Fire and Polymers

ACS SYMPOSIUM SERIES 425

Fire and Polymers

Hazards Identification and Prevention

Gordon L. Nelson, EDITOR
Florida Institute of Technology

Developed from a symposium sponsored
by the Macromolecular Secretariat
at the 197th National Meeting
of the American Chemical Society,
Dallas, Texas,
April 9–14, 1989

American Chemical Society, Washington, DC 1990

Library of Congress Cataloging-in-Publication Data

Fire and polymers: hazards identification and prevention

Developed from a symposium sponsored by the Macromolecular Secretariat at the 197th National Meeting of the American Chemical Society, Dallas, Texas, April 9–14, 1989.

Gordon L. Nelson, editor

p.   cm.—(ACS Symposium Series, 0097–6156; 425).

Includes bibliographical references.

ISBN 0–8412–1779–3

1. Polymers—Fires and Prevention—Congresses.

I. Nelson, Gordon L. II. American Chemical Society. Macromolecular Secretariat. III. Series.

TH9446.P65F57   1990
628.9'222—dc20

90–34195
CIP

The paper used in this publication meets the minimum requirements of American National Standard for Information Sciences—Permanence of Paper for Printed Library Materials, ANSI Z39.48–1984.

Foreword

The ACS SYMPOSIUM SERIES was founded in 1974 to provide a medium for publishing symposia quickly in book form. The format of the Series parallels that of the continuing ADVANCES IN CHEMISTRY SERIES except that, in order to save time, the papers are not typeset but are reproduced as they are submitted by the authors in camera-ready form. Papers are reviewed under the supervision of the Editors with the assistance of the Series Advisory Board and are selected to maintain the integrity of the symposia; however, verbatim reproductions of previously published papers are not accepted. Both reviews and reports of research are acceptable, because symposia may embrace both types of presentation.

Contents

Preface

Each year, fire causes some 6,200 deaths, 30,000 injuries, and $8 billion in property loss in the United States. Nationwide, 2.4 million fires, large and small, take their toll. Some, like the 1986 Dupont Plaza Hotel fire in Puerto Rico, involve large loss of both life and property. Clearly, fire is a major social issue.

Most fires involve the combustion of polymeric materials. Despite the involvement of polymers in fires, there have been no books on current research topics in this area of fire research in some years. Given the importance of fire and polymers as a social issue and the complexity and interest in the relevant basic science, Eli Pearce of Polytechnic University of New York and I cochaired a one-week symposium on fire and polymers. It was the first major symposium held on fire and polymers at an American Chemical Society National Meeting in about five years.

The symposium was planned as a state-of-the-art meeting, focusing on the basic science. Program areas included high heat polymers, fire performance of polymers, hazard modeling, mechanism of flammability and fire retardation, char formation, effects of surfaces on flammability, smoke assessment and formation mechanisms, and combustion product toxicity.

This book is based on that symposium. It is divided into five sections, designed to capture a representative segment of the latest research on fire and polymers. The book also includes some key contributions not presented at the symposium, all edited and organized for this book. The contributors are major figures in fire research, drawn from not only the United States but also Europe and Israel.

I gratefully acknowledge the work of Eli Pearce, dean of Arts and Sciences and director of the Polymer Research Institute, Polytechnic University of New York; Geoffrey N. Richards, director of the Wood Chemistry Laboratory, University of Montana; Barbara Levin of the National Center for Fire Research, NIST; Stanley Kaufman of Bell Laboratories; and Ronald L. Markezich of the Technology Center, Occidental Chemical Corporation. Their solicitation of papers was essential. I also acknowledge the donors of the Petroleum Research Fund, administered by the American Chemical Society, for support of travel for international speakers at the symposium.

GORDON L. NELSON
1988 ACS President
Florida Institute of Technology
Melbourne, FL 32902–2366

January 1990

FIRE   TOXICITY

Fire Toxicity

Fire creates a complex toxic environment involving flame, heat, oxygen depletion, smoke, and toxic gases. The nature of that environment is dependent not only upon the materials present but on the fire event, that is, the fire scenario. Materials have different toxic gas profiles under different conditions; therefore, toxic fire gas generation is not intrinsic to a material. Large fires in buildings constitute a severe toxic threat regardless of the materials being burned.

In the past, building codes in the United States included the phrase, "no more toxic than wood," in reference to fire gases from building materials. Such phrases have been deleted, because of the lack of either an accepted definition or test methodology to assess toxicity.

The basic goal of toxicity testing in a regulatory context is to allow society to make judgments about materials that will reduce risk in unwanted fires. While considerable discussion continues in the regulatory community, and laboratory toxicity tests find their way into liability litigation, these tests have not been accepted directly for use in the regulatory process in the United States. The basic reason for this is that laboratory tests assess toxic potency, not toxic hazard, of the combustion atmosphere, the latter dealing with the rate and amount of smoke, heat, and oxygen depletion generated, integrated with toxic potency. Toxic hazard is not directly obtainable from toxic potency; fire hazard assessment methodology is required.

While several states have attempted regulatory activity, the most visible approach, taken by the State of New York, has been the establishment of a data bank on toxic potency of building materials. The utility of such a data bank without available hazard analysis methodology is open to question.

A number of small-scale animal exposure tests have been developed to assess the potency of the toxic combustion products from combustible materials. Criticism of these tests relates to the relevance of the combustion module (a smoke generation apparatus) and the appropriateness of the animal model (rats or mice), particularly for irritant gases. Toxicity is only one of the several fire properties related to materials. All fire parameters are interrelated, that is, they are not independent variables.

In this section Levin and Gann provide a basic discussion of smoke toxic potency measurements. Recognizing that a large percentage of toxic potency is accounted for by a small number of gases, an "N" gas model has been developed to permit easier use of data in fire models.

Hydrogen chloride is liberated from polyvinyl chloride and other chlorine-containing synthetic polymers. The chapter by Hartzell and coworkers presents a state-of-the-art discussion of HCl toxicity using various animal models. Alarie and coworkers explore the role of hydrogen cyanide, another common fire gas, in fires.

Lundgren and Stridh discuss an analytical investigation of fire gases in full-scale and small-scale tests. Henderson presents a study of the toxicity of soot generated in the combustion process. In the final chapter Emmons discusses the state of fire modelling and the directions needed for a more comprehensive fire code.

2

Chapter 1

Toxic Potency of Fire Smoke

Measurement and Use

Barbara C. Levin and Richard G. Gann

Center for Fire Research, National Institute of Standards and Technology, Gaithersburg, MD 20899

Accurate measurement of the toxic potency of smoke is a
key to reducing human life loss in fires. This paper
summarizes the approaches taken in measuring toxic potency
and highlights four needed issues still to be researched.
Direct comparison of only toxic potency values is not a
valid means of determining the fire safety of materials and
is not sufficient for evaluating fire hazard. The paper
describes the N-Gas Model (a new method of assessing toxic
potency) and two approaches (in which toxic potency is one
of the factors) for assessing fire hazard: (a) HAZARD I, a
comprehensive tool for calculating the outcome of a fire,
and (b) a fire hazard index for comparing the contributions
of alternative materials to the toxicity component of fire
hazard.

As noted nearly five centuries ago, fires produce smoke; and as
learned this century, most of the fire deaths in this country
result from people breathing that smoke (1). Over the years, the
United States and Canada have had the worst fire loss records
among the industrialized countries which keep such records (2). At
present, the United States suffers 6,000 deaths and 30,000 reported
injuries per year (3). The annual property damage exceeds $7
billion, and the total cost of fire is over $50 billion (4).
Many factors affect fire hazard, or the threat to life safety
in a fire. These factors include:

- fire size,
- temperatures generated,
- vision impairment by soot and irritant gases,
- smoke toxicity,
- ease of escape,
- building design,
- time and clarity of alarm, and
- time of extinguishment.

There are various innovations that the fire science community can produce to reduce human losses. These include less flammable materials, early and accurate fire detection devices, and reliable suppression systems. Each of these approaches assists in keeping the fire from becoming large, an important factor since most fatalities in the United States occur some distance away from the room of fire origin from smoke generated from large fires (5).

As these strategies are brought to fruition, there remains one related issue: the determination of a smoke's potential harm per mass of material burned, i.e., the toxic potency of smoke. Accurate measurement of this key characteristic of fire smoke permits a more quantitative determination of the fire's toxic hazard which includes other factors as discussed below. Toxic potency assessment also tells us whether a small fire will produce smoke so toxic that only a small amount will kill. The presence of such "supertoxicants" has been a major topic of discussion within the fire community.

SMOKE TOXICITY MEASUREMENT

All fire smoke is toxic. In the past two decades, a sizable research effort has resulted in the development of over twenty methods to measure the toxic potency of those fire smokes (6). Some methods have been based on determinations of specific chemical species alone. Values for the effect (e.g., lethality) of these chemicals on humans are obtained from (a) extrapolation from pre-existing, lower concentration human exposure data or from (b) interpretation of autopsy data from accident and suicide victims. The uncertainty in these methods is large since:

- they presume which chemicals in the smoke are responsible for its toxicity,

- the data obtained from autopsies are often incomplete,

- autopsy data may have failed to account for metabolic changes in the body or blood (7), or

- the fatality may be the result of ingestion of the toxicant (under which case, an overdose is likely) rather than inhalation.

A second approach to the problem of toxic potency measurement has been to expose laboratory animals, usually rodents, to the smoke from the combustion of small samples of a burning material. Measurement of their response to the smoke leads to one of several biological endpoints, such as the LC_{50} (the concentration of smoke lethal to 50% of the test animals). In this approach, the animals respond to all the toxicants that are present in the smoke. It presumes that rodent mortality can be related to human mortality or, more simplistically, that the relative toxicity of the smokes will be similar in humans and rodents. However, since the relative contributions of the individual toxic chemicals in the smoke are not determined, a quantitative relationship between man and rodent is impossible using this approach.

Thus, each of these approaches has its drawbacks. While the number of methods under serious consideration has been reduced to a few, resolution of the drawbacks has not been forthcoming.

Recently, a third approach has been pursued that extracts the desirable features of the other methods and minimizes the difficulties (8-12). This has been called the N-gas method. It is based on the hypothesis that a small number ("N") of gases in the smoke account for a large percentage of the observed toxic potency. The lethality of each of these gases is determined for laboratory animals, e.g., rats. Similar measurements for combinations of these gases tell us whether the gases are additive (do not interact physiologically), synergistic, or antagonistic. The results of these mixed gas tests to date have been reduced to an algebraic equation which has been empirically determined for the exposure of rats to mixtures of CO, CO_2, HCN, and reduced O_2 (13).

$$\frac{m[CO]}{[CO_2] - b} + \frac{[HCN]}{d} + \frac{21 - [O_2]}{21 - LC_{50} O_2} \approx 1$$

where brackets indicate the actual atmospheric concentration of the gases. We have found that CO_2 acts synergistically with CO such that as the concentration of CO_2 increases (up to 5%), the toxicity of CO increases. Above 5%, the toxicity of CO starts to decrease again. The terms m and b define this synergistic interaction and equal -18 and 122000 if the CO_2 concentrations are 5% or less. For studies in which the CO_2 concentrations are above 5%, m and b equal 23 and -38600, respectively. The term d is the LC_{50} value of HCN and is 160 ppm for 30 minute exposures or 110 ppm for 30 minute exposures plus 24 hour post-exposure deaths. (Exposure to CO in air only produced deaths during the actual exposures and not in the post-exposure observation period; HCN did cause numerous deaths in the first 24 hours of the post-exposure period). The 30 minute LC_{50} of O_2 is 5.4% which is subtracted from the normal concentration of O_2 in air, i.e., 21%. In our pure and mixed gas studies, we found that if the value of the above equation is 1.1 ± 0.2, then some fraction of the test animals would die.

It has been shown that carbon dioxide also increases the toxicity of the other gases currently included in the model. For example, the 30 minute plus 24 hour LC_{50} value of HCN decreases to 75 ppm and that of O_2 increases to 6.6% in the presence of 5% CO_2. However, we empirically found that the effect of the CO_2 can only be added into this equation once. At this time, we have data on the effect of various concentrations of CO_2 on CO and only have information on the effect of 5% CO_2 on the other gases. Since CO is the toxicant most likely to be present in all real fires, we have included the CO_2 effect into the CO factor. As more information becomes available, the N-Gas equation will be changed to indicate the effect of CO_2 on the other gases as well.

Caution: The values used in the above equation are dependent on the test protocol, on the source of test animals, and on the rat strain. It is important to verify the above values whenever different conditions prevail and if necessary, to determine the values that would be applicable under the new conditions.

At this time, the above equation is for animal exposures of 30 minutes and includes deaths that occur within the following 24 hours.   Post-exposure deaths that occur after that are not predicted by the model.   Some data exists to expand this equation to other exposure times (11,13).

To measure the toxic potency of a given material with the N-Gas approach, a sample is combusted under the conditions of concern (e.g., non-flaming or flaming) and these principal components of the smoke measured.   Based on the results of the chemical analytical tests and the knowledge of the interactions of these four gases, an approximate LC_{50} value is predicted.   In one or two further tests, six rats are exposed to the smoke from a sample of such size that the smoke should produce an atmosphere that is predicted to be just below the observable lethal level.   The deaths of all of the animals then indicates the presence of unknown toxicants.   Survival of all the animals allows use of the approximate LC_{50} calculated from the gas mixture as a conservative lower limit for rats.   If more accuracy is needed, then a detailed LC_{50} can be determined.   Tables I-III (13) show some typical results using this method.   The good agreement (deaths of the animals when the N-gas values were above 0.9) indicates the high promise of this approach.   This method reduces the elapsed time and the number of test animals needed for the toxic potency determination and also indicates whether the toxicity is usual (i.e., the toxicity be explained by the measured gases) or is unusual (additional gases are needed to explain the toxicity).

Table I. Predictability of N-Gas Model Using CO, CO_2, and Reduced O_2[a]

N-GAS EXPOSURE VALUE	DEATHS WITHIN EXPOSURE	POST EXPOSURE	DAY OF DEATH
0.84	0/6	0/4	-
0.89	0/6	0/4	-
0.93	0/6	1/4	0
0.96	0/5	0/4	-
1.01	3/6	0/3	-
1.06	4/6	1/2	3
1.07	3/6	1/2	0
1.12	4/6	0/1	-
1.22	5/6	0/1	-

a. The information in this table was modified from reference 13. In all experiments, 6 rats were exposed.   The total number of rats in the post-exposure observation time changes because some rats were sacrificed for blood and some had died during the exposure.

Table II. Predictability of N-Gas Model Using CO, CO_2, and HCN[a]

N-GAS VALUE		DEATHS		
WITHIN EXPOSURE	POST EXPOSURE	WITHIN EXPOSURE	POST EXPOSURE	DAY OF DEATH
0.74	0.95	0/6	1/6	0
0.87	1.03	0/6	1/6	1
0.93	1.01	0/6	0/6	-
1.00	1.11	0/6	2/6	1,3

a. See legend of Table I.

Table III. Predictability of N-Gas Model Using CO, CO_2, HCN, and Reduced O_2[a]

N-GAS VALUE		DEATHS		
WITHIN EXPOSURE	POST EXPOSURE	WITHIN EXPOSURE	POST EXPOSURE	DAY OF DEATH
0.77	-	0/6	0/6	-
0.91	-	0/6	0/6	-
1.06	1.23	1/6	1/5	1
1.08	-	0/6	0/6	-
1.22	-	4/6	0/2	-

a. See legend of Table I.

The toxic potencies of many materials have been measured using a variety of the toxicity test methods. Comparison of toxic potency results between the various methods is, in most cases, meaningless. The frequent lack of agreement between methods is due to different methods of combustion, species of animals, and experimental apparatus (i.e., open or closed devices; also referred to as dynamic or static systems).

Most importantly, none of the methods have been sufficiently checked to assess how well they reproduce the gas yields or even the LC_{50} values from the appropriate segments of real-scale fire tests. To begin this process, a comparison procedure has been developed and a few materials have been checked using the NBS bench-scale combustor and the N-gas method (14,15).

Thus, there remain a small number of discrete, highly important technical issues to be researched in the field of toxic potency measurement:

• Addition of a few more key gases to the N-gas equation, so that the lethality (both within exposure and post-exposure) of most smokes is explicable;

- Determination of the proper combustion conditions of samples under small-scale laboratory conditions;

- Validation of small-scale results against the real-scale fire conditions of interest; and

- Extrapolation of the features of the N-gas equation from rodents to humans.

Much of this research could, in concept, be extended to sub-lethal effects. There is already some indication of a relationship between the smoke concentrations that cause death and those that result in physical collapse of the test animals (16). However, more subtle effects, such as decrease in human mental acuity, are expected to be very difficult to assess using rodent data.

USES OF TOXIC POTENCY DATA

There is general agreement among fire scientists and regulators alike that direct comparison of LC_{50} values (or any of the other toxic potency measures) is not a valid means of determining the relative fire safety of materials. Many other factors affect the true hazard that a commercial product presents in a fire. Thus, a product with a relatively high toxic potency might be considered acceptable if it is used in small quantities or if it is difficult to ignite. The same product might be considered unacceptable if used in large quantities in a room with a powerful ignition source. Thus, toxic potency numbers are but one factor in the calculation of the threat to people in fires.

There are two approaches that may be taken in integrating all of these factors. The first is a full calculation of the hazard from fires. This calculation requires the use of a computer and a sizable amount of expert judgment. The second approach involves a derived index. This is generally an algebraic combination of a few pieces of data leading to a value indicative of relative fire safety. An example of each approach is provided below.

Fire Hazard Assessment Methodology. HAZARD I is a set of pro-cedures combining expert judgment and calculations to estimate the consequences of a fire resulting from changes in such aspects as: materials fire properties, fire mitigation strategies, types of people present, and elements of building design (17). The procedures involve four steps:

- Defining the fire problem;

- Defining the fires of concern in the problem;

- Using the HAZARD I software to calculate the outcome of each of those fires, checking the sensitivity of the results to any assumptions made; and

- Evaluating the consequences of the calculated fires against a required performance level or traditional alternative.

The calculation of the fire's outcome in the third step includes the distribution of heat, smoke, and toxic gases throughout the building of concern. It allows the introduction of people into that building and monitors their movement in response to the fire. They may escape safely or fail to escape due to heat or the inhalation of toxic smoke. The benefits of changing some component of the defined fire problem is observed in the change in the number of deaths predicted, rather than by direct comparison of the toxic potencies of the different smokes. This mirrors the complexity of real-life fires.

A Fire Hazard Index. There are a number of situations where all that is needed is an indication of whether a change in a specific commercial product is beneficial or not. In these cases, one can presume that the people exposed to the fire and the building in which the fire exists are fixed. Moreover, it must be established that toxicity is the sole threat to escape, not smoke obscuration or heat.

With those simplifications, the key elements determining the toxic impact of the smoke are combined. One recent approach (18) proceeds as follows. To start, it is clear that the faster the already flaming material burns, the more mass is converted from solid to smoke. Second, the hazard is directly related to the toxic potency of the smoke. Since low LC_{50} values indicate high toxic potency, this parameter appears in the denominator. Third, the hazard depends on how swiftly the fire spreads, involving more and more material. Flame spread can be considered as a series of ignitions. Again, since a long ignition delay time (high resistance to ignition) reduces the hazard, it also appears in the denominator. These factors lead to the following equation:

$$\text{Toxic Fire Hazard Index} \quad \propto \quad \frac{\text{Mass Loss Rate}}{t_{ig} \times LC_{50}}$$

Note that this index only produces a relative number. Two products with widely different values of the index might be equally safe if, in fact, neither impedes escape. Conversely, two products with apparently similar values may produce different hazard levels if both products are close to the margin of safety. Thus, the scale for any index must be "calibrated", and it may well be different for each building or type of occupant. Generally, this will require a more complete hazard analysis and/or full-scale fire tests. Protocols for doing this are currently under consideration.

CONCLUSION

Knowing the impact of smoke toxic potency on escape from a fire is of sufficient importance that it has been the subject of research for over twenty years. As a result, we now have a realistic picture of proper contexts for the use of toxic potency data and a series of first-generation tools for measuring it. We also have a vision of the key technical issues to be resolved: developing a proper small-scale fire simulator, relating rodent results to people, and validating the small-scale data.

LITERATURE CITED

1. Birky, M. M.; Halpin, B. M.; Caplan, Y. H.; Fisher, R. S.;
    McAllister, J. M.; Dixon, A. M. Fire Fatality Study, Fire &
    Materials 1979, 3, 211-217.
2. America Burning: The Report of the National Commission on Fire
    Prevention and Control, Superintendent of Documents, U.S.
    Government Printing Office: Washington, DC, 1973.
3. Karter, M.J., Jr. Fire Loss in the United States During 1987,
    Fire Journal 1988, 82, 32-44.
4. Hall Jr., J.R. Calculating the Total Cost of Fire in the
    United States, Fire Journal 1989, 83, 69-72.
5. Gomberg, A.; Buchbinder, B.; Offensend, F.J. Evaluating
    Alternative Strategies for Reducing Residential Fire Loss - The
    Fire Loss Model, NBSIR 82-2551, National Bureau of Standards,
    Gaithersburg, MD, 1982.
6. Kaplan, H.L.; Grand, A.F.; Hartzell, G.E. Combustion Toxicol-
    ogy: Principles and Test Methods; Technomic Publishing Co.,
    Inc.: Lancaster, PA, 1983.
7. Levin, B.C.; Rechani, P.R.; Gurman, J.L.; Landron, F.; Clark,
    H.M.; Yoklavich, M.F.; Rodriguez, J.R.; Droz, L.; Mattos de
    Cabrera, F.; Kaye, S. Analysis of Carboxyhemoglobin and
    Cyanide in Blood from Victims of the Dupont Plaza Hotel Fire in
    Puerto Rico, J. Forensic Sciences (In press, January, 1990).
8. Babrauskas, V.; Levin, B.C.; Gann, R.G. A New Approach to Fire
    Toxicity Data for Hazard Evaluation, Fire Journal 1987, 81, 22-
    71. Also in ASTM Stand. News 1986, 14, 28-33.
9. Levin, B.C.; Paabo, M.; Gurman, J.L.; Harris, S.E.; Braun, E.
    Toxicological Interactions Between Carbon Monoxide and Carbon
    Dioxide, Toxicology 1987, 47, 135-164.
10. Levin, B.C.; Paabo, M.; Gurman, J.L.; Harris, S.E. Effects of
    Exposure to Single or Multiple Combinations of the Predominant
    Toxic Gases and Low Oxygen Atmospheres Produced in Fires,
    Fundamental and Applied Toxicology 1987, 9, 236-250.
11. Levin, B.C.; Gurman, J.L.; Paabo, M.; Baier, L.; Holt, T.
    Toxicological Effects of Different Time Exposures to the Fire
    Gases: Carbon Monoxide or Hydrogen Cyanide or to Carbon
    Monoxide Combined with Hydrogen Cyanide or Carbon Dioxide, In
    Proceedings of the Ninth Meeting of the U.S.-Japan Panel on
    Fire Research and Safety, Norwood, MA, 1987, NBSIR 88-3753,
    National Bureau of Standards, Gaithersburg, MD, 1988, p. 368-
    384.
12. Complex Mixtures: Methods for In Vivo Toxicity Testing, Complex
    Mixture Committee, National Research Council, National Academy
    Press, 1988.
13. Levin, B.C.; Paabo, M.; Gurman, J.L.; Clark, H.M.; Yoklavich,
    M.F. Further Studies of the Toxicological Effects of Different
    Time Exposures to the Individual and Combined Fire Gases-
    Carbon Monoxide, Hydrogen Cyanide, Carbon Dioxide and Reduced
    Oxygen, Polyurethane 88, Proceedings of the 31st SPI Con-
    ference, Philadelphia, PA, 1988, p. 249-252.
14. Braun, E.; Levin, B.C.; Paabo, M.; Gurman, J.; Holt, T.; Steel,
    J.S. Fire Toxicity Scaling, NBSIR 87-3510, National Institute
    of Standards and Technology, Gaithersburg, MD, 1987.

15. Braun, E.; Levin, B.C.; Paabo, M.; Gurman, J.L.; Clark, H.M.; Yoklavich, M.F. Large-Scale Compartment Fire Toxicity Study: Comparison with Small-Scale Toxicity Test Results, NBSIR 88-3764, National Institute of Standards and Technology, Gaithersburg, MD, 1988.
16. Levin, B.C.; Fowell, A.J.; Birky, M.M.; Paabo, M.; Stolte, A.; Malek, D. Further Development of a Test Method for the Assessment of the Acute Inhalation Toxicity of Combustion Products, NBSIR 82-2532, National Institute of Standards and Technology, Gaithersburg, MD, 1982.
17. Bukowski, R.W.; Peacock, R.D.; Jones, W.W.; Forney, C.L. Technical Reference Guide For Hazard I, NIST Handbook 416, Vol. II, National Institute of Standards and Technology, Gaithersburg, MD, 1989.
18. Babrauskas, V., Toxic Hazard from Fires: a Simple Assessment Method, Proceedings of conference on "Fire: Control the Heat, Reduce the Hazard, Queen Mary College Fire and Materials Centre and Fire Research Station, London, England, 1988, p 1-10.

RECEIVED November 20, 1989

Chapter 2

Toxicity of Smoke Containing Hydrogen Chloride

Gordon E. Hartzell[1], Arthur F. Grand, and Walter G. Switzer

Southwest Research Institute, P.O. Drawer 28510, San Antonio, TX 78284

Hydrogen chloride (HCl) is a relatively common fire gas toxicant about which there has been much speculation and controversy. It is formed from the combustion of chlorine-containing materials, the most notable of which is polyvinyl chloride (PVC). Hydrogen chloride is both a potent sensory irritant and also a pulmonary irritant. It is a strong acid, being corrosive to sensitive tissue such as the eyes. Assessment of the toxicity of HCl in humans is made particularly difficult due to the lack of a suitable animal model for its variety of toxicological responses. In spite of these difficulties, a clearer picture of the role of HCl in the toxicity of smoke is gradually being developed.

The effects of a toxicant may be characterized as being either lethal or sublethal. In spite of the common use of lethality as the index in combustion toxicology, the sublethal and possibly incapacitating toxic effects are most relevant to escape from a fire. When considered along with other factors such as pre-existing state of health and physical activity, sublethal effects will ultimately be the most useful to fire scientists and engineers in assessing toxic hazard. However, since toxicologists are limited with regard to sublethal response data on humans and nonhuman primates, attention must be given to the use of the rodent as an animal model, for both lethal and sublethal responses. In the case of the asphyxiant or narcosis-producing toxicants, carbon monoxide and hydrogen cyanide, it seems generally agreed that the rat is a reasonable model. With HCl and its irritant properties, choice of the appropriate animal model is quite complicated, particularly for sublethal effects. Rats, mice, guinea pigs and baboons have all been used, the latter only to a very limited extent, however.

There appears to be no question that the guinea pig is more sensitive to HCl than the rat or the mouse. Whether or not this greater sensitivity is meaningful in terms of human exposure is uncertain. Sublethal, respiratory responses of the guinea pig to the inhalation of HCl are claimed to be more comparable to those of humans,

[1]Current address: 3318 Litchfield Drive, San Antonio, TX 78230

0097–6156/90/0425–0012$06.00/0

i.e., initial sensory irritation, followed by a brief period of pulmonary irritation and then bronchoconstriction (1). Furthermore, it has been observed that the guinea pig, under conditions of moderate exercise, is more susceptible to HCl than when it is restrained (2). The tendency for severe bronchoconstriction, as well as laryngeal spasms and airways protective reflexes, by the guinea pig may cause it to be particularly sensitive to HCl, often leading to death either during or shortly after exposure. This is seen in the case of 30 minute (14-days postexposure observation) LC_{50} determinations, with the value for the guinea pig being about one-third that for the rat (3). There are those who would feel that the guinea pig is too sensitive, however.

A comparison of RD_{50} (respiratory depression) and LC_{50} values of HCl demonstrates that the mouse is also more sensitive than the rat to both the sensory irritant and also to the lethal effects of HCl. However, it has been claimed that the mouse may still be 7 to 10 times less sensitive than man and that a correction factor is required to extrapolate mice lethality data to man (4). The correction factor is based on the observation that HCl (or smoke from PVC) is about 7 to 10 times more lethal in mice exposed via a tracheal cannula, rather than through normal nose inhalation (5). The assumption was made that the respiratory tract of the mouse fitted with a tracheal cannula is comparable to that of a mouth-breathing human, an assumption which, to some, may be of questionable validity. It ignores the scrubbing capability of man's oral mucosa and marked differences in the tracheobronchial and pulmonary regions between the mouse and man (6). Furthermore, studies with the baboon, an animal species whose upper airway is recognized as a useful model for the human (7), refute the claim that man is more sensitive than the mouse to the lethal effects of HCl. These studies have shown unquestionably that baboons can survive short exposures to much higher concentrations of HCl than can the mouse (8).

Hydrogen chloride was found not to be physically incapacitating to baboons subjected to concentrations up to 17,000 ppm for 5 minutes (8). (The toxicant was reported, however, to cause an incidence of postexposure death at these high doses.) Comparable studies have not been conducted using actual PVC smoke. There have been questions as to the extent of respiratory dysfunction and susceptibility to infection caused by exposure to hydrogen chloride and PVC smoke. One study using baboons exposed to PVC smoke containing up to 4000 ppm HCl did not, however, indicate any significant residual effects on pulmonary function when tested at 3 days and again at 3 months postexposure (9).

Studies of the respiratory effects of both HCl and PVC smoke have been conducted, with marked differences being observed between rodents and nonhuman primates (9,10). In rodents (mice and rats), the response to both HCl and PVC smoke consists of a decrease in respiratory rate and minute volume, the typical response to a sensory irritant. In the anesthetized baboon, the responses both to PVC smoke and to HCl were similar, being characterized by an increase in respiratory rate and minute volume. Blood oxygen (PaO_2) values were reduced during exposure to both flaming and nonflaming PVC smoke; however, the hypoxemia was not as severe as that produced by exposure to a comparable concentration of pure HCl gas. Changes in pH and $PaCO_2$ values in PVC smoke-exposed animals were also less marked than in HCl-exposed animals.

Considerable controversy continues to exist as to what concentration of HCl is hazardous to man. Although numerous studies of the acute effects of HCl have been conducted with rodents, it is questionable whether lethality data from rodents can be directly extrapolated to man because of anatomical differences in the respiratory tract of the rodent and the primate. Interestingly, exposure doses (concentration × time) of HCl which cause postexposure lethality in rats are in the same range as those which have resulted in postexposure deaths of baboons, although the data for baboons are very limited and the comparison made is rather subjective (11). The lethal toxic potency of HCl with rats is actually only somewhat greater than that of carbon monoxide (12). Consideration of the exposure dose of carbon monoxide thought to be *hazardous* to humans would lead one, based simply on relative lethal toxic potencies, to suspect that exposure of humans to the range of 700 ppm or more of HCl for 30 minutes would be highly dangerous. It would be prudent to realize that HCl is dangerous to humans at concentrations well below those indicated by its lethal toxic potency (13,14).

Hydrogen Chloride in Smoke Toxicity Studies

Animal model lethality is the most common toxicity index used in the evaluation of materials for the toxicity of smoke (15). It is also the most common index used in toxic hazard modeling (16). There are two fundamental reasons for this: 1) Lethality is most easily incorporated into a standard test protocol; and, 2) Data are much more available for lethality than for sublethal effects. Extrapolation of lethality results to humans is more readily done and with greater confidence. For the purposes of current smoke toxicity testing methodology and toxic hazard analyses, a conclusion, based on the comparative data shown in Table I, is that the rat is a reasonable model for inhalation of HCl (3). Therefore, subsequent discussions will involve studies using rat lethality as the measured toxicological index. It should be added that, although most studies have been conducted with HCl, the effects discussed are likely to be experienced with irritants in general.

Table I. Lethal Toxic Potencies of Hydrogen Chloride
(LC_{50} Values in ppm)

Exposure Time (min)	Rats	Mice	Guinea Pigs	Baboons*
5	15,900[a]	13,745[b]		17,000
10	8,370[a]	10,138[c]		
15	6,920[a]		2,900[d]	10,000
22.5	5,920[a]			
30	3,800[a]	2,644[b]	1,350[d]	5,000
60	2,810[a]			

a) Hartzell (11)–includes 14 days postexposure
b) Darmer (17)–includes 7 days postexposure
c) Alarie (18)–includes 3 hours postexposure
d) Hartzell (3)–includes 14 days postexposure
* These values are not LC_{50}'s, but estimates based on conditions of animals after exposure. The 5-minute exposure resulted in postexposure deaths (8). The 15- and 30-minute exposures yielded subjects that survived indefinitely (11).
Source: Reprinted with permission from reference 19.

The consequences of exposure to atmospheres containing multiple toxicants have only recently begun to be realized. Although individual fire gas toxicants may exert quite different physiological effects through different mechanisms, when present in a mixture each may result in a certain degree of compromise experienced by an exposed subject. It should not be unexpected that varying degrees of a partially compromised condition may be roughly additive in contributing to incapacitation or death. This has been demonstrated in a number of studies, and is a key element in the assessment of toxic hazard from analytical data.

There is a further complication with combined toxicants which is more difficult to deal with. Each individual toxicant may have physiological effects other than that of its principal specific toxicity. These effects, particularly those involving the respiratory system, may alter the rate of uptake of other toxicants. For example, hydrogen cyanide is known to cause hyperventilation early in an exposure, with fourfold increases in the RMV of monkeys being reported (20). (Respiration eventually slows as narcosis results, however.) It has been suggested that this initial hyperventilation in primates may result in faster incapacitation from HCN than would be expected, along with more rapid COHb saturation should CO also be present. Carbon monoxide, in decreasing the oxygen transport capability of the blood, eventually results in a condition of metabolic acidosis. A slowing of respiration also results from the induced state of narcosis. Carbon dioxide is a respiratory stimulant which increases the rate of uptake of other toxicants, thus producing a faster rate of formation of COHb from inhalation of CO. Irritants, such as HCl, slow respiration in rodents due to sensory irritation, but increase RMV in primates from pulmonary irritation. With primates, there is also evidence of an irritant (HCl) causing bronchoconstriction which may interfere with oxygen reaching the alveoli for diffusion into the blood. With all these effects possible in the inhalation of mixtures of toxicants in real fire effluents, the situation is extremely complex. Very little research using toxicant combinations has been conducted using primates and the full seriousness of the combined effects on incapacitation and death of humans exposed to fire gas atmospheres is not fully understood. It is reasonable to say, however, that the effects described certainly do not favor the safety of those exposed.

In spite of the complexity of dealing with atmospheres containing multiple toxicants, considerable progress has been made in understanding some of the effects from studies using rodents. For example, it is fairly well agreed that carbon monoxide and hydrogen cyanide appear to be additive when expressed as fractional doses required to cause an effect (21,22). Thus, as a reasonable approximation, the fraction of an effective dose of CO can be added to that of HCN and the time at which the sum becomes unity (100%) can be used to estimate the presence of a hazardous condition.

In the case of mixtures of hydrogen chloride and carbon monoxide, empirical analysis of toxicological data shows that exposure doses leading to lethality of rats may also be additive (12). This is illustrated in Table II, in which data for mixtures of HCl and CO are expressed as fractional lethal doses. Summation of the fractional doses approximates unity for 50% lethality in each case. Although not confirmed with primates, these studies imply that hydrogen chloride may be much more dangerous than previously thought when in the presence of carbon monoxide or, conversely, carbon monoxide intoxication may be much more serious in the presence of an irritant. A rapid respiratory acidosis was seen in the blood of rats exposed to HCl, which when coupled with the metabolic acidosis produced by CO, resulted in severely compromised animals. This could be quite serious with real fires involving PVC, since carbon monoxide is almost always present from the burning of other materials. The effect may have significance with human exposure, such as prolonged hypoxemic conditions following rescue or escape. There is also some suggestion that the incapacitating effects of carbon monoxide may be enhanced in primates upon simultaneous exposure to HCl, the presence of which causes the blood PaO_2 to be decreased (23). This is probably the case with other irritants as well.

Table II. Summation of Fractional Effective (Lethal) Doses for
30-Minute Exposure of Rats to Mixtures of CO and HCl

| CO | | HCl | | ΣFED |
ppm	Fractional Lethal Dose	ppm	Fractional Lethal Dose	(50% Lethality)
6400*	1.0	–	–	1.0
5700*	0.89	600	0.16	1.05
5300*	0.83	1000	0.26	1.09
4150	0.65	1900*	0.50	1.15
3000	0.49	2100*	0.55	1.04
–	–	3800*	1.0	1.0

* Represents experimental LC_{50} value in the presence of a fixed concentration of
the cotoxicant.

It has recently been observed that there may also be additivity of fractional
doses between HCl and HCN (24). In Table III are shown 30-minute (14 days
postexposure) LC_{50} values for HCl and HCN, along with three LC_{50} values for HCN
in the presence of fixed concentrations of HCl.

Table III. Summation of Fractional Effective (Lethal) Doses for
30-Minute Exposure of Rats to Mixtures of HCN and HCl

| HCN | | HCl | | ΣFED |
ppm	Fractional Lethal Dose	ppm	Fractional Lethal Dose	(50% Lethality)
212*	1.0	–	–	1.00
219*	1.03	600	0.16	1.19
116*	0.55	1000	0.26	0.81
70*	0.33	1900	0.50	0.83
–	–	3800*	1.00	1.00

* Represents experimental LC_{50} value in the presence of a fixed concentration of
the cotoxicant.

Particularly striking was the incidence of postexposure deaths from concentrations of
the toxicants each of which alone would not be expected to result in any
postexposure lethality. Deaths were often several days after the exposure, as
illustrated in Table IV.

Table IV. Pattern of Lethality After Exposure of
Rats to HCl and HCN

| Average HCN (ppm) | Average HCl (ppm) | Exposure Number Dead | Postexposure Number Dead | | | Total Number Dead |
			Day 0	Days 1-4	Days 5-14	
170	–	0	0	0	0	0/6
–	1800	0	0	0	0	0/6
100	1900	0	0	0	6	6/6
180	1000	1	0	2	2	5/6

Although in most real multimaterial fires, exposure doses of CO produced normally far exceed those of other toxicants, significant combined effects have certainly been demonstrated with rodents in the laboratory. In addition to those described here with CO, HCN and HCl, such effects have also been reported due to CO_2 and to low oxygen when in combination with the narcotic toxicants (22,25). These all need to be studied further, preferably with nonhuman primates, in order to determine their impact on hazards to humans. Smoke atmospheres are likely to be much more hazardous than one would initially suspect from consideration of the concentrations of the individual toxicants taken separately. Perhaps, the major concern should not be so much the toxicity of HCl or HCN, but rather, the toxicity of combinations of these gases with CO, CO_2 and low oxygen as may be present in smoke.

Polyvinyl Chloride Smoke

Studies have been conducted to expose animals to a constant, known concentration of HCl as produced from the combustion of PVC (3,12). Concentrations of HCl were dependent upon other experimental parameters in addition to the amounts of PVC burned. Thus, the technique does not lend itself to determination of an LC_{50} based on PVC. Estimates can be made, however, by referring to published data which were obtained using the continuous combustion device (12).

The 30-minute LC_{50} values (14-days postexposure) based on HCl were determined for guinea pigs and for rats exposed to smoke produced from both flaming and nonflaming PVC. Data are shown in Table V. Both rodent models show the same trend, with guinea pigs being considerably more sensitive than rats and with the HCl produced from PVC having a greater "apparent" toxicity. This may be due to the presence of other irritants in addition to HCl, or perhaps to complications in respiratory penetration patterns involved with the complex gas/aerosol/particulate mixtures found in smoke.

Table V. Lethal Toxic Potencies of Hydrogen Chloride
Produced from Polyvinyl Chloride
(30-Minute LC_{50} Values in ppm)*

	Rat	Guinea Pig
HCl	3,800 (3,100-4,800)	1,350 (1,100-1,800)
HCl (PVC Nonflaming)	2,900 (2,200-3,700)	910 (660-1,500)
HCL (PVC Flaming)	2,100 (1,600-2,500)	650 (460-1,240)

* Values in parentheses are 95% confidence limits.

Modeling of Toxicological Effects of Fire Effluents

The apparent additivity of fractional doses of the major fire gas toxicants, including HCl, serves as the basis for modeling methodology being developed for use in the prediction of the toxic effects of fire effluents. Such methodology is beginning to provide useful tools for the assessment of toxic hazards in fires by combustion toxicologists and fire safety engineers. In addition to the estimation of toxic hazards from time-concentration data for individual toxicants, smoke concentration and toxic potency values can also be used for simulation of multimaterial fires (26). One such model, termed the fractional effective dose (FED) model, takes the form of Equation (1) for (n) materials. The total fractional effective exposure dose at any time (t) would be:

$$\sum_{i=1}^{n} \int_{t_{(bi)}}^{t} \frac{C_i - b_i}{K_i} \, dt \tag{1}$$

where C_i represents smoke concentration (from mass burning rate data) and K_i and b_i characterize a material's smoke toxicity (27). (K_i and b_i are the slope and intercept of a plot of LC_{50} vs. 1/time-of-exposure for the smoke produced by a burning material.) The time at which Equation (1) becomes unity (100%) is the time of exposure which would be expected to result in 50% effect. This summation of fractional effective doses for smoke from the burning of combinations of materials is justified on the basis of the apparent additivity of fractional doses of the major individual fire gases (21,22,28-30). Computer programs, using a variety of fire scenarios and material input data, have been developed for assessing potential toxic hazard for multimaterial fires by this method.

An obvious utility for the type of modeling described is to evaluate the effect of exchanging one material for another in a composite or an assembly or even the addition of a new material, perhaps one of high toxicity but with a low burning rate. It can be used to evaluate the contribution of a material that does not become involved until the later stages of a fire. The model has the potential of assessing the trade-offs of flammability vs. toxicity often encountered with the use of fire retardants.

A word of caution should be made with respect to estimating tenability for human exposures. Experimental LC_{50} values are determined using rat lethality as the bioassay. Judgement should be exercised for human tenability, with limits set considerably lower than FED summation at unity. An FED limit of, perhaps, 0.3 may be more appropriate for assessment of the potential escape of humans.

Acknowledgment

Work referenced as conducted at Southwest Research Institute was supported under U. S. National Institute of Science and Technology Grant Nos. NB83NADA4015 and 60NANB6D0635, Federal Aviation Administration Contract No. DTFA03-81-00065 and by the Vinyl Institute of the Society of the Plastics Industry, Incorporated. Portions of this text are reprinted with permission from *Advances in Combustion Toxicology*, G.E. Hartzell, ed., Vol. 1, Technomic: Lancaster, PA, 1988, p. 28-30.

Literature Cited

1.   Burleigh-Flayer, H.; Wong, K.L.; Alarie, Y. "Evaluation of the Pulmonary Effects of HCl Using CO_2 Challenges in Guinea Pigs," *Fund. Appl. Toxicol.* **1985**, *5*, 978-985.

2.   Malek, D.E.; Stock, M.F.; Alarie, Y. "Rapid Incapacitating and Lethal Effects of HCl in Guinea Pigs During Exercise," *Toxicologist* **1987**, *8*(4).

3.   Hartzell, G.E.; Grand, A.F.; Switzer, W.G. "Modeling of Toxicological Effects of Fire Gases: VII. Studies on Evaluation of Animal Models in Combustion Toxicology," *J. Fire Sciences* **1988**, *6*(6), 411-431.

4.   Kennah, H.E.; Stock, M.F.; Alarie, Y.C. "Toxicity of Thermal Decomposition Products from Composites," *J. Fire Sciences* **1987**, *5*(1), 3-16.

5.   Anderson, R.C.; Alarie, Y. "Acute Lethal Effects of Polyvinyl-chloride Thermal Decomposition Products in Normal and Cannulated Mice," *Abstracts*

of Papers, Society of Toxicology, Inc., 19th Annual Meeting, Washington, D.C., March 9-13, 1980.

6. Patra, A.L., "Comparative Anatomy of Mammalian Respiratory Tracts: The Nasopharyngeal Region and the Tracheobronchial Region," *J. Tox. Env. Health* **1986**, *17*, 163-174.

7. Patra, A.L.; Gooya, A.; Menache, M.G. "A Morphometric Comparison of the Nasopharyngeal Airway of Laboratory Animals and Humans," *The Anatomical Record* **1986**, *215*, 42-50.

8. Kaplan, H.L.; Grand, A.F.; Switzer, W.G.; Mitchell, D.S.; Rogers, W.R.; Hartzell, G.E. "Effects of Combustion Gases on Escape Performance of the Baboon and the Rat," *J. Fire Sciences* **1985**, *3*(4), 228-244.

9. Kaplan, H.L.; Anzueto, A.; Switzer, W.G.; Hinderer, R.K. "Acute Respiratory Effects of Inhaled Polyvinyl Chloride (PVC) Smoke in Nonhuman Primates," *The Toxicologist*, Abstracts of Papers, Society of Toxicology, Incorporated, 26th Annual Meeting, Volume 1, No. 1, p. 202, February 1987.

10. Kaplan, H.L.; Anzueto, A.; Switzer, W.G.; Hinderer, R.K. "Effects of Hydrogen Chloride on Respiratory Response and Pulmonary Function of the Baboon," *J. Tox. Env. Health* **1988**, *23*, 473-493.

11. Hartzell, G.E.; Packham, S.C.; Grand, A.F.; Switzer, W.G. "Modeling of Toxicological Effects of Fire Gases: III. Quantification of Postexposure Lethality of Rats from Exposure to HCl Atmospheres," *J. Fire Sciences* **1985**, *3*(3), 195-207.

12. Hartzell, G.E.; Grand, A.F.; Switzer, W.G. "Modeling of Toxicological Effects of Fire Gases: VI. Further Studies on the Toxicity of Smoke Containing Hydrogen Chloride," *J. Fire Sciences* **1987**, *5*(6), 368-391.

13. Henderson, Y.; Haggard, N. "Noxious Gases and the Principles of Respiration Influencing Their Action," Chem. Catalog. Co., New York, 1927, pp. 126-127.

14. *Chlorine and Hydrogen Chloride*, National Academy of Sciences–National Research Council, 1976, p 146.

15. Hartzell, G.E., "Assessment of the Toxicity of Smoke." In *Advances in Combustion Toxicology*, Technomic: Lancaster, PA, 1988; Vol. 1, pp. 8-18.

16. Bukowski, R.W.; Jones, W.W.; Levin, B.M.; Forney, C.L.; Stiefel, S.W.; Babrauskas, V.; Braun, E.; Fowell, A.J. "Hazard I. Volume I: Fire Hazard Assessment Method," NBSIR 87-3602, U.S. National Institute of Standards and Technology, 1987.

17. Darmer, K.I.; Kinkead, E.R.; DiPasquale, L.C. "Acute Toxicity in Rats and Mice Exposed to Hydrogen Chloride Gas and Aerosols," *Am. Ind. Hyg. Assoc. J.* **1974**, *35*, 623-631.

18. Alarie, Y. *Proceedings of the Inhalation Toxicology and Technology Symposium*, Leong, B.K., Ed.; Ann Arbor Science, 1980, 207-238.

19. Hartzell, G.E. (ed.). *Journal of Fire Sciences*, Vol. 7, Technomic: Lancaster, PA, 1989 (in print).

20. Purser, D.A.; Grimshaw, P.; Berrill, K.R. "Intoxication by Cyanide in Fires: A Study in Monkeys Using Polyacrylonitrile," *Arch. Env. Health* **1984**, *39*, 394.

21. Hartzell, G.E.; Stacy, H.W.; Switzer, W.G.; Priest, D.N. "Modeling of Toxicological Effects of Fire Gases: V. Mathematical Modeling Intoxication of Rats by Combined Carbon Monoxide and Hydrogen Cyanide Atmospheres," *J. Fire Sciences* **1985**, *3*(5), 330-342.

22. Levin, B.C.; Paabo, M.; Gurman, J.J.; Harris, S.E. "Effects of Exposure to Single or Multiple Combinations of the Predominant Toxic Gases and Low Oxygen Atmospheres Produced in Fires," *Fundamental and Applied Toxicology* **1987**, *9*, 236-250.

23. Kaplan, H.L., Personal Communication.

24.   Hartzell, G.E., Work in Progress.
25.   Levin, B.C.; Paabo, M.; Gurman, J.L.; Harris, S.E.; Braun, E.   "Toxicological
       Interactions Between Carbon Monoxide and Carbon Dioxide," *Proceedings of
       16th Conference of Toxicology, Air Force Aerospace Medical Research
       Laboratory*, Dayton, OH, October 1986; *Toxicology* **1987**, *47*, 135-164.
26.   Hartzell, G.E.; Emmons, H.W.   "The Fractional Effective Dose Model for
       Assessment of Hazards Due to Smoke from Materials," *J. Fire Sciences* **1988**,
       6(5), 356-362.
27.   Hartzell, G.E.; Priest, D.N.; Switzer, W.G.   "Modeling of Toxicological
       Effects of Fire Gases: II. Mathematical Modeling of Intoxication of Rats by
       Carbon Monoxide and Hydrogen Cyanide," *J. Fire Sciences* **1985**, *3*, 115-
       128.
28.   Tsuchiya, Y.; Nakaya, I.   "Numerical Analysis of Fire Gas Toxicity:
       Mathematical Predictions and Experimental Results," *J. Fire Sciences* **1986**,
       4(2), 126-134.
29.   Sakurai, T.   "Toxic Gas Test by the Several Pure and Mixture Gases Using
       Mice," *J. Fire Sciences* **1989**, *7*(1), 22-77.
30.   Purser, D.A.   "Toxicity Assessment of Combustion Products and Modeling of
       Toxic and Thermal Hazards in Fire," *SFPE Handbook of Fire Protection
       Engineering*, National Fire Protection Association, Quincy, MA, Section 1,
       1988, pp. 200-245.

RECEIVED November 1, 1989

Chapter 3

Role of Hydrogen Cyanide in Human Deaths in Fire

Y. Alarie, R. Memon, and F. Esposito

Graduate School of Public Health, University of Pittsburgh, Pittsburgh, PA 15261

Samples of smoke during fires have indicated that hydrogen cyanide is not of concern in fire deaths because the levels found were much below lethal levels.
However, blood analysis of cyanide in fire victims has
revealed high levels of this chemical in fire victims.
This study confirms previous studies that cyanide plays
a definite role in fire deaths. Thus, it cannot be ignored when investigating the principal toxicants involved in fire victims. The fact that hydrogen cyanide is
more potent and faster acting than carbon monoxide may
have further importance in fires involving large quantities of nitrogen containing polymers.

It is well known that hydrogen cyanide can be liberated during combustion of nitrogen containing polymers such as wool, silk, polyacrylonitrile, or nylons (1, 2). Several investigators have reported cyanide levels in smoke from a variety of fires (3, 4, 5). The levels
reported are much below the lethal levels. Thus the role of cyanide
in fire deaths would seem to be quite low. However, as early as 1966
the occurence of cyanide in the blood (above normal values) of fire
victims was reported (6). Since then many investigators have reported
elevated cyanide levels in fire victims (7-13). However, it has been
difficult to arrive at a cyanide blood level which can be considered
lethal in humans. In this report the results of cyanide analysis in
blood of fire victims are reported as well as the possibility that
cyanide may, in some cases, be more important than carbon monoxide as
the principal toxicant in fire smoke.

MATERIALS AND METHODS

Sixteen cities/jurisdictions in the U. S. and Canada have participated in this study as listed in Table I. Fire death reports were collected from January 1984 until March 1989. For each fire death, a
report from the fire department was submitted. An autopsy report was
submitted by the medical examiner or coroner. Laboratory reports on

0097–6156/90/0425–0021$06.00/0
© 1990 American Chemical Society

Table I. Participants in Fire Deaths Data Collection

Cities/Jurisdictions	Fire chiefs, Project Officers, Medical Examiners, Coroners, Pathologists, Analytical Chemists
Akron Ohio	J. Harris, R. Schueller, W. Cox, W. Hamlin
Birmingham Alabama	N. Gallant, O. D. Steadman, R. M. Brissie, C. Walls
Calgary Alberta, Canada	T. Minhinnett, C. Nickel, J. C. Butt
Columbus Ohio	D. Werner, S. Woltz, W. Adrion, J. Ferguson, P. Fardal
Dallas Texas	D. Miller, L. Mount, C. Petty, W. T. Lowry
Denver Colorado	M. Wise, N. Dinerman, G. Ogura, D. Wingeleth
Edmonton Alberta, Canada	R. Walker, T. F. Mansell, G. Jones
Miami Florida	K. McCullough, J. Gilbert, J. Davis, L. R. Bednarczyk
Newark New Jersey	S. Kossup, A. Freda, R. Goode, D. Barillo
Pierce County Washington	J. Burgess, E. Lacsina, W. Hamlin
Pittsburgh/Allegheny County, Pennsylvania	J. Harper, R. Jedrzejewski, J. Perper, C. Winek, F. Esposito
Seattle Washington	C. Harris, D. Vickery, D. Ready, W. Hamlin
Syracuse New York	T. F. Hanlon, M. S. Waters, E. Mitchell
Tacoma Washington	T. Mitchell, R. Beardsly, E. Lacsina, W. Hamlin
Wichita Kansas	J. E. Sparr, T. Austin, W. Eckert, W. Hamlin

blood analysis were submitted by the medical examiner and these data are reported here. A total of 369 fire fatalities were reported and 303 of them considered for statistical evaluation. The cases considered are those in which blood could be obtained within a few hours after death, and was refrigerated and analyzed within 24 hours in most cases or 48 hours. Cases in which the victims died after being treated are omitted. The reports for blood analysis contained information for the following items: carboxyhemoglobin, cyanide, ethanol, benzene, total hemoglobin, and methemoglobin as given in Table II. Proficiency blood samples were prepared for carboxyhemoglobin, cyanide and ethanol determinations. Five laboratories participated with the results of one survey presented in Table III.

RESULTS

The age group, sex and race distributions of the fire victims are given in Figures 1-3. The results for ethanol, carboxyhemoglobin and cyanide are presented in Figures 4-6. The methemoglobin level was below 3% in the majority of the cases and the benzene level below 0.5 mg/L in the majority of the cases. No value was found above 5.0 mg/L. No correlation was found between blood carboxyhemoglobin and ethanol levels or carboxyhemoglobin and cyanide levels as shown in Figures 7 and 8.

DISCUSSION

As shown in Figure 8, if we accept a level above 50% COHb level as probably lethal for humans, 148 cases had a COHb level exceeding it. This is slightly more than 50% of the total cases (292) evaluated for COHb. If we accept a level above 1.0 mg/L cyanide in blood as probably lethal for humans, 47 cases (18%) had a cyanide level exceeding it out of 255 total cases evaluated for this agent. As also shown in Figure 8, 39 victims had both carboxyhemoglobin levels and cyanide levels above the lethal levels given above while 16 of them had a cyanide level above the lethal level but carboxyhemoglobin below the lethal level. Cyanide alone, as a cause of death, is therefore not encountered as frequently as carbon monoxide.

The results presented in Figure 8 are very similar to the results of Anderson et al. (8) and both investigations show that a large number of victims died with less than 50% COHb and less than 1 mg/L of cyanide. We could conclude that in these cases both agents acting together will result in less COHb and cyanide at death as demonstrated in experimental animals exposed to CO and HCN in combination (12). However, if this was so, the cases with higher than 50% COHb and higher than 1 mg/L of cyanide become unexplainable. How can such high levels of CO and cyanide together be present, much above what we think is a lethal level for either one alone? At the same time we have results on fire victims showing practically no CO or cyanide. In some of the later cases cardiovascular or other diseases may be responsible (12). However, many of such cases reported here were healthy children as indicated by the autopsy reports. Thus other contributing

Table II. Blood Analysis of Fire Victims for 303 Total Cases

| | Distribution | | | | Detection |
	A	B	C	D	Level Value
Measurements					
Carboxyhemoglobin[a]	262	30	8	3	1%
Cyanide[b]	199	56	42	6	.01 mg/L
Ethanol[c]	111	165	23	4	.01%
Benzene[c]	51	30	216	6	.1 mg/L
Methemoglobin[a]	81	37	180	5	.1 %
Total Hemoglobin[a]	93	-	203	7	-

A: Cases reported above level of detection
B: Cases tested but below level of detection
C: Cases not tested
D: Cases unknown

[a]: Spectrophotometry
[b]: Microdiffusion and Spectrophotometry
[c]: Gas Chromatography

Table III. Comparison of Results
for Blood Levels of Carboxyhemoglobin,
Cyanide and Ethanol

| | Carboxyhemoglobin (%) Levels | | | Cyanide (mg/L) Levels | | | Ethanol (g/dL) Levels | | |
	1	2	3	1	2	3	1	2	3
Target Value	-	-	-	0	3.8	8.6	0.049	0.095	0.540
Laboratories									
A	19	36	71	0	4.7	9.2	0.045	0.10	0.595
B	16.4	35.7	72.5	0	3.7	6.6	0.045	0.096	0.578
C	20.6	40.4	76	0.13	4.1	7.3	0.034	0.082	0.71
D	19	35	72	0	4.1	6.5	0.045	0.09	0.557
E	17	36.9	73.1	0	4.0	8.0			
Statistical Analysis									
n	5	5	5	5	5	5	4	4	4
Mean	18.4	36.8	72.9	0.025	4.1	7.5	0.04	0.092	0.609
SD	1.7	2.1	1.9	0.06	0.36	1.1	0.005	0.008	0.066
% CV	9.2	5.8	2.6	223	8.8	14.8	13.0	8.5	10.8

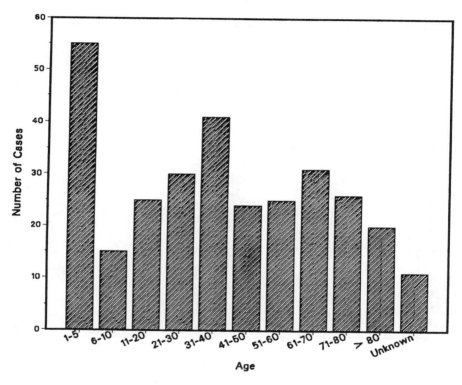

Figure 1. Distribution by age of fire deaths.

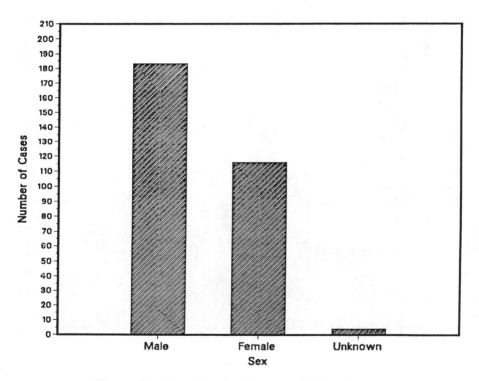

Figure 2.   Distribution by sex of fire deaths.

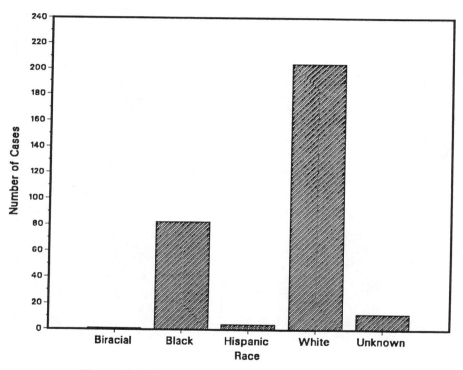

Figure 3. Distribution by race of fire deaths.

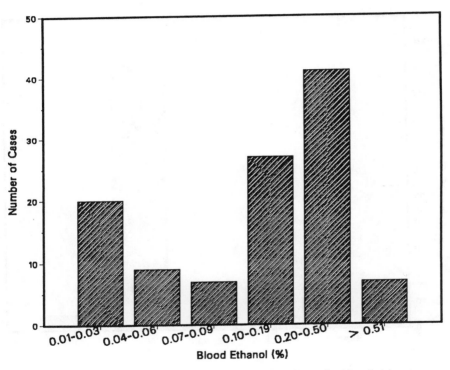

Figure 4. Levels of blood ethanol in fire death victims.

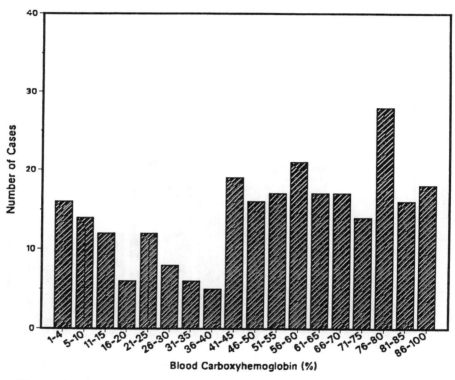

Figure 5. Levels of blood carboxyhemoglobin in fire death victims.

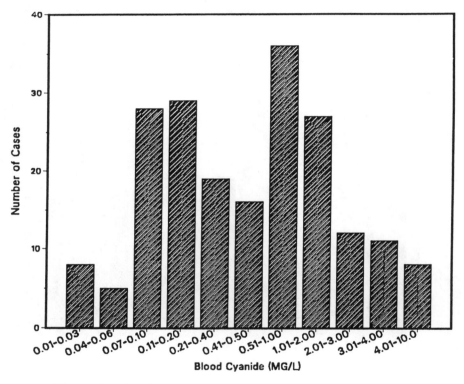

Figure 6.   Levels of blood cyanide in fire death victims.

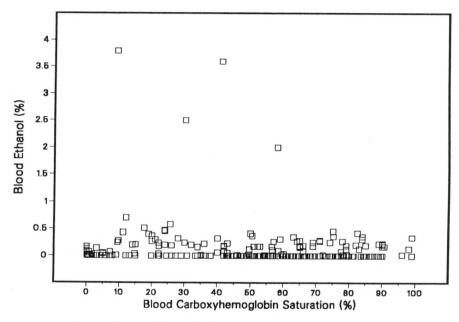

Figure 7. Relationship of blood ethanol levels and carboxyhemoglobin levels in fire death victims.

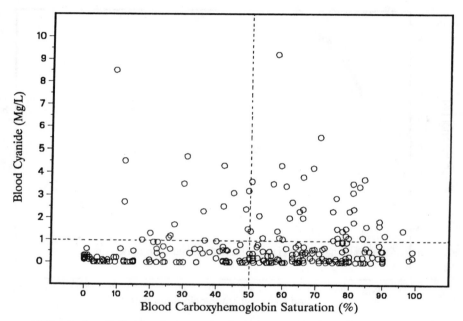

Figure 8.   Relationship of blood carboxyhemoglobin levels and
blood cyanide levels in fire death victims.

factors must be responsible when low COHb and blood cyanide levels
are present. Certainly low oxygen must be implicated in some of these
cases since the combination of low oxygen and cyanide has been shown
to be particularly effective in causing lethality with a blood
cyanide level of less than 0.6 mg/L (12). The former cases are much
more difficult to explain. However similar findings have been presented for mice exposed to smoke from some (not all) burning nitrogen containing polymers (12). Blood cyanide levels 3 times above the
lethal level found with exposure to HCN alone were reached (12).
With so many other toxicants present in fire smoke there must be some
which reduce the toxicity of HCN. Similar findings were also found
for cynomolgus monkeys exposed to smoke containing HCN vs. HCN alone
(14).

Establishing what the lethal level of blood cyanide is for humans is a difficult task. Based on extensive inhalation studies of
HCN in mice a level of 1 mg/L of blood was suggested (12). In rats,
a level of 2 mg/L of blood was found to be lethal (15). In limited
experiments with cynomolgus monkeys, an exposure concentration of
150 ppm for 30 min would result in 3 mg/L of blood cyanide and would
be lethal for these primates. Thus we have a range of 1 to 3 mg/L in
experimental animals. The lethal level, however, has been found to
be somewhat dependent upon exposure concentration and duration of exposure (12).

Can HCN by itself by responsible for death with inhalation of
fire smoke? While the data shown in Figure 8 indicate that this may
be so in a few cases, it is still a question to be explored. In one
case, analysis of blood from a victim who died in a small fire mainly involving a carpet, revealed a blood cyanide level of 1.2 mg/L and
a COHb level less than 5% saturation. When this carpet was burned in
experimental conditions a large amount of cyanide was generated and
mice exposed to the smoke reached 1.16 mg/L of blood cyanide with a
COHb level of 26% at death. Thus in this case cyanide was most probably responsible for this death with some contribution by carbon
monoxide and possibly other gases in the smoke (12). Recently, the
toxicologist investigating the deaths of 31 victims at the King's
Cross underground fire in London testified that 17 of the victims inhaled lethal quantities of cyanide gas (16). Therefore, we must review the measurements of atmospheric cyanide made in real fires (3,
4, 5). The methods used have been primitive and give a false sense
of security.

CONCLUSIONS

Nitrogen containing polymers have the potential to release cyanide
when thermally decomposed. This study demonstrates that cyanide
plays an important role in fire deaths.

ACKNOWLEDGMENTS

The collection of data on fire victims was initiated by the Foundation for Fire Safety and continued under Life Safety Systems Inc. We
thank Ms. Joyce Chidester for coordinating collection of the data.
Thanks are due to the individuals listed in Table I for their parti-

cipation and in particular to Dr. William Hamlin and Dr. A. Zebelman, Swedish Hospital Medical Center, Seattle WA 98104 for preparing stan -dard blood samples for analysis by the participating laboratories. They also performed blood cyanide analysis for several participating jurisdictions as given in Table I.

LITERATURE CITED
1. Cullis, C. F.; Hirschler, M. M. The Combustion of Organic Poly- mers.Clarendon Press, Oxford, 1981.
2. Alarie, Y. Ann. Rev. Pharmacol. Toxicol. 1985, 25, 325-347.
3. Gold, A.; Burgess, W. A.; Clougherty, E. V. Am. Ind. Hyg. Assoc. J. 1978, 39, 534-539.
4. Treitman, R. D.; Burgess, W. A.; Gold, A. Am. Ind. Hyg. Assoc. J. 1980, 41, 796-802.
5. Brandt-Rauf, P. W.; Fallon, L. F.; Tarantini, T.; Idema, C.; An- drews, L. Brit. J. Indust. Med. 1988, 45, 606-612.
6. Wetherell, H. R. J. Forensic Sci. 1966, 11, 167-173.
7. Clark, C. J.; Campbell, D; Reid, W. H. Blood Carboxyhemoglobin and Cyanide Levels in Fire Survivors. Lancet, 1981, 1332-1335.
8. Anderson, R. A.; Watson, A.A.; Harland, W. A. Med. Sci. Law 1981, 21, 288-294.
9. Anderson, R. A.; Watson, A.A.; Harland, W. A. Med. Sci. Law 1982, 22, 35-40.
10. Birky, M. M.; Halpin, B. M.; Caplan, Y. H.; Fisher, R. S.; McAl- lister, J. M.; Dixon, A. M. Fire and Materials 1979, 3 , 211- 217.
11. Birky, M.; Malek, D.; Paabo, M. J. Anal. Toxicol. 1983, 7, 265- 271.
12. Esposito, F. M.; Alarie, Y. J. Fire Sci. 1988, 6, 195-241.
13. Levin, B. C.; Rechani, P. R.; Gurman, J. L.; Landon, F.; Clark, H. M.; Yolavich, M. F.; Rodriguez, J. R.; Droz, B. S.; Kaye, S. J. Forensic Sci. in press.
14. Purser, D. A.; Grimshaw, P.; Berrill, K. R. Arch. Environ. Health 1984, 39, 401-408.
15. Levin, B. C.; Gurman, J. L.; Paabo, M.; Baier, L.; Procell, L.; Newball, H. H. Toxicologist 1986, 6, 59.
16. Dr. Toseland, quoted in The Times March 26, 1988.

RECEIVED November 1, 1989

Chapter 4

Chemical Analysis of Fire Effluents

B. Lundgren and G. Stridh

Swedish National Testing Institute, P.O. Box 857, S–501 15 Bors, Sweden

In the research project "Fire Hazard - Fire Growth in
Compartments in the early stage of Development (1) one
of the tasks was to investigate chemically the fire
effluents. The study was performed in the full scale
fire test room described in Nordtest Fire Test method
NT-FIRE 025 (2), a method similar to room tests
suggested by ASTM and ISO. Since many small scale
methods are believed to model different stages of
fire, it was determined to measure the emission of
fire effluents from three small-scale methods using
the same materials as in the full scale fire test. The
small scale test methods are; Nordtest NT-FIRE 004
(3), DIN 53436 (4) and a cone calorimeter test (5)
method similar to the NIST type of instrument. In the
full scale fire test thirteen different materials were
investigated. In the small scale fire tests six
different materials were investigated.
This presentation covers some of the basic data and
derived results are discussed. The gases species of
oxygen, carbon monoxide and carbon dioxide and nitrous
oxide have been measured for all the tests. In the
full scale fire tests hydrogen chloride and hydrogen
cyanides were measured. Hydrocarbons and their relative abundance were determined by collecting gas
samples on absorbent tubes for later analysis on a gas
chromatograph and a mass spectrometer.

In a joint research project in Sweden under the main title "Fire
hazard - Fire growth in compartments in the early stage of development (pre-flashover)" (1, 2) a number of different factors have
been studied. In the process of developing a full-scale fire test
method - "room-corner" configuration - for surface lining
materials, Nordtest NT-FIRE 025, the emission of smoke and gas was
studied. That study covers data from thirteen different single and

combined types of wall lining products. Some of these materials are today classified by using different small scale test methods. The discussion has for many years been concerned with the question, how well do the small scale fire test methods model real fire situations.

The purpose of this project was therefore to supply more information to answer this question. In the process of doing so the following tasks emerged
- to develop chemical analytical techniques
- to determine the different chemical species present
- to compare results for fire effluents from full scale and small scale fire tests

It was decided to take some of the thirteen materials through three small-scale fire tests and measure the emission of smoke and gases.

The small scale fire tests were, the Scandinavian "box test", Nordtest NT-FIRE 004, the DIN 53436-test and a cone calorimeter test. The study covered six different wall lining materials.

Test Methods and Materials

NT-FIRE 004 test method (3) is used for classification of wall lining materials in the Nordic countries. The burning behaviour, smoke emission and temperature increase are normally studied in NT-FIRE 004. The equipment consists of a box large enough to hold samples of wall lining materials, size 20 cm x 20 cm on four of the six walls. The floor and the back wall of the box have no lining material. Air inlet and exhaust chimmery are mounted on the back side wall. A gasburner is mounted on the floor plate having an energy output of 5 kW at 3.2 l/min of propane. The concentrations of gas species were measured with the aid of a diluting system inside the chimney of the "box" during the normal run of NT-FIRE 004.

The DIN 53436 - moving tube furnace (4). The equipment consists of a quartz tube, 100 cm long and 40 mm outer diameter. A round moving furnace is placed around the quartz tube. The furnace may give temperatures inside the tube of at least 600 °C. A tray 40 cm long holds the samples and a reference body which controls the inside temperature. The furnace being 10 cm long is moved at a specified speed over the sample heating continously new sections of the material during the experiment. The fumes from the material may be ignited with an igniter if necessary. The test has been performed at the Dantest Fire Technology laboratory under the following conditions: constant temperature 400 °C, ventilation rate 100 liter of air per hour and the oven moving in the counter air current mode over the samples.

The cone calorimeter used in this study (5) is a somewhat enlarged version of the model used at the National Institute of Standards and Technology in the United States. This particular equipment takes samples of size 20 cm x 20 cm mounted in a horisontal position on top of a load cell. Above the sample there is a cone heater and a spark ignitor. Gas samples are taken in fan ventilated exhaust duct mounted above the cone heater. The radiation used has been 50 kW/m^2 and free convection ventilation over the sample.

The ventilation rate in the fume hood and ventilation duct is approx 15 liter of air per seconds. The experiments have been performed with instrumentation at The Swedish Institute for Wood Technology Research.

The thirteen materials used in the over all study are presented in Table I. From the materials in Table I, numbers 2, 3, 12, 11, 6 and 8 were chosen for the study in the small scale fire tests.

Analytical Procedures

The first part of this study covered the development of measuring techniques for a large number of gases. After some testing and refinement of the gas analysis system we have studied the following species.

Carbon monixide	CO
Carbon dioxide	CO_2
Oxygen	O_2
Nitrous oxides	NO/NO_x
Hydrocarbons	$(CH)_n$
Hydrogen chloride	HCl
Hydrogen cyanide	HCN

In the full scale fire tests some additional gaseous species were studied specifically, i.e. formaldehyde. Not all gas species were studied in every test. Hydrogen cyanide and hydrogen chloride have only been studied in situations where evolution of these species were suspected. HCN and HCl have only been studied as collective (2, 5 or 10 minutes) samples for each fire test. It is most preferable to follow the concentrations with direct reading instruments. This has been the case for carbon monoxide, carbon dioxide, oxygen and in three out of four cases for nitrous oxide. Dräger tubes were used for measurements of nitrous oxides in the DIN 53436 test.

The amount of hydrocarbons present in the fire effluents have been measured in two different ways; 1) amount of non-burnt hydrocarbons 2) soot was separated from gas with a glassfilter and later extracted with cyclohexane. After the porous filter an absorbent glass-tube was connected with either charcoal or Tenax GC as the absorbent. Charcoal tubes were later extracted with carbon disulfide for analysis and Tenax tubes directly thermally desorbed to a gas chromatograph and a mass spectrometer.

For the small scale fire test methods it was possible to determine the mass of the sample burnt. In the full scale fire test this could not be done. To make gas emissions comparable between the fire models, the emissions of gases in the small scale fire tests have been reduced by the amount of material burnt in each case.

Results

Results full scale fire tests. The analytical results from six of the thirteen materials investigated in the full scale fire test are presented in Table II. The integrated amount of each gas from the start of the experiment until flash over in the room has occured is

Table I.   Tested Products

Products	Thick-ness (mm)	Density (g m^{-2})	Weight	Moisture content (%)	Application or other pretreatment
1 Insulating fiberboard	13	250	–	7.0	–
2 Medium density fiber board	12	600	–	5.9	–
3 Particle board	10	750	–	7.1	–
4 Gypsum plaster board	13	700	–	–	–
5 PVC wallcover-ing on gypsum plaster board	0.7 +13	–	240[1]	–	Glued
6 Paper wall-cover on gypsum plaster board	0.6 +13	–	200[1]	–	Glued
7 Textile wall-covering on gypsum plaster board	0.7 +13	–	370[1]	–	Glued
8 Textile wall-covering on mineral wool	0.7 +50	100[2]	370[1]	–	Glued
9 Melamine	13[3]	810[4]	–	6.7	Laminate was glued on both sides of the particle board
10 Expanded polystyrene	50	20	–	–	Glued to non-combustible silicate board
11 Rigid poly-urethane foam	30	30	–	–	–
12 Wood panel, spruce	11	530	–	10.0	–
13 Paper wall-covering on particle board	0.6 +10	–	200[1]	–	Glued

Notes:
1) The wallcovering only
2) The mineral wool only
3) The front laminate was of thickness 1.2 mm
4) The entire product

Table II.   Integrated Amount of Gases Generated up to
Flashover (f.o.) in the Full Scale Fire Test

	CO_2 (g)	CO (g)	NO_x (g)	$(CH)_n$ (g)	Time to f.o. [1] (sec)
Medium density fiber board	1950	380	nd	16	134
Particle board	2340	460	nd	12	150
Wood panel spruce	2730	410	nd	8	138
Rigid Poly- urethane foam	3510	350	d	8	14
Wall paper covering on gypsum plaster board	nd	nd	nd	nd	none [2]
Textile wall covering on mineral wool	4290	360	nd	10	55

d = detected, but to small for quantification
nd = not detected
1) The time until flame emerges through the doorway is defined as
flashover (f.o.).
2) No flashover within 600 seconds and not after increasing burner
output.

being presented. Flashover being defined as the moment when first
flame emerges through the opening of the door.
     The overall production of CO_2 during the experiment varies
between the samples within a factor of two. The amount of CO_2
produced did not correlate with time to flashover. The production
of CO increase very much up to flashover for most materials. High
concentrations of CO are produced directly before flashover is
reached. Wall paper covering on gypsum plaster board generated very
low concentrations of gases. It is therefore not possible to give a
fixed value. Particle board and fiber board seems to generate a
little more hydrocarbon than the rest of the materials. On the
other hand there is a significant difference in the time to flashover for the different materials. A closer look at the generation
schemes reveals increasing concentrations of CO and $(CH)_n$ just
before flashover, allowing more time for generation from the wood
fiber materials. The CO_2 production is fairly stable during the
whole of the experiment, for each of the materials.
     The NOx concentrations are very low. Only for rigid polyure-
thane foam could NOx be detected before flashover, the value still
too low to be specified. Recalculation of data through an estimation of average flow through the doorway just before flashover
leads to typical values of 2 %, 8 % and 500 ppm for CO, CO_2 and
$(CH)_n$ respectively for the materials studied. Both CO, CO_2
concentrations are reaching levels at which the gases have severe
effects on humans and in the case of CO become lethal to man within
one minute.

Results small scale fire tests. Initially we had in mind to make a comparison between NT-FIRE 025 and NT-FIRE 004 "box test" since most lining materials had been classified using that method. It turned out that the gases generated by the method NT-FIRE 004 needed a very high dilution before they could be fed into our direct reading instruments. Instrumentation used for smoke stack sampling, EMP-797, turned out to be suitable for our purpose.

During the experiments with NT-FIRE 004 a continous dilution of 1 to 20 was being used.

The sample sheets were weighed before and after each experiment and thus the material being burnt could be determined. The integrated amount of gas generated per gram burnt material is presented in Table III. The results are presented as average results of two similar experiments.

In each experiment with the cone calorimeter one piece of sample, 20 cm x 20 cm, was tested. During the test period three pieces of each type of sample were tested. The results presented in Table IV are therefore the average integrated amount of gases generated during sets of three experiments.

Table III.   Results from Nordtest NT-Fire 004

	CO_2 (mg/g)	CO (mg/g)	NO_x (mg/g)	$(CH)_n$ (mg/g)	Material burnt (g)
Medium density fiber board	501	94	0.22	0.15	375
Particle board	438	71	0.25	nd	333
Wood panel spruce	772	–	–	0.10	263
Rigid Poly-urethane foam	83	79	4.8	2.3	70
Wall paper covering on gypsum plaster board	919	46	0.46	2.0	123
Textile wall covering on mineral wool	938	159	0.65	3.7	92

nd = not detected

In calculating the data a dilution factor of 1/20 has been used for the EPM 797-system.

Table IV.   Results from the Cone Calorimeter

	CO_2 (mg/g)	CO (mg/g)	NO_x (mg/g)	$(CH)_n$ (mg/g)	Material burnt (g)
Medium density fiber board	1050	33	0.4	nd	184
Particle board	390	39	1.0	nd	156
Wood panel spruce	320	54	0.3	nd	99
Rigid Polyurethane foam	1650	220	6.0	nd	17
Wall paper covering on gypsum plaster board	1390	61	2.0	nd	33
Textile wall covering on mineral wool	4050	81	3.0	nd	21

nd = not determined

In calculating the data a dilution factor of 10.0 has been observed between the cone and the sampling line.

In running the DIN 53436 method hydrocarbon and hydrogen cyanide has only been determined qualitatively. The cyanide concentration has been determined four times during the 30 minute steady state combustion process. From these experiments the average concentration of emission has been estimated. The other results presented in Table V from DIN 53436 experiments have been measured in similar ways as for the other small scale test methods. It may be observed that the amount of material burnt in each experiment is smaller than in previous test procedures. The results presented are average values of two deteminations of each material.

Table V.   Results from DIN 53436 - Test Method

	CO_2	CO	NO_x	$(CH)_n$ 1)	HCN 2)	Material burnt
	(mg/g)	(mg/g)	(mg/g)	(mg/g)	(mg/g)	(g)
Medium density fiber board	610	305	<0.01	trace	0.3	8.2
Particle board	663	253	0.1	trace	1.2	8.3
Wood panel spruce	1020	246	0.02	trace	0.4	5.7
Rigid Poly- urethane foam	580	158	<0.01	nd	5.6	1.9
Wall paper covering on gypsum plaster board	296	74	0.05	nd	0.2	13
Textile wall covering on mineral wool	1160	220	0.3	nd	1.2	3.0

1) The measurement of hydrocarbon have only been made qualitatively
2) Hydrogen cyanides have been investigated using Dräger tube only. Thus the results are only qualitative.

A qualitative comparison of the results from the gas chromatography - mass spectrometer study of the different hydrocarbons from the wood materials, did not show significant differences between results from one method to the other. As far as can be judged it is mainly the amount of each component that differs between the small scale test methods and full scale fire tests.

Results from thermally desorbed samples taken from the fire effluents of the different materials in the full scale fire experiments are presented in Table VI. The different species of hydrocarbons have been grouped together and presented in three different cathegories; T, M and R representing Trace, Medium and Rich concentration. In this way it is possible to get an idea of the amount of contribution of different species of hydrocarbons to the fire effluents of each material. These results agree well in principle with results obtained by other researchers (6, 7).

Table VI. Organic Species found in the Fire
Effluents from the Fullscale Fire Test
NT-FIRE 025 for the Different Materials

Material	Chemical species						
	"1"	"2"	"3"	"4"	"5"	"6"	"7"
Insulating fibre-board	R	R	R	T	T	T	T
Medium density fibreboard	R	R	R	T	T	T	T
Particle board	R	M	M	R	T	T	T
Gypsum plaster board	R	R	R	T	T	T	T
PVC wallcovering on gypsum plast board	T	T	R	R	T	T	M
Paper wallcovering on gypsum plaster board	R	T	R	T	T	T	T
Textile wallcovering on gypsum plaster board	M	M	M	M	T	T	T
Textile wallcovering on mineral woll	M	M	M	M	M	T	T
Melamine faced particle board	T	T	T	T	T	T	T
Expanded poly-styrene	M	T	M	M	T	T	T
Rigid polyurethane foam	M	T	M	T	T	T	T
Wood panel, spruce	M	T	T	R	T	T	T
Paper wallcovering on particle board	T	T	T	T	T	T	T

1 = Alkanes, alkenes, alkynes ($C_1 - C_5$)
2 = Alkanes, alkenes, alkynes ($C_6 -$)
3 = Aromaties without heteroatoms
4 = Aromaties with heteroatoms
5 = Chlorinated hydrocarbons
6 = Aldehydes
7 = Alcohols

T = Trace, concentration < 1 mg/m^3
M = Medium, concentration 1 - 10 mg/m^3
R = Rich, concentration > 10 mg/m^3

Table VII.   Comparison between Methods for the Ratio
CO_2/CO

	Full scale	NT 004 "Box"	Cone Calorimeter	DIN 53436
Medium density filterboard	5.1	5.3	32	2.0
Particle board	5.1	6.1	10	2.6
Wood panel, spruce	6.7	–	5.9	4.1
Rigid Polyurethane foam	10.0	1.0	7.5	3.7
Wallpaper covering on gypsum plaster board	–	20	23	4.0
Textile wall covering on mineral wool	11.9	5.9	50	5.3

In the international discussions concerning standardisation of small scale toxicity tests it has been suggested how to compare different fire scenarios. Oxygen concentration and the ratio of CO_2/CO concentrations have been given as two of the major chemical indices for such a comparison in addition to temperature. The results from the full scale fire tests constitute accumulated concentration during a developing stage. Still it is of interest to compare those figures with the ones from the small scale fire tests on the same basis.

In Table VII the ratio of CO_2 and CO concentrations during the experiments are presented for each material and each testing procedure.

Discussion

A comparison of results for fire effluents from full scale and small scale fire tests has to be done in steps. A full scale fire is a developing event where temperature and major constitutions changes continously. A small scale fire test either take one instant of that developing stage and try model that or try to model the development in a smaller scale. On a priority one level rate of heat release, temperature, oxygen concentrations and the ratio of CO_2/CO concentrations have to be similar for a comparison. The full scale fire experiments reaches a temperature of 900 °C at the moment of flashover, while the small scale fire tests are reaching temperatures just above 400 °C for NT-FIRE 004 and the cone experiments. For the DIN 53436-method the temperature was set to 400 °C.

Keeping in mind the somewhat different approach for the production of fire effluents in the different methods used, one will find that the results presented in Table VII do not differ dramatically much from each other. The results for the cone calorimeter are however distinctly higher than the corresponding values for the other methods.

Oxygen concentration and the ratio of CO_2/CO are largely significant for the rate of ventilation in a fire. The full scale fire test and the box method, NT FIRE 004, are both representing developing stages of a fire. The cone calorimeter also represent a developing fire situation under very good ventilation conditions since air is driven over the sample partly by convection and partly by the draft from the fan in the exhaust duct. It is obvious that ventilation affects the ratio in Table VII for the cone calorimeter. The full scale fire test and the box-method demonstrate fairly good agreement between the two methods. The DIN 53436-method is a steady state procedure for generating nearly constant concentrations of fire effluents over at least a 30 minute period. We observe fairly consistent results from this test. One may also observe that the ratio CO_2/CO for the DIN-procedure is approximately half of that observed for the full scale fire test most probably due to a low ventilation rate. Monitoring of the oxygen concentrations during the experiments confirms these observations.

The low value of 1.0 in Table VII for the rigid Polyurethane foam in the box-methods may be explained by low ventilation rate. The experiment is over in less than 1 minute and most of the material is combusted during that period. The ventilation rate becomes to low to generate a "normal" ratio of CO_2/CO for fast burning materials. In this particular case the oxygen concentration become very close to 0 % oxygen.

In the cone calorimeter two materials, Rigid Polyurethane foam and Textile wall covering on mineral wool, give short dips in oxygen-concentration down to approx 10-11 % oxygen when ignition occurs. The remainder of the time the oxygen concentration is very close to ambient. Wood materials cause a decrease of oxygen concentration to 15-17 % over the whole experimental time.

In the DIN 53436 experiments the oxygen concentration is fairly stable over each experiment and varies between 13 to 18 % for the different materials. For the full scale fire test the oxygen concentrations stays close to the ambient almost all the way up to point of flashover.

Direct reading instruments may be used in the studies of fire effluents from different small scale testing methods. Moisture and high concentrations do create problems that has to be considered. Such problems may be overcome by using a dilution system such as the EPM 797-system. The instruments used must be fast responding since in certain situations the whole event is over within 1 minute. This also requires proper attention in the design of the sampling line.

When comparing results from one small scale method to the other it is observed that an increase in CO_2 in mg/g are not stoichiometrically balanced by a decrease in CO mg/g. Carbon takes

part in the formation of char and volatile hydrocarbons. The amount
of these species also varies considerably from one method to the
other. However improvement in precision of the results should be
possible.

The total amount of volatile hydrocarbons is very moderate
compared to the amount of carbon monoxide generated. It is therefore at this stage from a toxicological point of view less meaning
full to state the appropriate concentration of each specie.

Conclusions

Chemical measurements can be performed during developing stages of
fire using direct reading measurements and sampling techniques like
sorbent tubes.

Direct reading instruments are necessary for recording different stages of a fire in order to compare different methods and
materials to each other.

High precision in chemical measurements are needed for a good
evaluation of different fire tests.

Individual hydrocarbons may be sampled and analysed when the
information needed justifies the work to find the appropriate concentration.

A comparison of large scale and small scale methods require
records of the mass burnt for all methods. If no such information
the comparison is only qualitative.

This study demonstrate similar ranking of each material independent of test method used. At this stage it is premature to
choose one test as a better small scale model of full scale fires.
Each method needs further elaboration.

Acknowledgments

This project was financially supported by the Swedish Fire Research
Board and the National Testing Institute.

We are grateful to the Swedish Institute for Wood Technology
Research and the Dantest Fire Technology laboratory for their help
during this project.

Literature Cited

1. Sundström, B. Full Scale Fire Testing of Surface Materials
    Technical Report SP-RAPP 1986:45, Borås, Sweden, 1986; p 117.
2. Sundström, B. Room Fire Test in Full Scale for Surface Products
    SP-RAPP 1984:16, Borås, Sweden 1984.
3. Holmstedt, G. Rate of Heat Release Measurements with the
    Swedish Box Test, NT-Fire 004, SP-RAPP 1981:30, Borås, Sweden,
    1981.
4. Producing Thermal Decomposition Products from Materials in an
    Air Stream and Their Toxicological Testing: Part 1. Test Method
    53436. Deutsches Institut für Normung, Berlin, 1981.
5. Svensson, G., Östman, B., Rate of Heat Release for Building
    Materials by Oxygen Consumption, STFI-meddelande serie A no
    761, Stockholm, Sweden, 1982.

6. Fardell P. et al. Fire Research Station, BRE, United Kingdom, Private communication.

7. Levin, B.C., A summary of the NBS litterature Reviews on the chemical nature and toxicity of the pyrolysis and combustion from seven plastics: acrylonitrite-butadien-styrenes (ABS), nylons, polyesters, polyetylenes, polysterenes, poly(vinyl-chlorides) and rigid polyurethane foams, NB SIR 85-3267, 1986.

RECEIVED November 20, 1989

Chapter 5

Toxicity of Particulate Matter Associated with Combustion Processes

R. F. Henderson

Lovelace Inhalation Toxicology Research Institute, P.O. Box 5890, Albuquerque, NM 87185

Most of the concern for the toxicity of the atmospheres associated with fires has focused on vapors and gases. Vapors and gases are the components that are known to cause acute toxicity, and at high concentrations can lead to incapacitation and death. It is clear, how-- ever, that the smokes from fires also have particulate components in the form of soot and chemical reaction products, such as metallic oxides or ozonolysis prod-- ucts. The toxicity of these materials must also be considered.

Factors Influencing Toxicity of Particulate Matter

In evaluating the potential toxicity of airborne particles, it is important to consider where they will deposit in the respiratory tract, how long they can be expected to be retained there (or else-- where in the body if the particles clear from the lungs) and the inherent toxicity of the particles. All of these considerations are determined by the physical and chemical characteristics of the particles.

The site of deposition in the respiratory tract will depend on the aerodynamic diameter of the particle (Figure 1) ([1,2]). Particles greater than 10 µm aerodynamic diameter are not generally considered respirable. Particles in the 5–10 µm range will deposit mainly in the nasopharyngeal region while smaller particles will have signifi-- cant pulmonary deposition. Recent work indicates that a significant fraction of inhaled submicron particles deposit in the nose as well as in the lungs ([3,4]). Thus, particle size plays a key role in determining the portion of the respiratory tract that may be a target for toxicity of the particle.

Retention of particles in the lung will depend, in part, on their solubility in the milieu of the lung ([1]). Water soluble salts, such as most chlorides, will clear the lung rapidly, while

0097–6156/90/0425–0048$06.00/0

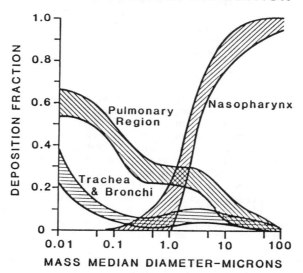

Figure 1. Effect of size on particle deposition in the respiratory tract. (Reproduced with permission from ref. 1. Copyright 1966 Pergamon.)

insoluble metallic oxides will be retained with half-times of clearance of hundreds of days (2,5). If the particle is of intermediate
solubility or contains a mixture of soluble and insoluble components, the clearance of the particle or its soluble fraction will be
influenced by the surface area available for dissolution. The
larger the surface area per unit mass, the higher the rate of
dissolution of the soluble components of the particles, and the
greater the potential for those components to interact with lung
tissue.

The inherent toxicity of the particle will depend on its chemical composition. For example, a particle of $CdCl_2$ can be expected
to be more toxic than NaCl because of the known toxicity of Cd^{+2}
(6). In the case of fibrous particles, the toxicity of the material will also depend on the size and shape of the fibers, with
long, thin fibers being the most toxic (7,8).

In summary, the type of airborne particle that is of most
toxicological concern is one that is small enough to have significant deposition in the lung, is insoluble enough in the fluids of
the lung to be retained in the lung for long periods of time, and is
inherently toxic to the lung tissue.

Potential Toxicity of Particles in Smokes

With the above factors in mind, what can be said about the potential
toxicity of the particulate matter in smokes? Each fire will have
smoke with a unique composition and this composition will vary with
time for a single fire. However, some generalities can be noted.
Fires can be expected to produce aerosols due to incomplete combustion and the condensation of volatile components. Most of these
particles that remain airborne in smokes will be in an aerodynamic
size range of submicron to micron particles (9). Thus, the smokes
will have respirable-sized particles and a significant fraction of
those particles, if inhaled, will be deposited in the deep lung
(2,5). Such small particles will also remain suspended in air
longer than larger particles and thus, will pose a potential exposure hazard for a longer period of time. In addition, small
particles have a large surface area per unit mass and, therefore,
the potential to adsorb or desorb more associated chemicals than
larger particles. Thus, the soot in smokes has the potential for
carrying various adsorbed toxicants into the deep lung where they
may be desorbed.

Second, many of the particles produced, such as soot and
metallic oxides, will have low solubility in the lung. Thus, once
deposited, these particles will remain in the lung for long periods
of time with a greater potential for exerting whatever inherent
toxicity they may have than would soluble particles (1,5). All of
these factors must be considered in evaluating the toxicity of
particulate matter in smokes.

Toxicity of Soot from a Controlled Combustion Process

One common particulate component of smokes is soot. For the
remainder of this paper, the toxicity of soot will be addressed.
Soot results from incomplete combustion of carbonaceous material

such as wood or fossil fuels. Variations in the toxicity of soot
should depend on the type and amount of chemicals adsorbed to the
soot. As a model, the soot produced from a controlled combustion
process--the burning of a standard fuel in a diesel engine--will be
considered. The toxicity of soot from the exhaust of a diesel
engine will then be compared to the toxicity of pure carbon black, a
soot with negligible adsorbed organic chemicals.

Physical/Chemical Characteristics of Diesel Soot. The diesel soot
discussed in this paper is from exhaust generated by a 1980,
5.7-liter Oldsmobile engine operating on a dynamometer using the
Federal Test Procedure urban driving cycle and burning a standardized certification fuel (D-2 Diesel Control Fuel, Phillips
Chemical Co.) (10,11). The carbon black is Elftex 12 from the Cabot
Corporation. The physical characteristics of the two soots are
similar. Both types of particles have volume median diameters of
~ 0.1 μm with surface areas per unit mass of 40 to 60 m^2/g (Wolff,
et al. Inhal. Toxicol., in press). Neither type of soot particle
is soluble in the lung. The distinguishing characteristic of the
diesel soot is that it contains 10-30% by weight of organic compounds (amount varies with the operating condition of the engine)
that are extractable into methylene chloride while pure carbon black
has negligible amounts of extractable organic material.
     The extractable organic matter associated with the diesel soot
has been chemically characterized and is known to have some biological activity (Table I) (12). Most of the mass of the material
extracted is in the form of high molecular weight aliphatic compounds, probably from unburned fuel and motor oil. The bacterial
mutagenic activity, however, is mainly associated with the aromatic
hydrocarbon fraction. Highly mutagenic nitroaromatic derivatives
would be expected to form from the interaction of aromatic compounds
in the fuel with the NO_2 formed during the combustion process. Such
nitroaromatic compounds have been detected in diesel fuel treated
with NO_2 and in the extracts of soot collected from the exhaust of
diesel engines (Table II) (13). Thus, diesel soot contains compounds that can interact with the genetic material, DNA. Organic
extracts of diesel soot have also been shown to be carcinogenic in
mouse skin assays (14). The next question is whether such compounds
are available to the lung tissue when diesel soot is inhaled and
deposited in the lung.

Table I. Extractable Organic Matter in Diesel Soot

Fractions from LH-20 Chromatography	Mass (%)	Bacterial Mutagenicity (%)
High MW aliphatic hydrocarbons	58	7
Aromatic hydrocarbons & derivatives	19	81
Polar compounds	23	12

SOURCE: Data from Bechtold et al. (12).

Table II. MS/MS Analysis of Nitroaromatics[a]

	Relative Intensity of Normalized Mass Spectra		
	A Fuel Aromatics + NO_2	B Filter Extract New Engine	C Dilution Tunnel Sediment Extract
Compounds			
Naphthalene-NO_2	+	+	
Biphenyl-NO_2	+++	++	+++
Fluorene-NO_2	++	++	+
Phenanthrene-NO_2	++	++	+
Dinitrobiphenyl			+
Pyrene-NO_2	+	+	+
Dinitrofluorene			+

SOURCE: Data from Henderson et al. (13).
[a]Analysis of nitroaromatics found by treating diesel fuel with NO_2
(column A) compared to nitroaromatics found in extracts of filters
of exhaust from a diesel engine (column B) or in extracts of
diesel soot deposited in a dilution tunnel of an animal exposure
system (13).

Bioavailability of Soot-Adsorbed Organic Chemicals. If the organic
compounds adsorbed to soot are available for direct interaction with
the DNA of the lung or if the compounds are available for metabolism
to DNA-reactive compounds by lung microsomal enzymes, the potential
genotoxicity of the diesel particles could be significant, particularly in view of the persistence of the insoluble soot particles
in the lung. In one study performed in vitro, ^{14}C-benzo(a)pyrene
(BaP), a typical aromatic compound produced in combustion processes,
was adsorbed onto diesel soot and the ability of lung or liver
microsomes to facilitate removal of the ^{14}C-BaP was observed (Figure
2) (15). Transfer of BaP from the soot to the microsomes was found
to be dependent on the lipid content of the microsomes with greater
amounts of BaP found in association with the more lipid lung microsomal fractions than with the liver microsomes. After two hours of
incubation, the amount of BaP transferring from the soot to the
microsomes leveled off at approximately 3% of the BaP originally on
the soot, and of that, only 1-2% was metabolized (Table III). This
indicates that soot-adsorbed chemicals are not readily available to
the lung, but that microsomes or the lipids in the milieu of the
lung may facilitate a very slow removal of a small fraction of the
chemicals. This slow removal, however, could be significant over a
long period of time.

Acute Toxicity of Diesel Soot. The acute toxicity of diesel soot
deposited in the lung is low compared to that of toxic particles
such as Ga_2O_3 (16) or nickel salts (17). In rats that had similar
lung burdens of diesel soot or Ga_2O_3 (0.5-0.6 mg/g lung), there was
no evidence of an inflammatory response to the soot, as evaluated by
bronchoalveolar lavage, while there was a strong inflammatory
response to the Ga_2O_3 (18,19). Less than 0.5 mg/g lung of Ni_3S_2,

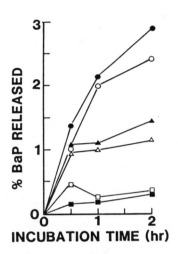

Figure 2. Bioavailability of soot-adsorbed benzo(a)pyrene (BaP). The transfer of ^{14}C-BaP from diesel soot to microsomes was measured. Microsomal protein (0.5 mg) was incubated at 37° C with the [^{14}C]-benzo(a)pyrene-coated diesel particles. (●) lung microsomes and 0.2 mM NADPH; (○) lung microsomes; (▲) liver microsomes with 0.2 mM NADPH; (△) liver microsomes; (■) 0.5 mg albumin; (□) buffer (0.15 M phosphate buffer, pH 7.7, containing 3 mM $MgCl_2$ and 0.1 mM EDTA). The presence of NADPH, a cofactor necessary for BaP metabolism, did not affect the transfer of BaP from the soot. (Reproduced with permission from ref. 15. Copyright 1988 Elsevier.)

Table III. (^{14}C)Benzo(a)pyrene Metabolites Formed in
Incubation Medium and on Diesel Particles[a]

| BaP Metabolites | Particle-Associated BaP | | Free BaP |
	Particle (%)	Medium (%)	Medium (%)
9,10-diol	ND	ND	33.3-42.4
7,8,9,10-tetrol	ND	ND	12.8-14.0
3-hydroxy	ND	1.2-1.5	ND
B(a)P	100	98.5-98.8	44.8-52.7

ND = not detected.
[a][^{14}]BaP-coated diesel particles or [^{14}C]BaP were incubated at
37°C with liver microsomes (1 mg protein) and 0.2 mM NADPH for
1 h. The data indicate the percent of the BaP present as parent
compound or as individual metabolites in each of the fractions.
SOURCE: Reprinted with permission from ref. 15. Copyright 1988 Elsevier.

NiSO$_4$ and NiCl$_2$ instilled into rat lungs also produced an inflammatory response (17). However, analysis of bronchoalveolar lavage
fluid from rodents exposed to diesel exhaust containing 3.5 mg
soot/m^3, 7 h/day for 2, 12 or 17 days indicated no influx of
inflammatory cells (20). Thus, the diesel soot, at lung burdens of
0.5 mg/g lung, does not produce an acute inflammatory response.

Chronic Toxicity of Inhaled Diesel Soot in Animal Studies. Numerous
studies on the chronic toxicity of inhaled diesel engine exhaust
have been reported. Studies conducted in the United States (National Institute of Occupational Safety and Health, Environmental
Protection Agency, Southwest Research Institute, General Research
Laboratories, Lovelace Inhalation Toxicology Research Institute), in
Germany (Fraunhofer Institute for Aerosol Research), Switzerland
(Batelle-Geneva Research Institute) and Japan (Japan Automobile
Institute) have been summarized in the proceedings of an international meeting on the subject (14).
     One of the studies at the Fraunhofer Institute clearly indi-
cated that the toxicity resulting from chronic inhalation of diesel
engine exhaust was due to the particulate component of the exhaust
and not the gases (21). Rats were exposed by inhalation over most
of their life span to filtered or unfiltered diesel exhaust.
Exposures were 19 h/day, 5 days/wk with soot concentrations of 4
mg/m^3. All of the measures of toxicity determined, including
decreases in body weight, alveolar clearance, and various measures
of lung function, as well as the induction of lung tumors, were
observed only in animals exposed to the unfiltered exhaust.
     In a study conducted at the Lovelace Inhalation Toxicology
Research Institute (ITRI), rats were exposed for up to 30 months, 7
h/day, 5 days/wk, to diesel exhaust containing 0, 0.35, 3.5, or 7.1
mg soot/m^3 of air. The diesel engine exhaust was generated as
indicated in the section of this paperon "Physical/Chemical
Characteristics of Diesel Soot." The lowest exposure concentration,
0.35 mg soot/m^3, is directly relevant to some occupational exposures
and is 10 to 100 times higher than any current or anticipated
environmental exposures. Observations of the animals were made at
6-mo intervals and included measures of dosimetry (mg soot/g lung),

microdosimetry (level of DNA adducts), clearance of secondarily
introduced tracer particles from the lung, inflammatory responses,
immune responses, respiratory function and histopathology. The
gaseous and particulate contents of the diesel exhaust are shown in
Table IV (22).

Table IV. Diesel Exhaust Composition

	Control (Air-Exposed)	Low	Medium	High
Particles (mg/m^3)	0.01	0.35	3.5	7.1
CO_2 (%)	0.2	0.2	0.4	0.7
CO (ppm)	1	3	17	30
Hydrocarbons (ppm)	3	4	9	13
NO_2 (ppm)	0	0.1	0.3	0.7
NO (ppm)	0	0.7	6	10

SOURCE: Data from Henderson et al. (13).

The accumulation of particles in the lung with time (Figure 3)
indicated that there was a greater lung burden in animals exposed to
the two higher exposure levels than would be expected based on the
deposition and clearance rates observed in the low level-exposed
rats (23). This suggested that chronic exposure to the high levels
of particles had impaired the normal clearance mechanisms of the
lung. This hypothesis was confirmed by studies on the ability of
the soot-laden lungs to clear inhaled ^{134}Cs-labeled fused aluminosilicate particles (FAP) from the lungs (Figure 4). The two high
level-exposed groups of animals cleared the secondary particles with
a half-time twice that of the control and low level-exposed rats
(23).
There was a detectable response of the lung to the inhaled
diesel soot only at the two highest exposure concentrations. The
inflammatory response detected in the bronchoalveolar lavage fluid
was exposure concentration-dependent and, in general, increased with
time of exposure and increasing lung burden (Figure 5) (22). Part
of the soot was cleared to the lymph nodes; these sites had increased total cells and increased antigen-specific antibody forming cells in response to subsequent immunization (24). Respiratory
function changes were consistent with development of dust pneumo-
coniosis at the two higher exposure concentrations (25). The
histopathology observations indicated a progressive inflammatory,
proliferative and fibrotic lung disease in animals exposed at the
two higher concentrations, with a small but significant increase in
tumors (Table V) (11).
In summary, these life-span studies in rodents suggest that
large quantities of diesel soot deposited in the lung are toxic.
However, there was no life-span shortening at any level of exposure. For noncarcinogenic endpoints there appeared to be a
threshold relationship with no significant responses at the low
exposure concentration and progressive alterations of many
parameters at the higher exposure concentrations when lung burdens

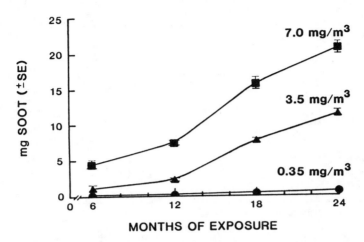

Figure 3. Accumulation of diesel soot in lungs exposed to diesel engine exhaust containing 0.35, 3.5, or 7.0 mg soot/m^3. Lung burdens at the lower exposure concentration did not exceed 1.0 mg/g lung (Data from ref. 23).

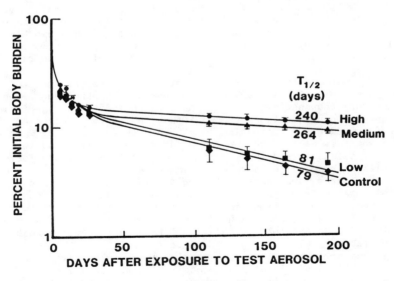

Figure 4. Clearance of ^{134}Cs-fused aluminosilicate particles from lungs of rats exposed to diesel exhaust (Data from ref. 23).

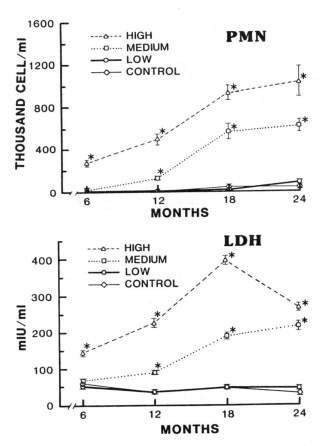

Figure 5. Pulmonary inflammatory response to chronic diesel exhaust exposure as measured in bronchoalveolar lavage fluid. The total amount or activity of material removed from the lung has been normalized to the weight of control lungs. Inflammatory response is indicated by influx of neutrophils (PMN). Cytotoxicity is indicated by extracellular lactate dehydrogenase (LDH). *(Continued on next page.)*

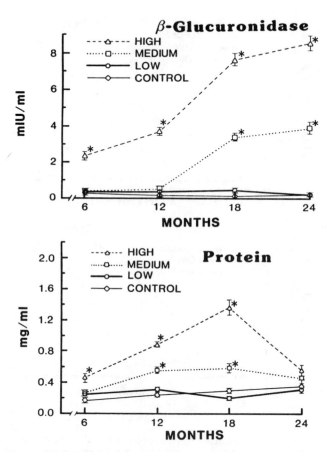

Figure 5. Continued. Phagocytosis is indicated by the release of the lysosomal enzyme, β-glucuronidase. Increased alveolar/capillary permeability is indicated by the increase in protein in the lavage fluid. (Reproduced with permission from ref. 22. Copyright 1988 Academic.)

Table V. Percentages of Rats with Lung Tumors
Following 30 Months Exposure to Diesel Exhaust

Exposure Group	Adenomas	Adenocarcinomas + Squamous Cell Carcinomas	Squamous Cysts Only	All Tumors
High	0.4	7.7[a]	4.9[a]	12.8[a]
Medium	2.3[a]	0.5	0.9	3.6[a]
Low	0	1.3	0	1.3
Control	0	0.9	0	0.9

[a]Difference from control significant at $p < 0.05$.
SOURCE: Reprinted with permission from ref. 23. Copyright 1987 Academic.

of particles were so high as to impair normal clearance mechanisms.
The observation of greatest interest was the induction of tumors in
the highest exposure groups. One must question whether this result
is due to the organic material associated with the diesel soot or
whether the tumors arise, in part, as a result of the overloading of
the lung with particles and the resulting chronic inflammatory
response. To determine if the organic material on the soot could
interact with DNA, shorter-term (12 wk) studies (Bond et al.
Toxicology, in press; Bond et al. In Assessment of Inhalation
Hazards: Integration and Extrapolation Using Diverse Data, 1989, in
press) were conducted and indicated that the diesel soot caused an
increase in the total DNA adducts (DNA altered by covalent binding
of reactive organic compounds), but the increase was equal for all
exposure levels (Figure 6). This may be because even the low lung
burdens saturate some step necessary for the formation of the
adducts. This finding is unexpected based on the fact that the
organic chemicals adsorbed to the soot are only slowly available.
This paradox will be addressed again in the discussion of the
formation of DNA adducts in response to exposure to carbon black.
The regional distribution of the DNA adducts in the respiratory
tract of diesel exhaust-exposed rats appears to agree with the known
deposition pattern of submicron particles, with the highest
concentration of adducts in the nasal and the pulmonary tissue
(Figure 7) (Bond et al. In Assessment of Inhalation Hazards: Integration and Extrapolation Using Diverse Data, 1989, in press).
  To further investigate the role of genetic vs epigenetic
mechanisms in the induction of the tumors, one can compare the
response of rats to diesel soot to that elicited by the pure form of
soot, carbon black.

Comparison of Toxicity of Diesel Soot and Carbon Black

In studies conducted at the ITRI, rats were exposed to 3.5 and 10 mg
particles/m^3 air of either Elftex 12 carbon black or diesel exhaust.
These exposures were 7 h/day, 5 days/wk for 12 wk (Wolff et al.
Inhal. Toxicol., in press; Bond et al. In Assessment of Inhalation
Hazards: Integration and Extrapolation Using Diverse Data, 1989, in
press). Dosimetry (mg particles/g lung), microdosimetry (DNA
adducts), pulmonary inflammation, and histopathology of the lungs of

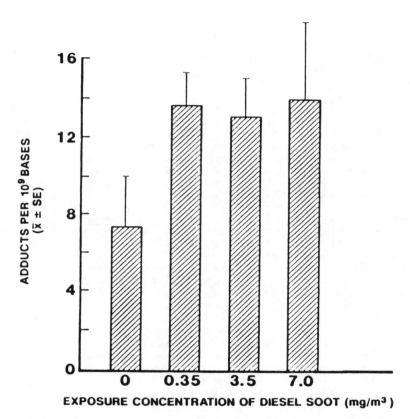

Figure 6. DNA adduct concentrations in rat lungs exposed to diesel exhaust containing 0.35, 3.5, or 7 mg/m^3 soot. 7 h/day, 5 days/wk for 12 wk. Quantitation was by the ^{32}P-postlabeling technique (Bond et al. *Toxicology,* in press; Bond et al. In *Assessment of Inhalation Hazards: Integration and Extrapolation Using Diverse Data,* 1989, pp 315–324).

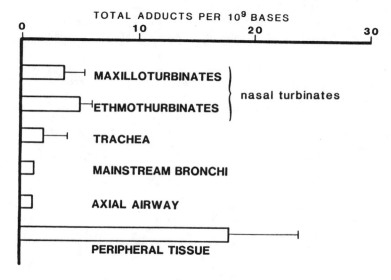

Figure 7. Regional distribution of DNA adducts in the respiratory tract of diesel exhaust-exposed rats (Bond et al. In *Assessment of Inhalation Hazards: Integration and Extrapolation Using Diverse Data,* 1989, pp 315–324).

rats exposed to the two types of soot were compared. The presence
of adsorbed organic compounds did not affect either the accumulation
or the clearance of the particles from the lung and the lung burdens
were identical in animals exposed to either particle. The micro--
dosimetry studies showed fewer total adducts in lung tissue from the
carbon black--exposed rats than from the diesel exhaust--exposed rats,
but the adducts in the diesel exhaust--exposed rats were not dose
dependent, an observation in agreement with earlier findings (Figure
8) (Bond et al. In Assessment of Inhalation Hazards: Integration
and Extrapolation Using Diverse Data, 1989, in press). Also, the
degree of difference in the concentration of the total adducts in
the DNA of rats exposed to the two types of particles was not as
great as one would expect, based on the degree of difference in the
extractable organic compounds associated with the particles. This
observation, combined with the fact that there was not a significant
increase in the number of lung tumors observed in the long--term
studies at the lowest exposure level, suggests that it is not the
concentration of the total DNA adducts that is of importance, but
rather it is more likely to be the level of specific DNA adducts
that will correlate with the induction of tumors.

The inflammatory responses, which were small, and the histo--
pathological responses to the two types of particles were identical,
indicating that the adsorbed organic compounds do not influence the
acute response to the diesel soot. The remaining question is whether
long--term exposures to carbon black would induce the same incidence
of tumors as did long--term exposures to diesel soot. Those studies
are underway at the ITRI and at the Fraunhofer Institute.

Epidemiology Studies of Diesel Soot Toxicity

In addition to animal studies, epidemiology studies by Garshick et
al. (26,27) point to an association between exposure to diesel
exhaust and lung cancer. In a case--control study in railroad
workers, workers 64 yr of age or younger at death and with 20 yr of
work in a diesel--exhaust--exposure job had an increased relative odds
of lung cancer of 1.4 (26). The same authors reported a retrospec--
tive cohort study with a large (55,407) cohort of U.S. railroad
workers (27). A relative risk for lung cancer of 1.45 was found for
workers with the longest history of diesel exhaust exposures. Based
on these and other epidemiology studies, as well as animal research,
the International Agency for Research on Cancer (IARC) has desig--
nated diesel exhaust as an animal carcinogen and a probable human
carcinogen (IARC. Monograph on the Evaluation of the Carcinogenic
Risk of Chemicals to Humans, 1989, in press).

Conclusions

Knowing the results of epidemiology studies on the association
between diesel exhaust exposure and lung cancer, and animal studies
on the toxicity of two types of soot, what can we conclude concern--
ing the toxicity to humans of soot associated with fires? Combus--
tion processes produce small, respirable--size soot particles. The
composition of the soot from fires can vary widely with the condi--

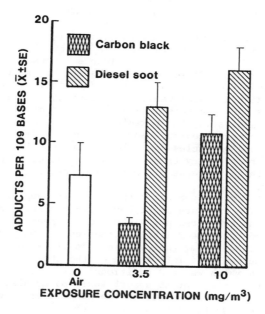

Figure 8. Accumulation of DNA adducts in rats exposed to carbon black or diesel soot for 12 wk. Quantitation was by the ^{32}P-postlabeling technique (Bond et al. In *Assessment of Inhalation Hazards: Integration and Extrapolation Using Diverse Data,* 1989, pp 315–324).

tions of the fire. Such soot, if inhaled and deposited in the deep
lung, can be expectd to be retained in the lung for a long time.
Small lung burdens (less than 1.0 mg soot/g lung), of the type of
soot generated by the combustion of diesel fuel, do not appear to
elicit adverse effects. Thus, for acute exposures, such as in
fires, when lung burdens would not be expected to reach 1 mg/g lung,
toxicity due to a similar type of soot particle would be of little
concern. Repeated inhalation of high concentrations of soot could
lead to accumulation of soot in the lung, impaired clearance of
particles, enhanced accumulation of soot, and eventually to pulmonary fibrosis and cancer. However, firefighters, the persons most
likely to have repeated exposures to soot from fires, are supplied
with protective breathing equipment that should prevent any
overexposure to soot.

Little is known about the toxicological significance of materials that may be adsorbed to the soot in various types of fires.
The studies described above provide valuable information concerning
the toxicity of soot in diesel engine exhaust, but similar studies
have not been done for soot from fires. The small respirable soot
particles represent an ideal vehicle for carrying various adsorbed
toxic substances into the deep lung and this potential must be
considered when evaluating the toxicity of fire-generated soot.
Finally, in the present review, there has been no discussion of the
potential that some particles generated by the chemical reactions in
smokes might be respiratory tract irritants that could produce acute
changes in breathing patterns or bronchoconstriction, just as many
gases and vapors do (28).

In summary, fires represent a particularly complex problem for
toxicologists. The exposure atmospheres are a mixture of many
substances, and no two fires are alike. One aspect of the mixture
that has received relatively little attention is the particulate
component of smokes. From studies on carbon black and on soot in
diesel exhaust we now have information on the chronic toxicity of
soot. An area in which we need additional information is the potential for fire-generated soot to adsorb toxic materials, and then to
deposit and desorb such materials in the lung. For this information
we must await the results of future research.

Acknowledgments

The author gratefully acknowledges the support of the U.S. Department of Energy, Office of Health Effects Research, Contract No.
DE-AC04-76EV01013, for the writing of this manuscript and for
supporting part of the research reviewed.

Literature Cited

1.   Task Group on Lung Dynamics. Health Phys. 1966, 12, 173-207.
2.   Schlesinger, R. B. In Concepts in Inhalation Toxicology;
     McClellan, R. O.; Henderson, R. F., Eds.; Hemisphere Publishing
     Corporation, 1989; pp 163-92.
3.   Yamada, Y.; Cheng, Y. S.; Yeh, H. C. Inhal. Toxicol. 1988,
     Premier Issue, 1, 1-11.

4. Cheng, Y. S.; Yamada, Y.; Yeh, H. C.; Swift, D. L.   J. Aerosol
    Sci. 19, 741-51.
5. Snipes, M. B. In Concepts in Inhalation Toxicology; McClellan,
    R. O.; Henderson, R. F., Eds.; Hemisphere Publishing
    Corporation, 1989; pp 193-7.
6. Commission of the European Communities, Directorate General
    Scientific and Technical Information and Information Management,
    Luxenbourg. Criteria (Dose/Effect Relationships) for Cadmium;
    Pergammon Press: Elmsford, NY, 1978.
7. Stanton, M. F.; Layard, M.; Tegeris, A.; Miller, E.; May, M.;
    Kent, E. J. Natl. Cancer Inst. 1977, 58, 587-603.
8. Wright, G. W.; Kuschner, M. In Inhaled Particles IV; Walton,
    W. H., Ed.; Pergammon Press: New York, 1977; pp 455-73.
9. Swift, D. L. In Aerosols in Medicine, Principles, Diagnosis
    and Therapy; Moren, F.; Newhouse, M. T.; Dolovich, M. B., Eds.;
    Elsevier Science Publishers, 1985; pp 53-76.
10. Wolff, R. K.; Henderson, R. F.; Snipes, M. B.; Griffith, W. C.;
    Mauderly, J. L.; Cuddihy, R. G.; McClellan, R. O. Fundam.
    Appl. Toxicol. 1987, 9, 154-66.
11. Mauderly, J. L.; Jones, R. K.; Griffith, W. C.; Henderson,
    R. F.; McClellan, R. O. Fundam. Appl. Toxicol. 1987, 9, 1-13.
12. Bechtold, W. E.; Dutcher, J. S.; Li, A. P.; Royer, R. E. In
    Inhalation Toxicology Research Institute Annual Report
    1981-1982; 1982; LMF-102, UC-48, pp 106-9.
13. Henderson, T. R.; Royer, R. E.; Clark, C. R.; Harvey, T. M.;
    Hunt, D. F. J. Appl. Toxicol. 1982, 2, 231-7.
14. Ishinishi, N.; Koizumi, A.; McClellan, R. O.; Stöber, W.
    Carcinogenic and Mutagenic Effects of Diesel Engine Exhaust;
    Elsevier Science Publishers, 1986.
15. Leung, H. W.; Henderson, R. F.; Bond, J. A.; Mauderly, J. L.;
    McClellan, R. O. Toxicology 1988, 51, 1-9.
16. Wolff, R. K.; Griffith, W. C.; Henderson, R. F.; Hahn, F. F.;
    Harkema, J. R.; Rebar, A. H.; Eidson, A. F.; McClellan, R. O.
    J. Toxicol. Environ. Health 1989, 27, 123-38.
17. Benson, J. M.; Henderson, R. F.; McClellan, R. O.; Hanson,
    R. L.; Rebar, A. H. Fundam. Appl. Toxicol. 1986, 7, 340-7.
18. Henderson, R. F. In Toxicology of the Lung; Gardner, D. E.;
    Crapo, J. D.; Massaro, E. J., Eds.; Raven Press, 1988; pp
    259-60.
19. Henderson, R. F. In Concepts in Inhalation Toxicology;
    McClellan, R. O.; Henderson, R. F., Eds.; Hemisphere Publishing
    Company, 1989; pp 415-30.
20. Henderson, R. F.; Leung, H. W.; Harmsen, A. G.; McClellan,
    R. O. Toxicol. Lett. 1988, 42, 325-32.
21. Stöber, W. In Carcinogenic and Mutagenic Effects of Diesel
    Engine Exhaust; Ishinish, N.; Koizumi, A.; McClellan, R. O.;
    Stöber, W., Eds.; Elsevier Science Publishers, 1986; pp 421-39.
22. Henderson, R. F.; Pickrell, J. A.; Jones, R. K.; Sun, J. D.;
    Benson, J. M.; Mauderly, J. L.; McClellan, R. O. Fundam. Appl.
    Toxicol. 1988, 11, 546-67.
23. Wolff, R. K.; Henderson, R. F.; Snipes, M. B.; Griffith, W. C.;
    Mauderly, J. L.; Cuddihy, R. G.; McClellan, R. O. Fundam.
    Appl. Toxicol. 1987, 9: 154-66.

24. Bice, D. E.; Mauderly, J. L.; Jones, R. K.; McClellan, R. O.
     Fundam. Appl. Toxicol. 1985, 5, 1075–86.
25. Mauderly, J. L.; Gillett, N. A.; Henderson, R. F.; Jones,
     R. K.; McClellan, R. O. Ann. Occup. Hyg. 1988, 32, 659–69.
26. Garshick, E.; Schenker, M. B.; Muñoz, A.; Segal, M.; Smith,
     T. J.; Woskie, S. R.; Hammond, S. K.; Speizer, F. E. Am. Rev.
     Respir. Dis. 1987, 135, 1242–8.
27. Garshick, E.; Schenker, M. B.; Muñoz, A.; Segal, M.; Smith,
     T. J.; Woskie, S. R.; Hammond, S. K.; Speizer, F. E. Am. Rev.
     Respir. Dis. 1988, 137, 820–5.
28. Alarie, Y. Food Cosmet. Toxicol. 1981, 19, 623–6.
29. Bond, J. A.; Wolff, R. K.; Harkema, J. R.; Mauderly, J. L.;
     Henderson, R. F.; Griffith, W. C.; McClellan, R. O. Toxicol. Appl.
     Pharmacol. 1988, 96, 336–346.

RECEIVED November 1, 1989

Chapter 6

Toxic Hazard and Fire Science

Howard W. Emmons

Division of Applied Sciences, Harvard University, Cambridge, MA 02138

The manner in which toxicological knowledge must work together
with the knowledge of human behavior, fire dynamics, and
chemistry to produce an acceptable level of fire safety is
proposed. A hypothetical example illustrates what must be done
with adequate accuracy in order to design fire safety to a
performance code. The example may give the impression that this
can already be done. In fact, each computer code used contains
dozens of assumptions, some very crude, so that the accuracy of
present predictions are unacceptably low.

The ultimate objective of the study of toxicity of fire-produced toxic
agents is the design and construction of a fire-safe environment. There
are many ways to accomplish this aim. We might develop materials which,
when heated, produce no toxic gases. We might use materials which do not
burn.

These fanciful solutions to the fire problems of society are already
possible. We could make everything from steel, reinforced concrete, and
ceramics, a solution that is uncomfortable, unesthetic, and unacceptable.
We thus commit ourselves to a balancing of risk, cost, and desirability.
We would like our built environment to satisfy a performance code which
states that "all occupants of this building will be able to safely exit or
reach a safe refuge area no matter where or when a fire starts."

This performance code may seem equally fanciful. It is the purpose
of this paper to discuss what is necessary to reach that goal and to show
the progress to date.

We can see what is necessary by looking at more mature engineering
fields to see how safety is assured. Building structural design, for
example, satisfies a performance code which avoids building collapse by
requiring the safe support of specified floor loads for various purpose
rooms. The structural engineer selects an appropriate detailed design by
use of scientifically derived and carefully validated formulas.

Fire safety design requires the same kind of approach. We must
develop the necessary scientific quantitative understanding of fire so as
to be able to predict the level of a building's fire safety. Again,
looking at structural engineering, we see the use of simple formulas for
beams, columns, joints, reenforcing rods, etc., which permit quantitative
evaluation of structural safety. The phenomena of fire also results in

0097–6156/90/0425–0067$06.00/0

formulas--sometimes not so simple--by which fires ignite, grow, decay, and extinguish. However, in the fire case, the interactions of physical, chemical, psychological, and physiological effects are so complex and numerous that, until the advent of the modern computer, comprehensive quantitative predictions of fire safety were far beyond our computing capacity.

What needs to be predicted to assure fire safety? Occupants must be alerted to the existence of a fire in their building. We must be able to predict what they will do; verify the alarm, try to find and extinguish the fire, notify others of danger, help handicapped, and move along an escape route. As they delay their escape more, the fire continues to grow. Thus, by the time that they walk, run, crawl along the escape route, there may be hot, toxic gas along the way. If the exposure is too high, they may be incapacitated and perhaps later, die.

To predict a safe escape from a building, we must know where the occupants may be, predict their behavior, predict their exposure to hot, toxic gases, and predict the effect of these gases. Before any of these predictions are possible, we need to predict the fire. The fire location, ignition, and growth controls the time at which the installed alarm system detects and announces an emergency. The fire and the details of the building design controls the time history of the gas temperature and composition on the escape route and these details must be predicted by scientifically-based fire safety engineering. With this information, it will be possible to predict the time of the alarm and the hazards on various escape routes. It will then be possible to predict whether or not all building occupants can safely escape.

This fire safety engineering program may sound as fanciful as some of my first suggestions. It was considered impossible 30 years ago. In fact, 30 years ago, it was impossible. But in 1989, it is not only possible, but all parts of the required science are making significant progress toward the minimum necessary level, and present computer codes can predict many, but by no means all, of the essential parts of this program (albeit at unacceptably low accuracy).

Illustrative Example

To make clear how far we have come along this road, let us compute a specific fire. Consider an apartment house with various apartments opening onto a 44m (144.4 ft) long corridor, Figure 1. A fire occurs in a 4 x 5 x 2.4m (13x16x8 ft) room 30 meters (98.4 ft) from the open end of the corridor. A family, father, mother, young boy, and baby, are asleep in a suite of rooms at the closed end of the corridor. The room where the fire occurs contains a bed, polyurethane mattress, an upholstered polystyrene frame, polyurethane foam padding and fabric chair and a wooden dresser. The fire starts in the upholstered chair. The door to the fire room is closed until the photoelectric detector in the room alarms and the occupant after ten seconds leaves the room and leaves the door open. The occupants in the distant suite, after various delays, move down the corridor to escape.

At present, there is no one computer fire code sufficiently comprehensive to compute this fire, including the people's response. In fact, no combination of present codes can solve this problem with the required engineering accuracy. To get an approximate illustrative solution to this case, a number of different computer fire codes must be used in succession and hand fit data transferred from one to the next. The computer programs used to make this (low accuracy) prediction and some of their often severe limitations will be indicated.

The fire in the chair grows proportional to the heat feedback from all sources, which change with time. The flames and hot gases rise to the

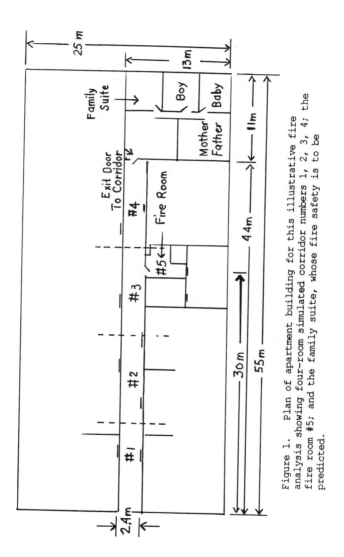

Figure 1. Plan of apartment building for this illustrative fire analysis showing four-room simulated corridor numbers 1, 2, 3, 4; the fire room #5; and the family suite, whose fire safety is to be predicted.

ceiling. A photoelectric detector at the ceiling alarms. The fire room
occupant awakes, runs out of the room and leaves the door open. The flames
from the chair, the hot layer of gas and the gradually heating ceiling, all
radiate to the bed and dresser. These items heat up and at their ignition
temperature start to burn. As they burn, radiation from their flames
enhance the burning rate of all burning objects and their plumes add to the
hot ceiling layer. Since the door has been left open, hot ceiling layer
gas, as soon as it is deep enough, flows out at the top of the door and
cold air flows in at the bottom. Soon the rate of pyrolysis in the room is
so high that all flammable pyrolysis gases cannot burn in the room; the
room fire becomes oxygen starved. All of these phenomena were computed
using the computer fire code FIRST. (1) (The computation took about ten
minutes for 2,000 seconds of fire time on a VAX 8600.)

There is, as yet, no code which will compute when the ceiling layer
with excess pyrolyzate will ignite and burn, nor the time when flames will
come out of the door into the corridor, nor the buoyant flow down the
corridor. However, the computer fire code FAST, (2) starting with a given
gaseous fuel release rate in a room, computes the distribution of the
resultant gases through a series of interconnecting rooms. Since FAST only
provides for five rooms and no corridors, the present problem is solved by
dividing the corridor into four, 11m (36 ft) long rooms separated by vents
the full width of the corridor and 2m (6.5 ft) high. FAST calculates the
resultant hot layer depth, temperature and composition in each room (four
of which are simulated parts of the corridor). (This calculation took 12
hours on an IBM-PC-AT.)

Since the hot layer in the fire room, because of O_2 starvation,
contains up to 20 mass percent fuel, and neither FIRST nor FAST can burn
this fuel in the corridor (which is what really happens), FAST was told to
burn all of the fuel in the fire room so that all the available energy is
released, even though in the wrong place, and is distributed by FAST to the
four (simulated corridor) rooms.

It was assumed that the alarm which sounded in the fire room was
barely audible in the suite of rooms. Father and mother slowly awaken and
prepare to move (70 seconds). The computer program EXITT (3) was used
which sends father to investigate the fire and then to awaken the boy,
mother to get her baby and all to go down the corridor to escape.

Finally, the computer program HAZARD (4) can compute the incapacitation or death of persons during an escape attempt. This program was not
used because a more detailed calculation is needed for the future and will
be illustrated here.

Results of Illustrative Example -- The Fire Dynamics

The rate of heat release in the fire room is shown in Figure 2. The smoke
at the ceiling is sufficient to sound the alarm at 109.2 seconds. The
chair burns out early. The bed and dresser, after being ignited at about
300 seconds, are soon limited in heat release by the limited oxygen
available. The pyrolysis rate is shown in Figure 3. The layer of hot
smokey gas at the ceiling has a maximum temperature of $1000°$ K, a maximum
fuel content of 20 percent, and a maximum carbon monoxide content of 5500
ppm, as shown in Figure 4.

The flow of hot gas out of, and fresh air into, the fire room is
shown in Figure 5. The fire room occupant wakes up and escapes at 121
seconds, 12 seconds after the alarm. He leaves the door open so fire gases
start to flow out. The outflow increases rapidly to 1.2 kg/sec while the
inflow, somewhat later, reaches 1 kg/sec. The outflowing hot layer gas
carries its fuel content which reaches a maximum of about .22 kg/sec.
When this fuel is burned in the corridor at the open door, 4.4 Mw is being
released. Compare this to the maximum of 1.68 Mw in the fire room. The

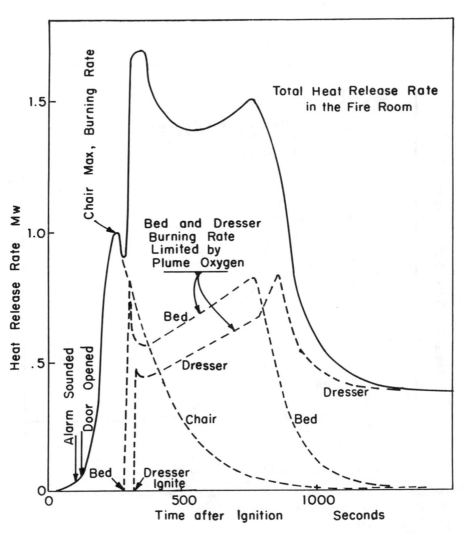

Figure 2.   The rate of heat release, Mw, predicted for the fire room by
computer fire code FIRST. The resultant heat release consists of contributions from an upholstered chair, a bed, and a dresser. These
latter two show severe burning limitations by the limited oxygen supply.

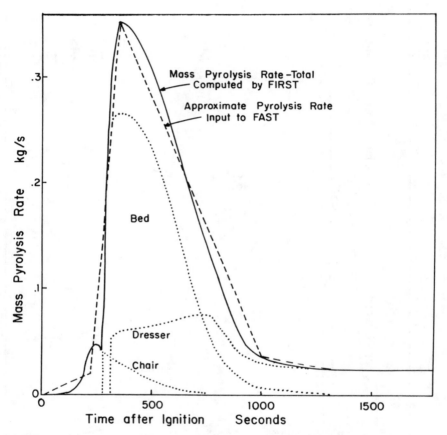

Figure 3.   The mass pyrolysis rate, kg/sec, predicted by FIRST:
Total (———), Chair, Bed, Dresser (· · · ·), Approximation input to
FAST (— — — —).

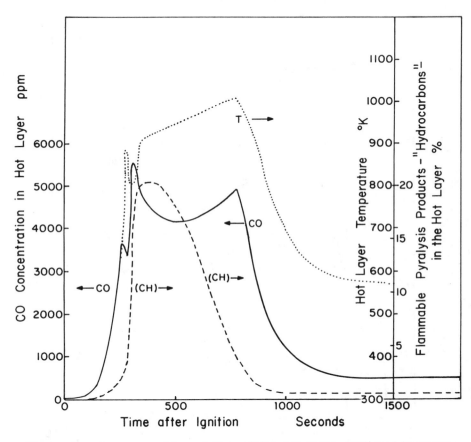

Figure 4.   Some properties of the ceiling hot layer predicted by FIRST:
Temperature   (· · · ·) K;   CO concentration   (————) ppm;   Unburned
fuel pyrolyzate   (— — — —)%.

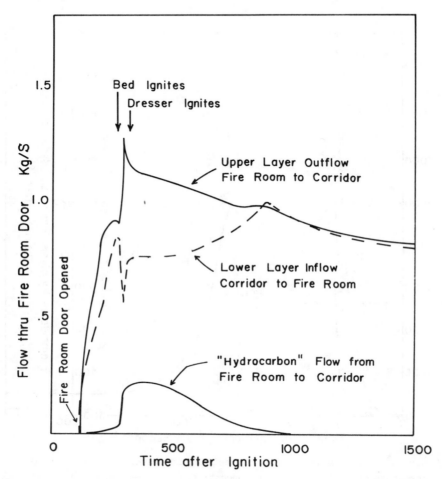

Figure 5.   Flow through fire room door. kg/sec, predicted by FIRST:
Upper hot layer outflow (————) containing unburned flammable
pyrolyzate (————); Lower cold layer inflow (— — — —).

sudden change in flow rates at 300 seconds may look wrong, but is consistent with the frequently observed pulse of gas which occurs in real fires as new heated fuel rapidly ignites.

On Figure 3 is shown (— — — — —) the mass pyrolysis rate used in computing the combustion product distribution by FAST. FAST ignores oxygen starvation so it will burn all the pyrolyzed fuel in the fire room instead of burning the excess pyrolyzate in the corridor. The hot toxic gases flow out along the corridor ceiling leaving a relatively cool, nontoxic layer of air returning at the floor. The depth of this cold layer computed by FAST is shown in Figure 6. The times and positions at which the cold layer is deeper than 1.2 meters are shown (\ \ \ \ \ \ \). In these positions and times, an able adult can move at 1.3 m/sec to escape. Only early in the fire (before about three minutes) is the cold layer deep enough to avoid crawling.

The temperature of the hot layer in the corridor is shown in Figure 7. In the corridor, the hot gas is untenable for an upright person next to the open fire room door after about 160 seconds. After about four minutes, the radiation from the hot layer would be too high to permit a person to pass. Furthermore, in actual fact, the temperature would be higher than that calculated at the fire room door, because of the fuel which would burn in the corridor, thus providing flame temperature radiation in addition to the hot layer temperature computed here.

The carbon monoxide in the corridor is shown on Figure 8. The data on CO production, especially during oxygen starvation, is very inadequate. These computed values are probably too small. It will, nonetheless, serve as an illustrative example. Figures 6-8 constitute the hazard maps for the escape route, in this case, the corridor. The time line for the fire is given in Table I.

The Occupant Actions

In the present example, the only fire detector is in the fire room. When it alarms, the fire room occupant is awakened immediately, but occupants in the suite can hardly hear it. They are slow to respond. The program EXITT knows where all persons are, decides mother will get baby, seeks the shortest path to get there and calculates the time at 1.3 m/sec. Mother and baby then follow the shortest route to the corridor door. The times required for these actions are given in Table II.

Simultaneously, the Father goes to the corridor door to investigate the fire. It is bad. He decides to return to awaken his son and they then go to escape down the corridor. Again, Table II gives the time line.

Having arrived at the corridor, can the suite occupants make it to safety? Mother with baby arrives at time 286.4 seconds (Table II). At this time, the cold layer at the floor is only .9 meter deep (Figure 6). The corresponding hot layer temperature is about 335° K (62° C) (Figure 7), but the cold layer temperature is the original low value. Also, the CO is about 1000 ppm in the layer above. To escape, the mother must crawl taking 84.6 seconds to reach the open end and safety. The line A-A on Figures 6-8 show the conditions during her escape.

The father and boy encounter conditions on line B-B. They must also crawl in order to avoid breathing CO of 5000 ppm and a maximum temperature of 625° K (352° C), which is lethal. They will all escape, if they can remain low enough and are not incapacitated by radiation from the hot layer over their heads. Mother encounters about 450° K (177° C) overhead for about 61 seconds, while father and boy would have to endure over 500° K (227° C) overhead for a minute with about 600° K (327° C) for 17 seconds. Mother and baby may make it to safety, but father and boy will probably succumb to the high temperature radiation.

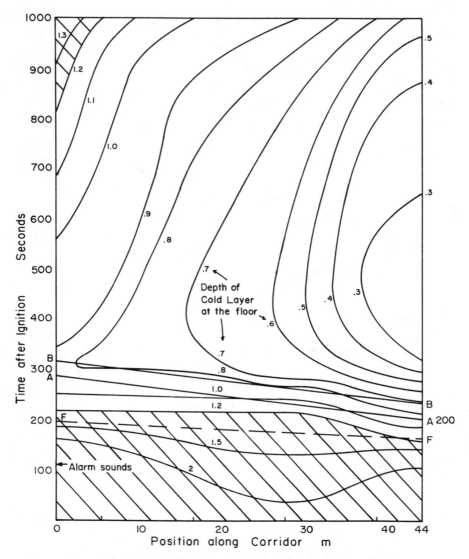

Figure 6.   Height above the floor of the interface between the upper
(hot) and lower (cold) layers in the corridor, m, predicted by FAST:
(A-A)   Mother-baby escape route;   (B-B)   Father-boy escape route;
(\ \ \ \)   Region where running is possible;   (F-F) Latest possible
escape route without crawling.

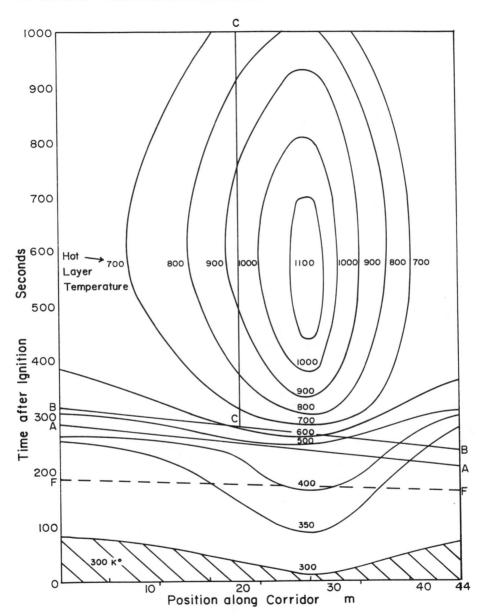

Figure 7. Hot layer temperature in corridor, K, predicted by FAST: (A–A) Mother-baby escape route; (B–B) Father-boy escape route; (C–C) Path of overhead radiation temperature exposure for person, if heat incapacitated at 20 m from exit; (F–F) Latest possible escape route without crawling; (\ \ \ \) Temperature not yet changed from ambient.

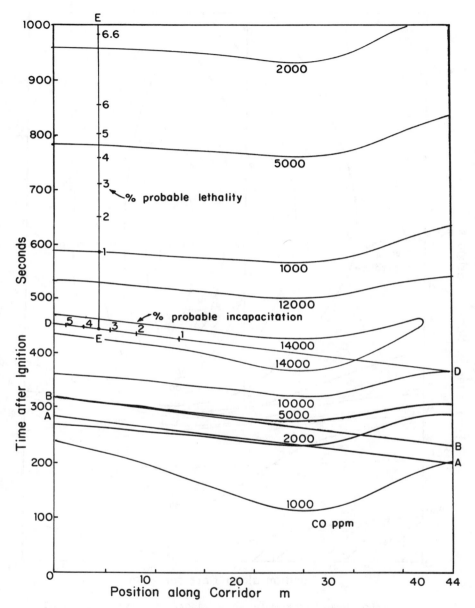

Figure 8.   Carbon monoxide in corridor, ppm, predicted by FAST:
(A-A)   Mother and baby escape route;   (B-B)   Father and boy escape
route;   (D-D)   Escape route through toxic but cooled layer.   Numbers
show probability of incapacitation location.   (E-E)   Shows CO exposure
for any person incapacitated (3.4%) 5 meters from the corridor exit.
Numbers show probability of death location.

Table I. Time Line of Fire

time-seconds	
0	Ignition of chair (flaming starts)
109.2	Photoelectric detector in fire room alarms
121	Fire room occupant leaves and leaves door open
160	Hot layer at end of corridor falls to 1.2 meters from floor
185	Hot layer in corridor by fire room door becomes incapacitating T = 350° K (77° C)
277.4	Bed heated to 520° K (247° C) ignites
278.3	Bed fire oxygen starved -- flames probably come out the fire room door at about this time (hot layer fuel = 9.85%)
310	Maximum CO in the fire room, 5500 ppm
314.1	Dresser heated to 600° K (327° C) ignites
315.5	Dresser oxygen starved
350	Maximum heat release rate in room, 1.65 Mw
390	Minimum depth of cold layer in fire room, .5 m
400	Maximum CO in corridor 14,500 ppm
480	Minimum depth of cold layer in corridor .29 m
580	Maximum temperature in corridor at fire room door, 1180° K (907° C)
610	Maximum temperature at corridor closed end, 710° K (437° C)
770	Maximum temperature in fire room, 1013° K (740° C)
772.5	Bed ceases to be oxygen starved
845.3	Dresser ceases to be oxygen starved
953	Bed reduced to 1Kg
~3000	Dresser reduced to 1Kg

Table II. Time Line of Occupants

time-seconds		
109.2		Alarm sounds in fire room. Father and mother vaguely hear it and are slow to awake.
179.2	(109.2 + 70)	Father and mother partially dressed and start to move.
183.8	(179.2 + 4.6)	Mother gets baby.
188.8	(183.8 + 5)	Mother starts to exit with baby.
193.2	(179.2 + 14)	Father arrives at corridor door to investigate fire.
201.8	(188.8 + 13)	Mother arrives at corridor to escape with baby.
203.2	(193.2 + 10)	Father concludes the fire is real and returns to get boy
212.9	(203.2 + 9.7)	Father arrives in boy's room.
222.9	(212.9 + 10)	Father awakens boy and starts to exit.
232.6	(222.9 + 9.7)	Father and boy arrive at corridor door.
286.4	(201.8 + 84.6)	Mother with baby crawls to exit door.
317.2	(232.6 + 84.6)	Father and boy crawl to exit door.

So long as they stay out of the hot layer, the exposure to toxic gas is negligible. We might note, however, if Father and/or boy is incapacitated by heat just at the end of the hottest overhead gas, i.e., at 20 meters (66 ft) from the open end of the corridor, they would fall into the cold layer at the floor and still receive little CO. However, they would encounter temperature radiation conditions along line C-C in Figure 7. The maximum overhead temperature ranges up to 950° K (677° C), more than enough to set their clothing afire. They would not survive unless firefighters with breathing apparatus and a fog nozzle got into the corridor within a minute or so.

To continue the illustration of hazard considerations, let us suppose that the CO composition of Figure 8 is encountered floor to ceiling on an upper floor where the gas had already been cooled by heat transfer to building and contents below so that no high temperature hazard exists.

Suppose a person tried to escape at 365 seconds (Figure 8) where the CO encountered is 10,000 ppm and he/she travels at .5 m/sec along line D-D. In this process, the person would encounter CO, shown as D-D in Figure 9. This is the exposure dose.

The Fractional Effective Dose (FED) of toxicant received by an exposed person is defined by Hartzell (5,6) as

$$ FED = \sum_i \int_{t(b_i)}^{t} \frac{C_i - b_i}{K_i} \, dt \qquad\qquad i = specie \qquad\qquad (1) $$

where for

	Incapacitation			Lethality	
	K (ppm)(sec)	b (ppm)		K (ppm)(sec)	b (ppm)
CO	2.2×10^6	233	} FEDI	6.2×10^6	1778 } FEDL
HCN	4.2×10^4	92		1.9×10^5	66

Assuming the above coefficients from rat data apply to humans, a person will have received an IC50 or LC50 dose, if the corresponding FED=1. When our knowledge of toxic hazard is complete, there will be a single algorithm for both incapacitation and death since these effects follow each other. However, for now, the FED approach which requires two separate calculations, is the latest advance.

For persons following path D-D, Figure 8, they will have received an FEDI = 0.525 for incapacitation and an FEDL = .170 for lethality by the time they reach the corridor exit. The exposure is far less than a dose for 50 percent effect. What, if anything, can be said about the escapees condition? Figure 10 is drawn following Hartzell et al.'s suggestive figure (Figure 6 in Ref. 7). By this figure, FEDI = .525 corresponds to 5 percent incapacitation, while FEDL = .17 corresponds to a completely negligible probability of death. For each individual person, the probability of escape is 95 percent, with the percent probability of being incapacitated along the escape path D-D, Figures 8 and 9, varying as indicated by the small numbers.

Once incapacitated, the person falls to the floor and continues to breath CO from the changing fire gases. If a person falls 5 meters (16 ft) from the open end of the corridor, further exposure occurs along the line E-E in Figure 8. This exposure is also plotted in Figure 9. At the incapacitation time (444 seconds, 5 m along corridor), the lethal dose is

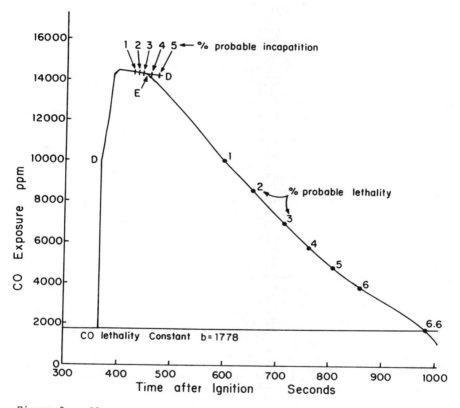

Figure 9.   CO exposure experienced on escape route: (D-D) numbers are probability of incapacitation; (E-E) CO exposure of incapacitated person 5 meters from corridor exit. Numbers are probability of death. Line at 1778 ppm below which CO makes no contribution to death.

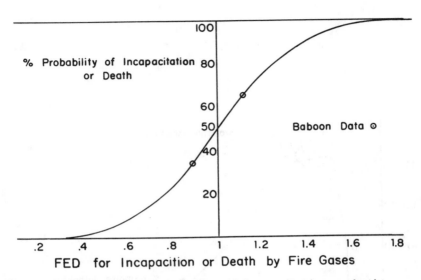

Figure 10.   The percent probability of incapacitation or death as
dependent on the corresponding Fractional Effective Dose (FED) defined
by Equation 1.

only FEDL =   .17. As the FEDL increases along E-E, the probability of
dying increases as shown by the small numbers along E-E.
        As a last illustration, return to the original problem of Figure 1
and ask how much time the family in the suite has, if they are to be able
to safely run at 1.65 m/sec to exit at the open end of the corridor.   In
order to run, the program HAZARD assumes that the hot-cold interface is at
least 1.2 meters from the floor. The latest they could leave the suite is
at 160 seconds following the dashed line F-F in Figure 6.
        In order to get out of the suite by 160 seconds, they must respond
within 160 - 109.2 = 50.8 seconds after the alarm sounds. This is not much
time. If there had been a self-closing door on the fire room, the corridor
would have been passable for a much longer period and the family could
easily escape.

Conclusions

The illustrative examples could be produced only by using both fire
dynamics and human factor information, which contains many crude approximations to the real world and omits completely many important effects.
However, it is clear that real progress is being made toward attaining a
sufficiently accurate predictive understanding of fire and its consequences
so that a performance code can eventually be attained. The computer fire
codes need to be made more comprehensive. There needs to be a mechanism
set up to evaluate the validity of computer fire codes for use with a legal
performance code, just as is done with all other legal codes.
        The toxicological "facts" I have used goes beyond present validated
knowledge and thus indicates directions that future work might take to
produce the data that can actually be used in a computer fire model.

As soon as a really Comprehensive Computer Fire Code, including human reactions and hazard effects, is available, we will be able to obtain toxicity data for humans. Every fire in which there are deaths is a toxicity test run with no control. Surely, of the 6000 or so such fires in the U.S. every year, a few hundred will be in sufficiently well-defined conditions that a comprehensive fire code will be able to predict where the bodies should be found. If the bodies are not found where expected, the rat data can be modified appropriately.

Literature Cited

1. Mitler, H.E. and Rockett, J.A. Users Guide to FIRST, A Comprehensive Single-Room Fire Model, NBSIR 87-3595, 1987.

2. Walton, W.D.; Baer, S.R.; and Jones, W.W. Users Guide for FAST, NBSIR 85-3284, 1985.

3. Levin, B.M. EXITT--A Simulation Model of Occupant Decisions and Actions in Residential Fires, User's Guide and Program Description, Appendix B, Reference 4.

4. Bukowski, R.W.; Jones, W.W.; Levin, B.M.; Forney, C.L.; Stiefel, S.W.; Babrauskas, V.; Braun, E.; and Fowell, A.J. HAZARD 1, Vol. 1, Fire Hazard Assessment Method, NBSIR 87-3602, 1987.

5. Hartzell, G.E.; Priest, D.N.; and Switzer, W.G., *J. of Fire Sciences* 1985, **3**(2): 115-128.

6. Emmons, H.W., *Fire Safety Journal* 1987, **12**, 183-189.

7. Hartzell, G.E.; Grand, A.F.; Kaplan, H. L.; Priest, M.S.; Stacy, H.W.; Switzer, W.G.; and Packham, S.C. Analysis of Hazards to Life Safety in Fires: A Comprehensive Multidimensional Research Program, Year 1, Final Report, SwRI Project 01-7606, NBS Contract NB83NADA4015.

RECEIVED November 17, 1989

FIRE RETARDANTS AND FIRE-RETARDANT COMMODITY PLASTICS

Fire Retardants and Fire-Retardant Commodity Plastics

The interest in fire retardants is as old as the Egyptians. Man's experiments with and understanding of fire retardant materials continues to evolve and develop.

The chapter by Hindersinn provides a short history of fire retardance in polymers. Weil extends that discussion by looking into promising new areas of technology. Drews and coworkers provide an example of an analyses of chemical reactions of fire retardants under pyrolytic conditions, to understand its detailed chemistry. Lewin, Mango, Shen, Wilkie, Cusack, Costa, and coworkers discuss very specific fire retardant systems.

What is particularly noteworthy is the diversity of chemistry which leads to improved fire retardant performance of polymers. As more is learned about how specific systems lead to fire retardant behavior at a microscopic level, the more fire retardant chemistry can be tailored to individual polymers to maximize total polymer performance.

Chapter 7

Historical Aspects of Polymer Fire Retardance

Raymond R. Hindersinn

Hindersinn Associates, 4288 Lower River Road, Youngstown, NY 14174

The history of polymer fire retardance is reviewed
from its inception with the early Egyptians to the
most recent developments in intumescent fire
retardants and inherently fire retardant polymers.

EARLY HISTORY

The control of polymer flammability, which has enjoyed
considerable success in the last forty years, had its
beginnings in antiquity with the first early attempts to
reduce the flammability of natural cellulosic materials
such as cotton and wood (1-3). Some of these early
developments are summarized in Table I. Perhaps the
earliest reference to this development was reported by the
Greek historian Herodotus (484-431 BC) who noted that the
Egyptians were imparting a degree of fire retardance to
wood by soaking it in alum (potassium aluminum sulfate).
About two centuries later, the Romans "improved" the
process by adding vinegar to the mixture. Military
applications for fire retardant wood were subsequently
reported in the first century B.C. by Vitruvius where
early siege towers were protected against incendiaries by
a thick coating of clay reinforced with hair. An
"incombustible cloth" was subsequently developed in the
17th century for Parisian theater curtains by treating
canvas with a mixture of clay and gypsum. The first
patent on a fire retardant treatment for wood and textiles
was issued to Wyld in 1735.

0097–6156/90/0425–0087$06.00/0
© 1990 American Chemical Society

A basic scientific investigation of fire retardancy, however, remained to be initiated by Gay-Lussac in France at the request of King Louis XVIII in 1821 who was again interested in reducing the flammability of theater curtains. This researcher noted that the ammonium salts of sulfuric, hydrochloric and phosphoric acids were very effective fire retardants on hemp and linen and that the effect could be improved considerably by using mixtures of ammonium chloride, ammonium phosphate and borax. This work has withstood the test of time and remains valid to this day. Thus the basic elements of modern fire retardant chemistry had been defined early in recorded history and remained the state of the art until early in the twentieth century. The most effective treatments for cellulosic materials being concentrated in Groups III, V and VII elements.

In 1913, the renowned chemist William Henry Perkin became interested in the problem of reducing the high flammability of a then popular fabric known as "flannelette" and in the process of his work first defined the most important requirements for a fire retardant fabric which included properties such as durability, feel, non-poisonous nature, low cost and printability after treatment. The Perkins treatment consisted of impregnating the fabric with aqueous solutions of sodium stannate and ammonium sulfate. A subsequent heat treatment converted the chemicals to insoluble stannic oxide which was believed to be the active retardant. The Perkins process did not win popular favor and further scientific work on fire retardance remained dormant until World War II when new developments in synthetic polymers ushered in a new era of fire retardant chemistry.

MODERN FIRE RETARDANT DEVELOPMENTS

The advent of synthetic polymers was of special significance since the water soluble inorganic salts defined up to that time were of little or no utility in these largely hydrophobic materials. Modern developments therefore were concentrated on the development of polymer compatible permanent fire retardants. Although a multitude of individual products have since been developed, Table II attempts to list the most significant developments with the largest impact on the direction of fire retardant chemistry.

CHLORINATED PARAFFIN AND ANTIMONY OXIDE. The demands of the armed forces in World War II for a fire retardant, waterproof treatment for canvas tenting led to the development of a combination treatment containing a chlorinated paraffin (CP), antimony oxide and a binder (4, 5). This was the first definition of the halogen-antimony synergistic combination which has since been shown to be so effective in many fire retardant polymer

TABLE I
Early Historical Fire Retardant Developments

Development	Date
• Alum used to reduce the flammability of wood by Egyptians.	About 450 B.C.
• Romans used a mixture of alum and vinegar on wood.	About 200 B.C.
• Mixture of clay and gypsum used to reduce flammability of theater curtains.	1638
• Mixture of alum, ferrous sulfate and borax used on wood and textiles by Wyld in Great Britain.	1735
• Alum used to reduce flammability of balloons.	1783
• Gay-Lussac reported a mixture of $(NH_4)_3 PO_4$, NH_4Cl and borax to be effective on linen and hemp.	1821
• Perkin described a FR treatment for cotton using a mixture of sodium stannate and ammonium sulfate.	1912

TABLE II
Most Important Modern Developments in Polymer Fire Retardance

1) Chlorinated paraffin, antimony oxide and a binder as a treatment on canvas.

2) Chlorine containing unsaturated polyesters.

3) Filler-like retardants.

4) Oxygen index method of evaluating relative polymer flammability.

5) Intumescent fire retardant systems.

6) Inherently fire retardant polymers.

products and the first introduction of organic halogen
compounds in place of the inorganic salts previously in
vogue.

REACTIVE FIRE RETARDANTS. This new treatment was then
immediately applied to polyvinyl chloride (PVC) and
unsaturated polyesters; both products which entered the
development stage during World War II. The new products
enjoyed a degree of success in PVC, because of the need
for a plasticizer to allow processing of that polymer.
The plasticizing nature of the CP, however, not only
reduced the desirable physical properties of the polyester
laminates but tended to "wash out" in many environments in
which these products were being used thus reducing or
eliminating the desirable fire retardant properties for
which they had been added in the first place. These major
deficiencies of CP rapidly led to the conclusion that a
reactive fire retardant system would be preferred which
would be chemically reacted into the polyester at some
stage of the polyester synthesis and/or fabrication of the
final product and thus confer permanent fire retardant
properties to the final product.
        The first fire retardant polyester containing a
reactive fire retardant monomer was introduced by the
Hooker Electrochemical Corporation in the early 1950's
containing chlorendic acid as the reactive monomer (6).
This pioneering development rapidly led to the
introduction of variety of reactive halogen and phosphorus
containing monomers, such as tetrabromophthalic anhydride,
chlorostyrene and tetrabromobisphenol A, which found
application in a wide variety of condensation polymer
systems.

FIRE RETARDANT FILLERS. The next major fire retardant
development resulted from the need for an acceptable fire
retardant system for such new thermoplastics as
polyethylene, polypropylene and nylon. The plasticizer
approach of CP or the use of a reactive monomer were not
applicable to these polymers because the crystallinity
upon which their desirable properties were dependent were
reduced or destroyed in the process of adding the fire
retardant. Additionally, most halogen additives, such as
CP, were thermally unstable at the high molding
temperatures required. The introduction of inert fire
retardant fillers in 1965 defined two novel approaches to
fire retardant polymers.
        One of these products was a thermally stable
insoluble chlorocarbon prepared from cyclooctadiene and
hexachlorocyclopentadiene. The diadduct, dodecachlor-
dodecahydrodi-methanodibenzo [a, e] cyclooctene, has the
structure given in I and is commonly referred to as a
cycloaliphatic chlorine compound (7).

The extreme insolubility of this high melting thermally stable hydrocarbon allowed it to be compounded into most thermoplastics without decomposition or discoloration. Its high chlorine content (65.1% by weight) and filler-like properties not only increased the heat distortion and flexural modulus of the original polymer, without the degradation of such important properties as electrical and water resistance, but was essentially non-migrating at elevated temperatures and in aqueous environments. Although since displaced by more effective aromatic bromocarbons in polyolefins such as polystyrene and ABS, the product still finds considerable utility in fire retardant nylon compositions.

The other fire retardant filler, hydrated aluminum oxide (or alumina) (8), exerts its fire retardant effect in polymer compositions by dehydrating under flame conditions and preventing burning by injecting large amounts of non-flammable water vapor into the atmosphere adjacent to the heated polymer surface. Self extinguishment is conferred by cooling the surface and displacing the oxygen necessary for continued flammability. Because of its relatively low decomposition temperature (245-320°C) hydrated alumina finds its greatest utility in polymer compositions requiring low processing temperatures, such as polyesters, where its low cost, hydrophobicity and reinforcing properties can be used to great advantage. Its effectiveness is also limited when applied to polyolefins because the large quantities. required for effective fire retardance make processing difficult or impossible. The major advantages of this fire retardant system are low smoke production and no hydrogen halide off-gases produced during pyrolysis on fire exposure.

OXYGEN INDEX TEST METHOD. Prior to the introduction of the Oxygen Index Test (OI) in 1966 by Fenimore and Martin (9), fire retardant polymer research was dependent upon a plethora of test methods with variable degrees of stringency for fire retardant evaluation. Not only were test results related only to a burning or self-extinguishing set of observations, but compositions passing one set of test conditions could not easily be related to behavior in a more demanding set of conditions. Additionally, the required test specimens varied from films, coatings, foams and rigid plastics, all of which affected the degree of ignitability of the specimen.

The OI test is carried out on a film specimen in a variable mixture of oxygen and nitrogen with that gas mixture being determined in which a small change in gas composition alters the burning characteristics of the material from self extinguishing to a burning classification. The oxygen concentration at this point was reported as a ratio or index of oxygen concentration to the total concentration of oxygen and nitrogen in the mixture multiplied by a hundred. The OI for air under this system is reported as 21. As indicated in Table III, all materials could now be assigned an OI value that would relate its relative flammability to any other material provided that the material was capable of burning in pure oxygen. Such an assignment was extremely useful in fire retardant research activities since the simple determination of a composition's oxygen index allowed an estimate of its relative flammability. Of particular interest is the relative position of carbon on this scale since its high OI value of 65 has often been used to decrease the flammability of more flammable polymers.

Subsequent research on the test has also shown that the temperature of the test conditions could be increased to allow the assignment of OI values to fire resistant materials not easily flammable in pure oxygen at room temperature (10, 11). Fenimore and co-workers (12) also used the oxygen index test conditions to study the mechanism of fire retardant systems by comparing the OI in oxygen/nitrogen and nitrous oxide/nitrogen oxidizing atmospheres.

Despite its usefulness as a research tool, however, the test conditions cannot be related to real fire situations and large scale testing is still required to determine the fire resistance of materials in actual field conditions.

INTUMESCENT FIRE RETARDANT SYSTEMS. As previously mentioned, the relatively high OI value for elemental carbon in Table III has led to the recent development of FR additive systems for many highly flammable polymers which obtain their fire retardant effect by catalyzing the pyrolysis of the polymer backbone into carbonaceous char or by supplying the carbonaceous ingredients in the additive mixture. Thus the polymer is then converted from a composition with an OI of 20 into an OI approaching that of carbon, well above the value of 25-30 required for most fire retardant systems. This approach to polymer fire retardance not only requires a lower level of additives to obtain a required degree of fire retardance but often reduces the large volumes of heavy smoke evolved during exposure of the halogen/antimony oxide compositions to flame conditions. The large quantities of undesirable halogen halide that accompanies the pyrolysis of these latter compositions is also eliminated.

Although the use of intumescent combinations of a polyhydric organic compound, an acid forming catalyst and a gas forming component have long been known to form excellent fire protective coatings for flammable substrates (13), the incorporation of one or more of these components into the basic polymer composition to confer an intumescent character when exposed to flame temperature is only a recent development. The chemical composition of the polymer substrate seems to be an important variable if not the most important variable in determining the effectiveness of this approach to polymer fire retardance and directly determines the loading level and number of ingredients required for effective results. Some of the intumescent systems and their effectiveness in the recommended polymer substrates are summarized in Table IV together with representative results of a commercial halogen/antimony oxide composition for comparison. The first system, designated "Melabis" for convenience, contains all three components of the intumescent coating compositions incorporated into a single water insoluble thermally stable additive that can be conveniently compounded into polypropylene and injection molded without premature decomposition. As can be seen, a 20% fire retardant level by weight is sufficient to confer a VO rating by the UL 94 flammability test (14) while a 48% loading is required for the same degree of fire retardance with the conventional cycloaliphatic chlorine/antimony oxide system. By comparison, only one weight percent or less of an aromatic metal sulfonate (an acid forming component) is required to attain a fire retardant rating by the ASTM D635 test conditions (15, 16) to the highly aromatic polycarbonate substrate.

The third composition in Table IV seems to be related to the aromatic sulfonate/polycarbonate technology just discussed with some modifications being necessary in order to compensate for the aliphatic nature of the polypropylene (17, 18) substrate. In this case the aromatic sulfonate is replaced with a metal salt (preferably magnesium stearate). A silicone oil and or gum has been added to enhance the intumescent character and a small amount of inert filler and decabromodiphenyl oxide is included probably to improve the molding characteristics of the total composition. Fire retardant compositions with a good surface char can be obtained at total loadings only about half that required for the halogen/antimony oxide composition.

THERMALLY STABLE POLYMERS

No discussion of polymer fire retardance would be complete without at least a brief mention of the highly aromatic polymers, all of which are very difficultly flammable if they burn at all (19). Although the low flammability of phenolic and furane resins are well known, these thermally

TABLE III
Limiting Oxygen Indexes of Various Materials

Material	OI (%)
Polyoxymethylene	15
Candle	16
Polymethylmethacrylate	17
Polypropylene	17
Polystyrene	18
Chlorinated Polyether	23
Polycarbonate	27
Polyphenylene Oxide	29
Polyvinyl Chloride (No Plasticizer)	45
Polyvinylidene Chloride	60
Carbon	65
Polytetrafluoroethylene	95

SOURCE: Data from ref. 9.

TABLE IV
Some Representative Intumescent FR Systems

FR System	Recommended Application	Approx. Loading Level Required for V-0 Designation Via UL-94	Oxygen Index
MELABIS	Polypropylene	20	-
Aromatic Sulfonates	Polycarbonates	1*	-
Magnesium Stearate/ Silicone/Talc	Polypropylene	21.8	30
Cycloaliphatic Chlorine Antimony Oxide	Polypropylene	48	26

* This composition has only a SE rating by ASTM D 635

stable linear polymers are unique because of their controlled chemical structure yields exceptional properties without the need for fillers and/or reinforcements or a long curing cycle for their high temperature properties and thermal resistance. The high cost of the finished polymers and their often specialized fabrication techniques, however, will limit their utility for the foreseeable future to specialized applications where economics are of secondary importance. The thermal and flammability characteristics of only a few representative commercial polymers are summarized in Table V. The oxygen indexes of these unmodified polymers are well above that necessary for fire retardant ratings and most have VO classifications under UL94 test conditions. Some will also withstand direct flame conditions for appreciable lengths of time without losing their coherence and without emitting large quantities of smoke or unusually noxious gases.

TABLE V: Some Thermally Stable Polymers

Chemical Structure	Mp. °C	Tg °C	Oxygen Index	UL-94 Rating
	427	-		
	334	-	35	VO5V
	-	190	30	-
	285	88-93	46-53	VO
	421	369	42	VO

SOURCE: Data from ref. 17.

Literature Cited

1)   F. L. Browne, <u>Theories of Combustion of Wood and It's</u>
      <u>Control</u>. Report #2136, Forest Products Laboratory,
      Forest Service, U.S. Department of Agriculture,
      Madison, Wisconsin; (December 1958).
2)   C. F. Cullis and M. M. Hirschler, <u>The Combustion of</u>
      <u>Organic Polymers</u>, Clarendon Press, Oxford, England,
      (1981), p. 229.
3)   D. L. Chamberlain, <u>Mechanism of Fire Retardance in</u>
      <u>Polymers</u>, W. C. Kuryla and A. J. Papa, Eds., Marcel
      Dekker, Inc., New York, New York, (1978), p. 110.
4)   R. W. Little, Ed., <u>Flameproofing Textile Fabrics</u>, ACS
      Monograph, No. 104, Reinhold Publishing Company, New
      York, (1947), p. 410.
5)   J. W. Lyons, <u>The Chemistry and Uses of Fire</u>
      <u>Retardants</u>, Wiley-Interscience, New York, (1970).
6)   P. Robitschek and C. T. Bean, <u>Ind. Eng. Chem.</u>, 1954,
      <u>46</u>(8), 1628.
7)   R. R. Hindersinn and J. F. Porter, U.S. Patent
      3,598,733, 1971.
8)   W. J. Connolly and A. M. Thornton, Mod. Plastics,
      1965, <u>43</u> (2), 154.
9)   C. P. Fenimore and F. J. Martin, Mod. Plastics, 1966,
      <u>44</u>(3), 141.
10)   D. E. Stuetz, A. H. DiEdwardo, F. Zitomer and B. P.
      Barnes, J. Pol. Sci., Polym. Chem., 1975, <u>13</u>,585.
11)   D. E. Stuetz, Symposium on the Flammability
      Characteristics of Polymeric Materials, University of
      Utah, June 21-26, 1971.
12)   C. P. Fenimore and G. W. Jones, <u>Combustion and Flame</u>,
      1968, <u>10</u>(3), 295.
13)   H. L. Vandersall, <u>J. Fire Flammability</u>, 1970, <u>2</u>, 97.
14)   Y. Halpern, D. Mott and R. Niswander, <u>Ind. Eng.</u>
      <u>Chem., Prod. Res. Dev.</u>, 1984, <u>23</u>, 233.
15)   V. Mark, U.S. Patent 3,940,366, 1976.
16)   A. Ballistreri, G. Montaudo, E. Scamporrino, C.
      Puglisi, D. Vitahini and S. Cucinella, J. Pol. Sci.;
      Part A: Pol. Chem., 1988, <u>26</u>, 2113.
17)   R. Bush, Plastics Eng., 1986, 29.
18)   R. Frye, U.S. Patent 4,387,176, 1983.
19)   A. Klein, <u>Plastics Design Forum</u>, 1988, 95.

RECEIVED November 1, 1989

Chapter 8

Prospective Approaches to More Efficient Flame-Retardant Systems

Edward D. Weil, Ralph H. Hansen, and N. Patel

Polytechnic University, Brooklyn, NY 11201

Possible routes to major advances in flame retardant efficacy are surveyed with reference to reported instances of unusually effective systems, and with reference to theory. In flame studies, some highly efficient quenching species have been noted. Char formation is a major factor in flame retardancy, and there are several approaches for improving the yield and integrity of char or for producing barrier layers containing elements other than carbon. The role of condensed phase oxidation is discussed in reference to the possibility of finding inhibitors. Examples of flame retardant systems based on oxidative dehydrogenation, Friedel-Crafts, and other catalytic modes are examined. Polyblending may prove to be a useful approach to flame retardancy, but results are difficult to predict. Some favorable ("synergistic") interactions between flame retardants can be expected to lead to efficient systems.

Currently available flame retardants must often be employed at concentrations at which they have adverse effects on the other properties of flame-retarded plastics or textiles.

It is our intention to point out clues, mostly from the literature, some from our own work, which suggest approaches to new flame retardant systems with greatly increased efficiency. Both vapor phase and condensed phase mechanisms will be considered.

Vapor Phase Flame Retardants

Flame retardants currently in use which operate by inhibiting vapor phase flame chemistry may be far from optimum. Those flame retardant systems which evolve hydrogen chloride, and perhaps even those which evolve hydrogen bromide, may be acting by little more than a physical effect (1). Some of our own work on tris(dichloroisopropyl) phosphate in polyurethane foams also suggests a physical mode of action (2).

Although there seems little doubt that antimony trihalides play a chemical role in inhibition of free radical chain reactions in the flame zone, a comparison

0097–6156/90/0425–0097$06.00/0

of antimony trichloride as a flame inhibitor in model hydrocarbon flames shows certain other inhibitors with at least an order of magnitude greater activity, iron carbonyl and chromyl chloride for example (3) (Table I).

Table I. Comparison of Concentration of Inhibitor Needed for 30% Reduction of Burning Velocity of Hexane Air Mixture

Inhibitor	Weight % Needed
Antimony trichloride	1.64
Phosphorus oxychloride	0.96
Tetraethyl lead	0.18
Iron pentacarbonyl	0.10

Model flame studies in hydrogen-oxygen flames show that vaporized Mg, Cr, Mn, Sn or U salts are active as radical recombination catalysts at about 1 ppm (4). It was not determined whether this effect was produced by homogeneous reactions in the vapor phase or by the presence of fine particles.

As suggested by such experiments on model flames, it appears possible that vapor phase flame retardants better by one or two orders of magnitude than the best present systems may be found for use in plastics.

Efficient Physical Barriers to Heat, Air and Pyrolysis Products

Carbonaceous char barriers may be formed by the normal mode of polymer burning, and besides representing a reduction in the amount of material burned, the char may act as a fire barrier. The relationship of char yield, structure, and flame resistance was quantified by Van Krevelen (5) some years ago. For polymers with low char-forming tendencies, such as polyolefins, one approach to obtain adequate char is to add a char-forming additive. Such additives generally bear a resemblance to intumescent coating ingredients (6, 7).

Thermal calculations suggest that the char barrier approach can be highly efficient if optimized. Funt and Magill (8) showed that a 1 mm layer would keep an underlying substrate from reaching ignition temperature when the external fire atmosphere was at 743 °C, and a 2.7 mm layer would suffice when the fire atmosphere was at 1500 °C (Table II).

Table II. Effect of a Closed-Cell Char Foam in Preventing a Substrate from Reaching Ignition Temperature (300 °C)

Thickness (cm)	External Temp. (°C)
0.01	342
0.1	743
0.27	1500
1.0	4600

Intumescent layers of such thicknesses are not difficult to achieve, but let us consider some limitations to this approach as well as some clues to improving intumescent char or char-like barriers.

Permeability of Carbonaceous Foam

To perform optimally, the char, or similar barrier should be continuous, coherent, adherent and oxidation-resistant. It should be a good thermal insulator (which implies closed-cell character) and it should have low permeability to gases, to liquid pyrolysate, and to molten polymer. Moreover, the char must be formed in a timely manner before the polymer is extensively pyrolyzed.

Gibov et al. (9) showed that combustion vapors and air could penetrate through a typical char layer. Capillarity served to bring molten polymer to the surface where it could pyrolyze and burn. One answer to this problem is obviously to create a closed cell foam. Gibov et al. showed that the incorporation of boric acid and ammonium phosphate helped minimize penetrability of the char (Fig. 1).

In a more macroscopic way, cracking or exfoliation of the char can occur to cause exposure of the underlying material to burning. In rigid polyurethane foams, this can sometimes be seen as a massive falling away of the char, exposing the foam underneath to renewed burning (10). In intumescent coatings, small cracks can appear, or pieces of the char can flake off. A general way to improve the coherence of char may be to apply one of the principles used in formulating improved intumescent coatings (11), namely, to include in the formulation a high aspect-ratio inorganic filler as a "bridging agent".

Oxidative Destruction of the Char Layer and Its Prevention

The char layer from a burning polymer, while it exerts protective action, is itself vulnerable to oxidation. This can manifest itself either during flaming combustion as a constant destruction of the char as it forms, or as afterglow. Means for prevention of this undesired char destruction have been reported. In studies on preventing combustion of carbon fibers, incorporation of borates, phosphates, or low melting glasses has been shown to be effective (12, 13).

However, other types of barriers besides carbonaceous char have been shown to function in flame retardancy. In brief, these include the following:

1. Glassy coatings using low melting glasses

2. Glassy foams

3. Carbonaceous foams with substantial noncarbon content

4. Fluorocarbon films and coatings

5. Metallic surface coatings

The idea that borates may form a glassy film has long been in the literature, supported by visual observation of the melting behavior of many borates. The deliberate synthesis of efficient low melting glasses as fire barriers was undertaken by Kroenke, Myers and Licursi (14, 15) at B. F. Goodrich. Some efficient glass forming flame retardants were found, based on sulfate glasses. A particularly efficient foamed glassy material was based on ammonium borate (16).

Chars having a substantial silica content were formed by fire exposure of siloxane-carbonate polymers, and were shown by Kambour (17) to be excellent fire barriers, resistant to oxidation and with good insulating ability (Fig. 2).

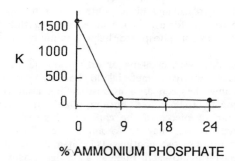

Figure 1. Effect of ammonium phosphate on penetrability (Darcy Constant) of char from novolak pyrolysis.

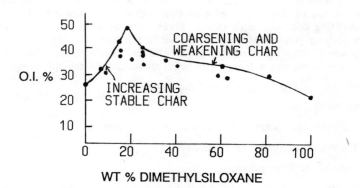

Figure 2. Oxygen index of silicone/bisphenol-A polycarbonate block polymers of varying silicone block content.

The importance of thermally stable char was shown to be critical in a further study of siloxane block polymers by General Electric researchers. (18). The char enhancing action of magnesium soaps and a special silicate silane (19) and likewise the char stabilizing action of lead salts (20) were demonstrated in polyolefins by General Electric investigators.

An efficient flame retardant effect was demonstrated with 2-mil zinc coatings on polyphenylene oxide-polystyrene blends (Noryl) by Nelson (21). The action may relate to enhanced char formation by chemistry specific to this blend. However, other metal coatings on some other polymers also appeared to contribute a measurable flame retardant effect.

Free Radical Inhibitors in the Condensed Phase

Studies by Steutz et al. (22), Burge and Tipper (23), Cullis et al. (24) and Brauman (25) on combustion of polyolefins show that surface (condensed phase) oxidation plays a significant part in flaming degradation. Steutz et al. showed that oxygen penetrated well beneath the surface of a burning polypropylene (Fig. 3). The contribution of surface oxidation to the degradation of polystyrene under conditions relevant to burning has also been shown recently (26). In principle, therefore, an antioxidant system in the condensed phase should have a flame retardant effect.

Experimental evidence for this theoretical prediction is hard to find. Most antioxidants and free radical scavengers are not effective at the temperatures of the oxidatively-pyrolyzing surface. At these temperatures, initiation of oxidative chain reactions is very rapid and any sacrificial antioxidant is rapidly destroyed. However, it is possible that in a formulation "on the edge" between propagation and extinguishment, the use of a free radical inhibitor could favor extinguishment. We were able to show small but statistically significant oxygen index elevations by means of anti-oxidants in marginally flame retarded polyethylene (27) (Table III).

Table III. Effects of Some Antioxidants on the Oxygen Index of Marginally Flame Retarded Polyethylene

Antioxidant (each at 2%)	Oxygen Index
none	22.1 ± 0.4
2,6-di-tert-butyl-p-cresol	23.6
octyl-N-phenyl-alpha-naphthylamine	24.2
N,N'-diphenyl-p-phenylenediamine	24.2
phenothiazine	23.8

An interesting proposal has been made by Czech researchers that triaryl phosphates may act in the condensed phase of burning polystyrene as free radical scavengers (28). If this is in fact one of the modes of action, triaryl phosphates are not very efficient in this system. New types of high temperature antioxidant systems based on polyconjugated aromatic structures (29) or based on reducing agents such as metals (30) have been suggested by Russian workers; these may offer leads to antioxidants effective at the surface of burning plastics.

One of the present authors (31) has developed a series of additives which combine the features of both free radical inhibitors and flame retardants of the tetrabromophthalimide or chlorendic imide type with hindered phenol antioxidant structures such as the following compounds:

We have recently evaluated the chlorendic imide/hindered phenol for its effect on the oxygen index of polyethylene, and we found only a miniscule increase, not considred statistically significant, in comparison to the same loading of chlorine as chlorendic anhydride. We believe that if the antioxidant approach to flame retardancy is to be successful, special high temperature antioxidant structures must be designed for this purpose.

Catalytic Modes of Flame Retardant Action

Certain catalytic modes have been well exploited in flame retardant systems, namely the dehydrating action of compounds which yield strong acids under flaming or smoldering conditions. Friedel-Crafts and other acid catalyzed condensation reactions have been exploited to increase char. These mechanisms don't work very well for polymers of mainly hydrocarbon character. Are there other modes of catalysis which might work better?

In principle, if a polyolefin could be made to undergo oxidative dehydrogenation to form water and char, its heat of combustion could be reduced to about one-third the heat of its complete combustion to carbon dioxide and water. Not only would this afford a drastic reduction in flammability (in view of the cooperative effects of reducing the heat of combustion while providing a char barrier), but in the optimal case, the only combustion product would be a harmless vapor, water. In view of these potential benefits, it is worth reviewing the dehydrogenation/oxidative dehydrogenation mode of catalysis to see what might be done. Regarding this concept, the literature is sparse, but some encouragement is found.

One very efficient system which may work this way was reported by Chien and Kiang (32) who found that 1.5% chromium, introduced by the Étard reaction, raised the oxygen index of polypropylene to 27 (Fig. 4) and char formation was promoted. The hypotheses as to mode of action included the idea that dehydrogenation catalysis might be involved.

Cullis and Hirschler (33) found that zinc acetylacetonate and cobalt acetylacetonate at 1% in polypropylene afforded self-extinguishing properties by the ASTM D635 test (Table IV). These additives appear to be catalytic pro-oxidants which enhanced the carbon yield.

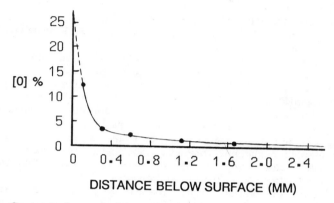

Figure 3. Content of oxygen in condensed phase of burning polypropylene at varying distances below the surface.

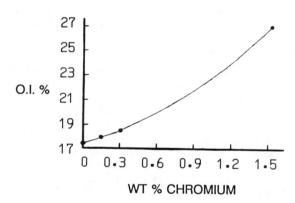

Figure 4. Effect of chromium content on oxygen index of polypropylene.

Table IV. Flame Retardant Effect of Some Metal Compounds in Polypropylene at 1% by the ASTM D635 Test

Compound	Charring	Drip	Fire Rating
Nickel AcAc	slight	yes	burns
Cupric stearate	slight	yes	burns
Cobalt AcAc	yes	no	self-ext.
Zinc AcAc	yes	no	self-ext.

Studies at the International Tin Research Institute showed that 2.5% zinc stannate strongly enhanced the flame retardant action of ATH in ethylene-acrylic rubber, and enhanced the char yield (34) (Fig. 5).

Hitachi Cable Ltd. (35) has claimed that dehydrogenation catalysts, exemplified by chromium oxide–zinc oxide, iron oxide, zinc oxide, and aluminum oxide–manganese oxide inhibit drip and reduce flammability of a polyolefin mainly flame retarded with ATH or magnesium hydroxide. Proprietary grades of ATH and $Mg(OH)_2$ are on the market which contain small amounts of other metal oxides to increase char, possibly by this mechanism.

Working with polymethyl methacrylate, Sirdesai and Wilkie (36) have shown that certain phosphine–platinum complexes undergo oxidative insertion reactions and thus catalyze crosslinking leading to flame retardance. This catalyst is expensive and not particularly efficient, but serves as a lead.

Probably the most efficient flame retardant system ever discovered for a polymer is platinum, which at 1 ppm flame retards silica-filled silicones and increases unburned residue (Fig. 6). In a very thorough study by MacLaury at GE (37), platinum was shown to exert a catalytic action to induce coupling between chains and with the filler. The detailed mechanism is still uncertain; nevertheless, the remarkable efficacy of platinum in this system supports the idea that very efficient f.r. agents may be designed by using catalysis principles.

Polyblends of Flame Retardant with Non-flame-retardant Plastics

The principles needed to design a polymer of low flammability are reasonably well understood and have been systematized by Van Krevelen (5). A number of methods have been found for modifying the structure of an inherently flammable polymer to make it respond better to conventional flame retardant systems. For example, extensive work by Pearce et al. at Polytechnic (38, 39) has demonstrated that incorporation of certain ring systems such as phthalide or fluorenone structures into a polymer can greatly increase char and thus flame resistance. Pearce, et al. also showed that increased char formation from polystyrene could be achieved by the introduction of chloromethyl groups on the aromatic rings, along with the addition of antimony oxide or zinc oxide to provide a latent Friedel-Crafts catalyst.

However, from a commercial standpoint, modifying the polymer mainly for improved flame retardancy is usually done reluctantly, since other properties usually suffer and cost is generally increased. The present trend for developing improved polymers is to utilize polyblending. Can a polyblending approach achieve efficient flame retardancy?

Commercial examples are notably PVC-ABS blends and the blends of polyphenylene oxide with polystyrene. In the case of PPO-PS blends, it has been shown that the good char forming ability of the PPO greatly helps flame

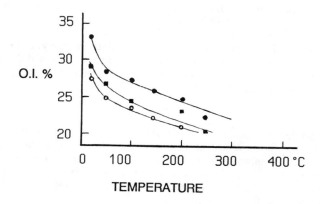

Figure 5. Effect of tin compounds on oxygen index vs. temperature of ethylene-acrylic rubber containing 50% ATH. Key: ●, 2.5% $ZnSn(OH)_6$; ■, 2.5% SnO_2; ○, no tin.

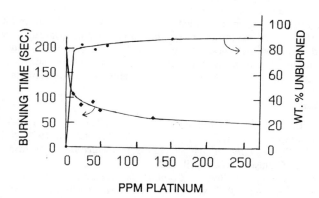

Figure 6. Effect of platinum on burning time and unburnt residue of silicone rubber containing 44 % silica.

retardancy, requiring only the addition of a triaryl phosphate which serves to flame retard the polystyrene pyrolysate which reaches the vapor phase (40).

Polytetrafluoroethylene has an oxygen index of 95%, and is relatively impervious to gases. The use of a low level of finely divided PTFE as an anti-dripping additive in flame retardant polycarbonates is described in the patent literature, and is in commercial use (41).

Can we expect highly efficient flame retardant systems by blending a small amount of a very flame retardant high-charring polymer with a large amount of a very flammable low-charring polymer? We know from the somewhat analogous area of textile blends that results can deviate radically from the "rule of averages" - in the wrong direction for blends of f.r. cotton and PET - and that new burning chemistry can even be elicited (42)! Perhaps some favorable cases remain to be discovered.

"Synergism"

The quantitative relationship of flammability of a polymer with respect to the concentration of flame retardant is usually not linear, and there is no logical reason to expect combinations of different flame retardants to show a linearly additive result either (43). The actual result is often found to be "synergistic" or "antagonistic", or in regression analysis terminology, the interaction term is often found to be statistically significant.

Some painstaking and elegant studies by Antia, Cullis and Hirschler (44) show what can be accomplished by optimizing flame retardant compositions. For example, by "contour-mapping" the oxygen index of ABS as a function of the concentration of decabromodiphenyl oxide, antimony oxide and ferric oxide, they discovered that the highest achievable oxygen index was with mixtures having all three additives present within a certain defined range of composition.

Careful attention to quantitative activity vs. concentration relationships, to the effect of interaction terms in combinations (using computerized regression analysis and experimental design), and careful observation of the manner in which one mode of action supports and reinforces another, seems likely to lead us to the next generation of highly efficient flame retardant systems.

Literature Cited

1. Larsen, E. R. J. *Fire Flamm./Fire Retardant Chem.* 1974, 1, 4-12; ibid. *1975*, 2, 5-19; Larsen, E. R.; Ecker, E. L. *J. Fire Ret. Chem. 1979*, 6 182-192.
2. Weil, E. D. in *Polyurethane Technology Conference* (preprints); Clemson University, Clemson, SC, April 1987.
3. Morrison, M. E.; Scheller, K. *Combustion and Flame* 1972, *18*, 3-12.
4. Bulewicz, E. M.; Padley, P. J. *13th Intl. Symposium on Combustion* (proceedings); Combustion Institute: Pittsburgh, PA, 1971; pp. 73-80.
5. Van Krevelen, D.W. *Chimia* 1974, *28*, 504: *Polymer* 1975 *16*, 615; *Properties of Polymers*; Elsevier: Amsterdam, 1976: pp. 525-536.
6. Halpern, Y.; Mott, M.; Nisdwander, R. H. *I&EC Prod. Res. & Dev.* 1984 *23*, 233-8.
7. Bertelli, F.; Locatelli, R. (to Montedison), U.S. Pat. 4 579 894, 1986; Landoni, G.: Fontani, S.: Cicchetti, O. (to Montedison), U.S. Pat. 4 336 182, 1982.

8. Funt, J.; Magill, J. H. *J. Fire Flamm.* 1975, *6* 28.
9. Gibov, K. M.; Shapovalova, L. N.; Zhubanov, B. A. *Fire & Materials* 1986, *10*, 133-5.
10. Weil, E. D.; Jung, A. K.; Aaronson, A. M.; Leitner, G. C. *Proc. 3rd European Conf. on Flamm. and Fire Ret.* Bhatnagar, V. M., Ed.; Technomic Publ.: Westport, CT, 1979.
11. Anderson, Jr., C. E.; Dziuk, Jr., J.; Mallow, W. A.; Buckmaster, J. *J. Fire Sci.* 1985, *3* 161-194.
12. McKee, D. W. *Carbon* 1986, *24*, (6), 737-741.
13. Strife, J. R.; Sheehan, J.E. *Ceramic Bulletin* 1988, *67*, (2), 369-374.
14. Myers, R. E.; Licursi, E. *J. Fire Sci.* 1985, *3* 415 431.
15. Kroenke, W. J. *J. Mat. Sci.* 1986, *21*, 1123-1133.
16. Myers, R. E.; Dickens, Jr., E. D.: Licursi, E.; Evans, R. E. *J. Fire Sci.* 1985, *3*, 432-449.
17. Kambour, R. P.: *J. Appl. Polym. Sci.* 1981, *26*, 861- 877.
18. Kambour, R. P.; Klopfer, H. J.; Smith, S. A. *J. Appl. Polym. Sci.* 1981, *26*, 847-859.
19. Frye, R. B., *Preprints, Amer. Chem. Soc. Div. of Polymeric Materials* Fall 1984, *51* (2); pp. 235-239.
20. MacLaury, M. R.: Schroll, A. L. *J. Appl. Polym. Sci.* 1985, *30*, 461-472.
21. Nelson, G. *Proc. Intl. Conf. on Fire Safety*, *13*, Product Safety Corp., Jan. 1988, pp. 367-378.
22. Steutz, D. E.; DiEdwardo, A. H.; Zitomer, F.; Barnes, B. P. *J. Polym. Sci., Poly. Chem. Ed.* 1975, *13*, 585-621; *ibid.* 1980 *18*, 967-1009.
23. Burge, S. J.; Tipper, C. F. H. *Combustion and Flame* 1969, *13*, 495.
24. Cullis, C.F. *J. Anal.Appl. Pyrolysis* 1987 *11*, 451-463.
25. Brauman, S. K. *J. Polym. Sci., Part B, Polym. Phys.* 1988, *26*, 1159-1171.
26. Martel, B. *J. Appl. Polym. Sci.* 1988, *35*, 1213-1226.
27. Weil, E. D. *Plastics Compounding*, Jan.-Feb. 1987, 31-40.
28. Tkac, A. *J. Polym. Sci., Polym. Chem. Ed.* 1981, *19*, 1495-1508.
29. Berlin, A. A. *Vysokomol. Soedin.* 1971. *A13* (2)(Engl. trans. ed.), 309-331.
30. Shustova, O. A.; Gladyshev, G. P. *Russian Chem. Rev.* 1976, *45* (9) (English trans.ed.), 865-882.
31. Hansen, R. H. (to Canusa Coating Systems Ltd.) U. S. Pat. 4 464 240, 1984; 4 465 571, 1984; 4 514 533, 1985.
32. Chien, J. C. W.; Kiang, J.Y. *Macromol.* 1980, *13*, 280-8.
33. Cullis, C. F.; Hirschler, M. M. *Eur. Polym. J.* 1984, *20* (1), 53-60.
34. Cusack, P. paper presented at 1988 Meeting of Fire Retardant Chemical Assn., March 20, 1988.
35. Shingyoji, K.; Inada, T.; Sato, M. (to Hitachi Cable Co.), Japan. Pat. 62/181, 1987.
36. Sirdesai, S. J.; Wilkie, C. A. *Polymer Preprints, Am. Chem. Soc., Div. Polym. Chem.* 1987, *28* (1) 149.
37. MacLaury, M. R. *J. Fire Flamm.* 1979, *10*, 175-198.
38. Khanna, Y. P.; Pearce, E. M. in *Flame Retardant Polymeric Materials*, Vol. 2, Lewin, M.; Atlas, S.: Pearce, E. M., Eds.; Plenum: New York, 1975; pp. 43-61.
39. Pearce, E. M. in *Contemp. Topics in Polym. Sci.*, Vol.5; Vandenburg, E. J., Ed.; Plenum: NY, 1984; pp. 401-413.

40. Carnahan, J.; Haaf, W.; Nelson G.: Lee, G.; Abolins, V.; Shank, P., in *Proc. 4th Intl. Conf. Flammability and Safety; Product Safety Corp.:* San Francisco, Jan. 1979.
41. Thomas, L.S.; Ogoe, S.A. paper presented at the March 1985 meeting of the Fire Retardant Chemical Assn., Houston, TX.
42. Miller, B.: Martin, J.R.; Meiser, Jr., C. H.; Gargiullo, M. *Textile Res. J.* 1976, *46* (7), 530-8.
43. Weil, E. D. in *Flame Retardancy of Polymeric Materials,* Vol. 3: Kuryla, W. C.; Papa, A. J., Eds. Marcel Dekker, Inc.: New York, 1975: pp. 186-243.
44. Antia, F. K.; Cullis, C. F.; Hirschler, M. M. *Eur. Polym. J.* 1982, *18,* 95-107.

RECEIVED January 5, 1990

Chapter 9

Ternary Reactions among Polymer Substrates, Organohalogens, and Metal Oxides in the Condensed Phase under Pyrolytic Conditions

M. J. Drews, C. W. Jarvis, and G. C. Lickfield

School of Textiles, Clemson University, Clemson, SC 29634

The results reported here represent part of an extensive and continuing investigation concerning the antimony oxide/organohalogen synergism in flame retardant chemistry and in particular into the mechanisms of volatile antimony species formation during pyrolytic degradation and combustion, and the role of the polymer substrate in these reactions. Mixtures of decabromodiphenyl oxide (DBDPO) with antimony oxide, inert fillers, a variety of active metal powders and two different polymer substrates (high density polyethylene and polypropylene) were pyrolyzed isothermally in the TGA and the volatile DBDPO pyrolysis products characterized by capillary gas chromatography (CGC) and CGC/MS. Based on the results of these and other supporting experiments, mechanisms are proposed to account for the observed product distributions.

The antimony oxide/organohalogen synergism in flame retardant additives has been the subject of considerable research and discussion over the past twenty-five years (1-17). In addition to antimony oxide, a variety of bismuth compounds and molybdenum oxide have been the subject of similar studies (18-20). Despite this intensive investigation, relatively little has been conclusively established about the solid state chemical mechanisms of the metal component volatilization, except in those cases where the organohalogen component is capable of undergoing extensive intramolecular dehydrohalogenation.

In the earlier literature, several different mechanisms have been proposed to account for the antimony volatilization which occurs during the combustion of polymer substrates in the presence of antimony oxide and an organohalogen.

0097–6156/90/0425–0109$06.25/0
© 1990 American Chemical Society

These reaction sequences can be separated into two categories according to whether or not the organohalogen component can undergo intramolecular dehydrohalogenation.

For those organohalogen compounds which can undergo intramolecular dehydrohalogenation, reaction sequence [1] has been proposed as the principal route to the generation of volatile antimony containing species (3, 5, 7, 18, 20).

$$R-CH_2CHX-R \longrightarrow R-CH=CH-R + HX \uparrow$$
$$6 \ HX + Sb_2O_3 \longrightarrow 2 \ SbX_3 + 3 \ H_2O \uparrow$$

[1]

Examples of additives and polymer substrates which could react via [1] would be chlorinated paraffin waxes, hexabromocyclododecane (HBCD) and poly(vinylchloride) or poly(vinylbromide).

For those organohalogen compounds which cannot readily undergo intramolecular dehydrohalogenation, two alternative reaction sequences, [2] and [3], for the generation of volatile antimony containing species have been proposed (9, 21, 22).

$$R-CH_2CH_2-R \longrightarrow 2 \ R-CH_2 \cdot$$
$$R-CH_2 \cdot + R'-X \longrightarrow R-CH_2-R' + X \cdot$$
$$X \cdot + R-CH_2CH_2-R \longrightarrow HX + R-CH_2\overset{.}{C}H-R$$
$$6 \ HX + Sb_2O_3 \longrightarrow 2 \ SbX_3 \uparrow + 3 \ H_2O \uparrow$$

[2]

$$Sb_2O_3 + R-X \longrightarrow \{SbO \cdots X \cdots R\}^*$$
$$n \ \{SbO \cdots X \cdots R\}^* \longrightarrow SbX_3 \uparrow + \{R_2O\}^*$$

[3]

Examples of additives to which either or both of these reaction sequences could apply are decabromodiphenyl oxide (DBDPO), tetrabromobisphenol-A (TBBPA) and the chlorinated cyclopentadiene adducts.

There is substantial data to support reaction sequence [2] and some data suggesting sequence [3] (21, 22) may occur under certain pyrolysis conditions, but no evidence has been published for [3] in the absence of a large excess of polymer substrate. In the case of [2] and [3] it has been suggested that these reactions would be dependent on the carbon-halogen bond strength in the organohalogen component (9).

Presumably other metal oxide compounds, such as bismuth or molybdenum oxide, could undergo similar reactions under the appropriate conditions.

More recently, based on the results of an extensive series of small scale degradation studies, two additional mechanisms for the volatilization of antimony from antimony oxide/organohalogen flame retardant systems have been proposed (23,24). Of these two proposed mechanisms, [4] and [5], [4] does not involve HX formation at all and [5] suggests an important role for the direct interaction of the polymer substrate with the metal oxide prior to its reaction with the halogen compound.

$$Sb_2O_3 + Polymer \longrightarrow Sb^\circ + [Polymer-O-]$$

$$Sb^\circ + R-X \longrightarrow [SbX_nO_yR_z] + [R] \qquad [4]$$

$$Sb_2O_3 + Polymer \longrightarrow [Sb_2O_3-Polymer]^*$$

$$[Sb_2O_3-Polymer]^* + R-X \longrightarrow [SbX_nO_yPolymer] + [R\cdot]^* \qquad [5]$$

Mechanism [4] was based on studies involving the direct reaction of antimony metal with DBDPO in the absence of a hydrogen source. The data from these experiments clearly show that if the oxide is reduced to the metal, direct interaction with DBDPO would occur, and that this is a specific and highly exothermic reaction. However, no direct evidence for the presence of metallic antimony in mixtures containing antimony oxide, a polymer substrate and an organohalogen compound was obtained.

Mechanism [5] was based on the results obtained from multi-step sequential pyrolysis experiments in an inert atmosphere (23). This mechanism [5] differs from [3], primarily in that [5] was proposed to be surface catalytic in nature, and that the reaction between the oxide particle surface and the organohalogen was considered only as the first step, initiating the process leading to the eventual formation of volatile antimony species.

The results reported here represent only a small part of an extensive and continuing investigation designed to elucidate, in more detail than had previously been established, the mechanisms of volatile antimony and other metal species formation during the pyrolytic degradation and combustion and the roles of the polymer substrate and organic halogen sources in these reactions. Because of its commercial and potential mechanistic (since it contains no C-H bonds) significance, much of the emphasis of this work has been focused on the solid state chemistry which occurs with DBDPO as the halogen source.

Experimental

The thermal analysis, tube furnace, and other apparatus previously employed in this work, as well as the procedures for measuring the antimony volatilization gravimetrically have been described elsewhere (23-25). For the collection of the volatile organics from the pyrolysis experiments, the use of various trap designs and a variety of solvents was investigated. For the majority of the capillary gas chromatographic (CGC) and CGC/MS analysis procedures, an Ace Glass Works glass microimpinger containing 20.0 ml of solvent (benzene, xylene or toluene) connected by a ball joint to the purge outlet of the pyrolysis apparatus (DuPont Instruments 951 TGA module) was used.

The CGC analysis of the volatile degradation products were performed using a Perkin-Elmer Sigma 2000 capillary gas chromatograph. The column used was either a fused silica 0.25 micron, bonded methyl silicone (10 m, 0.25 mm I.D.) or a methyl/5% phenyl silicone (15 m 0.25 mm I.D.) bonded phase. The carrier gas was helium and the capillary column head pressure was maintained at 20 psi. The make-up gas for the pulsed electron capture detector (ECD) was 95% Ar/5% methane supplied at a flow rate of 60 ml/min.

All of the injections were 1 microliter and the capillary column split ratio was nominally set at 200:1. Column and program times and temperatures were varied depending on the volatile products being analyzed.

All of the initial CGC organobromine degradation products assignments were based on the use of relative retention time standards. The relative retention times for the partially brominated degradation products were established either by the synthesis of the expected degradation products or by the use of selective debromination reactions.

In the latter stages of this investigation, CGC/MS was employed to confirm the assignments of the relative retention time standards. The instrumentation used in the majority of these experiments consisted of a Hewlett-Packard 5890A caplillary gas chromatograph interfaced directly to a HP-5970B mass selective detector (MSD). All of the data acquisition and analysis were done using a HP-59970C Chemstation and software. A limited number of additional experiments were carried out using a HP-5988A CGC/MS system. The capillary column employed for the mass spectrometry studies was a HP, 25m, methyl/5% phenyl silicone bonded phase, 0.25 mm ID fused silica with a film thickness of 0.33µm. The chromatographic condition used for these experiments were as similar to those employed for the prior CGC analysis as the instrumentation would permit.

For the decabromodiphenyl oxide (DBDPO) pyrolysis reactions, two different procedures were used to synthesize the series of brominated diphenyl oxides and dibenozofurans employed as the relative retention time standards: $AlBr_3/Br_2$ in ethylene dibromide and Fe^o (metal)/Br_2 in tetrachloroethylene. The rate of the initial bromination steps in the former reaction was so rapid that only the higher degree of bromination adducts could be isolated. The rate of the Fe^o/Br_2 reaction was found to be much slower, especially during the initial stages, and these reactions yielded a broader range of relative retention time reference peaks.

In a typical bromination experiment 1.0 g of diphenyl ether (or dibenzofuran), 1.0 g of Fe^o powder and 15 ml of Br_2 were added to approximately 250 ml of tetrachloroethylene. The mixture was gently heated with stirring for a period of up to two weeks. At intervals, approximately 20 µl of the reaction mixture was withdrawn, diluted to 3 ml with benzene and the resultant solution analyzed chromatographically. The assignment of degrees of bromination to specific retention times was based on following the progression of the very slow reaction, and the assumption that the longest observed retention times correspond to the fully brominated species. This assumption could be confirmed for the DBDPO but not absolutely for the octabromodibenzofuran (OBDBF), since no pure OBDBF standard was available.

An analogous series of dibenzodioxin relative retention time standards were prepared from dibenzodioxin (DBD) using Fe^o/Br_2 and the same procedure used to brominate the ether and the furan. The DBD itself was prepared by refluxing ortho-chlorophenol with NaOH.

The relative retention times employed in the study of the HBCD pyrolysis reactions were obtained by the use of selective chemical debromination. The HBCD was selectively debrominated using Zn^o (metal) powder in a refluxing ether solution. A small amount of acetic acid was used as the catalyst.

All of the organohalogen compounds studied were commercial products obtained from various manufacturers and used as received. Only the DBDPO was purified further by recrystallization for some of the chromatography and thermal analysis experiments. Samples of antimony trioxide and antimony pentoxide were also obtained from commercial sources. The ultrapure antimony trioxide, bismuth trioxide, bismuth metal, antimony metal, dibenzofuran and diphenyl ether were all obtained from Aldrich Chemicals. The poly(propylene) (PP) resin was 0.7 mfi, food grade from Novamont and the poly(ethylene) was unstabilized, high molecular weight, HDPE from American Hoechst.

Results and Discussion

On the basis of the data obtained from the early thermal analysis and tube furnace pyrolysis experiments performed during the initial phases of this investigation, it became apparent that in order to establish the principal reaction pathways to the generation of volatile antimony species, the volatile degradation products of the DBDPO itself would need to be characterized (24, 25).

To accomplish this objective a capillary gas chromatograph was used and an analytical protocol for the analysis of the volatile degradation products of DBDPO developed. While there has recently been considerable activity with respect to the characterization of the thermal degradation products of chlorinated and brominated aromatics (26-30), including DBDPO (29,30), at the time this work was first undertaken during late 1984, neither an analytical method nor analytical standards were available for the study of the thermal degradation products of DBDPO. Therefore, several sets of relative retention time standards were prepared corresponding to the anticipated partially brominated degradation products of DBDPO (24). These relative retention time standards included a series of partially brominated reaction products based on diphenyl oxide, dibenzofuran and dibenzodioxin.

In order to confirm the relative retention times established for DBDPO using only CGC, additional sets of partially brominated diphenyl oxides and dibenzofurans were synthesized using the Fe^O/Br_2 procedure. The course of these reactions was followed by both CGC and CGC/MS. As a result, it was possible to simultaneously confirm the previous relative retention time peak assignments as well as to correlate the retention times between the two instruments. Some of the pertinent comparative retention time data obtained from these experiments is summarized in Table I. Upon completion of the individual reactions, a cocktail containing both partially brominated furans and diphenyl oxides was mixed. A typical CGC chromatogram and a CGC/MS total ion chromatogram for this cocktail are shown in Figures 1 and 2, respectively.

On the basis of the results obtained from these combined CGC and CGC/MS experiments it was possible to positively confirm all of the relative retention times assignments which had previously been made using CGC analysis alone.

All of the DBDPO and HBCD thermal degradation experiments reported here were carried out isothermally with a nitrogen purge in a thermal gravimetric analyzer (TGA) at temperatures between 390 and 410° C for DBDPO and 240° C for HBCD, respectively.

Table I.   Comparison of the Isomer Retention Ranges
             Obtained from CGC and CGC/MS Analysis of the
             Relative Retention Time Standards

Rentention Time Range, min

Bromine#	Diphenyl Oxides CGC[1]	CGC/MS[2]	Dibenzofurans CGC[1]	CGC/MS[2]
3	11	19	13-15	22-23
4	15	23-24	19-20	27-28
5	19-21	27-27	23-24	32-35
6	22-24	31-34	27-28	40-44
7	25-27	37-40	33-34	60
8	29-30	45-51	44-45	98-100
9	35-36	70-75	------	-------
10	48-50	112-115	------	-------

[1] Perkin-Elmer Sigma 2000,15m,methyl/5% phenyl
    silicone column
[2] HP-5980-A,25m, methyl/5% phenyl silicone column.

The samples pyrolyzed included the pure organohalogen compound, simple mixtures of the organohalogen and a second component representative of a possible additive or reaction product present during the thermal degradation of a flame retarded polymeric material, such as Sb_2O_3 or a polymer substrate, and ternary mixtures containing all three components. After some preliminary experimentation with a variety of mixing techniques, it was concluded that the most reproducible results were obtained by preparing the sample mixtures directly in the quartz sample pans for the TGA pyrolysis experiments. The rationale for the choice of temperature ranges and experimental conditions employed have been described in detail elsewhere (24,25).

When pure DBDPO was pyrolyzed at $390°$ C and the volatile products analyzed by CGC, 98% was recovered. The only reaction of note appeared to be the degradation of the nonabromodiphenyl oxide which had been detected as an impurity present in the starting sample.

These results are quite different from others which have been reported in the literature concerning the high temperature (>600° C) thermal and thermal oxidative degradation of DBDPO, in which >90% degradation was reported (28). We confirmed that temperature was the principal cause for these different observations, by reproducing the 600° C isothermal pyrolysis of DBDPO in which less than 10% unreactive DBDPO was recovered.

As this work progressed, it became convenient for comparative purposes to express the pyrolysis results for a specific pyrolysis experiment in terms of the extent of the observed reaction of the initial organohalogen component content recovered chromatographically. The extent of reaction data as determined by the CGC analysis of the volatile reaction products for the pyrolysis of some representative simple mixtures of DBDPO are summarized in Table II. As illustrated by these data, the results obtained for the DBDPO/Sb_2O_3 mixture suggest that, in the absence of a polymer substrate, Sb_2O_3 exhibits the same extent of reaction as observed for other inert fillers such as glass beads or alumina.

Previous thermal analysis studies had indicated that while Sb_2O_3 did not react directly with DBDPO, there was some evidence that the reaction of a polymer substrate with the Sb_2O_3 generated a species which was very reactive (23), and that this product was antimony metal (Sb^0). Therefore, simple mixtures of DBDPO with powdered antimony, bismuth and zinc metals (mole ratio of bromine to metal of 3:1) were pyrolyzed and the extent of reaction determined by CGC.

As shown by the data for these mixtures also presented in Table II., under the conditions employed here, all of these metals reacted directly with DBDPO. From a mechanistic perspective, that for all three metals the principal DBDPO volatile reaction product was observed to be the octabromodibenzofuran (OBDBF) was considered to be the most important result of these experiments. The observed reactivity of the metal towards the DBDPO decreased in the order Zn^0 > Sb^0 > Bi^0. Of these data, only the rather large observed difference in the relative reactivities of Sb^0 (~38% OBDBF) and Bi^0 (~9% OBDBF) was unexpected based upon the thermodynamics for the formation of their respective halides.

While several additional experiments indicated that the low melting point of Bi^0 (271° C) may have effected the observed extent of reaction, other factors such as significant differences in the surface areas of the powdered metals cannot be ruled out.

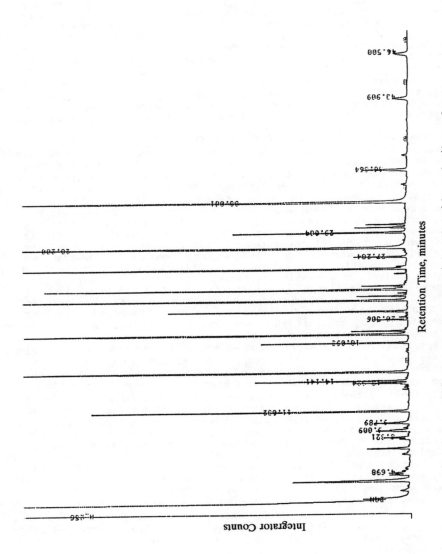

Figure 1.   CGC Chromatogram of Diphenyl Oxide and Dibenzofuran
Bromination Reaction Product's Cocktail

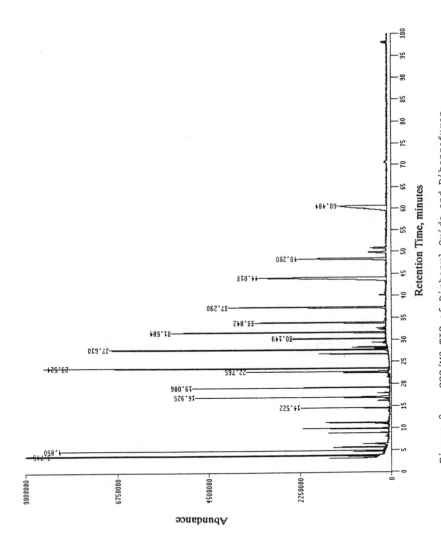

Figure 2.   CGC/MS TIC of Diphenyl Oxide and Dibenzofuran Bromination Reaction Product's Cocktail

By analogy to the utilization of Zn^o in the Wurtz-Fittig reaction (31), Scheme [6] illustrated for Zn^o is proposed to account for the observed reaction between DBDPO and any other active metals such as Sb^o or Bi^o.

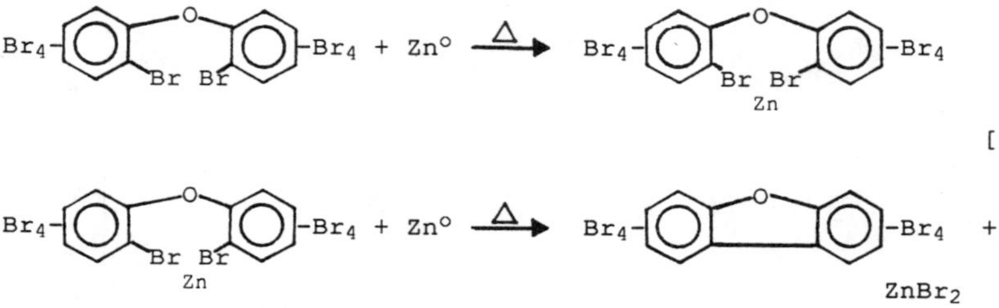

[6]

The two polymer substrates investigated as part of the study of DBDPO mixtures were polypropylene (PP) and linear high density polyethylene (HDPE). While both PP and HDPE decompose by similar random chain scission, radical mechanisms, chain transfer occurs much more readily during the pyrolysis of PP because of the presence of the tertiary hydrogens. In addition, only primary chain end radicals are formed when the HDPE chain cleaves homolytically. Therefore, a comparison of the PP/DBDPO and the HDPE/DBDPO mixtures' volatile product distributions was undertaken.

In comparing the results of these experiments as summarized in Tables III and IV, it was found that the extent of reaction was much greater for the HDPE/DBDPO mixtures (only ~8% undegraded DBDPO) than for the corresponding PP/DBDPO samples (~64% undegraded DBDPO). In addition, the peak area fraction for the octa-ether was almost an order of magnitude greater, the furans' areas three times and the unassigned peaks' area fraction more than five times that measured for the PP/DBDPO mixtures (≈20% as compared to ≈3%). One possible mechanism which would be consistent with the observed data, especially considering the magnitude of the difference in the observed extent of degradation between HDPE and PP, is the chain transfer reaction shown in Scheme [7].

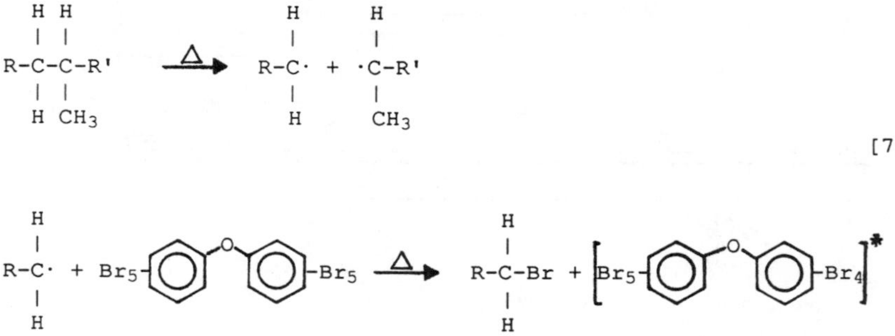

[7]

Further evidence that the principal degradation pathway involves chain radical attack on the DBDPO as shown in [7] and not halogen radical attack as would occur in [2] includes the relatively small quantity of HBr formed (23,25) as well as the large number of

Table II.   Summary of Volatile Product CGC Analysis for the Extent Reaction of Simple Mixture Pyrolyzed at 390°C in the TGA

Mixtures	DPO^1_{10}	$DPO^1_{9,8}$	$DPO^1_{7,6}$	$DPO^1_{5,4}$	$DBF^2_{8,7}$	DBF^2_{6-4}
$DBDPO/Sb_2O_3$	94.7	1.6	0.7	---	1.6	0.1
$DBDPO/AL_2O_3$	92.8	3.3	---	---	0.6	---
$DBDPO/GB^3$	91.8	3.1	---	---	0.1	---
$DBDPO/Bi^O$	81.6	5.3	---	---	8.7	---
$DBDPO/Zn^O$	52.5	2.8	0.4	---	10.1	14.1
$DBDPO/Sb^O$	58.3	3.0	---	---	38.4	---

[1] Brominated diphenyl oxides
[2] Brominated dibenzofurans
[3] Glass Beads

Table III.   Summary of the Volatile Product GC Analysis for the DBDPO/PP Mixture Pyrolyzed at 390°C in the TGA

Br #	DPO^1	DBF^2	PHENOLS	BENZENES
			Component Peak Area, %	
3	-----	-----	-----	-----
4	-----	-----	-----	-----
5	-----	-----	0.3	-----
6	0.2	0.3	-----	0.3
7	0.4	2.8	-----	-----
8	2.9	2.5	-----	-----
9	23.6		-----	-----
10	64.3		-----	-----

[1] Brominated Diphenyl Oxides
[2] Brominated Dibenzofurans

brominated alkyl fragments detected by the ECD detector during the CGC analysis.

From a mechanistic perspective, one of the most important aspects of the results of these experiments is the further confirmation of the formation of volatile metal halide species by a reaction pathway, as illustrated by Scheme [6], that does not include the formation of HX.

In this regard, it should be noted at this point that one of the products identified by CGC/MS from these pyrolysis reactions was $SbBr_3$. Furthermore, the data presented concerning the importance of the polymer substrate in the degradation of the DBDPO and the proposed chain radical transfer mechanism [7] would suggest that the condensed phase chemistry could be much more important in antimony oxide/organohalogen flame retardant systems than had been previously thought.

Several series of ternary mixtures containing DBDPO with PP and a variety of metals, metal oxides and other third components have been pyrolyzed in the TGA at $390°$ C and their volatiles analyzed by CGC. The CGC data from these experiments were found to fall into three distinct categories with respect to the observed extent of reaction. In the first class, Class I, the extent of reaction was found to be not significantly greater than that measured for the PP/DBDPO mixture alone.

The data for the Class I components (Fe^o, 100 mesh-size glass beads and chromatographic grade aluminum oxide) are summarized and compared to that from the simple PP/DBDPO mixture data in Table V. As illustrated by the data in Table V, in this class, ≈50% of the total volatile peak area corresponded to undegraded DBDPO and the principal degradation product was found to be the nonabromo ether (≈33%). Of these results, only the Fe^o (metal) data was surprising since it implied that Fe^o was acting as an inert filler and not forming significant quantities of iron halides under these experimental conditions. During the review of this manuscript it was suggested that the Fe^o may have been surface oxidized, while the TGA was carefully purged and all of the metal samples were reagent grade or ultrapure, surface oxidation cannot be ruled out as an explanation for the observed lack of reactivity for the Fe^o samples.

The second class, Class II, was represented by Zn^o and molybdenum (III) oxide (MoO_3). These data are summarized and compared to the results for the simple PP/DBDPO mixture in Table VI. As shown by the data in Table VI, only ≈20% of the total product peak area was accounted for by the undegraded DBDPO in this class.

In addition, ≈20% of the product peak area was assigned to furan formation as compared to ≈10% for Class I. Perhaps, the most significant results from this class are those for the MoO_3 mixture since these data represent a clear indication that this metal oxide can undergo chemistry similar to that observed for Sb_2O_3 as has been suggested but seldom supported in the past literature.

The third class, Class III, contained those mixtures exhibiting the greatest extent of reaction. Included in this class were Sb_2O_3, Sb^o(metal), Bi^o(metal), bismuth trioxide (Bi_2O_3) and antimony trisulfide (Sb_2S_3). These data are summarized and compared to those from the simple PP/DBDPO mixture in Table VII.
In these ternary mixture experiments, unreacted DBDPO accounted for ≈4% or less of the total volatile peak product areas.

Table IV.   Summary of the Volatile Product CGC Analysis
           for the DBDPO/PE Mixture Pyrolyzed at 390° in the TGA

Bromine #	Peak Area, %	
	Diphenyl Oxides	Dibenzofurans
5	---	0.8
6	1.7	6.5
7	5.4	8.1
8	25.8	1.1
9	21.9	---
10	7.5	---

Table V.   Summary of the Volatile Product CGC Analysis
           for the Class I Extent of Reaction Ternary
                    Mixtures at 390°C

Mixtures	Component Peak Areas, %					
	DPO^1_{10}	$DPO^1_{9,8}$	$DPO^1_{7,6}$	$DPO^1_{5,4}$	$DBF^2_{8,7}$	DBF^2_{6-4}
PP/DBDPO/Fe^o						
	56.4	30.6	.1	.1	8.1	0.5
PP/DBDPO/Glass beads						
	49.6	35.8	.1	.1	8.4	0.2
PP/DBDPO/AL$_2$O$_3$						
	49.5	36.2	.1	.1	9.8	1.1
PP/DBDPO						
	64.3	26.5	0.6	.1	5.3	0.3

[1] Brominated diphenyl oxides
[2] Brominated dibenzofurans

Table VI. Summary of the Volatile Product CGC Analysis for
the Class II Extent of Reaction
Ternary Mixtures at 390°C

Mixtures	DPO^1_{10}	$DPO^1_{9,8}$	$DPO^1_{7,6}$	$DPO^1_{5,4}$	$DBF^2_{8,7}$	DBF^2_{6-4}
		Component Peak Areas, %				
PP/DBDPO/Zn^o						
	26.4	43.6	1.0	0.1	12.3	5.7
PP/DBDPO/MoO$_3$						
	15.2	36.4	4.2	0.7	8.6	13.7
PP/DBDPO						
	64.3	26.5	0.6	0.1	5.3	0.3

1 Brominated diphenyl oxides
2 Brominated dibenzofurans

Table VII. Summary of the Volatile Product CGC Analysis for
the Class III Extent of Reaction Ternary
Mixtures at 390°C

Mixtures	DPO^1_{10}	$DPO^1_{9,8}$	$DPO^1_{7,6}$	$DPO^1_{5,4}$	$DBF^2_{8,7}$	DBF^2_{6-4}
		Component Peak Areas, %				
PP/DBDPO/Sb$_2$O$_3$						
	3.3	29.3	8.0	1.4	9.0	22.4
PP/DBDPO/Bi$_2$O$_3$						
	1.4	8.7	9.2	7.5	2.9	16.3
PP/DBDPO/Sb^o						
	2.0	16.7	28.8	0.1	9.3	12.2
PP/DBDPO/Bi^o						
	0.8	4.4	11.3	8.0	2.8	22.7
PP/DBDPO/Sb$_2$S$_3$						
	3.2	19.0	6.7	5.7	4.2	18.2
PP/DBDPO						
	64.3	26.5	0.6	0.1	5.3	0.3

1 Brominated diphenyl oxides
2 Brominated dibenzofurans

The most significant characteristic of these data, as compared to that from Class I and II, was the overall increase in component peak areas for the lower degree of bromination furans (4 to 6 bromines accounting for 12-20% as compared to ≈2% for Class I and a maximum of 14% for Class II) and ethers (6 and 7 bromines accouting for 7-29% as compared to ≈1% for Class I and a maximum for 4% for Class II).

In addition, for the Class III components, at least ≈30% and as much as ≈60% of the peak areas could not be assigned to known relative retention times. In comparing the results obtained within this class, the greatest overall extent of DBDPO reaction was observed for the mixture containing PP/DBDPO/Bi°. Also, somewhat surprisingly, there was essentially no difference observed between the extent of reactions for the DBDPO/PP/Sb$_2$O$_3$ and the DBDPO/PP/Sb$_2$S$_3$ mixtures. These results would suggest that Sb$_2$S$_3$ should be at least as effective an additive in DBDPO/oxide flame retardant formulations as Sb$_2$O$_3$. In addition, if the sulphur itself was vapor phase active, Sb$_2$S$_3$ could be more effective than Sb$_2$O$_3$.

The extent of reaction data for the ternary mixtures PP/DBDPO/Sb$_2$O$_3$ and PE/DBDPO/Sb$_2$O$_3$ are summarized and compared to their respective polymer/DBDPO simple mixtures in Table VIII. As was observed earlier, a greater extent of reaction was found for the PE ternary mixture than for the PP ternary mixture, and both of the ternary mixtures showed significantly greater extent of reaction than the respective simple mixtures. If the data summarized in Table VIII are compared to those in Tables III and IV, it would appear as if the presence of the Sb$_2$O$_3$ tended to reduce the differences observed between the two polymer systems. However, since the extent of reaction for the PE ternary mixture was so great that only ≈50% of the peak areas could be assigned it is difficult to discern with certainty that this is truly the case. In any case, it is clear that these two similar polymers exhibit significant differences in the extent of the DBDPO reaction and these results emphasize the importance of the polymer substrate degradation chemistry with respect to the formation of the volatile metal species.

Several experiments considered as pertinent to the interpretation of the ternary mix data with respect to the possible role of potential species formed in-situ, such as SbBr$_3$, were carried out with HBCD and HBCD mixtures. The organohalogen HBCD is prepared by the bromination of cyclododecatriene and nominally contains bromines at the 1, 2, 5, 6, 9 and 10 positions on the aliphatic ring.

In the study of the pyrolysis of HBCD using bromine ion selective electrodes, it was found that an average of three bromines per molecules was lost on heating above the decomposition temperature (25). These data, and the organic literature, suggested that one product of this elimination reaction should be a cyclododecatriene ring containing three vinyllic bromines which would be relatively difficult to remove, if the proposed elimination mechanism was correct. When this same reaction was analyzed by the use of CGC, it was found that the pyrolysis of HBCD by itself at 240° C yields a variety of products with different degrees of bromination (13% -Br$_6$, 7% -Br$_5$, 17% -Br$_4$, 31% -Br$_3$, 19% -Br$_2$ and 2% -Br$_1$) and not only the proposed triene.

Of these different products the triene was found to be the principal product and the average loss of bromine was in fact observed to be 3. However, clearly a large percentage of the debromination did not occur solely by the proposed concerted elmination (25). However, GC/MS analysis of the degradation products identified $SbBr_3$ as one of the principal products from a simple mixture of HBCD and Sb_2O_3.

To evaluate the effects of species such as $SbBr_3$ on the degradation of HBCD additional experiments were performed which compared the CGC volatile product distributions, the time to 50% weight loss (isothermal TGA) and the temperature at 50% weight loss (dynamic TGA) for mixtures of HBCD with Sb^O, Sb_2O_3, glass beads and $SbBr_3$. Of these mixtures, as shown by the data in Tables IX and X, the HBCD/$SbBr_3$ was the most reactive mixture as evidenced by the shortest time to 50% weight loss and the greatest extent of reaction based on the total CGC area counts. In a second experiment, Bi_2O_3 was pyrolyzed at 280°C with HBCD.

Based on the GC/MS results obtained from the HBCD/Sb_2O_3 mixture, it was assumed that one of the products formed under these conditions with Bi_2O_3 would be $BiBr_3$, via the direct reaction of the Bi_2O_3 with the HBr formed from the decomposing HBCD. Both the TGA weight loss and residue data were consistent with the formation of a significant quantity of $BiBr_3$ under the conditions employed. Confirmation of the formation of $BiBr_3$ by GC/MS was unsuccessful. The $BiBr_3$ formed "in-situ" was then subsequently pyrolyzed at 390° with a mixture of PP/DBDPO. This ternary mixture was found to yield the most complete extent of reaction data for any system which had been studied as only 0.3% of the total volatile product peak area was found to be undegraded DBDPO. However, the nature of these experiments makes it difficult to truly assess the significance of these data and any future work with this system would have to be carried out with well characterized $BiBr_3$.

The overall results obtained from the HBCD mixture experiments were interpreted in terms of a catalytic effect of strong Lewis acids such as $SbBr_3$ or $BiBr_3$ on the degradation of the HBCD by facilitating the loss of halogen. In the polymer/oxide/DBDPO ternary mixtures the in-situ formation of these species should produce a similar effect on the halogen loss from the system.

In addition, since $BiBr_3$ is significantly less volatile than $SbBr_3$ (bp. 441°C vs. 288°C), in a ternary mix reaction at 390°C, the $BiBr_3$ would be expected to remain in the reaction mixture for longer periods of time than the $SbBr_3$. Consequently, it could appear to be a more effective catalyst than $SbBr_3$ under these experimental conditions and therefore, yield a higher overall extent of reaction.

These data suggest that the formation of species such as SbX_3 may not only be important because of their volatility and flame inhibition, but also for the role they play in the solid state degradation chemistry.

<u>Summary</u>

From the results of the small scale thermal analysis experiments previously reported (23,25), it was concluded that the antimony volatilization and bromide release observed for ternary mixtures containing organobromine compounds, which did not undergo intermolecular dehydrohalogenation, could not be accounted for solely on the basis of HBr formation during degradation.

Table VIII.   Summary of Volatile Product CGC Analysis for
the Extent of Reaction of PP and PE Ternary
Mixtures Pyrolyzed at 390°C in the TGA

Mixtures	Component Peak Areas, %					
	DPO^1_{10}	$DPO^1_{9,8}$	$DPO^1_{7,6}$	$DPO^1_{5,4}$	$DBF^2_{8,7}$	DBF_{6-4}
$DBDPO/PP/Sb_2O_3$						
	3.3	29.3	8.0	1.4	9.0	22.4
$DBDPO/PE/Sb_2O_3$						
	0.6	16.2	8.5	6.7	2.2	15.5

[1] Brominated diphenyl oxides
[2] Brominated dibenzofurans

Table IX.   Summary of the Volatile Product CGC Analysis
for the HBCD Mixtures Pyrolyzed Isothermally in the
TGA at 240°

Bromine #	Mixture Peak Area, %				
	Neat	Sb^O	Sb_2O_3	$SbBr_3$	Glass Beads
1	2.3	----	----	----	----
2	19.3	21.8	4.3	28.4	3.2
3	31.1	15.2	9.7	----	41.5
4	17.3	51.4	36.8	18.6	45.5
5	7.1	----	----	2.2	----
6	12.7	11.8	46.9	----	----
Count [1]	4.5	0.3	0.4	0.1	2.5

[1] Total integrator counts for the entire chromatogram
multiplied by (10^{-6}).

Table X.   Summary of the Time to 50% Weight Loss Data
            for the HBCD Mixtures Pyrolyzed Isothermally in
            the TGA at 240°

Sample	Time to 50% Weight Loss, min.
HBCD	11.5
HBCD/Sb°	2.6
HBCD/Sb$_2$O$_3$	6.8
HBCD/SbBr$_3$	2.3
HBCD/Glass Beads	11.4

From the CGC and CGC/MS analysis of the thermally initiated
reactions of HDPE and PP with DBDPO and a variety of other
components reported here, it has been shown that attack of polymer
chain radicals on the DBDPO is the most probable first mechanistic
step in the ternary mixture pyrolytic degradation of DBDPO.

Experiments with antimony metal and DBDPO clearly demonstrated
for the first time that antimony volatilization was possible in the
complete absence of any source of hydrogen at polymer pyrolysis
temperatures.

On the basis of the results obtained from additional experiments
with other metals it was concluded that in the presence of active
metal such as Sb° or Bi° or Zn° direct attack on the DBDPO would
readily occur.   However, all attempts to directly measure the extent
of Sb° formation from Sb$_2$O$_3$ during polymer degradation were
unsuccessful.

From the single component, simple mixture and ternary mixture
extent of reaction data presented, it was concluded that the extent
of DBDPO reaction in the ternary mixtures studied, which was
measured by CGC and CGC/MS analysis, was greater than could be
accounted for on the basis of the reactivity of the individual
components initially present.   Taken together these data suggest
that a reaction intermediate or product was formed during the
degradation of the ternary mixtures which was either: (a) more
reactive towards DBDPO itself or (b) a powerful catalyst for the
subsequent dehalogenation of the DBDPO.

Data obtained from studies on the dehydrohalogenation of HBCD
with metal oxides and metal halides clearly show the catalytic
effect of strong Lewis acids on this reaction.   These results
suggest that any similar metal halide or oxyhalides could exert a
similar effect on the dehydrohalogenation of other halogenated
organics formed during the course of the solid state degradation
including the original halogen source itself.

Finally, the results reported here clearly demonstrate the
utility of utilizing the flame retardant additives as sensitive
probes into studying the course of the solid state chemistry which
occurs during pyrolysis.

Acknowledgment

The authors would like to gratefully acknowledge the Center for Fire Research of the National Institute of Standards and Technology for their support of this work through Grants NB79NADA0011, NB80NADA1042 and 60 NANB4D0033.

Literature Cited

1. J.W. Lyons,The Chemistry and Use of Fire Retardants, Wiley-Interscience, New York, 1970.
2. S. Backer, G.C. Tesoro, T.Y. Toong, and N.A. Mousa, Textile Fabric Flammability, MIT Press, Cambridge, Massachusetts, 1976.
3. J.J. Pitts, "Antimony-Halogen Synergistic Reactions in Fire Retardants,"J. Fire Flamm., 3, 51 (1972).
4. C.P. Fenimore and G.W. Jones, "Mode of Inhibiting PolymerFlammability," Combust. and Flame, 10, 295 (1966).
5. J.W. Hastie, "Molecular Basis of Flame Inhibition," Journal of Research of the National Bureau of Standards - A. Physics and Chemistry, $77A$, 733 (1973).
6. J,W. Hastie and C.L. McBee, "Mechanistic Studies of Halogenated Flame Retardants: The Antimony-Halogen Systems,"in Halogenated Fire Suppressants, R.G. Gann, ed., ACS Symposium Series, 16, 1975.
7. S.K. Brauman and A.S. Brolly, "Sb_2O_3 - Halogen Fire Retardance in Polymers I. General Mode of Action," Journal of Fire Retardant Chemistry, 3, 66 (1976).
8. I. Touval, "The Use of Stannic Oxide Hydrate as a Flame Retardant Synergist," J. Fire and Flamm., 3, 130 (1972).
9. S.K. Brauman, "Sb_2O_3 - Halogen Fire Retardance in Polymers II. Antimony-Halogen Substrate Interaction, "Jounal of Fire Retardant Chemistry, 3, 117 (1976).
10. S.K. Brauman, "Sb_2O_3 - Halogen Fire Retardance in Polymers III. Retardant-Polymer Substrate Interaction," Journal of Fire Retardant Chemistry, 3, 138 (1976).
11. S.K. Brauman, N. Fishman, A.S. Brolly, and D.L. Chamberlain, "Sb_2O_3-Halogen Fire Retardance in Polymers IV. Combustion Performance, "Journal of Fire Retardant Chemistry, 3, 225 (1976).
12. F.K. Antra, C.F. Cullis, and M.M. Hirschler, "Binary Mixtures of Metal Compounds as Flame Retardants for Organic Polymers," European Polymer J., 18 96 (1982).
13. T. Handa, T. Nagashima, H. Yamamoto, S. Miyanishi, N. Ebihara, and S. Orihashi, "The Synergistic Effect of Sb_2O_3 and Other Metal Oxide on PVC," J. of Fire Retardant Chemistry, 8, 171 (1981).
14. T. Handa, T. Nagashima and N. Ebihara , "Synergistic Action of Sb_2O_3 with Bromine-Containing Flame Retardants in Polyolefins. II. Structure-Effect Relationships in Flame Retardant Systems," J. of Fire Retardant Chemistry, 8, 37 (1981).

15. L. Costa, G. Camino and L. Trossaarelli, "Thermal Degradation of Fire Retardant Chloroparaffin - Metal Compound Mixtures - Part I. Antimony Oxide,"Polym. Degradation and Stability, 5, 267 (1983).
16. R.G. Gann, R.A. Dipert and M.J. Drews, "Flammability, "Mark - Bikales - Overberger - Menges: Encyclopedia of Polymer Science and Engineering, Vol. 6, Second Edition, John Wiley and Sons, Inc.N.Y., N.Y., 154 (1986).
17. R.C. Nametz, "Bromine Compounds for Flame Retarding Polymer Compositions," FRCA Proceedings, Callaway Gardens, Pine Mountain, Georgia, March 28-30, 1984, 56 (1984).
18. L. Costa, G. Camino and L. Trossaarelli, "Thermal Degradation of Fire Retardant Chloroparaffin - Metal Compound Mixtures - Part II. Bismuth Compounds," Polym. Degradation and Stability, 5, 355 (1983).
19. L. Costa, G. Camino and L. Trossaarelli, "Thermal Degradation of Fire Retardant Chloroparaffin - Metal Compound Mixtures - Part III. A Comparison of Behaviors of Antimony Trioxide and Bismuth Compounds," Polym. Degradation and Stability, 5, 367 (1983).
20. R.M. Lum, D. Edelson, W.D. Reents, Jr., W.H. Stunes, Jr.,and L.D.escott, Jr., "New Insights into the Flame-Retardance Chemistry of Poly (Vinyl Chloride)," Nineteenth Symposium (International) on Combustion, Combustion Institute, Pittsburgh, PA, 807 (1982).
21. S.K. Brauman and I.I. Chen, "Influence of the Fire Retardant Decabromodiphenyl Oxide - Sb_2O_3 on the Degradation of Polystyrene, "J. of Fire Retardant Chemistry, 8, 28, (1981).
22. E.R. Wagner and B.L. Joesten, "Halogen-Modified Impact Polystrene, Quantification of Preflame Phenomena," J.of Applied Polym. Sci., 20, 2143 (1976).
23. T. L. Gilstrap, "A Study of Polymer/Antimony Oxide Interactions During Pyrolysis," MS Thesis, Clemson University, Clemson, S. C., 1984.
24. M.J. Drews, C.W. Jarvis and G.C. Lickfield, "Degradation and Combustion Products in the Presence of Antimony Oxide," FRCA Conference Proceedings, Fall 1986, Kiawah Island, South Carolina, 133(1986).
25. M.J. Drews, C.W. Jarvis, and E.A. Leibner, Organic Coatings and Plastics Preprints, 43, 181 (1981)
26. C. Rappe and R. Lindahl , "Formation of Polychlorinated Dibenzofurans (PCDFs) and Polychlorinated Dibenzo-p-Dioxins (PCDDs) from the Pyrolysis of Polychlorinated Diphenyl Ethers,"Chemosphere, 9, 351 (1980).
27. C. Rappe,"Analysis of Polychlorinated Dioxins and Furans,"Environ. Sci. and Tech., 18, 78A (1984).
28. H. Thoma, S. Rist, G. Hauschultz and O. Hutzinger, "Polybrominated Dibenzodioxins and -Furans from the Pyrolysis of Some Flame Retardants," Chemosphere,15,649 (1986).
29. H. R. Buser, "Polybrominated Dibenzofurans and Dibenzo-p- dioxins:Thermal Reaction Products of Polybrominated Diphenyl Ether Flame Retardants," Environ. Sci.and Tech.,20, 404 (1986).

30. H. Thoma, G. Hauschulz, E. Knorr and O. Hutzinger, "Polybrominated Dibenzofurans (PBDF) and Dibenzodioxins (PBDD) from the Pyrolysis of Neat Brominated Diphenylethers, Biphenyls and Plastic Mixtures of these Compounds,"Chemosphere, 16, 277(1987).

31. J. March, "Advanceed Organic Chemistry:Reactions, Mechanisms,and Structure, "McGraw-Hill, New York, N.Y., 1968.

RECEIVED January 18, 1990

Chapter 10

A Novel System for the Application of Bromine in Flame-Retarding Polymers

Menachem Lewin[1], Hilda Guttmann, and Nader Sarsour

School of Applied Science and Technology, Hebrew University and Israel Fiber Institute, P.O.B. 8001, Jerusalem, Israel 91 080

Experiments pertaining to a new system for the application of bromine to flame retardant polypropylene and foamed polystyrene are described. The FR compound, ammonium bromide, is formed in the amorphous regions of the polymer phase by the interaction of bromine sorbed on the polymer and ammonia, sorbed subsequently. Gaseous nitrogen which is also produced, expands and brings about the rearrangement of the chains to produce a porous structure. The ammonium bromide produced is finely divided and imparts FR properties to the polymer. Amounts of NH_4Br in the polymer are determined by the concentration of the bromine in the bromine water, temperature, time and structure of the polymer. 5.2% of NH_4Br containing PP fabric gave an O.I. value of 24.2 and the effectivity, e.g. slope of plot of O.I. vs % Br was 1.24, as compared to 0.6 for aliphatic bromine. Treatment of foamed beads of PS enabled production of materials containing up to ca 300 % of NH_4Br, with O.I. values of up to 39, the effectivity decreasing with increasing % NH_4Br. The effectivity was significantly higher for PS samples expanded after encapsulation of NH_4Br. The results are discussed in the light of the behavior of NH_4Br in pyrolysis and of the specific thermal properties of the cellular polymers.

Bromine is applied to flame retarding (FR) polymers in a number of systems, which are divided into two major groups, according to the way in which the FR agent interacts with the polymer [1]. In one group, e.g. the reactive systems, the bromine is chemically bound with the polymer. The introduction of the bromine into the polymer molecule can be carried out by copolymerizing a bromine-containing

[1]Current address: Polymer Research Institute, Polytechnic University, Brooklyn, NY 11201

monomeric unit with the main monomers to form the polymer, as in
the case of copolymerizing dibromoterephtalic with terephtalic acid
and ethylene glycol to form poly(ethylene terphtalate) [2]. Another
approach is the direct bromination of the polymer as in the case of
the bromination of the lignin in wood [3] or of polystyrene [4].
The introduction of bromine can also happen by graft polymerization
[5].

The second group comprises the mixing of the polymer with non-
reactive bromine compounds. This can be done by adding the compound
to the melt in the extruder, by topical deposition or coating of
the polymer with a suitable formulation containing the compound, as
well as by exhaustion of dissolved compounds from a solvent bath
[1]. Cases are also known in which the bromine compound was encapsulated in the polymer in order to avoid deleterious interaction
[6].

In the present study, a new way of introducing a non-reactive
bromine-containing compound, ammonium bromide, is discussed. The
treatment, which can be carried out for many thermoplastic polymers
[7,8] in the solid state, is performed in two consecutive stages:
1. immersion of the polymer (fiber, film, beads, etc.) in an aqueous
bromine solution for a predetermined time, during which the bromine
is sorbed on the polymer; 2. treatment of the bromine-containing
polymer with an ammonia solution, in which the bromine is exother-
mically reduced by the ammonia in the amorphous regions of the polymer, without a chemical interaction with the polymer, according to
the overall equation:

$$3 \ Br_2 + 8NH_3 = 6NH_4Br + N_2 \tag{1}$$

This treatment, termed "porofication", produces two major effects:
1. the production of NH_4Br as an ultrafine powder distributed and
trapped between the chains; 2. the simultaneously-formed nitrogen
gas expands, creating sizeable local pressures, bringing about the
repacking of the chains and the formation of pores, thus changing
the structure of the amorphous regions of the polymer. This happens
without impairing the strength properties [1].

The application of this system for flame retarding, first reported in this paper for polypropylene (PP) and foamed polystyrene
(PS), utilizes the trapped or encapsulated NH_4Br. The flame retarding properties of NH_4Br were already reported upon in the
literature [9 - 11] and appeared to be highly effective, although
no quantitative assessment of the FR activity and comparisons with
other bromine-containing FR agents are given. Its use was limited
by it low temperature of decomposition into ammonia and hydro-
bromic acid, of ca 220°C, which was, in most cases, lower than the
polymer processing temperatures, causing corrosion of equipment and
pollution. The introduction of the NH_4Br into the polymer after
the main high temperature processing stages may overcome the major
obstacles mentioned above and enable its use.

Experimental

Materials: PP fabric prepared from highly oriented film-slit fibers.
PS foam: 3 samples provided by Israel Petrochemical Industries, Haifa,
were used. Sample 1: commercial foamed 5 mm beads containing a

bromine-based flame retardant, with an O.I. of 22.8; density –
0.014 g/cm^3. Sample 2: commercial 3 mm beads of foamed PS not containing flame retardant with an O.I. of 18.4, and density of
0.0145 g/cm^3. Sample 3: unfoamed 1 mm beads, containing 6% pentane.
The specimens of this sample were steam foamed after the porofication treatment and deposition of the NH$_4$Br, to beads of density of
0.0145 g/cm^3 for testing. Pure liquid bromine and a 25% NH$_3$ solution were obtained from Frutarom Electrochemical Industries, Haifa.
A G.E.   O.I. tester was used for the flammability measurements.
The foamed PS samples were placed in a meshed-wire bag, made of
stainless steel and mounted in the O.I. cylinder. The measurements
were accurate and reproducible within 0.1 O.I. units.

        Bromine solution was prepared by shaking liquid bromine in
water at room temperature in closed glass vessels and suitably
diluted. The details of the bromine and ammonia stages were given
in a previous publication [8].

        PP and PS accurately weighed samples were placed in small
knitted cotton bags, weighted with glass beads and immersed in
bromine-water solutions in stoppered glass vessels, placed in a
thermostat for varying time intervals. The cotton bags were then
removed from the bromine solution, lightly squeezed and immersed
in the ammonia solution, similarly placed in a thermostat for 15
minutes, with shaking. The polymer was then thoroughly rinsed in
running water for 30 minutes, removed from the cotton bags and dried
in air.

        The analytical methods used were described previously [7,8].
The amounts of NH$_4$Br deposited in the polymer were determined, in
most cases, from weight increase. In the case of small concentrations of NH$_4$Br, the polymer was dissolved in a solvent and the remaining NH$_4$Br determined argentometrically. Density was determined
by weighing known volumes of polymer.

Results and Discussion

Polypropylene. 3 series of experiments, in which the first stage
was performed with Br$_2$ solutions of 3 different concentrations, at
28°C and various time intervals are summarized in Figure 1. The
amounts of NH$_4$Br found in the polymer increase with Br$_2$ concentration to a maximum concentration of 5.7% of NH$_4$Br. The O.I. values
of the samples (Fig. 1) appear to correlate closely with the NH$_4$Br.
Table I shows the average effectivity, e.g. the slope of the plot
of O.I. vs % Br for the PP fabric to be 1.24. This value is seen in
Table II to be considerably higher than the value of 0.6 reported by
van Krevelen [12] for aliphatic and aromatic bromine compounds and
of 0.5 found by Green [13] for aliphatic Br compounds. According to
Green [13], 10% of aliphatic bromine are needed to produce a PP with
an O.I. of 23.5, as compared to 4.6% Br from NH$_4$Br, which produced
an O.I. of 24.2 in the present experiments.

        In Figure 2, taken from reference [12], we inserted curve 5,
which contains our present results. It is evident that the NH4Br
is more effective for PP than aliphatic or aromatic bromine throughout the 1 – 6% bromine range when no synergists are added.

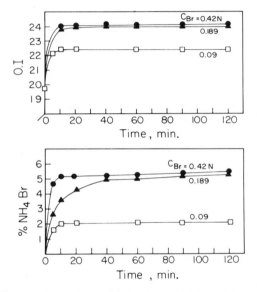

Figure 1. Weight % of NH$_4$Br and O.I. of treated PP fabric vs. time of contact with three Br$_2$ water solution at 28 °C.

Table I. Change in O.I. with Br
Content for PP Fibers

Br Wt.%	O.I.	Effectivity Δ O.I. / %Br
1.63	22.13	1.55
2.9	23.84	1.46
3.26	23.88	1.31
4.10	24.02	1.08
4.24	24.08	1.06
4.65	24.24	1.0
	Av.	1.24

Table II. Effectivity of Various Br Compounds
in Flame Retarding PP

FR Compound	ΔO.I./%Br	Reference
Aliphatic Br	0.6	12
Aliphatic Br	0.5	13
Aromatic Br	0.45	12
NH_4Br by Porofication	1.24	this paper

It is of interest to note that similar results were previously obtained in this laboratory for cellulose. A part of these
results is reproduced in Table III. In these experiments, chromatographic grade paper strips were quantitatively treated with several
flame retardants, including NH_4Br in a series of concentrations,
dried in air and tested by igniting at 3 different angles. The % Br
needed to prevent ignition of the sample was determined. It is evident that the effectivity of the Br in NH_4Br is more than twice
higher than in tris(2,3 dibromopropyl) phosphate and in tris(tribromophenyl)phosphate [14].

The high flame retarding effectivity of NH_4Br as compared to
organic bromine-containing materials has not been adequately discussed in the literature. It is logically explained by the following
data: 1. decomposition of NH_4Br yields HBr directly, which is considered to be the desirable favored pyrolysis product of a polymer-
flame retardant system [15]; 2. the dissociation energy of NH_4Br
to HBr and NH_3 is much lower than the bond dissociation energies
of bromine in organic compounds [15,16,17]. The dissociation con-

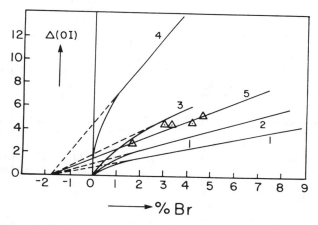

Figure 2. Flame retardancy of PP: (1) aromatic Br; (2) aliphatic Br; (3) aliphatic and (4) aromatic Br with Sb_2O_3 as synergist. (Curves 1–4 reprinted with permission from ref. 12. Copyright 1977 John Wiley and Sons. Curve 5, this paper; data from Table I.)

stant (4.58×10^{-4} at 300°) and the degree of dissociation (16.1 at 330°C) increase at 320°C to 4.97×10^{-4} and 38.7, respectively, and decrease at higher temperatures. Sizeable amounts of HBr are therefore readily available when PP and PS begin to decompose at 300°C [18]. At this temperature, NH_4Br sublimes and exists in a gaseous phase.

Table III.[*] Flammability Limits of Bromine Compounds on Cellulose

Compound	Angle	%Compound	%Br	%P	%N
Tris(2,3 dibromo-	90	24.8	17	1.1	
propyl)phosphate	45	20.5	14.1	0.91	
	0	8.6	5.9	0.38	
Tris(tribromo-	90	27.7	19.2	0.83	
phenyl)phosphate	45	19.3	13.4	0.58	
	0	7.8	5.4	0.23	
Hexabromocyclo-	90	34.5	all test strips burned		
dodecane	45	31.4	23.4		
	0	12.7	9.4		
2,4,6-tribromophenol	0	30.7	22.3 all test		
			strips burned		
2,4,6-tribromoaniline	0	31.6	23.0 all test		
			strips burned		
Ammonium bromide	90	8.8	7.2		1.3
	45	7.0	5.7		1.0
	0	3.4	2.8		0.49

* From Reference 14.

Polystyrene Foam. 2 typical experiments in which samples 1 and 2 were treated for various time intervals with a $0.42NBr_2$ solution at 28°C are presented in Figures 3 and 4. Amounts of NH_4Br of up to 287% of the weight of the PS foam (densities up to 0.053 g/cm^3) were deposited in specimens of sample 1. O.I. values of above 38 were obtained. Similar results were obtained with sample 2, with O.I. values of up to 36 and lower NH_4Br contents. The amounts of NH_4Br generally increased with the temperature and time of the bromine sorption stage (Table IV and V).

The dependence of O.I. on % NH_4Br in foamed PS (sample 1) prepared with several concentrations of Br_2 is shown in Figure 5. It can be clearly seen that the FR effectivity decreases with the increase in wt % of the NH_4Br. A similar result is seen for sample 2. This decrease can tentativley be explained by the large excess of bromine in the polymer, which is not fully utilized through the gaseous FR mechanism typical of bromine compounds. This excessive NH_4Br can be looked upon as a filler.

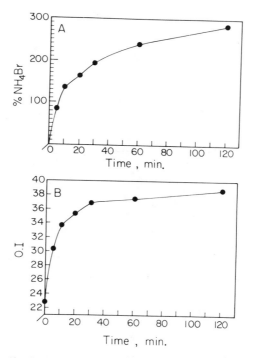

Figure 3. Weight % of NH₄Br (A) and O.I. (B) in PS foam, sample 1 vs time of contact with 0.42N Br₂ at 28 °C.

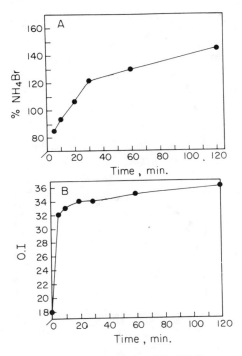

Figure 4. Weight % of NH₄Br (A) and O.I. (B) in PS foam, sample 2 vs. time in 0.42N Br₂ at 28 °C.

Table IV. Effect of Temperature of Br_2 Sorption on % Deposited
NH_4Br and O.I. In. Conc. of Br_2: 0.31N; Sample 1.

Temp., °C	% NH_4Br			O.I.		
	10 min.	30 min.	120 min.	10 min.	30 min.	120 min.
5	87	109	128	30.3	31.8	32.5
28	97	126	178	31.2	32.6	35.8
40	109	147	225	32.1	34.02	37.1

Table V. Effect of In. Br_2 Concentration on % NH_4Br Deposited
and O.I. Temp. 28°C; Sample 1.

In. Conc. N	% NH_4Br			O.I.		
	10 min.	30 min.	120 min.	10 min.	30 min.	120 min.
0.089	18	26	34	24.1	25.6	26.2
0.189	56	77	87	28.2	29.6	30.2
0.31	97	126	178	31.2	32.6	35.8
0.42	135	195	287	33.6	36.9	38.7

Figure 6 summarizes results of a series of experiments in which
the bromine sorption and ammonia stages were carried out on unfoamed
PS beads, containing 6% of absorbed gaseous pentane. The porofication treatment did not affect the pentane and the smaples could be
expanded by steam in the usual manner, to obtain foam densities of
0.0145 g/cm^3. It is noteworthy that the amounts of NH_4Br deposited
are much smaller than in the case of the foamed samples 1 and 2.
The maximum amount, 35 wt.%, of NH_4Br was already obtained after
ca 30 minutes' treatment in bromine water of 0.42N at 28°C. The
corresponding maximum O.I. value of 30.5, measured on the specimens
after the expansion indicates a much higher FR effectivity than in
Samples 1 and 2 (see Table VI). This may be due to several reasons.
The structure of the foam may be different in the former due to the
porofication treatment being carried out prior to the foaming. It
is possible that the linings (membranes) between the cell compartments in the foam have different insulation properties due to
different density or thickness [19]. It has been found in previous
work [8] in these laboratories that the porofication treatment produces very small pores down to the range of 0.01 - 0.001 microns.
Thus, the presence of such pores, filled with nitrogen or air and
NH_4Br in the membranes between the larger foam cells is indicated.
This will, however, bring about a much finer and more homogeneous
distribution of the NH_4Br in the foam. It will concentrate more in
the membranes, rather than being deposited inside the cells as is
probably the case when the porofication is carried out on the expanded PS. The bromine during the sorption stage diffuses and
accumulates into the cells and reacts with ammonia in the cells
rather than in the membrane. The large amounts of NH_4Br formed

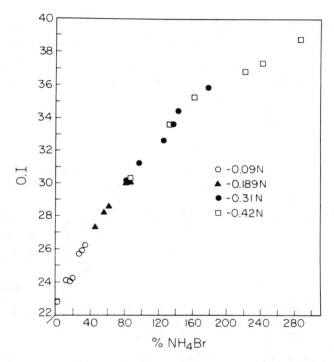

Figure 5. O.I. vs. weight % NH_4Br in foamed PS (sample 1) treated with four concentrations of aqueous Br_2 at 28 ° C.

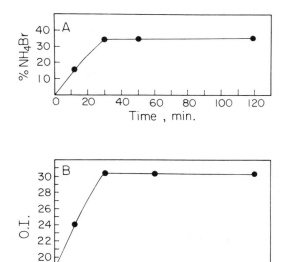

Figure 6. Weight % of NH_4Br (A) and O.I. (B) in unfoamed PS beads (sample 3) vs. time in Br_2 solution.

in samples 1 and 2 point to this possibility. The Br present in
high local concentrations in the cell will tend to be less effective
in combustion than the Br in the membrane. It will have to diffuse
after sublimation during pyrolysis through the cell and into the
membrane before initiating the flame inhibition cycle.

Table VI. Flame Retarding of PS Foam with Ammonium
Bromide Produced in the Polymer

Sample	Range of NH_4Br present % wt.	Range of O.I.	Effectivity $\dfrac{\Delta\,O.I.}{C_{Br}}$
(1)	0 – 40	22.8–27.2	0.085
5 mm	40 – 145	27.2–34.3	0.068
beads	145 – 287	34.3–38.9	0.032
(2)	14 – 26	24–26.9	0.28
3 mm	22 – 56	26.3–30.1	0.137
beads	85 – 145	32.1–36.2	0.089
(3) unfoamed beads	0 – 35	18.6–30.5	0.40

On the other hand, the presence of the small pores in the membranes (cell walls), due to the porofication treatment before foaming, increases the thermal insulation value of these membranes.
According to Woolley [19], this can have a marked effect on the
growth of fire in a compartment, due to the conservation of heat.
For a given heat flux, the time t_s needed to attain a given
surface temperature is given by

$$t_s \sim K \, \rho \, C \qquad\qquad (2)$$

where K- heat conductivity, ρ- density and C is the surface heat.
T_s will be by two orders of magnitude lower for a PS foam of
a density of 0.02 g/cm^3 than for a solid PS with a density of 1.05
g/cm^3, while the K increases only by a factor of 2.35 and C decreases by 15%. A decrease in the density and heat conductivity due
to formation of small pores in the membranes may thus partly offset
the simultaneous increase in the effectiveness of the Br in the membrane.
The above considerations, based on equation 2, also explain the
higher FR effectivity of Br in solid PS as compared to foamed PS,
shown in Table VII. It appears, however, that the NH_4Br deposited
according to the present study is almost as effective in foamed PS
as aromatic Br compounds in solid PS.
It should be realized that flammability of foams is a complex
subject area and the "mechanism by which cellular polymers with
different physical forms (cell sizes, etc.) lose heat at high temperatures have received surprisingly little attention" [19]. The

above discussion cannot be considered as conclusive and much additional work appears to be needed in order to elucidate further the flammability behavior of cellular polymers treated by various FR methods.

Table VII. ($\triangle$ O.I./%Br) of Various Bromine Compounds in Flame Retarding Polystyrene

Type of PS	FR Compound	$\triangle$O.I./%BR	Ref.
Solid PS	Aliphatic Br	1.1	12
Solid PS	Aromatic Br	0.3	12
Impact PS	Aromatic Br	0.27	13
Foamed PS			
Treated before foaming	NH_4Br	0.40	this paper
Treated after foaming	NH_4Br	0.032-0.28	this paper

Literature Cited

1. Lewin, M. In Chemical Processing of Fibers and Fabrics; Lewin, M.; Sello, S.B., Eds. Marcel Dekker, New York, 1984; Vol. 2, Part B., p. 1-141.
2. Masai, Y; Kato, Y; Fukua, N. U.S. Patent 3 719 727, 1973.
3. Lewin, M. U.S. Patents 3 150 919, 1964; 3 484 340, 1969; 3 547 687, 1970; British Patent 940 575, 1960.
4. Tress, J.J.; Heilman, W. U.S. Patent 3 050 476, 1962.
5. Stannet, V.T.; Walsh, W.K.; Bittencourt, E.; Liepins, R.; Surles, J.R. In Fiber Science; Lewin, M., Ed., J. Appl. Polym. Sci., Appl. Polym. Symp. 1977, 31, 201.
6. French Patent 2 119 743, 1971.
7. Lewin, M.; Guttmann, H. U.S. Patent 4 066 387, 1978.
8. Lewin, M.; Guttmann, H.; Naor, Y. Polymer-Bromide Interactions. I. Interactions with Polyacrylonitrile. Macromol. Sci.-Chem. 1988, A25, 1367-1383, and unpublished results.
9. McMaster, E.L.; Eichhorn, J.; Nagle, F.B. U.S. Patent 3 058 927, 1962.
10. Nagle, F.B.; Eichhorn, J. U.S. Patent 3 133 037, 1963.
11. Longstreet, M.O.; McMaster, E.L.; Nagle, F.B. U.S. Patent 3 108 016, 1963.
12. Van Krevelen, A. In Fiber Science; Lewin, M. Ed.; J. Appl. Polym. Sci., Appl. Polym. Symp., 1977, 31, 269-292.
13. Green, I. In Flame Retardant Polymer Materials. Lewin, M.; Atlas, S.M.; Pearce, E.M. Eds.; Plenum, New York, 1982, Vol. 3, 1-36.
14. Lewin, M.; Sello, S.B. In Flame Retardant Polymeric Materials. Atlas, S.M.: Pearce, E.M. Eds.Plenum: New York, 1975, Vol. 1, 19-136.
15. Petrella, R.V. In Flame Retardant Polymeric Materials. Lewin, M.; Atlas, S.M.; Pearce, E.M. Eds. Plenum: New York, 1978, Vol. 2, 159-201.
16. Mellor, J.W. Inorganic and Theoretical Chemistry. Longmans, Green & Co.: London, 1952, Vol. II, 590-595.

17. Jolles, Z.E., Ed., <u>Bromine and its Compounds</u>. E. Benn, 1966.
18. Mark, H.F.; Atlas, S.M. In <u>Flame Retardant Polymeric Materials</u>.
     Lewin, M.; Atlas, S.M.; Pearce, E.M. Eds., Plenum: New York,
     1975, Vol. 1, 1-19.
19. Woolley, W.D. In <u>Fire and Cellular Polymers</u>. Buist, J.M.;
     Grayson, S.J.; Woolley, W.D., Eds. Elsevier Appl. Sci:
     London, 1986, 37-41.

RECEIVED November 1, 1989

Chapter 11

Flame-Retardant Latices for Nonwoven Products

P. A. Mango and M. E. Yannich

Air Products and Chemicals, Inc., Allentown, PA 18195

Nonwoven products ranging from medical disposables to automotive fabrics are required to meet specific flammability standards. These fabrics are generally composed of cellulosic and/or synthetic fibers which are flammable. Additionally, polymer coatings are applied to the fabric to impart properties such as strength, abrasion resistance and overall binding. It is the purpose of this paper to describe the various polymer coatings commonly used in the nonwovens' industry and their effect on flammability of the substrates. Additionally, the effect of flame retardant additives, commonly used in latex formulations, will be discussed.

As nonwovens penetrate more markets traditionally held by other materials, the need for flame retardancy is frequently encountered. In hospitals, nurses' caps, surgeon's masks, surgical drapes and insulation gowns are required to meet flammability standards. In automobiles, nonwoven carpets, high loft padding and fiberfill must meet federal and industry flammability standards. In building materials, nonwoven drapes, blinds, wall coverings, air filters, and furniture fiberfill are all subject to local flammability ordinances. These are a few of the applications that the nonwovens' manufacturer must establish an effective flame retardant product for.

Air Products, a manufacture of latex binders, has completed a comprehensive study of flame retardants for latex binder systems. This study evaluates the inherent flammability of the major polymer types used as nonwovens binders. In addition, 18 of the most common flame retardants from several classes of materials were evaluated on polyester and rayon substrates. Two of the most widely recognized and stringent small scale tests, the NFPA 701 vertical burn test and the MVSS-302 horizontal burn test, are employed to measure flame retardancy of a latex binder-flame retardant system. Quantitative results of the study indicate clear-cut choices of latex binders for flame retardant nonwoven substrates, as well as the most effective binder-flame retardant combinations available.

Latex Binders Used in Flame Retardant Nonwovens

There are several polymer systems widely employed as nonwoven binders:

* acrylics
* ethylene-vinyl chloride copolymers (EVCL)
* vinyl acetate-ethylene copolymers (VAE)
* polyvinyl chloride and copolymers (PVC)
* styrene-butadiene copolymer (SBR)
* polyvinyl acetate homopolymers (PVAC)

0097–6156/90/0425–0145$06.00/0
© 1990 American Chemical Society

Each is used to some degree in nonwovens requiring flame retardancy, even though some actually impart flammability where others are inherently flame retardant. Table I summarizes the characteristics of some representative polymers widely used on rayon and polyester nonwoven substrates. Table II further identifies binder physical and chemical properties. Each binder was evaluated for its inherent flame retardancy on polyester and rayon as well as its flame retardancy in combination with commonly available flame retardants.

Flame Retardants for Nonwovens

There are literally hundreds of flame retardants commercially available for nonwoven polyester and rayon. These can be subdivided into durable and non-durable. In this paper, non-durable means water soluble in room temperature water. Durable means able to withstand at least five washes in hot water with detergent. Flame retardants with performance somewhere in between, often called semi-durable, were not utilized in this study. Table III is a compilation of the flame retardants included in the study. Compositions include:

- halogen containing compounds (with or without antimony)
- phosphorus containing compounds
- hydrated compounds
- other inorganic salts

The different classes of chemicals act in different ways to inhibit combustion. Prior to combustion of a nonwoven, several stages(1) occur:

- substrate temperature elevated by heating
- decomposition of substrate
- evaporation of decomposition products
- oxidation

Each class of flame retardant attacks the combustion process at different stages. For example, the hydrated compounds (alumina trihydrate, sodium silicate) contain bound water which is released upon heating and acts to reduce the temperature of the substrate. The phosphorus containing compounds(2) are thought to decompose to phosphoric acids which form stable nonvolatile products preventing libration of flammable materials. These are sometimes called solid phase flame retardants. A third mechanism for stopping combustion is to retard oxidation. So called vapor phase flame retardants, primarily the halogenated compounds with or without antimony, decompose to form free radicals which react with flammable matter. Without antimony, hydrogen halides are the precursor species; with antimony, antimony trihalide is the precursor. Hydrogen halides alone form noncombustible gases which oxygen starve the oxidation, act as a heat sink, and interfere with the free radical chain reaction. Antimony trihalides volatilize and are stronger inhibitors of the same free radical reaction. Because fire retardant mechanisms are not fully understood, this study developed empirical data - no correlation to or elucidation of mechanism was attempted.

Flame Retardant Test Pertinent to Nonwovens

There are many flame retardant tests used with some frequency for nonwovens. This study uses two of the most widely utilized and stringent tests: NFPA-701, a vertical flame test and MVSS-302, a horizontal test.

Table I. Performance Properties of Selected Binders on Rayon and Polyester

LATEX	TENSILE STRENGTH DRY	WET	HAND	RESILIENCY	HEAT SEALABILITY	UV STABILITY
EVCL	Good(R) Excellent(P)	Good(R) Excellent(P)	Moderate	Good	Excellent	Average
Acrylic (Soft)	Good(R) Poor(P)	Excellent(R) Poor(P)	Soft	Average	Average	Excellent
Acrylic (Moderate)	Good(R) Poor(P)	Good(R) Poor(P)	Moderate	Average	Average	Excellent
PVC	Poor(R) Good(P)	Poor(R) Good(P)	Stiff	Good	Average	Average
PVC (Plasticized)	Poor(R) Good(P)	Poor(R) Good(P)	Moderate	Good	Excellent	Average
VAE (Moderate)	Excellent(R) Excellent(P)	Excellent(R) Good(P)	Moderate	Average	Poor	Average
VAE (Very Soft)	Excellent(R) Good(P)	Poor(R) Poor(P)	Very Soft	Poor	Average	Average
SBR	Good(R) Poor(P)	Good(R) Poor(P)	Soft	Excellent	Good	Poor

Note: R - Rayon
 P - Polyester

Table II. Physical and Chemicals Properties of Selected Binders

Binder	Tg(°C)	Ionic Nature	pH	Viscosity (CPS)	% Solids
EVCL	+30	Anionic	8.0	100	50
Acrylic (Soft)	-22	Nonionic	3.0	550	46
Acrylic (Moderate)	0	Anionic	3.0	100	50
PVC	+69	Anionic	10.5	20	55
PVC (Plasticized)	+14	Anionic	9.8	43	56
VAE (Very Soft)	0	Anionic	5.5	700	52
VAE (Moderate)	-25	Nonionic	5.5	200	52
SBR	-14	Anionic	5.7	110	50

NFPA-701 (large and small scale): These tests are sanctioned by the National Fire Protection Association, a volunteer organization of fire protection professionals. Figure 1 illustrates the set up. The vertical alignment of flame and sample allows maximum heat and flammable gases to feed the flame front of the sample, assisting further combustion.

MVSS-302: This test, Motor Vehicle Safety Standard 302, is a Federal Standard for materials used in interiors of automobiles. High loft polyester pads in door panels, dash boards, and behind upholstery are key targets in the nonwovens area. Figure 2 illustrates this test set up. The horizontal alignment of the sample, with flame impingement at a 90° angle, allows heat and flammable gases to transfer away from the moving flame front.

A third test often cited, Oxygen Index (O.I.) is an ASTM test which measures the flammability of the base polymers. The flammability of the polymer film may alone differ significantly from the flammability of the same polymer on a nonwoven substrate. Theoretically, the higher O.I. for a sample film, the higher its nonflammability.

Table III. Flame Retardants for Nonwovens

Retardant	No	Manufacturer	Composition	Durability
Caliban P-44	1	White Chemical	Decarbromodiphenyl oxide/ Antimony Trioxide/Water	Durable
Caliban P-95	2	White Chemical	Antimony Trioxide/Water	Durable
ND-74	3	White Chemical	Inorganic Salts Organic Buffer	Nondurable
Fyarestor 100	4	Pearsall Chemical	Bromochlorinated Paraffins	Durable
Fyarestor 330	5	Pearsall Chemical	Bromine-Phosphorus Compound/Water	Nondurable
ID-28-70A	6	Pearsall Chemical	Bromine-Phosphorus Compound/Water	Nondurable
N-22	7	Nyacol Products, Inc.	Pentabromodiphenyl oxide/ Antimony Pentoxide/Water	Durable
A-1530	8	Nyacol Products, Inc.	Antimony Pentoxide	Durable
N-24	9	Nyacol Products, Inc.	Bromochloro/Antimony Pentoxide/Water	Durable
X-12	10	Spartan	Blend of Organic Salts	Nondurable
CM	11	Spartan	Blend of Organic Salts	Nondurable
Fyran J3	12	Crown Metro Inc.	Ammonium Sulfamate/Water	Nondurable
Diammonium Phosphate	13	-	Diammonium Phosphate	Nondurable
Delvet 65	14	Diamond Shamrock	Chlorinated Paraffins/Water	Durable
Santicizer 141	15	Monsanto	2-ethylhexyl diphenyl phosphate	Durable
Hydral 710	16	Alcoa	Aluminum Trihydrate	Durable
Firebrake ZB	17	U.S. Borax	Zinc Borate	Durable
CRF-NW	18	Chem-Cast	Tricresyl Phosphate	Durable

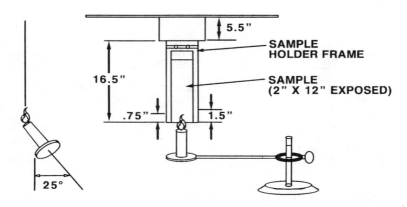

Figure 1.   NFPA-701 test configuration.

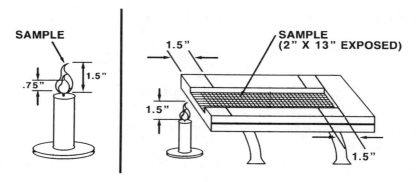

Figure 2.   MVSS-302 test configuration.

Flame Retarding Nonwovens

In flame retarding nonwovens, the contribution of components may not be additive. Rather, the interaction of binder, flame retardant, and substrate is critical in the performance of the flame retardant nonwoven. Similarly, the flammability of a binder film or the flammability of a flame retardant coated woven cloth often do not predict the flame retardancy of the same binder or flame retardant on a nonwoven substrate of rayon or polyester. Actual data on a nonwovens substrate is the only accurate measure of a system's flame retardancy. For this study, two widely used substrates were selected. The first, lightweight rando rayon, is representative of material used in nurse caps, surgeon's masks, and miscellaneous coverstock. This material is constructed of 1 1/2 denier fiber, weighs 1 1/2 ounces per square yard, and is relatively dense web. Rayon as a material is water absorbent, burns rather than melts, and is readily flammable. This fiber ignites around 400°C(3) and has an oxygen index of about 19.0. Certain binders adhere well to rayon while others do not. Apparently, this lack of affinity for the substrate affects flame retardancy, as will be demonstrated later.

The second substrate utilized was a high loft polyester, representative of material used in air filters, furniture padding, and automobile interior pads. The material used here is constructed of 1 1/2 denier fiber, weighs 1 1/2 ounces per square yard, and is relatively open in construction. Polyester as a material is water resistant and flammable. This fiber ignites at around 520°C, but melts at about 265°C. In many flame retardant tests, the polyester will melt and "shrink" away from the flame, giving spurrious results. Once again, not all binders have an affinity for this substrate, and flame retardancy is affected.

Each substrate was saturated or sprayed with binder or binder-flame retardant to attain a 40% dry add-on. The high loft polyester was tested using the MVSS-302 protocol as this type material often is utilized in applications requiring this test. The rando rayon was tested using NFPA-701 protocol, as this test is one of the most stringent used with nonwovens.

Flammability of Latex Binder on Substrate

The first experimental program undertaken was a study of the flammability of latex binders alone on nonwoven substrates. Table IV summarizes the results of this study. From composition, the halogen containing binders would be expected to perform best as flame retardants. Logically, the more halogen, the better the performance. Empirically, however, this expectation is only supported by the oxygen index values on stand alone binder films. Once coated on a substrate, especially polyester, this supposition is not borne out. Airflex 4530, an ethylene-vinyl chloride copolymer containing 46 weight % chlorine is superior to the approximately 56 weight % containing PVC. The simplistic hypothesis that flame retardant performance of a material correlated with halogen content is not supported. A second finding is that oxygen index does not correlate well with actual flame retardancy on substrates. Evidently, the liberation of hydrogen chloride and the resultant formation of combustion inhibiting free radicals is rate dependent on the molecular environment. Surrounded by other polymer chains, the rate is rapid. When absorbed and bonded to a substrate, this rate may be slowed. While the actual cause is not completely understood, the effect is empirically shown in Table V. A final conclusion drawn from Table V is that for tests less stringent than MVSS-302 on polyester, a binder alone can be sufficient to attain flame retardancy. For more stringent tests or more flammable substrates such as rayon or cotton, none of the binders investigated here provide flame retardancy alone.

Flammability of Binder-Flame Retardant Systems on Polyester

The second experimental program conducted was an evaluation of binder-flame retardant combinations on high loft polyester. Each of the eighteen flame retardants in Table III were screened for flame retardant performance with a latex binder on polyester. Six of the best performing flame retardants were selected for more extensive testing, the results of which appear in Table V. Two new findings are noted in this table; TACKY, indicating a plasticization of the polymer by the flame retardant which causes severe surface tackiness, and FLOCKS, indicating the latex binder and flame retardant, at concentrations used here, are not compatible and severe flocculation of latex or flame retardant occurs. With these new limitations, further observations are:

- EVCL's are compatible with all flame retardants tested, and perform well with all but one.

Table IV. Flammability of Latex Binders on Nonwoven Substrates

Latex	NFPA 701 (Rando Rayon)	MVSS 302 (High Loft Polyester)	Oxygen Index (Film Only)
EVCL	Total Burn	Pass (3.3 in/min)	25.0
PVC	Total Burn	Fail (4.1 in/min)	46.0
PVC (DOP Plasticized)	Total Burn	Fail (4.6 in/min)	23.4
Acrylic (Soft)	Total Burn	Fail (9.1 in/min)	19.4
VAE (Moderate)	Total Burn	Pass (3.9 in/min)	17.4
VAE (Very Soft)	Total Burn	Fail (12.5 in/min)	17.4
SBR	Total Burn	Fail (11.1 in/min)	17.2
PVC (Self Crosslinking)	- - -	Fail (6.3 in/min)	17.7

NFPA 701: Pass: <5.5 inch burn (average of 10) with no single sample >6.5 inch burn.

NFPA 302: Pass: <4 in/min burn rate of <2 in. burned before self-extinguishing (S.E.).

Table V. Effectiveness of Flame Retardants (High Loft Polyester)

Binders	White Chemical Caliban P-44	White Chemical ND-74	Witco Fyarestor 330	Witco ID-28-70A	Nyacol N-22	Diammonium Phosphate
EVCL	Pass (2" S.E.)	Pass (2" S.E.)	Pass (4"/MIN)	Pass (2" S.E.)	Pass (2" S.E.)	Fail (5" S.E.)
VAE (Very Soft)	Pass (2" S.E.)	Fail (3" S.E.)	Fail (4.5" S.E.)	Pass (2" S.E.)	Fail (4.1"/MIN)	FLOCKS
VAE (Moderate)	Pass (2" S.E.)	Pass (2" S.E.)	Fail (3.5" S.E.)	Pass (1.5" S.E.)	TACKY	FLOCKS
PVC (DOP Plasticized)	Pass (2" S.E.)	Fail (3.5" S.E.)	FLOCKS	FLOCKS	Fail (4" S.E.)	FLOCKS
PVC	Fail (4.5" S.E.)	Fail (3.5" S.E.)	Fail (3.5" S.E.)	Pass (1.5" S.E.)	Pass (2.2"/MIN)	Pass (2.5"/MIN)
Acrylic (Soft)	Fail (4.8" S.E.)	Fail (4" S.E.)	Fail (2.5" S.E.)	Pass (2" S.E.)	Pass (2.9"/MIN)	FLOCKS
PVAC (Self-crosslinking)	Fail (4.5"/MIN)	Fail (6.7"/MIN)	Pass (1" S.E.)	Fail (3" S.E.)	Pass (2.1"/MIN)	FLOCKS

CONDITIONS: 40% Binder Add-On
40% Flame Retardant (of total solids)

Polyester: 1.5 oz., 1.5 denier

Flame Test: Horizontal Burn Test (MVSS 302)
Pass: <4 in/min burn rate or <2 in. burned before self-extinguishing (S.E.).

- Two of the other binders, VAE and PVC, pass with three of the six flame retardants (which is the next best performance).
- Addition of a dioctyl phthalate plasticizer (DOP) increases flammability and incompatability of PVC with flame retardants.
- Acrylics, in general, are flammable materials with or without flame retardants.

A significant advantage to performing well with a wide range of flame retardants is formulating flexibility. There are many factors which limit the choice of flame retardants asided from flame retardant performance and compatability. For example, environmental constraints (no antimony to the sewer, no ammonia in the workplace) and compatability constraints (shorter than normal shelf life with certain emulsions) may limit the choice.

Another issue is the PVC dilemma: to plasticize or not to plasticize. If PVC is not plasticized, it shows some latitude in flame retardant acceptability and good performance in flame testing, but does not film form well and is stiff. If PVC is plasticize (most commonly with DOP), flammability increases and incompatability with additives proliferates. Additionally, plasticizer volatilization and diffusion can present performance and environmental problems. A compromise may be reached by blending a plasticized and unplasticized PVC. Approaches utilizing internal plasticizers, monomers copolymerized with VCM, have shown varied success. Using ethylene as the comonomer, as is done with EVCL's gives excellent results. Incorporating an acrylate may solve some problems, but significantly raises flammability.

An additional advantage illustrated in Table V is the potential to use less flame retardant with certain binders to achieve desired performance. This can lead to advantages in strength and hand, or cost advantages in lower added costs.

Flammability of Binder-Flame Retardant Systems on Rayon

Another experimental program was conducted with binder-flame retardant combinations on rando-rayon. The eighteen flame retardants were paired with a latex binder and screened for performance in an NFPA-701 flame retardant test. Seven of the best performing flame retardants were chosen for further study. Additionally, since all seven happened to be non-durable, the best performing durable flame retardant was added at a higher add-on and flame retardant level. More than coincidence is probably responsible for the easy penetration and deposition of the non-durable (water soluble) flame retardant. The durable (non-water soluble) flame retardants, restricted to surface deposition are not as effective. The results of the testing, using the stringent NFPA-701 test protocol, are summarized in Table VI. Once again, several conclusions can be made:

- EVCL's pass with five of the eight flame retardants, and is very close on two others. None of the flame retardants were incompatible with the EVCL's.
- The next best latex, PVC, passed with three of the eight.
- In this test, one VAE, the acrylic, and the plasticized PVC latexes failed with all of the retardants.

Conclusions

The performance of several latex binders in flame retardant testing of nonwoven polyester or rayon substrates with and without added flame retardants has been investigated. Correlation of coating flammability (i.e. by oxygen index) to actual performance on a substrate is poor. Results generated on both rayon and polyester

Table VI. Effectiveness of Flame Retardants (Rando Rayon)

Flame Retardant Binders	Witco Fyarestor 330	Spartan CM	Crown Metro Fyran J3	Diammonium Phosphate	White Chemical ND-74	White Chemical P-44*
EVCL	Pass (5.3")	Pass (5.4")	Pass (4.9")	Fail (5.7")	Pass (5.1")	Pass (5.3")
VAE (Very Soft)	Fail (6.0")	Fail (8.6")	TOTAL BURN	Fail	Fail (8.9")	TOTAL BURN
VAE (Moderate)	Fail (6.0")	Fail (7.1")	Fail (8.1")	Fail (8.3")	Fail (9.9")	Pass (5.3")
PVC (DOP Plasticized)	FLOCKS	Fail	Fail (7.8")	FLOCKS (7.8")	Fail (6.6")	- -
PVC (Unplasticized)	Fail (6.0")	Pass (5.5")	Fail (5.9")	Fail (5.9")	Pass (5.5")	Pass (4.9")
Acrylic (Soft)	Fail (8.7")	Fail (9.9")	TOTAL BURN	Fail (8.1")	Fail (8.4")	TOTAL BURN
SBR	Fail (6.3")	Fail (9.9")	TOTAL BURN	TOTAL BURN	Pass (5.5")	- -

Conditions: 40% Total Solids Add-On
40% Flame Retardant (of total solids)

Rayon: 1.5 oz., 1.5 denier

Flame Test: Vertical Burn Test (NFPA-701)
Pass: <5.5 inch burn (average of 10 samples), with no samples >6.5 inch burn.

*60% Total Solids Add-on 75% Flame Retardant

substrates have demonstrated a clear performance advantage in flame retardancy for an ethylene-vinyl chloride copolymer containing 46% chlorine over all other considered latexes, with and without added flame retardants. A postulated advantage for PVC latex, containing about 56% chlorine, was not observed and plasticized PVC demonstrated poor flame retardant tolerance. Acrylics and SBR, with and without flame retardants, exhibited little flame retardancy.

<u>Literature Cited</u>

1 Mischutin, V., <u>A New FR System for Synthetic/Cellulosic Blends</u>, Textile Chemist and Colorist, March 1985, Vol. 7, No. 3.
2 Weil, E. D., <u>Flame Retardants for Nonwovens</u>, Tappi Notes: 1985 Nonwovens Binders: Advanced Chemistry and Use Seminar, September 30 - October 2, 1985.
3 Kuryla, W. C., and Papa, A. J., <u>Flame Retardancy of Polymeric Materials</u>, Marcel Dekker, Inc., 1975, Vol. 3, p 161.

RECEIVED November 1, 1989

Chapter 12

Zinc Borate as a Flame Retardant, Smoke Suppressant, and Afterglow Suppressant in Polymers

Kelvin K. Shen[1] and T. Scott Griffin[2]

[1]United States Borax and Chemical Corporation, Los Angeles, CA 90010
[2]United States Borax Research Corporation, Anaheim, CA 92801

Zinc Borate ($2ZnO \cdot 3B_2O_3 \cdot 3.5H_2O$) is a unique multifuntional fire retardant. It can function as a flame retardant, smoke suppressant, afterglow suppressant, as well as an anti-tracking agent in polymers. It has been used extnesively as a fire retardant in PVC, polyolefin, nylon, polyester, epoxy, acrylic, urethane, rubbers, etc.

The use of bromine or chlorine halogen sources to impart fire retardancy in polymers is well known in the plastics industry. The halogen source can be physically blended into the resin (additive approach) or chemically built into the polymer backbone (reactive approach). In order to enhance fire retardancy, antimony oxide is usually used as a synergistic additive. In recent years, considerable effort has been expended to find either partial or complete substitutes for antimony oxide. This effort has been spurred by the desire to achieve smoke reduction, better cost/performance balance, to obtain a product with low tinting strength, and by the concern for the possible toxicity of antimony oxide. One of these substitutes is a special form of zinc borate with a molecular formula of $2ZnO \cdot 3B_2O_3 \cdot 3.5H_2O$, known in the trade as FIREBRAKE ZB (1, 2).

Zinc borates with different mole ratios of $ZnO:B_2O_3:H_2O$ can be readily prepared by reacting zinc oxide with boric acid (Fig. 1). Among all these zinc borates, the compound with a molecular formula of $2ZnO \cdot 3B_2O_3 \cdot 3.5H_2O$, is the most commonly used fire retardant. This article will review recent research results on the use of this particular form of zinc borate as a multifunctional fire retardant.

0097–6156/90/0425–00157$06.25/0

DISCUSSION

The major advantages of using the zinc borate can be summarized as follows:

1. High Dehydration Temperature - In contrast to other forms of zinc borates, the water of hydration of the zinc borate is retained up to 290°C, thus allowing it to be used in polymers requiring high processing temperatures. The proposed molecular structure for the zinc borate is depicted in Fig. 2. The high dehydration temperature can be explained by the absence of interstitial water in the crystal lattice.

2. Completely Reacted Material - The zinc borate contains no free zinc oxide. Thus, it does not have serious detrimental effects on the thermal stability of PVC or chlorinated paraffin as free zinc oxide does.

3. A Multifunctional Smoke Suppressant, Flame Retardant, and Afterglow Suppressant.

4. Low Tinting Strength - The zinc borate has a refractive index similar to most organic polymers, which results in the retention of considerable translucency. This provides for easy visual inspection of the finished products and allows the use of lower pigment loadings in formulations requiring deep shades.

5. Improving Electrical Properties - The zinc borate has good antitracking and antiarcing properties.

6. Low Toxicity - Unlike antimony oxide, the zinc borate is considered non-toxic.

The performance of the zinc borate as a fire retardant is summarized in the following:

ZINC BORATE IN HALOGEN-CONTAINING POLYMERS.
AS A FLAME RETARDANT. The zinc borate is an efficient synergist of organic halogen sources. In certain halogen-containing systems such as unsaturated polyester, epoxy (3), and rigid PVC, the zinc borate alone can outperform antimony oxide as shown by the Oxygen Index and UL-94 tests (Fig. 3, 4, and 5).

In other halogen-containing systems, such as flexible PVC and polyolefins, it is preferable to use the zinc borate in conjunction with antimony oxide for maximum performance. In flexible PVC, for example, the zinc borate alone is not very effective in the Oxygen Index test (Fig. 6), but a combination of the zinc borate and antimony oxide (1:1 ratio) outperforms equal weight of antimony oxide at a total loading of more than 10 phr (4). In the presence of alumina trihydrate (ATH), the beneficial effect of using a combination of the zinc borate and antimony oxide is dramatically increased (Fig. 7).

AS A SMOKE SUPPRESSANT. Antimony oxide performs well as a flame retardant in flexible PVC, however, as a gas phase flame retardant, its use can drastically increase the smoke production during PVC combustion as illustrated in Fig. 8. The zinc borate,

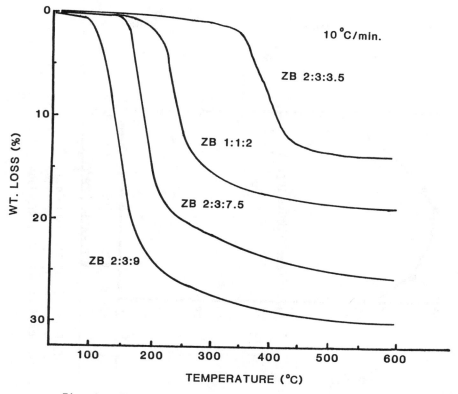

Fig. 1.   Thermogravimetric Analysis of Zinc Borates.

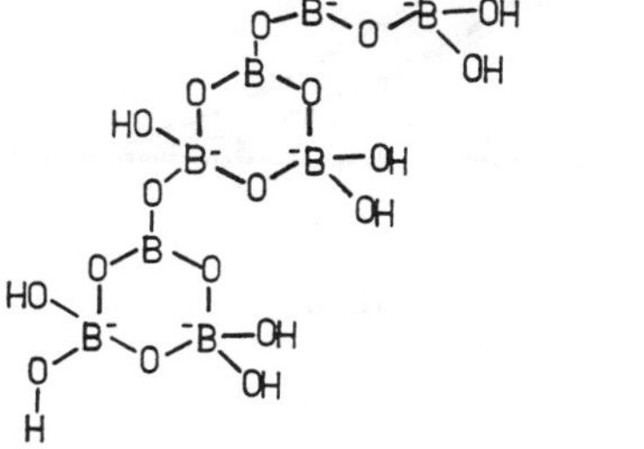

Fig. 2.   Proposed Molecular Structure for Zinc Borate $(2ZnO \cdot 3B_2O_3 \cdot 3.5H_2O)$.

160 FIRE AND POLYMERS

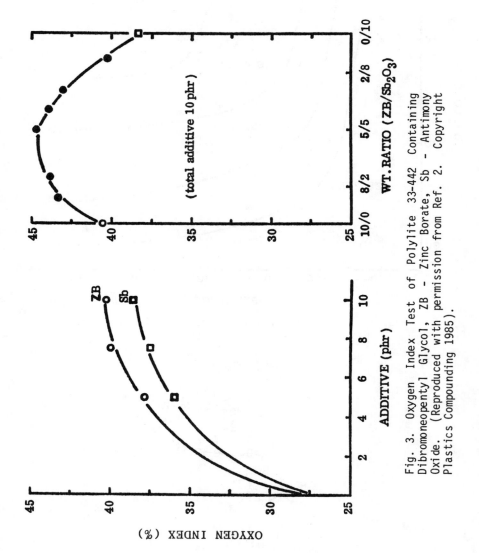

Fig. 3. Oxygen Index Test of Polylite 33-442 Containing Dibromoneopentyl Glycol, ZB - Zinc Borate, Sb - Antimony Oxide. (Reproduced with permission from Ref. 2. Copyright Plastics Compounding 1985).

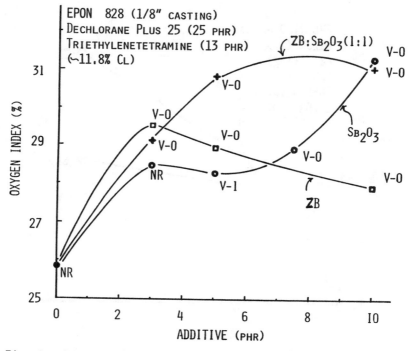

Fig. 4. Oxygen Index and UL-94 Tests on Epoxy (NR - not ratable, sample completely consumed). (Reproduced with permission from Ref. 2. Copyright Plastics Compounding 1985).

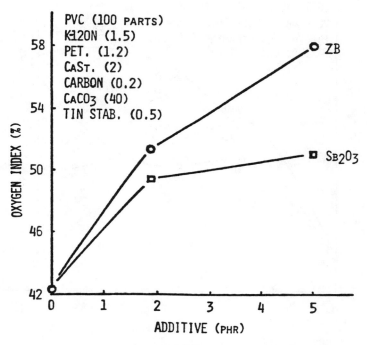

Fig. 5. Rigid PVC Conduit Formulation. (Reproduced with permission from Ref. 2. Copyright Plastics Compounding 1985).

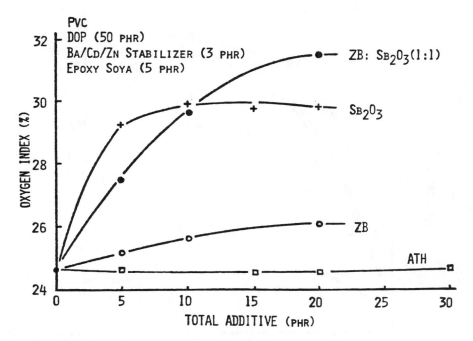

Fig. 6. Oxygen Index of Flexible PVC Formulations. (Reproduced with permission from Ref. 2. Copyright Plastics Compounding 1985).

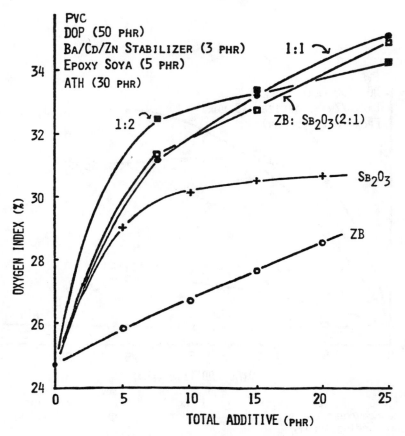

Fig. 7. Oxygen Index of Flexible PVC Formulations Containing
Alumina Trihydrate (30 phr). ZB - Zinc Borate. (Reproduced
with permission from Ref. 2. Copyright Plastics Compounding
1985).

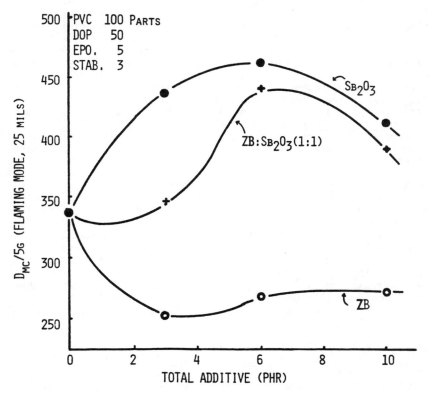

Fig. 8. NBS Smoke Test of Flexible PVC. ZB - Zinc Borate.

on the other hand, is an effective smoke suppressant. Used with antimony oxide, formulations with high ratios of zinc borate to antimony oxide yield the greatest smoke reduction (Fig. 9).

In the presence of ATH, the zinc borate reduces smoke effectively, even when used with antimony oxide at a 1:1 ratio. Fig. 10 shows this dramatic smoke reduction.

AS AN AFTERGLOW SUPPRESSANT. Due to its unique glass-forming ability (i.e., from the B_2O_3 moiety), the zinc borate can also function as an afterglow suppressant as illustrated in Table I (5). In addition, the zinc borate is also an effective antitracking and antiarcing agent(6). In a 1978 U.S. Patent, ICI disclosed the use of the zinc borate in combination with antimony oxide and Dechlorane Plus (an organic chlorine source) to give a V-0 reinforced nylon-6,6 (28% fiberglass) with high tracking resistance. While antimony oxide can decrease the tracking resistance, the zinc borate can improve the tracking resistance. (Table II). The water of hydration of the zinc borate, when released at high temperatures can diffuse and cool the arcs. The resulting anhydrous zinc borate undergoes vitrification and displays good electrical insulation properties.

MODE OF ACTIONS IN HALOGEN-CONTAINING POLYMERS. In halogenated polymers, such as flexible PVC, the zinc borate markedly increases the amount of char formed during combustion; whereas the addition of antimony oxide, a vapor phase flame retardant, has little effect on char formation. Analysis of the char shows that about 80-95% of antimony volatilized, whereas the majority of boron and zinc from the zinc borate remained in char (86% and 61%), respectively. (Table III). The fact that the majority of the boron remains in the condensed phase is in agreement with the fact that boric oxide is a good afterglow suppressant. The zinc species remaining in the condensed phase can alter the pyrolysis chemistry by catalyzing the dehydro-halogenation and promoting cross-linking, resulting in the increased char formation and a decrease in both smoke production and flaming combustion. The B_2O_3 moiety of the zinc borate can form a glassy layer inhibiting further oxidation of the char.

ZINC BORATE IN HALOGEN-FREE POLYMERS.

AS A FLAME RETARDANT AND SMOKE SUPPRESSANT. In recent years, the development of halogen-free fire retardant polymers has been a subject of considerable interest, because these polymers will not produce any corrosive and toxic hydrogen-halide gases during combustion. One of the common approaches to flame-retard a halogen-free polymer has been the addition of high loadings (100-200 phr) of ATH. ATH can undergo an endothermic dehydration by releasing 34% by weight of water in the temperature range of 220°-450°C. Recently, it was found that there are major advantages in partially replacing ATH with the zinc borate as a fire retardant in a variety of halogen-free polymer systems (7).

Fig. 11 illustrates the smoke suppression effect of the zinc borate when used in conjunction with ATH in a halogen-free ethylene-vinyl acetate (EVA) system. Sumitomo Electric, in a

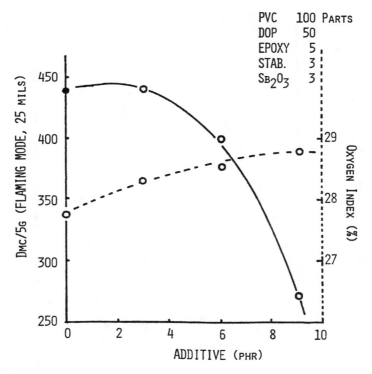

Fig. 9. NBS Smoke Test (——) and Oxygen Index Test (---)
of Zinc Borate in Flexible PVC.

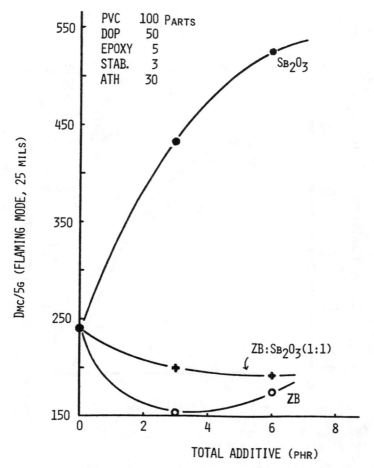

Fig. 10. NBS Smoke Test of Flexible PVC Containing Alumina Trihydrate. ZB - Zinc Borate.

Table I.   Fire Retardant Magnetic Polypropylene[a]

Compositions	Example			
	1	2	3	4
Ferrite (% by wt.)[b]	70	70	70	70
DBDPO[c]	35	35	35	–
Brominated Polystyrene	–	–	–	40
Sb_2O_3	12	12	12	14
Zinc Borate	–	20	40	10
Afterflame (sec.)	0	0	0	0-2
Afterglow (sec.)	115-127	22-28	10-15	3-5
UL-94 (1/16 in.)	HB	V-O	V-O	V-O

[a]Yokoda, K., et. al., U. K. Patent application GB 2,115,822A (1983, to Kanegafuchi Kagaku), U. S. Patent 4,490,498.

[b]All the other additives are parts per hundred parts resin by wt.

[c]Decabromodiphenyl oxide.

Table II.   Thermoplastic Nylon-6,6 Compositions[a]

Components	UL-94 (1.6 mm) 50%RH[b]	70C[c]	Tensile strength (MN/m^2)	CTR[d]
Zinc Borate (15 wt-%)	Fail	Fail	141	>600
Dechlorane 515 (10 wt-%)	Fail	Fail	147	300
Dechlorane 515 (10 wt-%) plus Sb$_2$O$_3$ (5 wt-%)	V-0	V-0	140	200
Dechlorane 515 (10 wt-%) plus Zinc Borate (15 wt-%) plus Sb$_2$O$_3$ (0.1 wt-%)	V-1	V-1	138	500
Dechlorane 515 (10 wt-%) plus Zinc Borate (15 wt-%) plus Sb$_2$O$_3$ (1wt-%)	V-0	V-1	137	475
Dechlorane 515 (10 wt-%) plus Zinc Borate (15 wt-%) plus Sb$_2$O$_3$ (2 wt-%)	V-0	V-0	-	<350

a.   Maslen, A.J., and W.H. Taylor, U.S. Patent 4,105,621 (1978, to ICI); nylon-6,6 with 28% glass fiber.
b.   Measured after conditioning at 50% RH and 23C for 48 hours.
c.   Measured after conditioning at 70C for one week.
d.   Comparative tracking resistance.

Table III.   Ignition of Flexible PVC at 560°C[a]

Sample (phr)	Percent Char	Percent of Original Elements in Char					
		Zn	B	Sb	Cl	C	H
No Additive	8.3	-	-	-	0.1	14.0	4.0
Sb$_2$O$_3$(10)	8.8	-	-	6.3	0.1	17.2	4.8
ZB[b](10)	18.6	60.6	85.8	-	2.3	24.7	7.9
Sb$_2$O$_3$(5)+ ZB (5)	16.3	64.2	73.8	19.9	1.0	24.7	7.0

[a]PVC - Geon 121 (100 parts), DOP (50), Epoxy Soya (5), Ferro 6V6A (3); 1/8 inch thick sample.

[b]Zinc Borate.

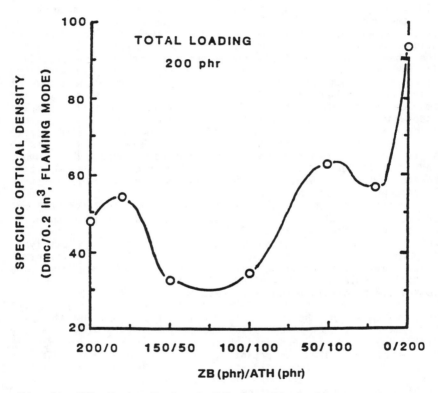

Fig. 11. NBS Smoke Test of Ethylene-Vinyl Acetate. ZB - Zinc Borate, ATH - Alumina Trihydrate.

1986 U.S. patent, disclosed that a highly filled combination of ATH and unspecified zinc borate in an EVA cable produced a much lower optical transmission signal loss than a conventional fluoropolymer in an optical composite cable application (Table IV).

More recently, General Electric reported the use of the zinc borate as a smoke suppressant in a halogen-free silicone-polyetherimide copolymer (8).

MODE OF ACTION IN HALOGEN-FREE POLYMERS. To quantify the zinc borate char forming performance, EVA samples containing ATH and/or the zinc borate were pyrolyzed at 500°C in an oven; and the residues were weighed and analyzed for carbon content (7). It was found that ATH (200 phr) catalyzed the combustion of EVA via a glowing combustion mode. As a result, it decreased the char yield, relative to the control without any additive, in the powdery residue. Partial replacement of ATH with the zinc borate (100 phr each) resulted in a tenfold increase of char and a change-over from the glowing combustion to the smouldering combustion mode. At 550°C (Table V), all four samples underwent a flaming combustion and showed low char yields. Interestingly, partial replacement of ATH with the zinc borate (100 phr each) resulted in a significant delay of the time to self-ignition. In addition, this combination produced a porous and hard residue, attributed to the sintering between the zinc borate and ATH. This was confirmed by the Scanning Electron Microscopy study of the residues (Fig. 12).

Differential Scanning Calorimetric (DSC) analyses of EVA samples in air showed that 50% replacement of ATH with the zinc borate at a total loading of 200 phr resulted in a remarkable 305 cal/g less of exothermicity in the oxidative-pyrolysis temperature range of 362-562°C (Fig. 13). Differential Thermogravimetric (DTG) analyses in air showed that the same partial replacement resulted in a delay of the peak "oxidative-pyrolysis weight loss rate" by 55°C (Fig. 14).

The increase in char yield, the decrease in exothermicity, and the delay of the peak oxidative-pyrolysis rate by the zinc borate indicate that it can alter the oxidative-pyrolysis chemistry of the polymer in the condense phase and render it less susceptible to combustion. More importantly, the zinc borate and ATH can form a porous and hard residue at temperatures above 550°C. This hard residue can protect the unburned polymer from being exposed to high temperatures and further combustion.

CONCLUSIONS

In halogen-containing polymers, the multifunctional zinc borate can function as a flame retardant, smoke suppressant, and afterglow suppressant.

In halogen-free polymers, the zinc borate in combination with ATH at high loadings can also function as a flame retardant and smoke suppressant by releasing water and forming a porous ceramic residue, which acts as a thermal insulator.

Table IV.   Flame Retardant Ethylene-Vinyl Acetate
Optical Composite Cable[a]

Components (phr)	Oxygen Index (%)	Maximum Smoke[b] Density (Dm)	Tensile Strength (kg/mm^2)
ATH[c](50)+ZB(50)	30	90	0.61
ATH(75)+ZB(75)	32	75	0.66
ATH(100)+ZB(100)	35	75	0.65
ATH(125)+ZB(125)	55	60	0.79
MgCO$_3$(50)+ZB(50)	31	88	0.59
MgCO$_3$(75)+ZB(75)	34	73	0.63
MgCO$_3$(100)+ZB(100)	36	55	0.63
MgCO$_3$(125)+ZB(125)	60	50	0.66

[a]Vinyl acetate content 60%; U.S. Patent 4,575,184 (1985 to Sumitomo Electric).

[b]Flaming mode, sample thickness unkown, D_m-maximum specific optical density.

[c]Alumina trihydrate.

Table V.   Pyrolyses of Halogen-Free EVA at 550°C in Air[a]

Composition (phr)	Percentage of Original Carbon Remaining in Residue (%)	Mode of Combustion (time)[d]	Residue Hardness
No additive	0.6	Flaming (3 sec.)	Very Soft
ATH[b](200)	0.7	Glowing (15 sec.) Flaming (1 min. 25 sec.)	Very Soft
ZB[c](200)	1.6	Flaming (1 min.)	Very Hard
ZB(100)+ATH(100)	0.9	Flaming (1 min. 40 sec.)	Very Hard

[a]Pyrolyses carried out in a forced-air oven for 6 minutes.

[b]Alumina Trihydrate.

[c]Zinc Borate.

[d]Time for combustion to start.

Fig. 12. Scanning electron micrographs (6000 x) of residues of air-pyrolyses of halogen-free ethylene-vinyl acetate, containing Zinc Borate (33%) and alumina trihydrate (ATH) (33%). At 500°C (top picture), a powdery residue was obtained; no sintering took place between the Zinc Borate (small particles) and ATH (large particles). At 550°C (bottom picture), a hard and porous residue was obtained; Zinc Borate acted as a sintering agent. A hard and porous residue is a good thermal barrier.

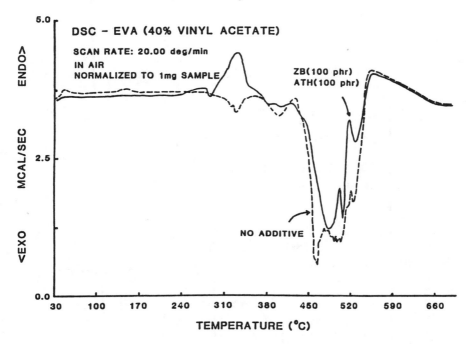

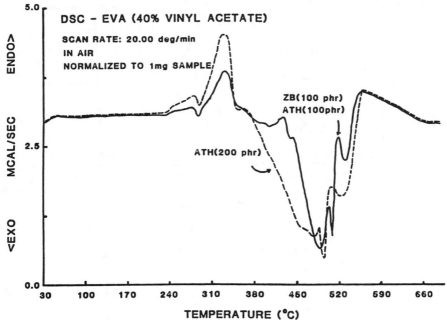

Fig. 13. Differential Scanning Calorimetric Analyses of Ethylene-Vinyl Acetate. ZB - Zinc Borate, ATH - Alumina Trihydrate.

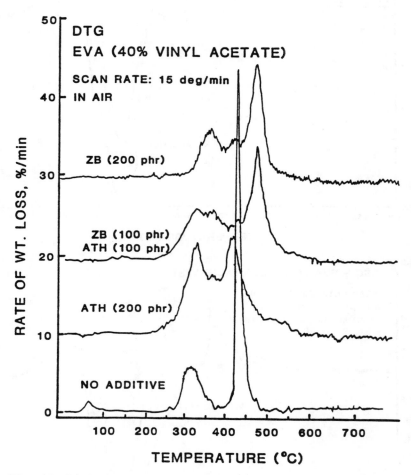

Fig. 14. Differential Thermogravimetric Analyses of Ethylene-Vinyl Acetate. ZB - Zinc Borate, ATH - Alumina Trihydrate. Zero values for the ordinates of success curves are at about 10%/min. intervals.

REFERENCES:

1.   Nies, N. P., and R. W. Hulbert, U. S. Patent Re. 27,424.
2.   Shen, K. K., Plastics Compounding, 8, No. 5, 66 (Sep./Oct. 1985).
3.   Shen, K. K., and R. W. Sprague, Journal of Fire Retardant Chemistry, 9, 161 (1982).
4.   Shen, K. K., Journal of Vinyl Technology, 4, No. 3, 120 (1982).
5.   Rakszawski, J. F., and W. E. Parber, Carbon, 2, 53 (1964).
6.   Shen, K. K., Fire Retardant Chemical Association Spring Conference, San Antonio, Texas (March 1989).
7.   Shen, K. K., Proceedings of the International Conference on Fire Safety, San Francisco, 12, 340 (1987).
8.   Cella, J.A., E. A. O'Neil, and D. A. Williams, U.K. Patent 2 193 216A (Feb. 3, 1988).

RECEIVED December 18, 1989

Chapter 13

The Design of Flame Retardants

Charles A. Wilkie

Department of Chemistry, Marquette University, Milwaukee, WI 53233

A detailed understanding of the course of a reaction between a polymer and an additive will permit one to use that information to design new flame retardants. The reaction between poly(methyl methacrylate), PMMA, and red phosphorus is described and that information used to determine that $ClRh(PPh_3)_3$ should be used as a flame retardant. The results of this investigation are then used to choose the next additive. A recurring theme is the efficacy of cross-linking as a means to impart an increased thermal stability.

The process of burning requires that large polymer chains be broken down to give smaller and smaller molecules. Eventually a stage is reached at which the molecules produced have sufficient volatility to escape the surface and enter the gas phase. One may choose to intervene either at an early stage in the decomposition process, and have a relatively small number of molecules present, or at some later stage when many molecules are present. It seems most advantageous to prevent the initial decomposition of the polymer and instead cause aggregation, and charring, of the polymer chains. Van Krevelen (1) has shown that there is a correlation between char residue and the oxygen index of a polymer, the higher the char, the lower the flammability. Parker (2) has noted that an increase in aromaticity yields high char residues which also correlate with the oxygen index.

It has been pointed out (3) that additives to protect polymers from heat and light are developed largely on a empirical basis. The basic understanding of the chemical functions of additives in polymeric systems is frequently unclear. Several rather fundamental questions about the course of a reaction between an additive and a polymer need to be answered and that information used to design suitable additives for the stabilization of polymers.

0097–6156/90/0425–0178$06.00/0

We separate the process of polymer burning into two stages, a fuel generation stage and a stage where these fuels are consumed. The generation of fuel typically does not require the presence of oxygen while the burning stage, of course, does require oxygen. In this work only the decomposition, and hence fuel generation, of the polymer is investigated. If one can find methods to prevent fuel generation, then flame retardancy of the polymer is a natural consequence. The objectives of this work are to understand how a particular polymer and an additive interact, i.e., at what site in the polymer does reaction occur and what functionality of the additive causes this reaction. If enough information on reactions of additives and polymers becomes available then this will permit the rational design of a suitable flame retardant.

There is a severe temperature requirement in these investigations. Since the polymer must be processed, it is necessary that no reaction between the additive and the polymer occur at or below the processing temperature.   The reaction must occur at some temperature which lies between the processing temperature and the point at which decomposition of the polymer occurs.

We note at the outset that our objective is not immediately to design flame retardants, rather it is to build up enough information that will permit the design of flame retardants. None of the materials that are described in this paper would be suitable additives, however the detailed reaction information that becomes available through this work will permit flame retardants to be designed.

In this paper we describe the reaction of poly(methyl methacrylate), PMMA, and red phosphorus and use that information to predict that the reaction of Wilkinson's catalyst, $ClRh(PPh_3)_3$, and PMMA may be a worthwhile investigation. Finally information from this reaction is utilized to identify other potential additives and the reaction of these cobalt compounds with PMMA is described. Part of the strategy that will be explored is a strategy of cross-linking to produce materials with greatly increased thermal stability.

Experimental

Chemicals used in this study were obtained from Aldrich Chemical Company. ^1H NMR spectra were obtained on a Varian EM-360, ^{13}C NMR spectra were obtained on a JEOL-FX-60Q or a GE QE-300. An Analect FX-6200 FT-IR spectrometer was used for infrared spectroscopy and a Hewlett-Packard Model 5980 with a model 5970 mass-selective detector was used for GC-MS. TGA measurements were performed on a Perkin-Elmer TGA-7 instrument.   HPLC measurements were obtained using a Waters Instrument.

All reactions were performed in evacuated sealed Pyrex tubes of about 80 cm^3 volume. Typically, a vessel is thoroughly evacuated on a high vacuum line (pressure < 1 x 10^{-5} Torr) for several hours, then it is sealed off from the line and placed in a muffle furnace preheated to the desired temperature. After two hours the tubes are removed and immediately cooled in liquid nitrogen. Care must be taken in these operations, tubes have been known to explode in the oven and upon removal. Typically, the material was dissolved in a solvent, separated by chromatography, and identified by spectroscopy. Some tubes were equipped with break-seals, upon removal from the oven these were resealed to the vacuum line and gasses were quantified by pressure-volume-temperature measurements and identified by infrared spectroscopy.

PMMA and Model Reactions.   Reactions between PMMA and red phosphorus were carried out with the ratio of PMMA to red phosphorus varying from 10:1 to 1:1, i.e., 1.0g PMMA and 0.1g phosphorus to 1.0g PMMA and 1.0g phosphorus, at temperatures ranging from 300°C to 375°C. The PMMA-rhodium system used a 1.0g sample of PMMA and 0.5g of ClRh(PPh$_3$)$_3$ at 260°C for 2 hours. Reactions with the model compound, dimethylglutarate, DMG, utilized a 1:1 stoichiometry, 0.17g DMG and 1.0g ClRh(PPh$_3$)$_3$ (4-7).

Preparation of Cobalt Hydrogenation Catalysts.   K$_3$Co(CN)$_6$ was prepared following Brauer (8), HCo[P(OPh)$_3$]$_4$ was prepared following an Inorganic Synthesis procedure (9), and K$_3$Co(CN)$_5$ was prepared following the procedure of Adamson (10).

Reaction of K$_3$Co(CN)$_6$ with PMMA.   A 1.0 g sample of PMMA and 1.0g of the cobalt compound were combined in a standard vessel and pyrolyzed for 2 hrs at 375°C. The tube was removed from the oven and the contents of the tube were observed to be solid (PMMA is liquid at this temperature). The tube was reattached to the vacuum line via the break-seal and opened. Gases were determined by pressure-volume-temperature measurements on the vacuum line and identified by infrared spectroscopy. Recovered were 0.22g of methyl methacrylate and 0.11g of CO and CO$_2$. The tube was then removed from the vacuum line and acetone was added. Filtration gave two fractions, 1.27g of acetone insoluble material and 0.30g of acetone soluble (some soluble material is always lost in the recovery process). The acetone insoluble fraction was then slurried with water, 0.11g of material was insoluble in water. Infrared analysis of this insoluble material show both C-H and C-O vibrations and are classified as char based upon infrared spectroscopy. Reactions were also performed at lower temperature, even at 260°C some char is evident in the insoluble fraction.

Reaction of HCo[P(OPh)$_3$]$_4$ with PMMA.   A 1.0g sample of PMMA and 1.0g of the cobalt compound were combined as above. After pyrolysis at 375°C for two hours the tube is noted to contain char extending over the length of the tube with a small amount of liquid present. The gases were found to contain CO, CO$_2$, hydrocarbon (probably methane), and 0.10g methyl methacrylate. Upon addition of acetone, 1.0g of soluble material and 0.19g of insoluble may be recovered. The infrared spectrum of the insoluble fraction is typical of char.

Reaction of K$_3$Co(CN)$_5$ with PMMA.   A 1.0g sample of PMMA and 1.0g of K$_3$Co(CN)$_5$ were combined as above. After pyrolysis for two hours at 375°C the tube was noted to contain char. The gases consisted of 0.07g CO, 0.23g CO$_2$, and 0.11g methyl methacrylate, these were separated by standard vacuum line techniques and identified by infrared spectroscopy. The tube was then opened and chloroform and acetone added. The soluble fraction had a mass of 0.36g while the insolubles were 1.10g. The insoluble fraction was slurried with water and filtered, 0.22g was insoluble. The infrared spectrum was typical of char.

Oxygen-Index Measurements.   The measurement of oxygen-index was performed by bottom ignition, as discussed by Stuetz (11).

Results and Discussion

The majority of investigations reported in the literature report on the efficacy of a material as a flame retardant but do not address the chemistry that gives rise to this retardant effect. The objective of this work is to come to an understanding of the course of the reaction between polymer and additive and to use that understanding of chemistry to choose new additives. Since the purpose of this work is to identify the chemistry that occurs between the additive and the polymer, unusually large amounts of the additive relative to the polymer are used. This permits one to more easily determine the course of the reaction.

PMMA - Red Phosphorus System.   The initial reaction that was investigated was that between PMMA and red phosphorus (4-5). Phosphorus was chosen since this material is known to function as a flame retardant for oxygen-containing polymers (12). Two previous investigations of the reaction of PMMA with red phosphorus have been carried out and the results are conflicting. Raley has reported that the addition of organic halides and red phosphorus to PMMA caused moderate to severe deterioration in flammability characteristics. Other authors have reported that the addition of chlorine and phosphorus compounds are effective flame retardant additives (12).

The initial observation is that PMMA is essentially completely degraded to monomer by heating to 375°C in a sealed tube while heating a mixture of red phosphorus and PMMA under identical conditions yields a solid, non-degraded, product as well as a lower yield of monomer. One may observe, by ^{13}C NMR spectroscopy, that the methoxy resonance is greatly decreased in intensity and methyl, methoxy phosphonium ions are observed by ^{31}P NMR. Additional carbonyl resonances are also seen in the carbon spectrum, this correlates with a new carbonyl vibration near 1800 cm^{-1} in the infrared spectrum and may be assigned to the formation of anhydride. The formation of anhydride was also confirmed by assignment of mass spectra obtained by laser desorption Fourier transform mass spectroscopy, LD-FT-MS.

Mass spectra (5) were obtained by the addition of potassium bromide to the sample as an ion source. Since the molecular weight of a methyl methacrylate monomer is 100, one expects to see peaks in the mass spectrum at $100n + 39$, where n is the number of monomer units and 39 is a potassium ion. In the low mass region, up to 1200, peaks are observed at $100n + 41$, while in the high mass region, 2100-2200, the peaks occur at $100n + 37$, in the intermediate mass region the peaks are at 100n + 39 as expected. This is clear evidence for hydrogen transfer reactions that occur between various ions. Indirect evidence for such hydrogen transfer has been reported (13). This investigation only detected PMMA oligomers up to 2200 molecular weight, there is no reason to think that higher oligomers may not be observed with improved instrumentation.

The evidence presented above clearly indicates that red phosphorus attacks PMMA at a carbonyl site with the elimination of methyl and methoxy and the formation of anhydrides. The evidence indicates that anhydride formation occurs via an intra-molecular process, this second bond along the polymer chain does provide added stability that will render depolymerization more difficult and the polymer more thermally stable. The suggested course of the reaction between red phosphorus and PMMA is delineated in Scheme 1.

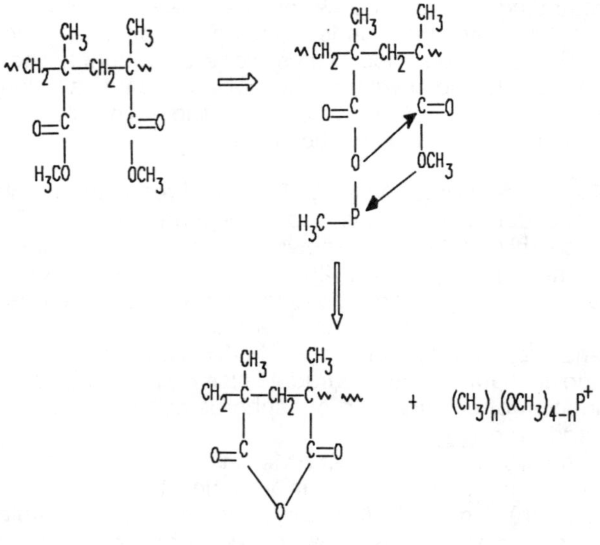

Scheme 1

The previous conflicting investigations may now be rationalized. Red phosphorus is known to thermally convert to white phosphorus, which will burn in air. If white phosphorus is formed, a fire is expected and no flame retardant activity will be observed. On the other hand, if the phosphorus reacts with the polymer as in Scheme 1, then thermal stabilization is expected. The efficacy of red phosphorus seems to be closely related to the efficiency of mixing of the additive and the polymer, when they are well-mixed the phosphorus will react with the polymer and lead to flame retardant activity, if the mixing is poor then the phosphorus will be converted to the white allotrope and burning will result. Since all of the work reported herein was carried out in sealed tubes under vacuum, the phosphorus must react and lead to stabilization of the polymer against molecular weight loss and fuel production, i.e. thermal stabilization.

The Importance of Cross-Linking. The observation that intramolecular anhydride formation leads to an increase in stability with respect to molecular weight loss and fuel generation suggests that cross-linking is an advantageous process for thermal stabilization of polymers. One stage of the analysis of this work was to determine that cross-linking should be a goal of this work and to begin to identify reagents that will effectively convert thermoplastic materials into cross-linked thermosets during the burning process and produce an intractable char. The reaction is shown below as Scheme 2.

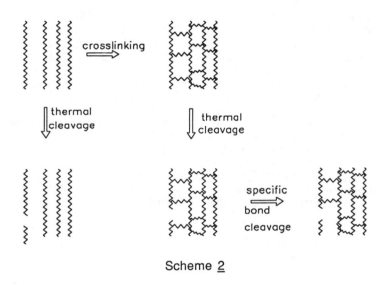

Scheme 2

On the left we show four polymer strands, all independent of each other, on the right these are now cross-linked. An input of thermal energy to a polymeric system (shown as cleavage in the figure) is sufficient to cause breakage of a bond to liberate a small molecule which is the source of fuel for polymer degradation. On the left side note that the cleavage of any bond will yield such small molecules. On the other hand, in the cross-linked structure (right-hand side), cleavage of any one bond is insufficient to produce small molecules, rather a second, and very specific, bond must also be broken. Since bond cleavage is a random process, it is unlikely that the specific bond will be cleaved without a great deal of thermal input, i.e., a very high temperature is required to produce small molecules. Thus the cross-linked structure is inherently less likely to produce fuel since much more heat is required to cause disruption of the molecule to produce the small molecules required for burning to occur.

PMMA - Wilkinson's Catalyst System. The reaction with phosphorus identifies the carbonyl as the site of reactivity in PMMA, accordingly an additive that will react at the carbonyl site should be used. Since

Wilkinson's catalyst, $ClRh(PPh_3)_3$, is known to react with acid halides with the elimination of the carbonyl, this reagent was chosen (14).

$$ClRh(PPh_3)_3 + RCOCl ==> RCl + ClRh(CO)(PPh_3)_2$$

The reaction between $ClRh(PPh_3)_3$ and PMMA produces both chloroform soluble and chloroform insoluble fractions (6-7). The soluble fraction contains a material which is very much like PMMA but also contains anhydride by FT-IR analysis. This polymeric fraction also contains rhodium, ligated by methyl and methoxy as well as triphenylphosphine. The chloroform insoluble fraction is about 25% of the total material and will not dissolve in any common laboratory solvent. Charring of PMMA under such conditions has not been previously seen. The char contains rhodium and is also found to retain chloroform, indicative of cross-linking. TGA analysis indicates that the chloroform may be driven off at about 150°C and the remainder is non-volatile at 600°C.

Films containing about 10% $ClRh(PPh_3)_3$ in PMMA were prepared and subjected to oxygen index, TGA, and DSC measurements. The oxygen index, bottom ignition (11), increases from about 14 for pure PMMA to about 20 for the rhodium compound in PMMA. TGA analysis indicates that about 25% of the sample is non-volatile at 600°C and the glass transition temperature increases by about 15°C by DSC.

In order to better understand the reaction of $ClRh(PPh_3)_3$ and PMMA model compound studies were begun. The model of choice is dimethylglutarate, DMG, since this provides a similar structure and a suitable boiling point for sealed tube reactions.

PMMA   DMG

The reaction of DMG and $ClRh(PPh_3)_3$ leads to the formation of several rhodium compounds, six of which were isolated and characterized. The compounds arise from oxidative insertion of the rhodium species into a carbon-oxygen bond of the DMG followed by ligand exchange reactions (7).

We believe that a similar reaction occurs between PMMA and $ClRh(PPh_3)_3$. A summary of the reaction is presented in Scheme 2.

The initial step is an oxidative addition of $RhCl(PPh_3)_3$ to a C-O bond of the ester moiety and produces rhodium-carbon and rhodium-oxygen bonds. Adjacent rhodium species can undergo further reaction with the formation of anhydride linkages. This anhydride formation may occur between adjacent pairs of reactants, between pairs in the same chain, or between pairs that are present in different chains. All of these reactions are observed, and in however the last reaction is the one of interest here since this leads to cross-linking and char formation. Rhodium is present in both the chary material and in the soluble fractions. From the reaction pathway in order for rhodium elimination to occur, two rhodium-inserted

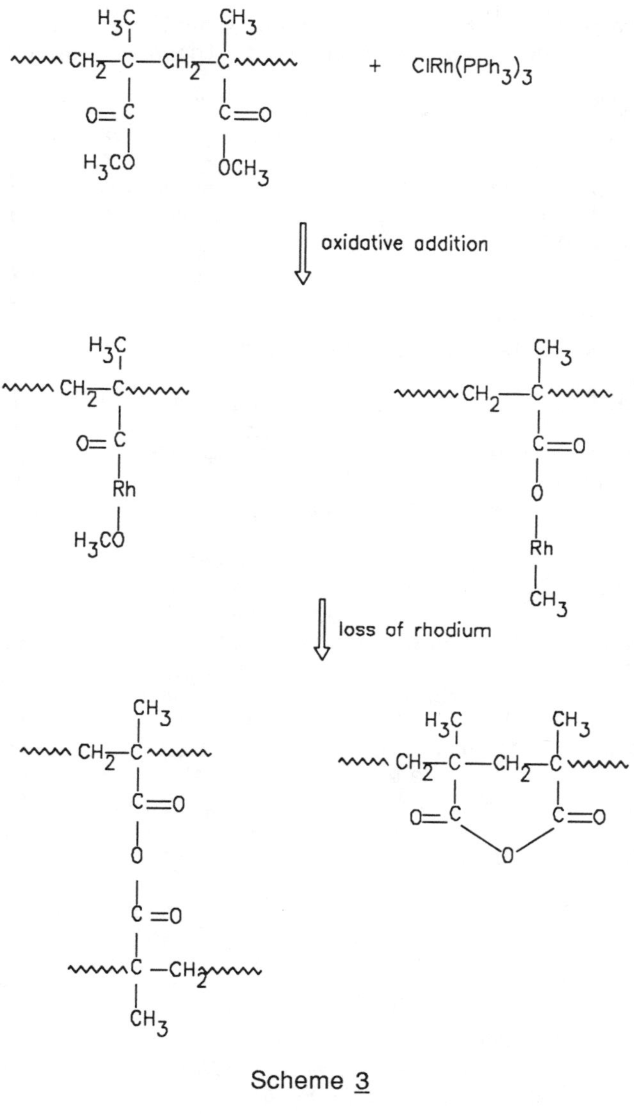

Scheme 3

moieties must be in close proximity. Since this cannot always occur, rhodium must be present both in the char as well as in the soluble fraction.

It is unlikely that ClRh(PPh$_3$)$_3$ will ever be useful as a flame retardant due to its red color, expense, and the potential toxicity associated with a heavy metal. An additional disadvantage of the rhodium system is the fact that char formation occurs at a temperature of 250°C, since this is near the processing temperature of PMMA char formation may occur during processing rather than under fire conditions. This discovery is nonetheless important because it allows one to delineate a set of requirements for flame retardants that function by this mechanism. These requirements are:

1. The compound must be able to effectively bind to carbonyl, thus it must be a transition metal compound.
2. The compound should be coordinatively unsaturated - capable of undergoing oxidative addition to the carbonyl.
3. Since Wilkinson's catalyst is a catalyst for a variety of reactions, a compound known to have catalytic activity may be desirable. This is a corollary of 2.
4. The compound should be colorless and inexpensive.

Extension to Cobalt Compounds. We believe that this set of requirements provide an excellent guide to the elucidation of compounds that will function as flame retardants for PMMA. Since Wilkinson's catalyst is best known as a hydrogenation catalyst and since a hydrogenation catalyst must possess all of the characteristics identified above, it seems reasonable to suggest that transition metal compounds that are known to function as hydrogenation catalysts may be useful flame retardants additives for PMMA. Cobalt compounds were selected for investigation since cobalt is in the same periodic family as rhodium and cobalt compounds may be expected to be inexpensive, possibly colorless, and possibly non-toxic. Two cobalt hydrogenation catalysts were identified, $K_3Co(CN)_5$ (15-16) and $C_3H_5Co[P(OMe)_3]_3$ (17). Both the cobalt(II) cyanide, $K_3Co(CN)_5$, and the cobalt(III) analogue, $K_3Co(CN)_6$, were selected for further study. The allyl cobalt phosphite was very air-sensitive and thus not suitable for these investigations. A related hydridocobalt compound, tetrakis-triphenylphosphitehydridocobalt, $HCo[P(OPh)_3]_4$, was chosen for study. We have found that the oxygen index increases by about 5 or 6 points, from 14 for neat PMMA to about 19 or 20 for all three of these compounds, $K_3Co(CN)_5$, $K_3Co(CN)_6$, and $HCo[P(OPh)_3]_4$.

Char formation and reduced monomer production are observed for all of these additives upon reaction with PMMA. Char formation increases as a function of temperature, for the hydrido cobalt compound, there is 5% char at 262°, 8.5% at 322°, 15% at 338°, and 19% at 375°C; the cobalt(III) cyanide produces 3% char at 338° and 11% at 375°C; the cobalt(II) cyanide yields 11% char at 375°C. At the highest temperature, 375°C, the amount of monomer formation is 22% for $K_3Co(CN)_5$, 11% for $K_3Co(CN)_6$, and 10% for $HCo[P(OPh)_3]_4$. Ideally one would hope to observe no monomer formation and complete char production. Such is not the case here, these materials probably have no utility as flame retardant additives for PMMA since monomer formation, even at a reduced level, will still permit a propagation of the burning process. While somewhat positive results for these three additives do not prove the validity of the hypothesis, we take this to be a starting point in our search for suitable additives, further work is underway to refine the hypothesis and to identify other potential hydrogenation catalysts and other additives that may prove useful as flame retardants for PMMA

It should be noted that other schemes for the production of anhydrides are also possible. We should also note that anhydride formation is not the only means of cross-linking PMMA chains, other schemes for cross-linking are also under investigation.

Conclusion

In this paper we have presented evidence to show that it is quite feasible to determine the detailed course of reaction between a polymer and an additive. Further, the understanding of this reaction pathway provides insight into new additives and schemes for the identification of efficacious flame retardant additives. Finally, we have elucidated schemes for the cross-linking of PMMA and have shown that the schemes do provide a route for flame retardation. It is imperative to realize that the purpose of this work is not to directly develop new flame retardants, rather the purpose is to expose the chemistry that occurs when a polymer and an additive react. This exposition of chemistry continually provides a new starting point for further investigations. The more that pathways for polymeric reactions are determined the more information is available to design suitable additives to prevent degradation of polymers.

Literature Cited

1. Van Krevelen, D. W., Polymer, 1975, 16, 615.
2. Parker, J. A.; Fohlen,G. M.; Sawko, P. M. cited in Pearce, E. M.; Khana, Y. P.; Raucher, D. "Thermal Characterization of Polymeric Materials," Turi, E. A., Ed., Academic Press, New York, 1981, p. 807.
3. Starnes, Jr., W. H. in "Developments in Polymer Degradation, Vol. 3," Grassie, N., Ed., Applied Science Publ., London, 1981, p. 135.
4. Wilkie, C. A.; Pettegrew, J. W.; Brown, C. E. J. Polym. Sci., Polym. Lett. Ed., 1981, 19, 409.
5. Brown, C. E.; Wilkie, C. A.; Smukalla, J.; Cody, Jr., R. B.; Kinsinger, J. A. J. Polym. Sci., Polym. Chem. Ed., 1986, 24, 1297.
6. Sirdesai, S.; Wilkie, C. A. J. Appl. Polym. Sci., 1989, 37, 863.
7. Sirdesai, S.; Wilkie, C. A. J. Appl. Polym. Sci., 1989, 37, 1595.
8. Brauer, G. "Handbook of Preparative Inorganic Chemistry," Academic Press. New York, 1965, p. 1541.
9. Levison, J. J.; Robinson, S. D. Inorganic Synthesis, 1972, 13, 105.
10. Adamson, A. W. J. Am. Chem. Soc., 1951, 73, 5710.

11. Stuetz, D. E.; Diedwardo, A. H.; Zitomer, F.; Barnes, B.P. J. Polym. Sci., Polym. Chem. Ed., 1973, 13, 585.
12. Peters, E. N. Flame Retardancy of Polymeric Materials, 1979, 5, 113.
13. Gritter, R. J.; Seeger, M.; Johnson, D. E. J. Polym. Sci., Polym. Chem. Ed., 1978, 16, 169.
14. Tsuji, J. Org. Syn. via Metal Carbonyls, 1977, 2, 595.
15. King, N. K.; Winfield, M. E. J. Am. Chem. Soc., 1961, 83, 3366.
16. Kwiatek, J.; Mador, I.L.; Seyler, J. K. J. Am. Chem. Soc., 1962, 84, 304.
17. Bleeke, J. R.; Muetterties, E. L. J. Am. Chem. Soc., 1981, 103, 556.

RECEIVED January 8, 1990

Chapter 14

Inorganic Tin Compounds as Flame, Smoke, and Carbon Monoxide Suppressants for Synthetic Polymers

P. A. Cusack[1] and A. J. Killmeyer[2]

[1]International Tin Research Institute, Uxbridge, United Kingdom
[2]Tin Research Institute, Inc., Columbus, OH 43201

In view of the current demand for novel, non-toxic flame- and smoke-suppressant systems for synthetic polymers, certain inorganic tin compounds have been evaluated as fire retardants in a number of plastic and elastomeric substrates. The results obtained indicate that tin compound additives, in particular, zinc hydroxystannate and zinc stannate, exhibit beneficial properties both in halogenated and halogen-free formulations. The tin compounds appear to act predominantly in the condensed phase by a char-promoting mechanism, and this leads to a significant decrease in the amounts of smoke and toxic gases evolved during polymer combustion. The observed carbon monoxide-suppression is particularly interesting, since CO inhalation is now known to be the cause of death in the vast majority of fire fatalities.

The last two decades have seen a major growth in the use of synthetic polymers as materials for construction, insulation, packaging, upholstery and transport applications (1). Unfortunately, this period has also seen a dramatic increase in the number of serious fires, and the number of deaths and injuries in fires remains appallingly high. Fire deaths are normally violent in nature, and smoke inhalation and not fire itself is the killer that accounts for over 80% of fire deaths (2). Therefore, recent advances in fire testing have placed great emphasis in developing products that have low flame spread properties and are low smoke producing.

The use of flame retardants in polymers has increased dramatically in recent years, in parallel to the growth of the plastics industry (1). Data for the U.S. consumption of these chemicals during 1985 are presented in Table I. Many of the existing commercial additives, however, have problems associated with their use. In particular, certain flame-retardant systems are known to cause an increase in the amount of smoke and toxic/corrosive

0097–6156/90/0425–0189$06.50/0

gases generated by plastics if they burn (4). In addition, a number
of the commercial fire retardants have been found to possess
undesirable toxicological properties themselves, (5,6) and there has
been considerable interest in finding new, safer chemical additives
for flammable materials.

Table I. Flame Retardants for Plastics
(USA Consumption, 1985)

Chemical	Metric tonnes
ADDITIVES:	
Alumina trihydrate	93,900
Antimony oxides	16,000
Boron compounds	5,000
Bromine compounds	17,000
Chlorine compounds	13,000
Phosphorus compounds	28,000
Others	8,100
REACTIVES	8,900
	TOTAL = 189,900

SOURCE: Data from the Fire Retardant Chemicals Association.
Reproduced with permission from ref. 3. Copyright 1985 Modern Plastics.

As a result, several inorganic compounds have found application
in this field, and alumina trihydrate, $Al(OH)_3$, is now by far the
highest volume flame retardant (3). Its use, however, is limited to
those polymers which can tolerate the exceptionally high loadings
required to be effective, without seriously affecting the mechanical
properties of the substrate (7).

At the present time, inorganic tin compounds find a relatively
small use in natural polymers, particularly as flame-resist treatments for woollen rugs and sheepskins (8,9). Although certain other
metal derivatives have received more attention, there has been much
interest recently in the potential use of tin chemicals as flame
retardants and smoke suppressants for synthetic polymers (10).

The purpose of this paper is to briefly review recent research
into the effectiveness and mode of action of tin compounds as fire
retardants in a number of halogenated and halogen-free, plastic and
elastomeric substrates.

TIN ADDITIVES

Anhydrous tin(IV) oxide ('Superlite' grade) and β-stannic acid
('Metastannic acid') were supplied by Keeling & Walker Ltd., Stoke-
on-Trent. 'β-stannic acid paste' was prepared at Chinghall Ltd.,
Milton Keynes, by dispersing Keeling & Walker's Metastannic acid, at
a level of 73% in a phthalate plasticiser. 'Colloidal tin oxide', a
25% aqueous dispersion of SnO_2, was supplied by Nyacol Products Inc.,
Ashland, Mass., U.S.A.

Zinc hydroxystannate and zinc stannate were synthesised at
I.T.R.I. according to previously reported procedures: (11,12)

$$Na_2Sn(OH)_6 + ZnCl_2 \xrightarrow{H_2O} ZnSn(OH)_6\downarrow + 2NaCl$$

$$ZnSn(OH)_6 \longrightarrow ZnSnO_3 + 3H_2O$$

The identities of the products were confirmed by infrared spectroscopy and x-ray powder diffraction patterns. Physical data on the inorganic tin additives studied are given in Table II.

TEST METHODS

Flammability. Flammability of the samples, in the form of thin strips of approximate dimensions of 120mm x 7mm x 3mm was determined by measurement of their oxygen indices (OI's) according to ASTM D2863, using a Stanton-Redcroft FTA module.
     It has been established that the OI values of polymers fall, in some cases dramatically, when the surrounding gas mixture is heated (1). Consequently, the OI test has been extended to materials at elevated temperatures, and the high temperature OI's of polymer samples were determined using a Stanton-Redcroft HFTA instrument. The temperature index (TI) of a polymer is defined as the temperature at which the test specimen just supports combustion in air.

Smoke density. Optical density measurements on the smoke evolved from burning plastic samples were carried out using an NBS Smoke Chamber. The samples, which measured 75mm x 75mm, with a thickness of 0.6 - 4mm, were burned in the 'flaming' mode in accordance with ASTM E662-79. Specific smoke density (Ds) values reported are the averages of three independent determinations.

Carbon Monoxide Evolution. Determination of the carbon monoxide evolved during combustion of polymer samples in NBS Chamber experiments was carried out using a Telegan CO Sensor (Type 3F). Quoted values are the numerical averages of three independant determinations.

HALOGEN-CONTAINING POLYMER SYSTEMS

The flame-retardant action of chlorine and bromine compounds, either as physically-incorporated additives to an organic polymer or as part of the polymer structure itself, is well established (1). Indeed, halogenated compounds find extensive commercial use as flame retardants, (3) and these are often used in conjunction with 'synergists', such as antimony trioxide or phosphorus derivatives (1). However, halogen-containing polymers generally evolve large amounts of smoke and corrosive gases during combustion, (4) and there is a great demand for novel smoke-suppressant formulations.

Polyester resins. The fire-performance characteristics of unsaturated polyester resins are of utmost importance in many application areas, particularly in the construction, transportation and electronics industries (13). Consequently, these plastics represent one of the major growth areas for fire retardants in recent years (14).

Table II. Physical Data on Tin Additives

Additive	Formula	Physical State	Tin Content(%)	Specific Gravity	Average Particle Size (μm)
Anhydrous tin(IV) oxide	SnO_2	white powder	78.8	6.9	0.3
β-stannic acid (hydrous tin(IV) oxide)	$SnO_2 \cdot xH_2O$	white powder	70.3	5.2	ca. 2
β-stannic acid- 'paste'	$SnO_2 \cdot xH_2O$	73% dispersion in a phthalate plasticiser	51.3	-	ca. 2
Colloidal tin(IV) oxide	$SnO_2 \cdot xH_2O$	25% aqueous dispersion	19.7	1.3	0.03
Zinc hydroxystannate	$ZnSn(OH)_6$	white powder	41.5*	3.3	ca. 1
Zinc stannate	$ZnSnO_3$	white powder	51.1**	3.9	ca. 1

* Zinc content = 22.8%
** Zinc content = 28.2%

Earlier studies at the ITRI have demonstrated the effectiveness of tin(IV) oxide, both in its anhydrous and hydrous forms, as a flame- and smoke-retardant additive for laboratory-prepared polyester resin formulations (15). In a recent study, carried out in collaboration with a major U.K. company, a number of inorganic tin additives have been incorporated into a commercial brominated polyester resin. Although this resin, which contains 28% by weight bromine, is intrinsically flame-retardant, giving samples with an OI of ca. 41 and which meet the UL94-VO classification, formulations with improved flame and smoke properties are sought.

Antimony trioxide, β-stannic acid (hydrous tin(IV) oxide), zinc hydroxystannate, and zinc stannate were incorporated into the commercial brominated polyester resin at levels of 1,2,5 and 10% by weight. No processing problems were encountered and the samples cured satisfactorily to give rigid, opaque strips.

The incorporation of Sb_2O_3 into the polyester leads to an interesting flammability effect: a progressive increase in OI is observed up to an additive level of 2%, above which the OI decreases dramatically (Figure 1). It has previously been reported (16,17) that the optimum atomic ratio of halogen: antimony in synergistic flame-retardant systems appears to depend upon the nature of the host polymer, and may often be far greater than the stoichiometric proportions (viz. 3:1). Indeed, the OI of polypropylene containing a chlorinated hydrocarbon passes through a very sharp maximum, with increasing Sb_2O_3, at a Cl:Sb ratio of ca. 25:1 (18). In the brominated polyester, the Br:Sb atomic ratio for optimum flame retardancy also appears to be about 25:1.

Of the tin additives studied, the anhydrous and hydrated zinc stannates, $ZnSnO_3$ and $ZnSn(OH)_6$ respectively, are considerably more effective flame-retardant synergists with the bromine present in the plastic than β-stannic acid (Figure 1). In line with this observation, oxidic tin-zinc systems have previously been found to exhibit superior flame-retardant properties to tin oxides alone (19-22). In addition, $ZnSnO_3$ gives higher values of OI than Sb_2O_3 at all incorporation levels studied, and, in fact, the 1% $ZnSnO_3$-containing plastics outperform samples containing 2% Sb_2O_3.

It is interesting to note that β-stannic acid is significantly more effective as a flame retardant than anhydrous SnO_2 at low additive levels (Figure 2) as previously reported in a number of halogen-containing polymer systems (15,22,23). In this connection, thermal analysis studies of ABS-decabromobiphenyl-tin(IV) oxide systems have shown that very little tin(IV) bromide or tin(IV) oxybromide is volatilised when anhydrous SnO_2 is used, but, if the tin is present as β-stannic acid, and if the atomic ratio of Br:Sn is higher than stoichiometric (viz. 4:1), then significant volatilisation occurs, giving rise to a vapour-phase flame-retardant action (23). This difference in reactivity between the two forms of tin(IV) oxide may be due to the difference in the surface areas of these materials: despite being a coarser powder in terms of its average particle size (Table II), hydrous SnO_2 has a surface area of at least an order of magnitude greater than that of the anhydrous material (24).

The advantage of using a paste dispersion of a flame-retardant additive in this polyester resin formulation is evident from the

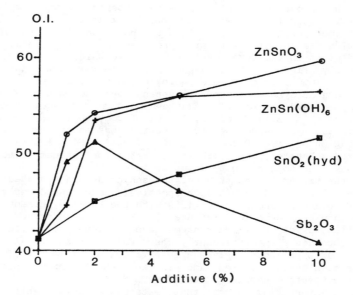

Figure 1. Relationship between the flammability and metal
compound level of brominated polyester resin samples.
(Reproduced with permission from Ref.34, Copyright 1989
 John Wiley & Sons Ltd.)

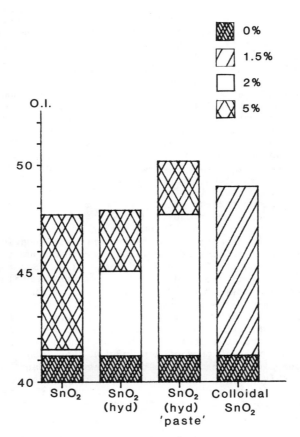

Figure 2.   Effect of incorporation level of tin oxide additives
on the flammability of brominated polyester resin.

data shown in Figure 2.   At incorporation levels of 2% and 5%, a
dispersion of β-stannic acid in a phthalate plasticiser (supplied by
Chinghall Limited, Milton Keynes, U.K.), gives a significantly
improved flame-retardant performance over that of powdered β-stannic
acid, presumably because of its more uniform distribution in the
polymer matrix.

A further improvement in flame-retardant efficiency is observed
when a colloidal dispersion of tin(IV) oxide is incorporated into
the polyester. At a 1.5% addition level, colloidal SnO_2 gives an
OI value (49.0) which is markedly higher than that obtained with 5%
loadings of either anhydrous SnO_2 (47.7) or β-stannic acid (47.9),
as powdered additives (Figure 2). In addition to its increased
flame-retardant ability, colloidal SnO_2 offers the further advantages
of translucency in the cured plastic, ease of incorporation and non-
settling in the resin prior to cure.

Evaluation of the smoke generated during flaming combustion of
the brominated polyester resins has been carried out in an NBS Smoke
Chamber. The data obtained, illustrated graphically in Figure 3 and
summarised in Table III, clearly indicate that both zinc
hydroxystannate and zinc stannate, when incorporated at levels of 2%
or 5%, significantly reduce both the total amount and the rate of
production, of smoke evolved from the burning brominated resin.
Although β-stannic acid and antimony trioxide have a slight
beneficial effect on the rate of smoke emission, neither additive
decreases the maximum optical density of the smoke, saturation
levels being reached in each case.

Furthermore, $ZnSn(OH)_6$ and $ZnSnO_3$ are found to be highly
effective as CO suppressants in the resin, when assessed in a novel
test assembly comprising the NBS Smoke Chamber in conjunction with
the continuous CO monitor (Figure 4 and Table III). Substantial
reductions in CO levels produced from zinc stannate-containing
samples in the NBS Smoke Chamber are evident at 3 minutes after
ignition and these effects continue throughout the tests, so that
maximum levels of CO, after 10 minutes burning, are also much lower
than for the base plastic. Antimony trioxide and β-stannic acid,
although showing some CO-suppressing ability, are considerably less
effective in this respect than either $ZnSn(OH)_6$ or $ZnSnO_3$ (Table III).

Hence, the zinc stannates have been shown to impart beneficial
properties to this polymer system, in terms of flammability and
smoke/carbon monoxide evolution, and the improvements in performance
are clearly superior to those exhibited by the commercially available
antimony trioxide.

Other Halogenated Systems. Inorganic tin compounds have been found
to exhibit flame-retardant and smoke-suppressant properties in a
number of halogen-containing synthetic polymer compositions. In
particular, Touval (22) reported that hydrous tin(IV) oxide is at
least as effective as antimony trioxide when incorporated as a
flame-retardant synergist with chlorine-containing compounds in
various thermoplastics including polyethylene, polypropylene and
acrylonitrile-butadiene-styrene (ABS plastic). Similarly, Fukatsu
(25) demonstrated the flame-retardant superiority of hydrous SnO_2
over Sb_2O_3, at levels of 0.5% - 3%, in a poly(vinyl alcohol)-poly
(vinyl chloride) blended fabric.

Recent studies at I.T.R.I., carried out in collaboration with

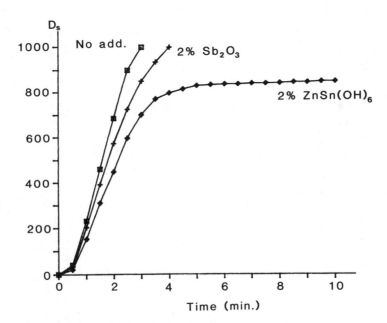

Figure 3. Effect of inorganic additives on the density of the
smoke evolved from brominated polyester resin in the NBS-type
Smoke Chamber.
(Reproduced with permission from Ref.34, Copyright 1989
 John Wiley & Sons Ltd.)

Table III.   Effect of Inorganic Additives on the Fire Properties of Brominated Polyester Resin

Additive	Flammability OI	Smoke Density				CO Evolution (ppm)			
		D_s(3min)*	Red^n(3min)	D_s(max)	Time(min) to $D_s = \infty$	Actual(3min)*	Red^n(3min)	Max.**	Red^n(max).
None	41.2	1000	-	∞	3.0	2225	-	5950	-
2% Sb$_2$O$_3$	51.2	851	14.9%	∞	4.0	1650	25.8%	4700	21.0%
" SnO$_2$(hyd)	45.1	879	12.1%	∞	4.0	1825	18.0%	5463	8.2%
" ZnSn(OH)$_6$	53.4	702	29.8%	850	-	1100	50.6%	3300	44.5%
" ZnSnO$_3$	54.2	737	26.3%	881	-	1390	37.5%	3645	38.7%
5% Sb$_2$O$_3$	46.2	1000	0%	∞	3.0	1917	13.8%	4783	19.6%
" SnO$_2$(hyd)	47.9	899	10.1%	∞	3.5	1690	24.0%	4967	16.5%
" ZnSn(OH)$_6$	56.0	629	37.1%	949	-	1013	54.5%	2813	52.7%
" ZnSnO$_3$	56.1	672	32.8%	∞	7.5	866	61.1%	2587	56.5%

* Measurements recorded 3 minutes after ignition of sample.
**Measurements recorded at end of experiment, i.e. 10 minutes after ignition of sample.

(Reproduced with permission from Ref. 34, Copyright 1989 John Wiley & Sons Ltd.)

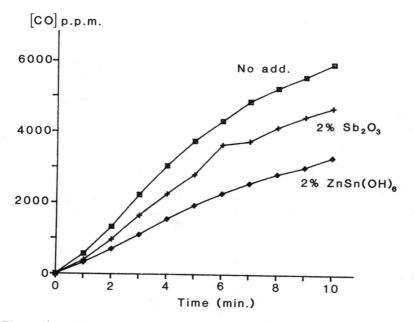

Figure 4.   Effect of inorganic additives on the emission of
carbon monoxide from brominated polyester resin in the NBS-type
Smoke Chamber.
(Reproduced with permission from Ref.34, Copyright 1989
 John Wiley & Sons Ltd.)

external organisations, have shown that certain inorganic tin
additives, including SnO_2, $ZnSn(OH)_6$ and $ZnSnO_3$, are effective flame-
and smoke-retardants in a range of chlorinated polymers, including
rigid PVC, (26) flexible PVC and neoprene.

Hence, it is apparent that certain inorganic tin compounds are
very effective flame retardants and smoke suppressants for
halogenated polymer formulations. Since these additives are
generally non-toxic, their potential use as partial or total
replacements for existing commercial flame retardants, such as
antimony trioxide, is thought to merit serious consideration.

HALOGEN-FREE POLYMER SYSTEMS

In recent years, there has been increasing concern about the
toxicity and corrosive nature of the smoke and gases generated during
the combustion of halogen-containing polymers (27). Consequently,
there is a considerable demand for plastics which comply to the
specification 'low smoke zero halogen' (LSOH) flame retardancy,
particularly for use in underground transport, shipping and power
stations. As a result, there has been much interest in the use of
non-toxic inorganic additives and fillers, and the market for these
compounds is likely to continue its rapid growth in the years ahead
(28).

Ethylene-acrylic Rubber. The use of PVC as a wire and cable
insulation material has declined in recent years, and it has been
replaced to a significant extent by halogen-free elastomeric
compositions. At the present time, such formulations are made
flame-retardant by the incorporation of alumina trihydrate (ATH).
Although ATH is essentially non-toxic and relatively inexpensive,
high addition levels are necessary for effectiveness, and this often
results in a marked deterioration in the mechanical properties of
the polymer (29). While recent advances in production technology
and surface coating have mitigated some of the problems associated
with high filler loading, (29) improved systems comprising
combinations of ATH with other additives are under investigation.

A collaboration has been undertaken with the Admiralty Research
Establishment, Poole, U.K., in which the fire-retardant properties
of a number of inorganic tin compounds in a non-halogenated, ATH-
filled ethylene-acrylic rubber formulation, are being assessed.
Preliminary results have indicated that a marked flame-retardant
synergism exists between certain tin compounds (at a 2.5% level) and
ATH (50% loading), and an increase in OI from 27.5 (for no tin
additive) to 33.0 was observed for the $ZnSn(OH)_6$ containing
formulations. It has been reported (30) that an ATH loading of 60%
is necessary to raise the OI of ethylene-acrylic rubber to a value
of 33, and that such a formulation meets the MOD Naval Specification
NES 518. Hence, it appears that lower total additive levels are
required for adequate performance when ATH - $ZnSn(OH)_6$ combinations
are used than when ATH is incorporated alone. This significant
reduction in the filler content may prove advantageous in terms of
the mechanical properties of the polymeric substrate.

Elevated temperature OI data (Figure 5) indicate that the tin-
containing elastomers retain their flame-retardant superiority up to
a temperature of 250°C above which the samples undergo extensive

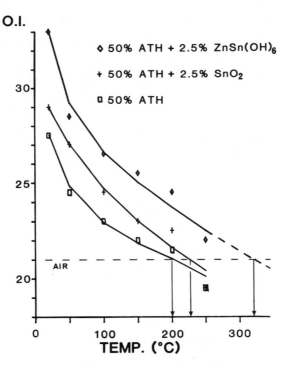

Figure 5. Effect of temperature on the flammability of
ethylene-acrylic rubber samples.

thermal degradation and determination of OI's becomes impractical.
It is of interest to note that the polymer containing $ZnSn(OH)_6$ does
not burn in air even at 250°C and, accordingly, this composition has
a temperature index of at least 50°C above that of the rubber
containing ATH alone. The OI and high temperature OI data therefore
provide substantial evidence as to the benefit of using $ZnSn(OH)_6$ as
a flame-retardant synergist with alumina trihydrate filler.

Other Non-halogenated Systems. Although oxidic tin compounds,
including hydrous SnO_2 and $ZnSn(OH)_6$, can impart a high degree of
flame-resistance to cellulosic substrates, such as cotton fabric (19)
and paper, they are generally ineffective as flame retardants in
halogen-free synthetic polymer formulations, unless the composition
has a relatively high inorganic filler content.

However, these tin additives can exhibit significant smoke-
suppressant properties in non-halogenated polymers. In a programme
of studies carried out at ITRI in collaboration with Borax
Research Limited, Chessington, U.K., a series of divalent metal
hydroxystannates and stannates were screened as smoke suppressants
in halogen-free glass-reinforced polyester (GRP) panels (31). The
best result obtained was for a 2phr (parts per hundred of resin)
incorporation level of $ZnSnO_3$, which gave a 46% reduction in smoke
density compared to the control sample.

Furthermore, Hirschler (32) has investigated the smoke-reducing
ability of a number of metal hydroxides and oxides, at incorporation
levels of up to 40 phr in acrylonitrile-butadiene-styrene (ABS)
copolymer. At the lowest additive level studied (10 phr), SnO_2 gave
a higher degree of smoke suppressancy (viz. 58%), than any of the
other compounds, and loadings of at least 30 phr were found to be
necessary to achieve comparable performance with either $Al(OH)_3$ or
$Mg(OH)_2$.

The effectiveness of inorganic tin compounds as flame and smoke
retardants, both alone and in combination with inorganic fillers, in
several halogen-free thermoplastic, thermosetting and elastomeric
substrates, is currently under investigation at the ITRI.

MECHANISTIC STUDIES

Although there have been many studies on the mode of action of fire
retardants generally, (16) the mechanistic behaviour of tin additives
is less clear, and may depend on several factors including the ratio
of halogen: tin in the system.

Thermal analysis has been used extensively to study in detail
the various individual stages occurring during the breakdown of
polymers under the action of heat, and can provide useful information
regarding the mode of action and the effectiveness of flame-retardant
additives. Simultaneous thermogravimetry (TG), differential
thermogravimetry (DTG) and differential thermal analysis (DTA), have
been used at the ITRI to investigate the thermal breakdown of
various tin-containing polymers and of the additives themselves.

Simultaneous TG/DTA studies of zinc hydroxystannate (Figure 6),
indicate that dehydration of this compound occurs during the
temperature range 190°-285°C, with a loss of 19.1% of its initial
weight, this corresponding to the loss of 3 moles of water:

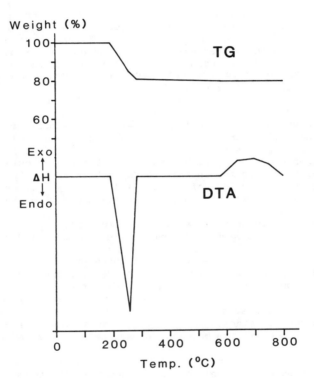

Figure 6.   Simultaneous TG/DTA traces of zinc hydroxystannate
in air.
(Reproduced with permission from Ref.34, Copyright 1989
 John Wiley & Sons Ltd.)

$$ZnSn(OH)_6 \longrightarrow ZnSnO_3 + 3H_2O$$

This thermal dehydration is accompanied by a large endotherm (absorption of heat), as previously reported (12). A secondary process occurs at higher temperatures (ca. 580°-800°C), accompanied by a broad exotherm, due to a solid state reaction in which no weight loss is observed: (12)

$$ZnSnO_3 \longrightarrow \tfrac{1}{2} Zn_2SnO_4 + \tfrac{1}{2} SnO_2$$

Extensive thermoanalytical studies have been carried out on the brominated polyester resin system. Figure 7 shows the differences in the thermal breakdown behaviour of the 2% $ZnSnO_3$-containing polyester, when the atmosphere is air as opposed to argon. Although there are slight differences in the initial decomposition stages, the most significant dissimilarity occurs at higher temperature where, in air, a third distinct weight loss occurs. This loss is accompanied by a large exotherm, indicative of an oxidative 'burn-off' of residual char.

The reactions of the polymer in air are obviously more relevant to its flammability than those in argon, and thermal analysis data for brominated polyester samples in air, are presented in Table IV. In the absence of any inorganic additive, the resin loses about 80% of its weight during the initial decomposition stage, this corresponding to the loss of styrene and HBr, (33) followed by char oxidation (13.7% loss) at a DTG max (temperature of maximum rate of weight loss) of ca. 528°C.

Incorporation of a 2% level of either $ZnSn(OH)_6$ or $ZnSnO_3$ into the resin, leads to marked differences in its thermal degradation profile. The initial decomposition comprises two distinct stages, the major step occurring at a lower temperature than that of the untreated polymer, which may be indicative of promotion of bromine volatilisation by the zinc stannates. Furthermore, the amount of residue burnt off in the char oxidation step is greatly increased in the tin-containing samples, and the temperature at which this process occurs is significantly higher than in the base resin. Hence, the tin additives may also be acting as condensed phase char promotors.

In contrast to these observations, a 2% addition of Sb_2O_3 to the polymer results in only a slight change in its thermal decomposition, with only a modest increase in char residue, which is, in turn, burnt off at a slightly higher temperature than that of the corresponding stage in the untreated resin (Table IV). These findings indicate that Sb_2O_3 has little condensed phase fire-retardant activity in this polyester system, and this is in agreement with the generally accepted observation that antimony additives operate primarily in the vapour phase by interfering with the free radical reactions associated with combustion processes (1,16).

Continuous monitoring of the carbon monoxide and carbon dioxide evolved during thermal decomposition of brominated polyester resin samples, has been carried out using a simultaneous thermal analysis-mass spectrometry technique. In order to allow measurement of the carbon monoxide evolved, the atmosphere chosen for these runs was 21% oxygen in argon, since the peak at 28 atomic mass units (amu)

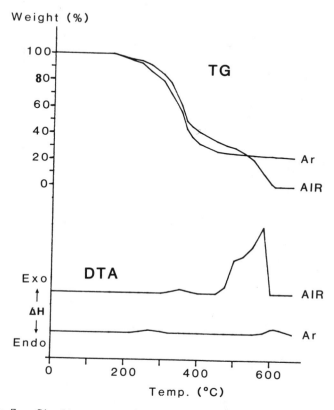

Figure 7. Simultaneous TG/DTA traces of brominated polyester resin containing 2% ZnSnO$_3$ in air and in argon. (Reproduced with permission from Ref.34, Copyright 1989 John Wiley & Sons Ltd.)

due to nitrogen (N$_2$) in air, would have obscured the CO peak, also at 28 amu. CO$_2$ was monitored at 44 amu. The results obtained (Table V) clearly show that the presence of 5% β-stannic acid in the resin substantially reduces both the total amounts of, and the rates of evolution of, CO and CO$_2$, produced during thermal degradation of the plastic.

Table IV. Thermal Analysis Data for Brominated Polyester Resin Samples*

Sample	Initial Degradation+		Char Oxidation++	
	Weight Loss (%)	DTG max (°C)	Weight Loss (%)	DTG max (°C)
No additive	80.7	361.6	13.7	527.9
2% ZnSn(OH)$_6$	48.4 15.7	332.1 414.3	35.5	572.5
2% ZnSnO$_3$	50.5 13.9	337.3 406.2	34.1	578.0
2% Sb$_2$O$_3$	78.5	372.3	18.5	558.4

* Experimental conditions: Atmosphere: air
$\quad\quad\quad\quad\quad\quad\quad\quad\quad$ Flow rate: 20cm^3/min
$\quad\quad\quad\quad\quad\quad\quad\quad\quad$ Heating rate: 10°C/min

+ Temp. range : ca. 180–450°C

++Temp. range : ca. 450–650°C

Table V. Analysis of Gases Evolved During Thermal
Degradation of Brominated Polyester Resin Samples

Sample*	Specific Emission**					
	Carbon Monoxide			Carbon Dioxide		
	I+	II++	Total	I+	II++	Total
No additive	65	146	211	78	375	453
5% SnO_2(hyd)	18	70	88	30	232	262

*Atmosphere : 21% oxygen in argon; heating rate = 20°C/min.

**Integrated gaseous production, normalised for initial sample weight (arbitrary units).

+First major decomposition stage (ca. 350-370°C).

++Second major decomposition stage (ca. 540-560°C).

Further insight into the mode of action of the individual additives is provided by simple combustion experiments, carried out in air. Table VI shows that the yield of involatile carbonaceous char, formed when brominated polyester is burned to completion, is more than doubled when brominated polyester is burned to completion, is more than doubled when 5% additions of either β-stannic acid or zinc stannate are made to the plastic, this observation being consistent with condensed phase behaviour. Elemental analysis of the residues suggests that, although in the case of β-stannic acid, only a small fraction of the tin is volatilised, a very significant proportion of both the zinc and the tin is volatilised from the zinc stannate-containing polymer, which may be indicative of vapour phase action. Interestingly, the extent of bromine loss is significantly reduced for tin-containing samples, particularly those containing $ZnSnO_3$. Antimony trioxide, which undergoes almost complete volatilisation in the polymer, shows little char enchancing behaviour and operates primarily in the gas phase. The apparent ability of zinc stannate to act in both the condensed and vapour phases may account for its overall flame-retardant superiority to the other inorganic additives studied.

CONCLUSIONS

1. Zinc hydroxystannate and zinc stannate are very effective flame-retardant synergists when incorporated at levels of 1-10% into a brominated polyester resin, and are, in general, markedly superior to other inorganic additives studied including, antimony trioxide, anhydrous tin(IV) oxide, and β-stannic acid.
2. The surface area and degree of dispersion in the polymer matrix of the fire-retardant additive has a pronounced effect on its efficiency. Colloidal tin(IV) oxide is significantly more effective, in terms of its flame-retardant ability, than powdered tin(IV) oxide or β-stannic acid.

Table VI. Residual Char Yields and Extents of Elemental Volatilisation from Brominated Polyester Resin Samples during Combustion in Air

Sample	Char Yield (%)	Elemental Volatilisation (%)				Primary Phase of Action*
		Br	Sn	Zn	Sb	
No additive	24.1	95.9	-	-	-	-
5% SnO_2(hyd)	53.0	85.1	18.4	-	-	condensed
5% $ZnSnO_3$	52.8	74.0	42.9	34.3	-	condensed + vapour
5% Sb_2O_3	33.5	94.4	-	-	93.0	vapour

*With regard to metallic element; bromine itself acts almost exclusively in the vapour phase (1,16).

(Reproduced with permission from Ref. 34, Copyright 1989 John Wiley & Sons Ltd.)

3. $ZnSn(OH)_6$ and $ZnSnO_3$, at levels of 2% or 5%, significantly decrease both the total amount and the rate of production, of smoke and carbon monoxide generated from brominated polyester resin, when burnt under limiting or flaming conditions. Antimony trioxide and β-stannic acid only exhibit marginal benefits in terms of smoke and CO suppression in this polymer system.

4. Thermoanalytical and related mechanistic studies indicate that the tin additives are primarily condensed phase flame retardants, and these can alter the pyrolysis of halogenated polymers by promoting the loss of halogen and enhancing cross-linking, to give an increase in the yield of thermally-stable carbonaceous char, with a concomitant decrease in the generation of volatile flammable products. This reduction in the amount of fuel supplied to the flame leads to the observed suppression of smoke and carbon monoxide. In addition, partial volatilisation of zinc and tin occurs during combustion of brominated resins containing $ZnSn(OH)_6$ or $ZnSnO_3$, and this indicates that these additives may contribute to vapour phase flame retardation, which may account for their superiority to tin oxide itself, where the extent of tin volatilisation is very low. Antimony trioxide has little effect on the thermal breakdown of the polymer and operates primarily in the gas phase.

5. Certain inorganic tin compounds are effective flame-retardant synergists when incorporated at a 2.5% level into a 50% ATH-filled ethylene-acrylic rubber composition. Tin-containing elastomer formulations retain their flame-retardant superiority at environmental temperatures up to 250°C, and samples containing 2.5% $ZnSn(OH)_6$ do not sustain combustion in air at this temperature.

6. In general, tin compounds do not exhibit flame-retardant properties in halogen-free polymer systems, unless the composition contains a high inorganic filler loading. However, tin additives often act as smoke suppressants in non-halogenated polymers.

7. Inorganic tin compounds, in particular, zinc hydroxystannate and zinc stannate, are effective fire retardants in a number of polymer systems. They appear to have advantages over certain existing commercial additives, namely:
    (a)   Non-toxicity
    (b)   No discoloration of substrate
    (c)   Effective at low levels
    (d)   Little apparent effect on mechanical properties
    (e)   Flame-retardant synergism with halogens and/or alumina trihydrate
    (f)   Smoke suppression
    (g)   Decrease in carbon monoxide production
    (h)   Wide range of applicability

ACKNOWLEDGMENTS

Messrs A.W.Monk and J.A.Pearce, and Miss S.J.Reynolds are gratefully acknowledged for experimental assistance. Miss J.Ratcliffe is thanked for assistance in the design of Figures for this paper.

LITERATURE CITED

1. Cullis,C.F.; Hirschler,M.M. The Combustion of Organic Polymers;
    Oxford University Press, Oxford, U.K.,1981.
2. Graham,R.A. Proc.Flame Retardants '87 1987, p 1/1.
3. Anon. Mod.Plast.Int. 1985, 15, 63.
4. Lindstrom,R.S.; Sidman,K.R.; Sheth,S.G.; Howarth,J.T. J.Fire
    Retard.Chem. 1974, 1, 152.
5. Sanders,H.J. Chem.Eng.News 1978, Apr. 24, 22.
6. Mischutin,V. Speciality Chemicals 1982, 2, 27.
7. McAdam,B.W. Speciality Chemicals 1982, 2, 4.
8. Ingham,P.E. Tin & Its Uses 1975, 105, 5.
9. Benisek,L. Br.Patent 1 385 399, 1975.
10. Blunden,S.J.; Cusack,P.A.; Hill,R. The Industrial Uses of Tin
    Chemicals; Royal Society of Chemistry: London, U.K., 1985.
11. Dupuis,T.; Duval,C.; Lecomte,J. Compt.Rend. 1963, 257, 3080.
12. Ramamurthy,P.; Secco,E.A. Can.J.Chem. 1971, 49, 2813.
13. Stepniczka,H.E. J.Fire Retard.Chem. 1976, 3, 5.
14. Nicholson,J.W.; Nolan,P.F. Fire & Mater. 1983, 7, 89.
15. Cusack,P.A. Fire & Mater. 1986, 10, 41.
16. Hirschler,M.M. Dev.Polym.Stab. 1982, 5, 107.
17. Antia,F.K.; Baldry,P.J.; Hirschler,M.M. Eur.Polym.J. 1982, 18,167.
18. Hindersinn,R.R.; Witschard,G. In Flame Retardancy of Polymeric
    Materials; Kuryla,W.C.; Papa,A.J., Eds.; Marcel Dekker: New York,
    1978; Vol.1, p 1.
19. Cusack,P.A.; Hobbs,L.A.; Smith,P.J.; Brooks,J.S. International
    Tin Research Institute Publ. No. 641 1984.
20. Busse,W.F. U.S. Patent 3 468 843, 1969.
21. Toray Industries Inc. Br. Patent 1 382 659, 1975.
22. Touval,I. J.Fire & Flammability 1972, 3, 130.
23. Donaldson,J.D.; Donbavand,J.; Hirschler,M.M. Eur.Polym.J. 1983,
    19, 33.
24. Anon. Tin Oxide and Metastannic Acid. Keeling & Walker Ltd. Data
    Sheets, Stoke-on-Trent, U.K., 1987.
25. Fukatsu,K. J.Fire Sciences 1986, 4, 204.
26. Cusack,P.A.; Smith,P.J.; Kroenke,W.J. Polym.Degrad.Stab. 1986,
    14, 307.
27. Woolley,W.D.; Fardell,P.J. Fire Safety J. 1982, 5, 29.
28. Anon. Plast. & Rubber Weekly 1987, Dec. 5, 9.
29. Brown,S.C.; Evans,K.A.; Godfrey,E.A. Proc.Flame Retardants '87
    1987, p 11/1.
30. Anon. Martinal as a Flame-retardant Filler for Cables.
    Martinswerk GmbH Data Sheet No. 2/2, Bergheim, West Germany,1986.
31. Cusack,P.A.; Smith,P.J.; Arthur,L.T. J.Fire Retard.Chem. 1980,
    7, 9.
32. Hirschler,M.M. Polymer 1984, 25, 405.
33. Haines,P.J.; Lever,T.J.; Skinner,G.A. Thermochimica Acta 1982,
    59, 331.
34. Cusack,P.A.; Monk,A.W.; Pearce,J.A.; Reynolds,S.J. Fire & Mater.
    1989, 14, 23.

RECEIVED January 17, 1990

Chapter 15

Mechanism of Thermal Degradation of Fire-Retardant Melamine Salts

L. Costa, G. Camino, and M. P. Luda di Cortemiglia

Dipartimento di Chimica Inorganica, Chimica Fisica e Chimica dei Materiali, Via Pietro Giuria 7, 10125 Torino, Italy

A thorough study of the mechanism of degradation of several fire retardant melamine salts is carried out using thermogravimetry, evolved gas analysis and spectroscopic characterisation of the products of degradation.
The salts can be classified in three main classes: salts which undergo thermal dissociation to acid and melamine; salts of strong acids which catalyse melamine condensation; salts of acids which react with melamine condensation products. Implications of the thermal behaviour of the salts in the mechanism of fire retardance is briefly discussed.

The increasing understanding of fire hazard in the case of organic polymeric materials has recently brought to general attention the importance of such factors as formation of dense smoke and of corrosive or toxic products in fire scenarios (1-3). Thus, the emphasis in the evaluation of fire retardant systems is now shifting from pure flame extinguishing parameters to a more comprehensive appraisal of hazard in fire retarded materials in which production of smoke, corrosive and toxic products is included. In this respect, some well established classes of fire retardants, as for example halogen-based systems, seem to be rather unsatisfactory. Most current developments in this field involve indeed halogen-free systems such as char forming or intumescent additives (4-7). These act in the condensed phase inducing the formation of char on the surface of the polymeric material which protects it from the action of the flame.

Melamine and its salts are widely used in formulations of fire retardant additives, particularly of the intumescent type (4-7). The role played by melamine structures in these additives is however not yet understood. The thermal behaviour is of paramount importance in studies of the fire retardance mechanism. It is known that melamine undergoes progressive condensation on heating with elimination of ammonia and formation of polymeric products named "melam", "melem", "melon" (8,9). The following schematic reaction is reported in the literature (10-12):

0097–6156/90/0425–0211$08.00/0
© 1990 American Chemical Society

$$2\ \underset{\text{melamine}}{C_3H_6N_6} \xrightarrow{-NH_3} \underset{\text{melam}}{C_6H_9N_{11}} \xrightarrow{-NH_3} \underset{\text{melem}}{C_6H_6N_{10}} \xrightarrow{-NH_3} \underset{\text{melon}}{C_6H_3N_9} \qquad (1)$$

While there is clear evidence that melam is the dimer of melamine in which two s-triazine rings are connected by a NH bridge (di 6,[2,4 diamino-1,3,5-triazine]amine) (10):

$$(2)$$

different structures have been proposed for melem (8-14), e.g.:

I   II

III

for which there is not yet a final proof supporting one of them. As far as melon is concerned, it is generally agreed that it is formed by elimination of ammonia between amino groups of melem, whichever its structure (8-13). The difficulties encountered in the assignment of structures to these products depends on the fact that they are either sparingly soluble (melamine, melam, melem) or insoluble (melon). Therefore characterisation is practically restricted to the use of UV and solid state IR spectrophotometry. (15-16) In particular, in spite of the uncertainty of the structure of melem and consequently of melon, there is a fairly good agreement in the literature on their IR spectrum (16) (Figure 1). It can be seen that melamine, melam, melem and melon all show the typical absorption at 795-815 cm^{-1} due to out of plane deformation of the s-triazine ring whereas they show clearly distinguishable features in the complex region between 1000-1700 cm^{-1} involving absorptions due to NH$_2$, NH, CN and s-triazine ring (16-17).

As far as fire retardance is concerned, it is to be noted that melamine heated in an open system (Figure 2) mostly evaporates unalterated above 250°C leaving a little residue (ca. 7%,) of condensation products which are more thermally stable than melamine itself: melam to about 350°C, melem 450°C, melon 600°C (18). The yield of condensation products can be increased by heating melamine in a closed system with however limitation by equilibrium with ammonia evolved. (11,18). Therefore, when melamine is incorporated in a polymer, in the absence of reactions with the matrix, it should leave the material when thermal degradation occurs, with negligible contribution to the charred residue possibly formed. In this case the contribution of melamine to fire retardance could be sought in a blowing action during the charring step of intumescent systems, if temperature of evaporation and charring conveniently match. Otherwise, typical gas phase actions could be performed by volatilised melamine such as dilution of flammable gases, etc.

A completely different situation may arise when melamine derivatives, e.g. melamine salts are introduced in the polymeric material, owing to the thermal behaviour of the salts which depends on the type of acid combined with melamine. We have characterised the thermal behaviour of a wide number of melamine salts and found that they may be classified into three main classes:

1. Salts which undergo thermal dissociation to acid and melamine
2. Salts of strong acids which catalyse melamine condensation
3. Salts of acids which react with melamine condensation products.

Examples of the three classes, including salts used in fire retardant applications, are shown and discussed here. The results obtained on heating in nitrogen are reported since they are identical to those obtained in air. On the other hand, the supply of combustible volatile products to the flame in polymer combustion is essentially due to the thermal degradation process. The role of oxygen in this step is still matter of debate and it might depend on the type of polymer and conditions of combustion such as presence or absence of flame. For example, in the presence of the flame, oxygen is present at the material surface in low concentration and might simply catalyse the thermal degradation process by heat resulting from limited thermal oxidation of the polymeric material.

Experimental.

Materials. Melamine and melamine salts were supplied by SKW, Trostberg, FRG. Melam was prepared by heating equimolar amounts of melamine and 2,4-diamino-6-chloro-s-triazine (19). Reference melam hydrobromide and nitrate were prepared by reacting melam with 20% solutions of HBr and HNO_3 respectively (20).

Thermal degradation. A Du Pont 951 thermobalance-1090 thermal analyser system was used connected to a U trap provided with entrance and exit stopcocks. Standard conditions were: 10 mg sample; heating rate, 10°C/min; nitrogen flow, 60 cm^3/min. Volatile products, swept by the purging nitrogen, condense out either on the glass envelope of the thermobalance emerging from the furnace (high boiling products) or in the U trap cooled at liquid nitrogen temperature (gaseous products). Degradation products were generally identified by Fourier Transform Infrared (FTIR, Perkin Elmer 1710) either in KBr pellets (residues and

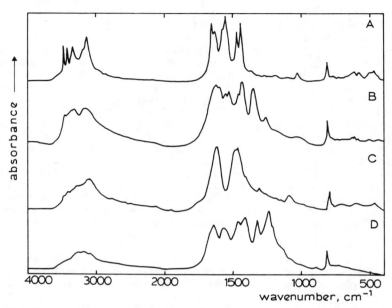

Figure 1. IR spectra of melamine (A), melam (B), melem (C), melon (D).

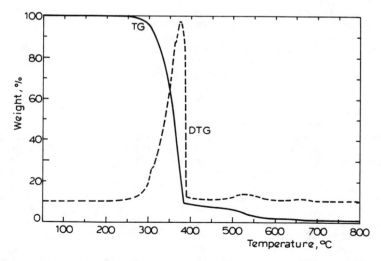

Figure 2. Thermogravimetric and differential thermogravimetric curves (TG and DTG) of melamine, heating rate 10°C/min; nitrogen, 60 cm³/min.

high boiling components) or in a gas cell after transfer from the U trap through a vacuum line. 31P solid state NMR was recorded on a Jeol JX-270 instrument equipped with magic angle spinning accessory using CP-Mass cross polarisation magic angle spinning with rotation speed of the sample (ω_R) between 3.5 - 4.2 kHz.

Thermogravimetry-evolved gas analysis (TG-EGA). Determination of water and ammonia evolved on heating was carried out by continuous monitoring of their content in the gas purging stream exiting the thermobalance. Water was measured by using a specific probe (Hygrometer III, Panametrics). Ammonia was measured by bubbling the gas stream into 0.1N sulphuric acid solution whose conductivity was continuously recorded. Conductivity variation was converted into ammonia evolved by means of calibration with titrated NH_4OH solutions.

Results and Discussion.

Salts which Undergo Dissociation to Acid and Melamine.

Cyanurate. The TG of dimelamine cyanurate (Figure 3) is characterised by three degradation steps as shown by the DTG curve. In the first (250-350°C), only melamine is volatilised as shown by IR. The weight loss in this step (30%) corresponds to elimination of one molecule of melamine per molecule of salt (calculated 33%). Indeed the IR of the residue after melamine loss (figure 4), is identical with that of reference melamine cyanurate. This behaviour was found to be typical of salts of polyfunctional acids owing to decrease of acidity with increasing number of acidic functions. Thus, the second molecule of melamine being less tightly bound, mostly tends to evaporate on heating (similarly to melamine heated alone) giving only a negligible amount of condensation.

Above its melting point (360°C) about 90% of melamine cyanurate is converted to volatile products (360-450°C, 2nd step) in which free melamine and melamine cyanurate were recognised by IR. This indicates that a competition takes place between evaporation of the unaltered salt and its thermal dissociation to melamine and cyanuric acid. Melamine behaves then as described above while cyanuric acid which, heated alone in TG volatilises completely above 300°C, is known to decompose to cyanic acid (8).

The IR of the residue left at 450°C is similar to that of condensation products obtained from TG of melamine and the yield (7%) is that expected on the basis of melamine content of the salt. The condensate from melamine cyanurate seems however to correspond to a somewhat larger degree of condensation, since it shows by IR a structure close to melon which is obtained from melamine at higher temperature (500°C) in TG. This suggests that cyanuric acid freed by heating may display a catalytic action on the condensation of melamine of the type discussed below, whose efficiency is however limited by decomposition to volatile cyanic acid. The melamine condensate undergoes then complete fragmentation to volatile products (HCN, $(CN)_2$, NH_2CN, high boiling fragments) on heating to 650°C (3rd step, Figure 3) as in the case of melamine (Figure 2). Summarising, the following scheme can account for the thermal behaviour of dimelamine cyanurate:

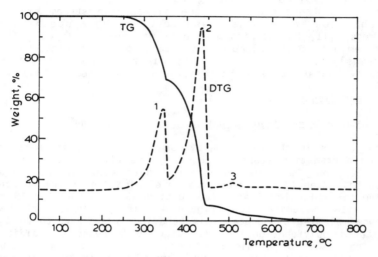

Figure 3. TG, DTG curves of dimelamine cyanurate. Conditions as Figure 2.

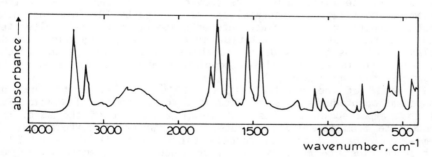

Figure 4. IR of residue of dimelamine cyanurate heated at 350°C: melamine cyanurate (by comparison with reference compound).

```
      dimelamine cyanurate
 step 1      ↓  250–350°C        ↗
    melamine cyanurate + melamine
 step 2      ↓  360–450°C                   ↗                        (3)
         melamine + cyanuric acid ──→ cyanic acid
         └──────────────┐        ↗
   evaporation       melon  +  NH₃
   step 3              ↓ >500°C
          decomposition
```

A behaviour similar to that of cyanurate is observed for the oxalate.

o-Phtalate. In the first step of volatilisation of melamine o-phtalate (220–280°C, Figure 5) a mixture of o-phtalic anhydride and o-phtalic acid is eliminated, as identified by IR. The residue (60%) shows the IR spectrum of free melamine. In this case, upon melting of the salt (240–280°C), the acid residue behaves as when heated alone where it distills as a mixture of o-phtalic acid and anhydride (21). The calculated amount of o-phtalic acid in the salt is 57% to be compared with 40% weight loss to 280°C. However the IR spectrum of the commercial o-phtalate we have used shows the presence of free melamine indicating that the acid was in defect. On the other hand, a few percent of the acid or of its degradation products may be left in the residue at 280°C since its subsequent thermal behaviour is not identical to that of free melamine. Indeed the DTG curve of the second step of degradation in which evaporation-condensation of melamine takes place (280–400°C, Figure 5) is more complex than the corresponding step of Figure 2. Moreover, a larger amount of melamine condensate is left from melamine o-phtalate at 400°C (13%) as compared to that calculated (ca. 6%) on the basis of melamine content of the salt. Finally, the condensate obtained at 400°C is characterised by an IR spectrum similar to that formed by heating melamine cyanurate but it seems more stable by about 100°C, degrading between 600–750°C.

The thermal degradation of melamine o-phtalate can be schematically represented as follows:

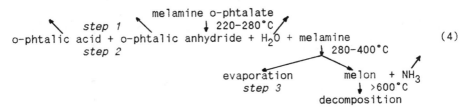

```
                      melamine o-phtalate
 ↖        step 1        ↘      ↓ 220–280°C     ↗
 o-phtalic acid + o-phtalic anhydride + H₂O + melamine           (4)
           step 2                          ↓ 280–400°C
                                      ↙              ↘              ↗
                      evaporation           melon  +  NH₃
                       step 3                 ↓ >600°C
                                      decomposition
```

Salts of Strong Acids which Catalyse Melamine Condensation.

Hydrobromide. A complex process takes place in the first step of weight loss of melamine hydrobromide (250–400°C) as shown by the DTG curve of Figure 6. The products evolved in this step were shown to be a mixture of melamine hydrobromide (IR) and ammonium bromide (IR and X-ray). The IR spectrum of the residue (55% at 400°C) which is compared with the original salt in Figure 7, is identical with that of reference melam hydrobromide. Furthermore, the IR of the residue

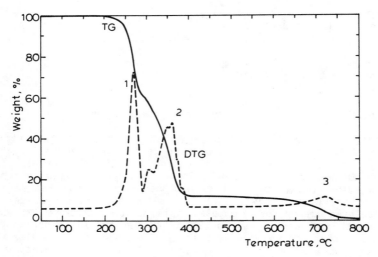

Figure 5. TG, DTG curves of melamine o-phtalate. Conditions as Figure 2.

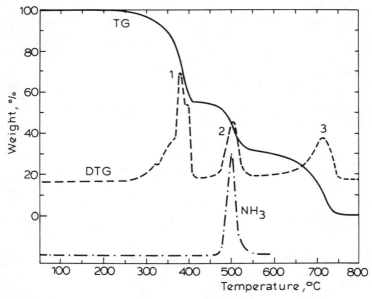

Figure 6. TG, DTG curves and rate of ammonia evolution (arbitrary units) for melamine hydrobromide. Conditions as Figure 2.

treated with 1% NaOH is that of melam. These data show that evaporation of melamine hydrobromide competes with the condensation of melamine to the dimer melam:

$$2 \text{ [melamine] } \cdot \text{HBr} \longrightarrow \text{ [melam] } \cdot \text{HBr} + \text{NH}_4\text{Br} \quad (5)$$

The by-product ammonium bromide is shown to volatilise in similar conditions in TG. This is a different behaviour altogether from that of melamine in whose thermal condensation melam can be isolated with difficulty only in carefully controlled experimental conditions (11,17). Indeed melem is the condensation product most easily isolated and this led some authors to doubt that melam could be an intermediate product of the condensation of melamine to melem and melon (19,22).

The high yield of condensation could be due to favoured nucleophilic attack when the amino group of melamine is protonated:

$$\text{[structure]} \longrightarrow \text{[structure]} \text{Br}^- \longrightarrow \text{C--NH--C} + \text{NH}_4\text{Br} \quad (6)$$

Quantitative condensation of melamine to melam on heating at 290°C in the presence of zinc chloride and acids was previously reported in the literature (20,23). The reason for limitation of the condensation process to the dimer has however not been discussed.

Melam hydrobromide undergoes thermal degradation between 450–550°C (2nd step, Figure 6) with complete elimination of bromine either through evolution of ammonium bromide, which is the major product, or melamine hydrobromide as shown by IR. Also a small amount of free ammonia is evolved in this step in which melon is formed as shown by the IR of the residue at 550°C (ca. 35% of original melamine hydrobromide).

These data suggest that, on heating, condensation of melam to melon occurs either through a reaction similar to Reaction 5 with elimination of ammonium bromide or through reaction between free amino groups giving ammonia. Formation of melamine hydrobromide, in this process, is in agreement with reported production of melamine during thermal condensation of melam (12). Degradation of melon takes place as expected (18) above 600°C (3rd step). The thermal behaviour of melamine hydrobromide can be described as follows:

$$
\begin{array}{c}
\text{melamine hydrobromide} \\
step\ 1 \quad \downarrow \quad 250\text{–}400°\text{C} \\
\text{evaporation} \nearrow \quad \text{melam hydrobromide} + \text{NH}_4\text{Br} \nearrow \\
step\ 2 \quad \searrow \downarrow \quad 450\text{–}550°\text{C} \\
\text{melon} + \text{NH}_3 + \text{melamine hydrobromide} + \text{NH}_4\text{Br} \\
step\ 3 \quad \downarrow \quad >600°\text{C} \\
\text{decomposition}
\end{array} \quad (7)
$$

Nitrate. The thermal behaviour of the nitrate (Figure 8), is qualitatively similar to that of the hydrobromide (Figure 6). Above its melting point (225°C), melamine nitrate undergoes competing evaporation and condensation to melam nitrate with evolution of ammonium nitrate (1st step, 250-370°C). Melamine nitrate and ammonium nitrate, which volatilises between 230-320°C in TG, were indeed identified by IR in the volatilised material. Moreover, the IR spectrum of the residue of this step of degradation (42%, Figure 9B) is identical to that of reference melam nitrate. Therefore in this step of degradation a reaction similar to Reaction 5 takes place in which HNO_3 is substituted for HBr.

Similarly to melamine hydrobromide, in the second step of weight loss (370-450°C, Figure 8), IR data show that melam nitrate evolves further ammonium nitrate, ammonia and melamine nitrate giving melon which decomposes completely to volatile products above 500°C. The degradation scheme is then:

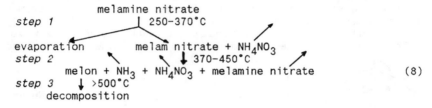

$$(8)$$

Sulphate. Melamine sulphate eliminates the water of crystallisation on heating between 100 – 230°C as shown by TG, DTG and water evolution curves of Figure 10 (1st step). The amount of water evolved corresponds to one molecule per molecule of salt (calculated weight loss: 7.4% experimental: 8%). The IR of the anhydrous salt is shown in Figure 11A in which the typical strong band of sulphate anion group at 1095 cm^{-1} is evident.

In the range 300-400°C (2nd step, Figure 10), the volatile products were identified as ammonia and water by using the specific detectors and melamine by IR. From the weight loss and the measured amount of water and ammonia (0.66 and 0.35 mole per mole of salt respectively) it can be calculated that at the utmost 15% of the melamine present in the sulphate should be volatilised. This shows that, in contrast to the above salts of strong acids (hydrobromide, nitrate) which tend to evaporate on heating (>250°C), melamine sulphate reaches decomposition temperature (>300°C) with liberation of melamine before vapor pressure allows significant volatilisation of the salt. At this temperature the freed melamine will then volatilise. However, similarly to the previous salts, also in the case of the sulphate, the loss of s-triazine structures is limited by the very efficient promotion of melamine condensation demonstrated by simultaneous evolution of ammonia. Indeed, it can be calculated that about 0.4 molecules of ammonia is eliminated per melamine molecule left in the condensed phase which is close to the amount corresponding to complete conversion to melam (0.5).

Limited decomposition of the sulphate and extensive melamine condensation overlap in this second step with condensation of sulphuric acid residues to pyrosulphuric structures. This is shown by water evolution and appearance of typical absorptions of pyrosulphate

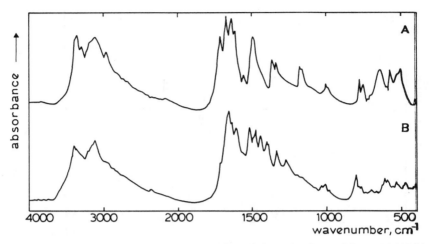

Figure 7. IR of melamine hydrobromide (A) and of residue at 400°C (B): melam hydrobromide (by comparison with reference compound).

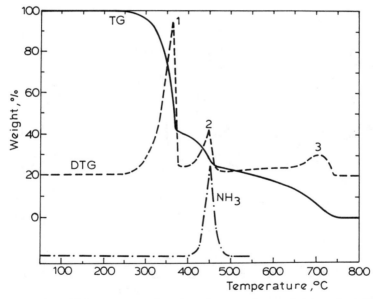

Figure 8. TG, DTG curves and rate of ammonia evolution (arbitrary units) for melamine nitrate. Conditions as Figure 2.

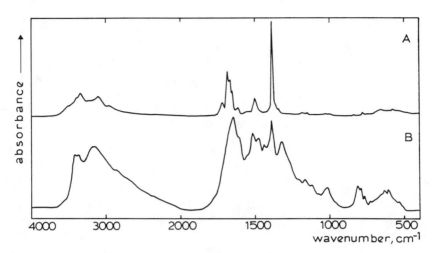

Figure 9. IR of melamine nitrate (A) and of residue at 370°C (B):
melam nitrate (by comparison with reference compound).

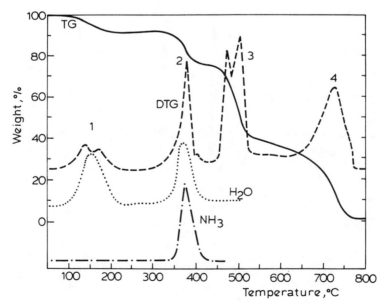

Figure 10. TG, DTG and rate of water and ammonia evolution
(arbitrary units). Conditions as Figure 2.

anion in the IR spectrum of the residue at 400°C (e.g.: 1020, 1050, 1225 cm^{-1}, Figure 11B) which were assigned by comparison with reference ammonium pyrosulphate. The reactions occurring in this step would thus involve a combined condensation of melamine to melam and of sulphuric acid to pyrosulphuric acid. For example, condensation of melamine could take place as in Reaction 5, followed by condensation to melam ammonium pyrosulphate:

$$2 \text{ melamine} \cdot H_2SO_4 \longrightarrow \text{melam} \cdot H_2SO_4 + NH_4HSO_4 \longrightarrow$$

$$2x \uparrow (9a) \quad [\text{dimelam}] \cdot H_2S_2O_7 + H_2O$$

$$(NH_4)_2S_2O_7 + H_2O$$

$$2x \uparrow (9b)$$

$$\longrightarrow [\text{melam}] \cdot HO-S(=O)_2-O-S(=O)_2-ONH_4 + H_2O \tag{9}$$

Condensation of sulphuric acid residues could also involve two molecules of melam sulphate giving dimelam pyrosulphate (9a). On the other hand, ammonium sulphate gives diammonium pyrosulphate in TG between 270-350°C (9b). The overall result of the combined condensation process would not change if condensation of sulphuric acid residues preceeds that of melamine. In this case dimelamine pyrosulphate would be the first product of reaction. The following condensation process leading to melam could then occur with formation of the three types of pyrosulphates discussed above. Formation of the melam salt of pyrosulphuric acid on heating melam sulphate was previously reported (20).

Ammonia is evolved here in the condensation process of melamine to melam whereas the respective ammonium salts are evolved in the case of the hydrobromide and nitrate. Reactions 9 and 9b show that ammonia should be neutralised by pyrosulphuric acid groups. In TG we have found that diammonium pyrosulphate eliminates ammonia above 350°C, with however complete decomposition:

$$(NH_4)_2S_2O_7 \longrightarrow 2NH_3 + 2SO_3 + H_2O \tag{10}$$

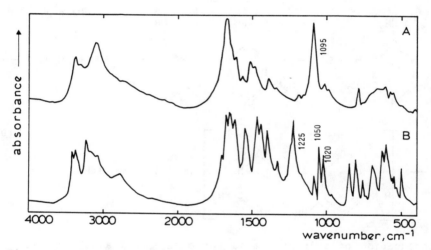

Figure 11. IR of anhydrous melamine sulphate (A) and of residue at 400°C (B): likely melam pyrosulphate.

Nevertheless, this reaction alone cannot account for the behaviour of melamine sulphate since it implies the elimination of one mole of SO_3 per mole of ammonia which would exceed by about 70% the experimental weight loss in this step of degradation. Moreover, we did not have evidence of relevant elimination of SO_3. On the other hand, water evolved (0.66 mole/g atom S) is larger than that calculated for Reaction 9 or 9a, 9b (0.50 mole/g atom S). Therefore   it seems that ammonium pyrosulphate groups undergo thermal scission in this step with evolution of ammonia while pyrosulphuric acid is partially stabilised by interaction with melam which is a thermally stable base. The SO_3 possibly evolved by partial decomposition of pyrosulphuric acid would decrease the amount of free melamine evolved calculated on the basis of TG. The calculated moles of ammonia eliminated per mole of condensing melamine would consequently decrease. However, since it was calculated that only 15% of original melamine would be the maximum amount volatilised, the ratio ammonia to melamine would fall in the range 0.35-0.40 which is in agreement with the above discussion once the experimental error is taken into account.

In the following third step (450-520°C), (Figure 10) the melam pyrosulphate is further condensed to melon with decomposition of the pyrosulphuric structure as shown by disappearance of typical absorptions of pyrosulphate or sulphate in the IR of the melon left as a residue. This last decomposes then above 520°C (4th step). The proposed degradation scheme for the sulphate is:

$$
\begin{array}{c}
\text{melamine sulphate} \cdot H_2O \\
step\ 1 \qquad \downarrow\ 100 - 230°C \qquad \nearrow \\
\text{anhydrous melamine sulphate} + H_2O \\
step\ 2 \qquad \downarrow\ 300 - 400°C \nearrow \qquad \nearrow \\
\text{melam pyrosulphate} + \text{melamine} + NH_3 + H_2O \\
step\ 3 \quad \downarrow\ 450 - 520°C \\
\text{melon} + \text{volatile products (likely: } NH_3,\ H_2O,\ SO_3) \\
step\ 4 \quad \downarrow\ > 520°C \\
\text{decomposition}
\end{array}
\tag{11}
$$

Salts of Acids which React with Melamine Condensation Products.

Phosphate. The phosphate, which is a most widely used fire retardant melamine salt, eliminates water in two successive steps with maximum rate at 280 and 320°C respectively (Figure 12). The solid state   31P NMR of melamine phosphate and of the residue of the two dehydration steps (300 and 330°C) are shown in Figure 13. In the case of the original salt (Figure 13A) the chemical shift anisotropy is axially symmetric and the isotropic peak is at 2.5 δ that is in the typical region of o-phosphates (24). The residue of the first step of degradation shows the isotropic peak centered at -8.6 δ (Figure 13B) that is at the chemical shift typical of pyrophosphates (24). Splitting into two peaks at -7.6 and 9.5 δ is likely to be due to asymmetry of the molecule owing to non equivalence of the melamine molecules in the pyrophosphate salt. In this case the chemical shift anisotropy is axially slightly asymmetric as shown by slightly different height of the spinning side bands. A small amount of the original o-phosphate is still present in this residue as shown by the small peak at 2.5 δ. The residue of the second degradation step shows the isotropic peak at -21.5 δ (Figure 13C) typical of middle groups of

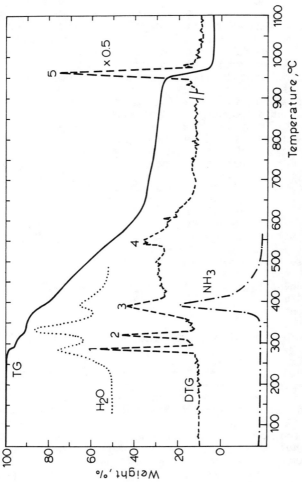

Figure 12. TG, DTG and rate of water and ammonia evolution (arbitrary units) for melamine phosphate. Conditions as Figure 2.

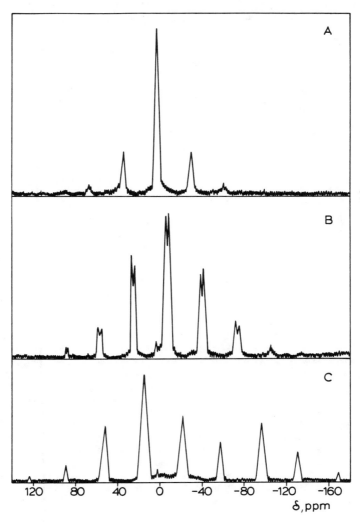

Figure 13. Solid state 31P NMR of melamine phosphate (A, ω_r =3.5 KHz), of residue at 300°C (B, ω_r =3.6 KHz:pyrophosphate) and 330°C (C, ω_r =4.0 KHz:polyphosphate).

linear polyphosphates (24) and the chemical shift anisotropy is strongly axially asymmetric. A degree of polymerization above 60 can be calculated on the basis of experimental sensitivity and absence of signals around -8 δ arising from end groups of polyphosphates (24). Thus, the dehydration processes involve condensation of melamine phosphate to pyrophosphate and polyphosphate:

(12)

The experimental weight loss in the first and second step (4.0 and 4.5% respectively) is in agreement with that corresponding to condensation to pyrophosphate (4.0%) and polyphosphate (4.2%, n 1). Furthermore, the IR spectra of melamine phosphate and of the residues at 300 and 330°C (Figure 14 spectra A, B and C respectively) show that besides the typical bands of phosphate salts (950-1300 cm^{-1}) which are present in the three spectra, a new absorption due to P-O-P bonds (ca. 890 cm^{-1}) appears in the spectra of the residues. The absorptions due to melamine salt structures (e.g. 780-790 and 1450-1750 cm^{-1}) are closely similar in the three spectra of Figure 14. Fire retardants based on melamine pyrophosphate and polyphosphate are reported in the literature (5) as well as methods for preparation of these salts (25-27).

The TG and DTG curves of Figure 12 show that melamine polyphosphate undergoes a complex degradation process between 330-650°C. In step 3 of the DTG curve (max. rate 390°C), water, ammonia and melamine are evolved. In this step the thermal behaviour of polyphosphate is somewhat similar to that of the sulphate in the same range of temperature (300-400°C). Indeed evolution of melamine indicates that thermal dissociation of polyphosphate giving free melamine takes place above 330°C. However, evaporation of melamine competes with its condensation as shown by evolution of ammonia.

Moreover, simultaneous evolution of water implies that condensation of melamine is combined with that of polyphosphoric structures. Assuming the reaction scheme proposed above, melam polyphosphate and ammonium polyphosphate groups should be formed from melamine polyphosphate:

(13)

It is known that ammonium polyphosphate tends to dissociate liberating ammonia above 300°C and the resulting free hydroxy groups condense giving crosslinked structures (ultraphosphate) with elimination of water (28,29):

(14)

Owing to overlapping of different processes in this range of temperature, quantitative considerations based on data obtained in programmed temperature TG cannot be made. However, it can roughly be estimated that up to 410°C, that is to the end of the process characterised by the narrow DTG peak 3 in Figure 12, about one mole of ammonia is evolved per mole of melamine. This would support the formation of melam by Reaction 13, followed by elimination of ammonia (reaction 14).

The residue of the third step of degradation of melamine phosphate (300-410°C) should then be the melam salt of ultraphosphoric acid. The IR spectrum of such a residue (Figure 15A) shows broad, poorly resolved absorption bands with however clear evidence of s-triazine ring absorption at 790-810 cm^{-1}. Crosslinking P-O-P bonds ("branching points") hydrolyse very fast in water to give phosphoric or polyphosphoric acid (30). Upon treatment with water of the ultraphosphate residue, a small fraction dissolved probably due to phosphoric or polyphosphoric acid molecules resulting from random hydrolysis of the branching points, whereas the salts of products of condensation of melamine are insoluble. The IR spectrum of the water-extracted residue (Figure 15B) shows slightly better resolved absorption bands than the original residue (Figure 15A). Typical absorptions of s-triazine ring (790-815 cm^{-1}) and of P=O groups (1080, 1250 cm^{-1}) are evident whereas identification of the melam structure is not straightforward. Nevertheless, the attribution of the IR spectrum of Figure 15B to melam phosphate or polyphosphate deriving from hydrolysis of melam ultraphosphate seems not unlikely if comparison is made with IR absorptions of melam (Figure 1B) and its salts (Figure 7,9,11) particularly in the regions 1500-1700 and 3000-3800 cm^{-1}. Melam polyphosphate is reported to be obtained on heating melam phosphate (20).

Above 410 °C the ultraphosphate undergoes a relatively slow degradation process which is completed at about 650°C with formation of a residue (35% of original phosphate) mostly stable to 950°C

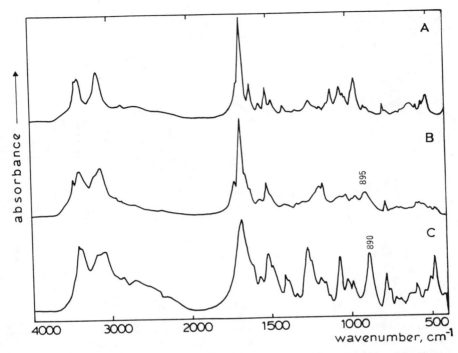

Figure 14. IR spectrum of melamine phosphate (A) and of residue at 300°C (B: pyrophosphate) and at 330°C (C:polyphosphate).

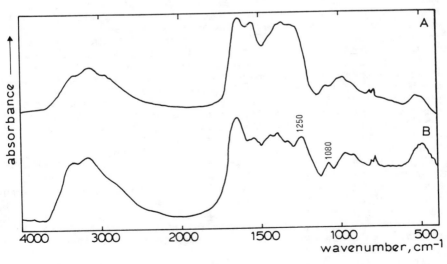

Figure 15. IR spectrum of residue of melamine phosphate heated to 380°C (A) and of this residue after water extraction (B).

(Figure 12). In this step the DTG curve shows a very broad peak with a narrower maximum superimposed (550°C) indicating the occurrence of different overlapping processes. This thermal behaviour cannot be explained on the basis of that of melamine condensation products or of ultraphosphates (e.g. ammonium salt). Indeed melamine condensate undergoes complete fragmentation to volatile products below 750°C [18] while ammonium ultraphosphate does so mostly below 700°C [29], in TG at 10°C/min. The presence of P in the material obtained at 650°C is shown by the solid state   31P NMR which however gives broad complex signals that could not be assigned to specific structures at this stage. The infrared spectrum (Figure 16A) shows broad absorptions among which those typical of s-triazine ring (810 cm^{-1}) and of NH$_2$ groups of melamine or its condensates (1650 and 3000-3600 cm^{-1}) are recognisable. A noticeable absorption begins to show at 2200 cm^{-1}.

Further modification of the residue takes place on raising the temperature above 650°C as shown by the low steady rate of weight loss (8% to 900°C, Figure 12). Indeed, the typical absorption of s-triazine ring and of NH$_2$ groups, still evident in Figure 16A, have disappeared in the IR spectrum of the material obtained at 900°C (Figure 16B). Here, a few broad absorption bands are present centered at 500, 950, 1280 and 2200 cm^{-1}. This material which is a white powder, undergoes a very rapid decomposition to volatile products with maximum rate at 960°C (step 5, Figure 12). While further characterisation is under way, it could be suggested that the product obtained at 900°C might be a highly thermally stable (PN) compound of the type obtained on heating phospham (NPNH) or phosphoryl nitride (OPN)$_n$ [31,32]. Sumarising, the overall degradation process of melamine phosphate follows the scheme:

$$
\begin{array}{ll}
& \text{melamine phosphate} \qquad \nearrow \\
\textit{step 1} & \downarrow \; 250\text{-}300°C + H_2O \\
& \text{melamine pyrophosphate} \\
\textit{step 2} & \downarrow \; 300\text{-}330°C \qquad \nearrow \\
& \text{melamine polyphosphate} + H_2O \\
\textit{step 3} & \downarrow \; 330\text{-}410°C \quad \nearrow \qquad \nearrow \\
& \text{melam ultraphosphate} + NH_3 + H_2O + \text{melamine} \\
\textit{step 4} & \downarrow \; 410\text{-}650°C \\
\text{(unidentified} & \text{phosphate-ultraphosphate-melamine type structures)} + \\
& \text{volatile products} \\
& \downarrow \; 650\text{-}940°C \\
& \text{loss of melamine structures (likely formation of PN)} \\
\textit{step 5} & \downarrow \; 940\text{-}970°C \\
& \text{decomposition}
\end{array}
\tag{15}
$$

Borate. Dehydration is the first reaction occurring on heating the borate as in the case of the phosphate but at a much lower temperature (130-270°C, step 1 Figure 17). Boric acid heated alone in the same conditions in TG, eliminates water between 80-300°C in two main overlapping steps with maximum rate at 170 and 190°C corresponding to formation of metaboric acid and boric anhydride respectively:

$$
H_3BO_3 \xrightarrow{-H_2O} HBO_2 \xrightarrow[-H_2O]{\times 2} B_2O_3
\tag{16}
$$

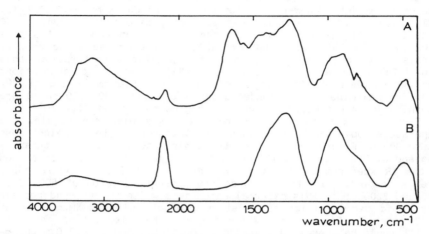

Figure 16. IR spectrum of residue of melamine phosphate heated to 650°C (A) and to 900°C (B).

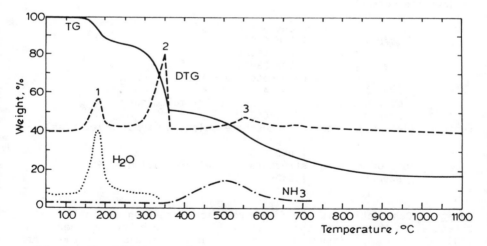

Figure 17. TG, DTG and rate of water and ammonia evolution (arbitrary units) for melamine borate. Conditions as Figure 2.

A physical mixture of boric acid and melamine shows this two step dehydration processes in TG whereas in the salt the two steps seem to merge. Indeed the dehydration process begins at higher temperature in the salt (130°C instead of 80°C) possibly because interactions of hydroxy groups with melamine molecules have first to be thermally broken. The overall weight loss at the end of the first step of degradation of the borate (15%, 270°C Figure 17) is in agreement with occurrence of Reaction 16 (calculated: 14.4%). The IR spectrum of the product of dehydration (Figure 18B) shows the typical absorptions of melamine with minor differences in relative intensity of some bands (see Figure 1A). Whereas the absorptions of OH groups present in the spectrum of the original salt (e.g. 2400,2900 cm^{-1}, Figure 18A), have disappeared as expected from occurrence of Reaction 16. The absorptions of boric anhydride (800(m), 1200(m), 1470(s) cm^{-1}) are difficult to recognise in Figure 18B owing to overlapping with the relatively strong absorptions of melamine. The dehydration process of melamine borate should then result in a physical mixture or adduct of boric anhydride with melamine.

In the range 270-350°C (2nd step, Figure 17) melamine volatilises as in the case of pure melamine (Figure 2) or of salts which free melamine at these temperatures, e.g. dimelamine cyanurate (1st step, Figure 3), melamine o-phtalate (2nd step, Figure 5). However, in the case of boric anhydride-melamine only about 50% of the melamine volatilises as it can be calculated from the weight loss in this step (35%, Figure 17) since boric anhydride is stable to >800°C in TG. Whereas more than 90% of free melamine can volatilise in the same range of temperature in the examples mentioned above. On the other hand free melamine is absent from the residue obtained when volatilisation stops as shown by the IR spectrum of Figure 19A in which, however, typical absorptions of the s-triazine ring (790-815 cm^{-1}) and of NH_2, NH groups (1600-1700 and 3000-3600 cm^{-1}) are recognisable in the otherwise complex absorption pattern. This residue was boiled with 5N HCl to eliminate soluble fractions such as boric anhydride. The extracted material shows an IR spectrum (Figure 19B) which could be consistent with a complex product of condensation of melamine. Indeed, the spectrum of Figure 19B is very close to that of the water extracted product obtained at the same temperature from melamine phosphate (Figure 15B) apart from the presence of minor absorptions between 900-1300 cm^{-1} (e.g. 1080 cm^{-1}, P=O) in this last . However, in the case of the phosphate, the formation of this residue occurred with evolution of ammonia indicating that it was issued from a condensation process involving NH_2 groups of melamine. Whereas, in the case of the borate, evolution of ammonia only begins at 350°C and takes place in the following third step (Figure 17). Although this point is to be further investigated it might be that condensation of melamine takes place in the boric anhydride-melamine system with however retention of ammonia by interaction with boric anhydride. Ammonia would then be released at higher temperature when such interactions are broken on heating. For example, ammonium pentaborate is shown to eliminate ammonia between 330-400°C in TG (33). The amount of ammonia released above 350°C corresponds to 0.6 mole per mole of melamine unvolatilised in the previous step between 270-350°C. This is close to ammonia evolved when the equivalent residue of Figure 15B is formed from the phosphate (0.40) and to the amount corresponding to

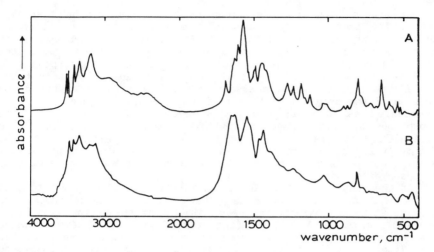

Figure 18. IR spectrum of melamine borate (A) and of residue at 280°C (B).

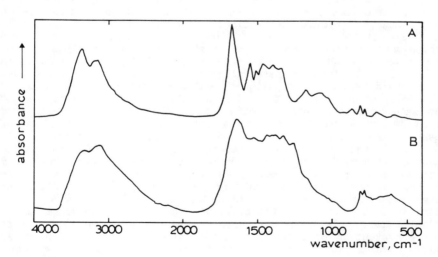

Figure 19. IR spectrum of residue of melamine borate heated to 380°C (A) and of this residue after acid extraction (B).

melam formation (0.50). A physical mixture of boric anhydride and melamine shows a TG identical to that of Figure 17, apart from the dehydration step.

Although a detailed identification of the residue obtained from the second step of degradation of melamine borate is not yet available, it is clear from Figure 19 that it contains s-triazine ring and amino groups besides boric anhydride. This material undergoes further modification on heating above 350°C with a relatively slow weight loss to 950°C (3rd step, Figure 17) to give a white product stable at least to 1100°C (17% of the original melamine borate). This step of degradation occurs with 32% weight loss of which only 2.6% is accounted for by ammonia. The content of boron in the material obtained at 950°C is 28% as determined by a method proposed for refractory boron nitride (34). The IR spectrum of this residue (Figure 20) shows a relatively sharp maximum at 1405 cm^{-1} which could be attributed to B-N bonds (17), and two overlapping broad maxima in the region of NH$_2$, NH stretching frequency (3000-3800 cm^{-1}) whereas absorptions due to the s-triazine ring (e.g. ca 800 cm^{-1}) are absent. Further broad absorptions are shown in Figure 20 in the regions 800-1300 and 1500-1700 cm^{-1}. Many papers and patents deal with preparation of refractory boron nitride (stable in inert atmosphere to 2700°C (35)) by heating boron oxy compounds with melamine generally above 900°C (e.g. ref. 36-41). The IR spectra reported in the literature for boron nitride (42,43) show a weak absorption at 809-813 cm^{-1} and a strong one at 1374-1389 cm^{-1}. The spectrum of Figure 20 could be partially consistent with that of a boron nitride since the strongest of the above absorptions is close to the strongest absorption in Figure 20. However, the IR spectrum of Figure 20 shows also relevant absorptions which are not attributable to boron nitride. On the other hand it was suggested (36) that on heating mixtures of orthoboric acid and melamine or other nitrogen containing compounds (e.g.: urea, cyanamide etc.) a BNO (B=31.6%, N=20.5%) residue is obtained at 600°C which could not be transformed into pure boron nitride on further heating to 1300°C; whereas the BNO residue gave boron nitride on heating in a stream of ammonia at 500-900°C. Although IR characterisation of the BNO was not shown, the B content of the residue of Figure 20 (28%) is close to that reported in the literature (36) for BNO (31.6%). Therefore, the residue obtained at 950°C from

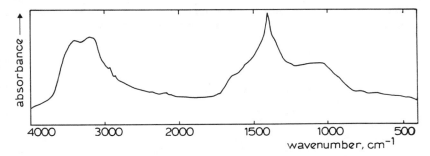

Figure 20. IR spectrum of residue of melamine borate heated to 950°C.

melamine borate could be similar to this BNO material rather than to pure boron nitride. While further characterisation work is in progress, it must be mentioned that also polyamino-borazines (35,44) are boron-nitrogen thermally stable products which might be formed in the above conditions and would account for the N-H absorption in the IR spectrum of Figure 20.

Summarising the data available at this stage, the thermal degradation of melamine borate can be represented by the scheme:

$$
\begin{array}{c}
\text{melamine borate} \\
step\ 1 \qquad \downarrow 130\text{-}270°C \\
\text{melamine-boric anhydride} + H_2O \qquad\qquad (17) \\
step\ 2 \qquad \downarrow 270\text{-}350°C \\
\text{melamine} + \text{unidentified (likely melamine condensate-boric anhydride)} \\
step\ 3 \qquad \downarrow 350\text{-}950°C \\
\text{loss of melamine structures (likely formation of BN or BNO)}
\end{array}
$$

Conclusion

This survey shows that melamine salts can display widely different thermal behaviour depending on the type of acid residue. As far as fire retardance is concerned the formation of volatile products and of thermally stable residue ("char") in the range 300-500°C in which most polymers undergo thermal degradation, is of paramount importance. First of all it is shown that in dimelamine salts one molecule of melamine behaves as free melamine volatilising almost quantitatively above 250°C. A similar behaviour is also shown by monomelamine salts of acids which tend to decompose on heating at relatively low temperature (200-400°C, e.g. o-phtalate, cyanurate, oxalate). These salts leave on degradation a small amount of residual char stable to about 500°C (<15%) which results from partial condensation of melamine to "melon".

Whereas acid catalysed condensation of melamine takes place on heating salts of relatively stable acids such as hydrobromide, nitrate. The corresponding ammonium salt is volatilised and melam salt is formed which undergoes further condensation to melon (ca. 20-35% at 500°C). The overall yield of the melamine condensation reaction is limited by competition with volatilisation of the original salt. A similar but somewhat more complex behaviour is shown by the sulphate. This salt undergoes on heating combined condensation of melamine to melam and sulphuric acid to pyrosulphuric acid with release of ammonia and water. Although condensation of melamine is limited in this case by the evaporation, about 80% of residue is obtained at 450°C. Further thermal modification leads to about 40% of melon-like product at 500°C.

Combined condensation of melamine and of the acid residue is also shown by the phosphate and borate. In these cases however, the polymeric dehydrated compound derived from the acid is thermally stable up to above 500°C allowing chemical reactions with melamine condensation products simultaneously formed. The resulting material is stable to 950°C (phosphate, ca 30% of original salt) or to above 1100°C (borate, ca. 20% of original salt).

These data are essential for understanding the role played by melamine salts both in gas or condensed phase during burning of polymeric materials. Indeed melamine salts may act either as a source of gases relatively stable to thermal oxidation occurring in the flame such as melamine, ammonium or melamine salts, water ("blanket" effect) or as a source of thermally stable char (char forming effect) depending on the type of salt. Moreover, the knowledge of the thermal behaviour of salts to be used as fire retardant additives will supply the rationale for selection of the most suitable salt for a given polymeric material. For example the melamine salt may be selected which evolves the gases or forms the char in the range of temperature in which the polymer degrades to volatile combustible products. In particular, in formulation of intumescent systems it is essential to obtain a suitable matching between gas evolution (blowing) and char formation. These considerations are strictly valid only if chemical interactions between the melamine salt and the degrading polymer and/or other additives do not take place on heating. However, even in these case, the results reported here may supply useful information for the understanding of such reactions.

Acknowledgments
The authors wish to thank Dr. R. Gobetto of the University of Torino, Italy for interpretation of solid state NMR data.

Literature cited

1. Kracklaurer J.; In Flame-Retardant Polymeric Materials.; Lewin M.; Atlas S.M.; Pearce E.M.; Eds.; Plenum Press: New York, 1978; Vol 2, p 285.
2. Kaplan H.L.; Grand A.F.; Hartzell G.E. Combustion Toxicology; Technomic: Lancaster, 1983.
3. Benjamin/Clarke Associates Fire Deaths; Technomic: Lancaster, 1984.
4. Vandersall H.L. J. Fire Flamm. 1971, 2, 97.
5. Kay M.; Price A.F.; Lavery I. J. Fire Retardant Chem. 1979, 6, 69.
6. Camino G.; Costa L. Rev. Inorg. Chem. 1986, 8, 69.
7. Camino G.; Costa L. Polym. Deg. and Stab. 1989, 23, 359.
8. Smolin E.M.; Rapoport L. In The Chemistry of Heterocyclic Compounds; Weissberger A. Ed.; Interscience: New York, 1959; Vol 13.
9. Bann B.; Miller S.A. Chem. Rev. 1958, 58, 131.
10. Takimoto M.; Yokoyama T.; Sawada M.; Yamashita M. Kogyo Kagaku Zasshi 1963, 66, 793; Chem. Abstr. 1964, 60, 5499f.
11. Takimoto M.; Funakawa T. Kogyo Kagaku Zasshi 1963, 66, 797; Chem. Abstr. 1964, 60, 5499g.
12. Takimoto M; Funakawa T. Kogyo Kagaku Zasshi 1963, 66, 804; Chem. Abstr. 1964, 60, 5499h.
13. May H. J. Appl. Chem. 1958, 349.
14. Schmidt A. Allg. Prakt. Chem. 1966, 17, 702; Chem. Abstr. 1966, 66, 46808k.
15. Takimoto M. Nippon Kagaku Zasshi 1964, 85, 159; Chem. Abstr. 1964, 60, 2937h.

16. Takimoto M. Nippon Kagaku Zasshi 1964, 85, 168; Chem. Abstr. 1964, 60, 2937f.
17. Colthup N.B.; Daly L.H.; Wiberley S.E. Introduction to Infrared and Raman Spectroscopy; Academic Press: New York, 1964.
18. Costa L.; Camino G. J. Thermal Anal. 1988, 34, 423.
19. Spriridonova N.V.; Finkel'shtein A.I. Khim. Geterotsikl. Soedin. Akad. Nauk. Latv. 1966, 126; Chem. Abstr. 1966, 65, 715f.
20. Gavrilova N.K.; Gal'perin V.A.; Finkel'shtein A.I.; Koryakin A.G. Zh. Org. Khim. 1977, 13, 669.
21. Handbook of Chemistry and Physics; Weast R.C., Ed.; The Chemical Rubbers Co.; 1971.
22. van der Plaats G.; Soons H.; Snellings R. In Proceedings of the Second European Symposium on Thermal Analysis; D. Dollimore, Ed.; Heyden: London, 1981; p. 215.
23. Gal'perin V.A.; Finkel'shtein A.I.; Gavrilova N.K. Zh. Org. Khim. 1971, 7, 2431.
24. Van Wazer J.R.; Callis C.F.; Shoolery J.N.; R.C. Jones J. Am. Chem. Soc. 1956, 78, 5715.
25. Vol'fkovich S.I.; Zusser E.E.; Remen R.E. USSR Patent 67 616, 1949; Chem. Abstr. 1949, 43, 3473g.
26. Vol'fkovich S.I.; Feldmann W.; Kozmina M.L. Z. Anorgan. Allg. Chem. 1979, 457; Chem. Abstr. 1980 92, 103417v.
27. Sheridan R.C. Inorg. Synth. 1982, 157.
28. Camino G.; Grassie N.; McNeill I.C. J. Polym. Sci. Polym. Chem. Ed. 1978, 16, 95.
29. Camino G.; Costa L.; Trossarelli L. Polym. Deg. and Stab. 1984, 6, 243.
30. Van Wazer J. Phosphorus and Its Compounds; Interscience: New York, 1958.
31. Toy A.D.F. In Comprehensive Inorganic Chemistry.; Trotman-Dickenson A.F. Ed.; Pergamon Press: Oxford ; Vol. 3, p. 451;
32. Miller M.C.; Shaw R.A. J. Chem. Soc. 1964, 3233.
33. Erdey L.; Gal S.; Liptay G. Talanta, 1964, 11, 913.
34. Erickson S.L.; Conrad F.J. Talanta, 1971, 18, 1066.
35. Greenwood N.N.; Thomas B.S., In Comprehensive Inorganic Chemistry.; Trotman-Dickenson A.F. Ed.; Pergamon Press: Oxford; Vol 3, p. 916.
36. O'Connor T.E. J. Am. Chem. Soc. 1962, 84, 1753.
37. Gorski A.; Podsiadlo S. Polish J. Chem. 1984, 58, 3.
38. Ogasawara T.; Koshida T.; Sasaki K. Chem. Abstr. 1986, 105, 81767s.
39. Koshida T.; Ogasawara T.; Sasaki K. Chem. Abstr. 1986, 105, 213266q.
40. Matsuda F.; Kato K. Chem. Abstr. 1986, 105, 63213d.
41. Sato T.; Ishii T. Chem. Abstr. 1986, 105, 211253j.
42. Miller F.A.; Wilkins C.H. Anal. Chem. 1952, 24, 1253.
43. Brame E.G. Jr.; Margrave J.L.; Meloche V.W. J. Inorg. Nucl. Chem. 1957, 5, 48.
44. Paciorek K.J.; Harris D.H.; Kratzer R.H. J. Polym. Sci. Polym. Chem. Ed. 1986, 24, 173.

RECEIVED November 1, 1989

FIRE   RETARDANCY
IN   ENGINEERING   PLASTICS

Fire Retardancy in Engineering Plastics

The term "engineering plastics" came to be used in the early 1970s to identify high performance, generally more expensive, synthetic polymers used for advanced applications. Many such materials have aromatic backbones. Flame retardancy in such systems many times involves subtle changes in chemistry to enhance char or to change volatiles rather than addition of generic additives.

This section begins with an overview by Tesoro. It is followed by a discussion of the use of brominated phosphate esters by Green. Whang and Pearce discuss improved performance in certain aromatic polymers. Factor presents a study using available analytical tools to elucidate mechanisms of char formation in aromatic polymers. And finally, in two chapters, Nelson and coworkers present a study of the effects of coatings on the fire performance of engineering polymers and insight into the mechanistic origins.

Chapter 16

Fire Resistance in Advanced Engineering Thermoplastics

Giuliana C. Tesoro

Department of Chemistry, Polytechnic University, Brooklyn, NY 11201

Engineering polymers generally comprise a high performance segment of synthetic plastic materials that exhibit
premium properties. In this paper, engineering thermoplastics developed for advanced applications, and particularly for enhanced thermal stability are considered.
The aromatic structure of these materials is of major
importance in their thermal response and in their inherent fire resistance according to established criteria.
Approaches to fire retardants that have been studied for
this class of polymers are reviewed, and illustrative
data on performance are presented.

Engineering polymers are generally understood to include those
plastic materials that may be shaped into functional parts of structural components and may replace metal. A recent list of engineering
thermoplastics (1-2) includes nylons, polyacetals, polyesters, polycarbonates, polyarylates, polysulfones, polyphenylenesulfide, and
polyetherimide. In this discussion, the focus will be on thermoplastic polymers developed primarily for enhanced long-term resistance to elevated temperatures in advanced applications. Representative structures of repeating units are shown in Figure 1. It is evident that the aromatic character of the macro-molecule is a common
denominator-consistent with the relationships of molecular structure
to thermal stability documented for a broad spectrum of high-temperature polymers (3). In the context of the thermal response behavior
of polymers such as those shown in Figure 1, it is not surprising
that their fire resistance has not been of major concern in most instances. This must be viewed with reference to specific criteria of
flammability behavior, and at this time, it is appropriate to select
measurements of oxygen index (ASTM D-2863), and of Underwriters Laboratory ratings (UL-94) as indicative of fire resistance for laboratory investigations and for evaluation of commercial candidates respectively. Some data for the polymers considered are summarized in
Table I, where the literature values shown refer to "neat resins" except where addition of a flame retardant (FR) is indicated for

0097–6156/90/0425–0241$06.00/0
© 1990 American Chemical Society

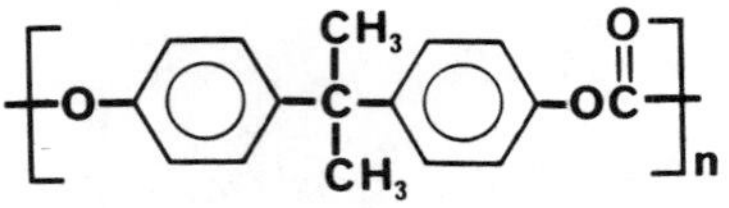

Poly[bisphenol-A]carbonate

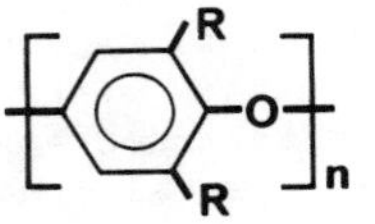

Phenylene ether based resin (PPO)

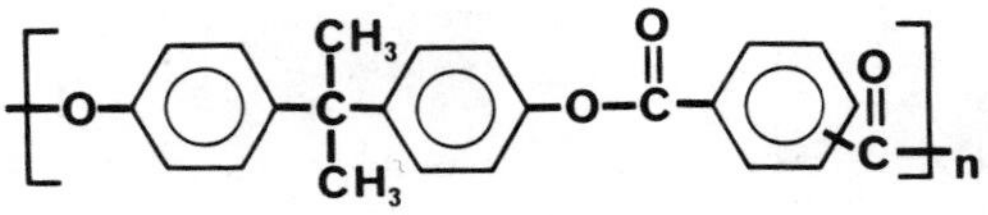

Polyarylate

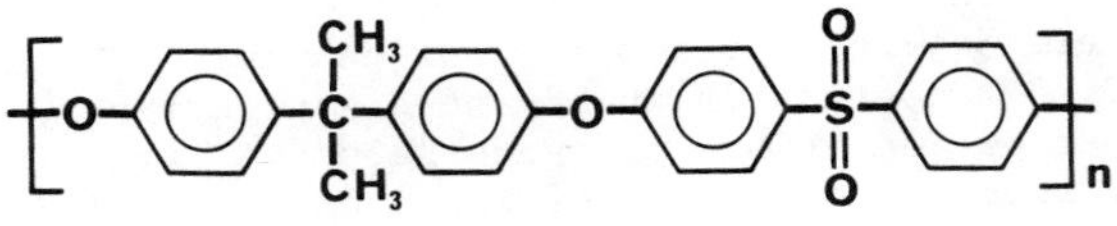

Polysulfone

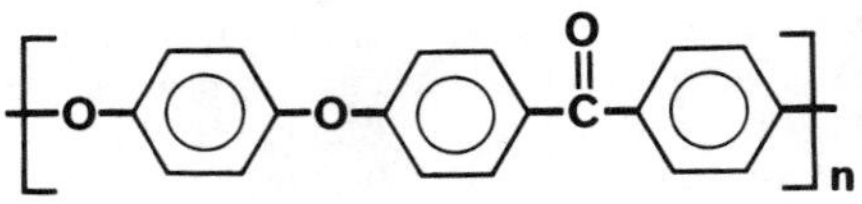

Polyether etherketone (PEEK)

Figure 1. Examples of repeating unit in advanced engineering
thermoplastics.

TABLE I

FIRE RESISTANCE OF SOME ENGINEERING THERMOPLASTICS*

Polymer	Oxygen Index Value	UL - 94 Rating	Reference
Polycarbonate (PC)			
LEXAN/1Ø1	25.Ø	V-2	(2) (5) (6)
LEXAN/92Ø (FR)	35.Ø	V-0	
Polyphenylene Oxide	32	-	(5) (6)
(PPO)			
NORYL 6P-275	24.Ø	-	
NORYL SE1ØØ (FR)	33.Ø	V-0	
Polysulfone			
UDEL-172Ø	32.Ø	V-0	(2) (4) (7)
Polyethersulfone	34-42	V-0	(2) (4)
(PES)			
Polyetherether Ketone	24-35	V-0	(2) (8)
(PEEK)			
Polyether Ketone	4Ø	V-0	(8)
Ketone			
(PEKK)			
Polyarylates			
ARDEL D-1ØØ	34	V-0	(2)

*LEXAN and NORYL are trademarks of the General Electric Co.

UDEL and ARDEL are trademarks of Amoco Performance Products, Inc.

specific commercial products where the desired V-0 (UL-94) rating was
not attained in the absence of FR additive. As a first approximation, the advanced engineering thermo-plastics considered have an oxygen index value >3$\emptyset$, and a UL-94 rating of V-0 (maximum and minimum
thickness in the intended application).

A discussion of test methodology is beyond the scope of the present paper. However, the fact that established tests do not accurately reflect the behavior of materials in fires has been widely
recognized (9), and the search for more meaningful techniques for the
evaluation of engineering materials has continued to be a valid research objective. The development of the cone calorimeter, a bench-
scale tool for the evaluation of fire properties of materials (10a)
at NBS, is of particular significance in this context.

Rate of heat release measurements have been attempted since the
late 1950's. A prominent example of instrument design for the direct
measurement of the sensible enthalpy of combustion products is the
Ohio State University (OSU) calorimeter. This has been standardized
by ASTM and a test method employing this technique (ASTM-E-906) is
part of a FAA specification for evaluation of large interior surface
materials.

Thermoplastics for aircraft interiors have been evaluated by
this technique (10b) in accordance with the FAA specification (peak
rate of heat release of 65 kilowatts per meter squared (Kw/m 2) or
less). In these tests (10b) Polyether sulfone demonstrated marginal
compliance. For Polyether imide (PEI) and PEI/Polydimethyl siloxane
copolymers peak heat-release rates were well below the specified
value. The overall trend suggested a possible correlation of peak
heat release values with aromatic carbon content in the polymers
evaluated.

For thermoplastic composites, results of flammability tests are
generally reported on the basis of oxygen index values and/or UL-94
ratings (e.g. (11-12). The general problems associated with composites and multicomponent systems have not been addressed in depth and
published data pertain primarily to specific glass-filled resins offered by manufacturers, or to composite systems designed to meet the
specifications of a particular end use.

APPROACHES TO FIRE RETARDANTS

Brominated Compounds

The traditional approach to fire retardation in most thermoplastics
has been the incorporation of brominated compounds with (or without)
antimony oxide as a synergist. This has been generally effective,
and compounds reported in the literature of the seventies have covered a broad range of structures and formulations designed to meet
the flammability requirements of polypropylene and thermoplastic
polyesters primarily. With concern for the toxicological and environmental factors associated with the use of TRIS (2,3, dibromopropyl) phosphate (TRIS), and eventually the ban of TRIS in 1977,
the use of decabromodiphenyl oxide was investigated for many applications (13-15). Work on analogues of TRIS also continued to elicit
interest (16-17). Other compounds of high bromine content have been
commercialized, for example octabromodiphenyl oxide and ethylene-bis-

tetrabromophthalimide. However, recent developments have focused primarily on oligomers of high bromine content designed to minimize effects of the additives on the processing and performance properties of the host polymer, and also to avoid outward diffusion in the system and consequent risks of environmental contamination. Figure 2 shows examples of structures for brominated flame retardants developed in recent years. Commercial products based on brominated polystyrene oligomers (19), on brominated PPO oligomers (18) and on tribromophenyl terminated tetrabromobisphenol A-carbonate oligomers (22) have been recommended primarily for nylon and polybutylene terephthalate, with additive levels of 15 to 25% required for a V-0 rating in the UL-94 test. The bromine content of the commercial products ranges from 50% to 70%, and the concepts underlying the developments reflect the need for essentially aromatic chemical structures that would be miscible with specific resins (non blooming), withstand relatively high processing temperatures, and would not impair resin performance at the concentrations required for flame retardant effectiveness.

This rationale is evident also in other bromine-containing oligomers, for example those based on brominated epoxy resins (20) and on poly-pentabromobenzyl acrylates (21).

A comparison of representative bromine-containing oligomeric flame retardants offered commercially is shown in Table II (adapted from reference 21). Some of these products are claimed to be "broad spectrum brominated flame retardants." As a practical matter, they are designed with miscibility with specific resins clearly in mind, and they can be employed to attain the desired V-0 rating in UL-94 tests of products that do not meet these requirements without added flame retardants.

Blends of flame retardant additives have been advocated as an approach to an optimum balance of properties in the finished products. For example, blends of tetrabromophthalate esters with decabromodiphenyl oxide or other flame retardants are reported to yield a V-0 rating in modified PPO and in polycarbonate resins without compromising melt processability or performance properties (23a-b).

In summary, new brominated flame retardants have essentially met the challenge for thermoplastics that do not meet current requirements (e.g. a V-0 rating) without additives. More stringent flammability requirements for advanced materials and applications, coupled with improved test methodology (e.g. 10a-b) may shift the focus of research to other approaches.

Phosphorus Compounds

Carnahan and co-workers have studied the effectiveness of phosphorus compounds as flame retardants for blends of poly-2,6 dimethyl-phenylene oxide and polystyrene (35/65), evaluating flammabilty by oxygen index measurements (24). The mode of action of the phosphorus containing species (Triphenyl phosphate, Triphenyl phosphine, Triphenyl phosphine oxide, p-Phenyl-phenol phosphate, Red phosphorus were included in the study) was investigated according to criteria designed to determine whether effectiveness in the gas phase or in the condensed phase dominated. The macroscopic criteria of chemical mechanism employed in this work for determining the mode of action of

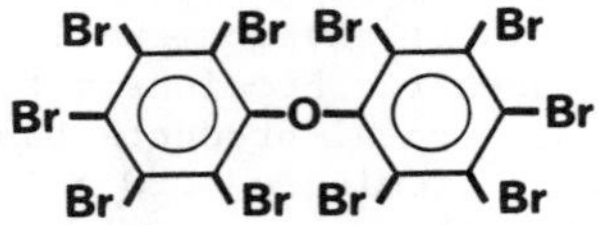

Decabromodiphenyloxide

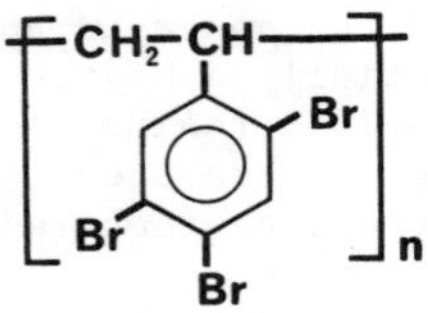

Brominated polystyrene

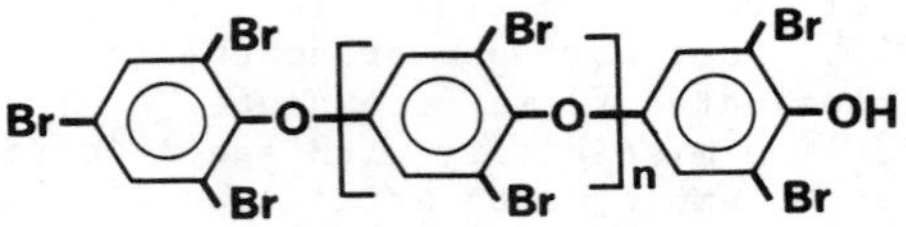

Poly-dibromophenylene oxide

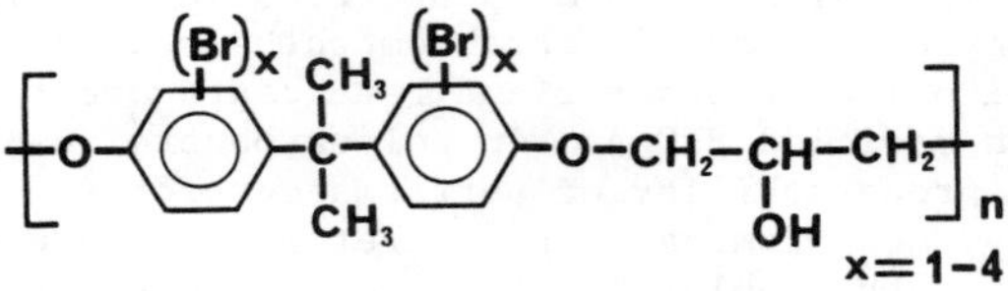

Brominated epoxy oligomer

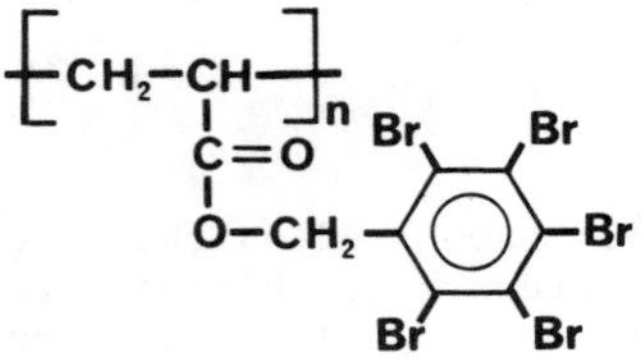

Poly-pentabromobenzyl acrylate

Figure 2.   Brominated fire retardants.

Table II. Properties of Polymeric Flame Retardants

Chemical Name	Poly(Pentabromobenzyl Acrylate)	Poly-dibromophenylene Oxide	Brominated Polystyrene	Brominated Epoxy
Tradename	FR-1025 (Ameribrom- Bromine Compounds)	PO-64P (Great Lakes Chemicals)	Pyrocheck 68 PB (Ferro)*	F 2400 (Makhteshim)
Appearance	White free flowing powder	Light brown powder	Off white pellets or powder	Yellow pellets
Molecular weight	30,000 - 80,000	6000	1000 - 12000	40,000
Bromine Content %	70.5	62	67	52
Specific Gravity, $g \cdot cm^{-3}$	2.05	2.25	2.10	1.81
Softening range, °C	205-215	210-240	215-225	140-160
Weight loss °C in air 5%	338	400	374	-
10%	339	460	-	375

*See also J.C. Gill - Proceedings, Fire Retardant Chemicals Ass. San Antonio, Texas, March 1989 pp. 17-33.

phosphorus flame retardants were based on a simple candle model of
polymer combustion proposed by Fenimore and Martin (25). This model
assumes no interaction of the polymer melt with the oxidant, and sufficient energy returned from the flame by radiation and conduction to
pyrolytically produce fuel gases. The oxygen index of polymer
treated with the organic phosphorus compounds increased from an initial value of 22 to a maximum of 34, depending on phosphorus content.
The experimental evidence showed that all species, with the possible
exception of red phosphorus, were active as gas flame poisons and not
in the condensed phase.

New Flame Retardants for Polycarbonates

Highly effective aromatic sulfonate salt fire retardant additives for
polycarbonate resins have been discovered by V. Mark at General Electric. Mark investigated a large number of inorganic and organic
salts of aromatic sulfonic acids (26) and found that flame retardant
effectiveness was attained at extremely low concentrations of the additive, such that the physical and mechanical properties of the polycarbonate were not impaired. The oxygen index of the resin increased
from a value of 25 to 35 (e.g. with 0.1% of sodium or potassium
2,4,5 trichlorobenzene sulfonate). UL-94 ratings of V-0 were obtained and commercial utility was demonstrated (e.g. Lexan 940 polycarbonate resin - General Electric) (26). The mode of action of the
salts has been studied for specific compounds (27) and, more recently, for polycarbonates of different structures and source (28).
Evidence was presented (28) that aromatic sulfonates present in catalytic amounts are capable of increasing the thermal degradation rate
of polycarbonate and promoting the formation of a carbon layer at the
burning surface. Thus the heat-insulating properties of the expanded
char (intumescence) are responsible for the observed flame resistance
of polycarbonates. Additional examples of flame retardant systems
that are effective in polycarbonates at very low concentration are
documented in the patent literature. Salts of carboxylic acids have
been claimed (29) and several systems are covered in patents issued
to Dow Chemical (30). These are briefly summarized below in Table
III.

Table III.   Selected Dow Chemical Patents on Flame
Retardants for Polycarbonates

Patent No. (Year)	Additive	Amount (%)	Oxygen Index ($\%O_2$)
4,254,015 (1981)	Metal salt of aromatic sulfonamide	0.05-0.2	29-40
4,335,038 (1982)	Metal perfluoroborate (+ Organosilane)	0.01-0.03 (0.01-0.05)	27-38
4,366,283 (1982)	Perfluorotitanate (metalate)	0.01-0.10	29-41
4,486,560 (1984)	Metal salt of aromatic sulfonamide (+ metal salt of perfluorometal- ate + halogenated organic compounds)	Total 0.01- 1.0	26-39

Extraordinary effectiveness has also been claimed for iron halides in polyphenylene oxide/polystyrene blends (40/60 to 60/40 (31), but much of the work on low concentration additives to date has been devoted to polycarbonates.

An example of a direct comparison of performance properties with and without added flame retardants for "New ignition resistant polycarbonate resins" is provided in a paper by workers at the Dow Chemical Company (32). The technology of the flame retardant additives is described as including

1. Organic alkali metal salts of the

$$\text{structure Ar} -- \overset{\overset{\text{O}}{\|}}{\underset{\overset{\|}{\text{O}}}{\text{S}}} -- \overset{-}{\underset{\overset{\|}{\text{R}}}{\text{N}}} \ \ \text{M}^{+}$$

   M = alkali metal
   Ar = aromatic group
   R = substituted aromatic group
2. A polymeric organobromine compound
3. Optionally, polytetrafluorethylene (PTFE) (to inhibit dripping)

Total additive loadings rarely exceed 1% of the finished resin. The criteria for ignition resistance are based on oxygen index and UL-94 ratings. The mode of action of the flame retardants is reported to be consistent with that of aromatic sulfonates as proposed by Webb (27).

Illustrative performance properties for a "general purpose polycarbonate," and for the same resin modified with the additive formulations "700" (without PTFE) and "800" (with PTFE) are summarized in Table IV (adapted from reference 32). It is clear that the objective of minimal effect on performance properties has been attained for this system. It is evident that flame retardant effectiveness attained with minimal levels of additive can provide optimum solutions to the problem of decreasing flammability without sacrifice in performance properties. Work documented to date suggests that in depth studies of thermal degradation such as reported for aromatic sulfonates in polycarbonates (28) would be rewarding for other systems.

Conclusions

In conclusion, the predominantly aromatic structure of engineering thermoplastics developed as thermally stable materials for advanced applications is such that flame resistance has not generally been a major concern. In some instances (notably in PPO/polystyrene blends and in polycarbonates), flame retardant grades have been developed to meet the requirements of specific applications. Approaches have ranged from those based on the traditional knowledge of flame retardant effectiveness (e.g. oligomers of high bromine content) to the exploration of new concepts (e.g. aromatic sulfonates). Recent work suggests that complex formulations merging several compounds and modes of action may attain the desired objective of flame resistance at additive concentrations which are sufficiently low to avoid change in the essential performance properties of the resins.

Table IV. Properties of Flame Resistant Polycarbonates

PROPERTY	TEST PROCEDURE	TEST DATA RANGE		
		WITHOUT PTFE	WITH PTFE	GENERAL PURPOSE PC
TENSILE STRENGTH (MPa)	ASTM D-638			
YIELD		55.5-65.5	55.1-62.0	56.5-65.5
BREAK		55.1-72.3	48.2-68.9	55.1-72.3
ELONGATION BREAK (%)	ASTM D-638	95 – 135	80 – 120	95 – 135
TENSILE MODULUS (MPa)	ASTM D-638	1929-2412	1929-2412	1929-2412
FLEXURAL MODULUS (MPa)	ASTM D-790	2067-2549	2067-2549	2067-2549
DEFLECTION TEMPERATURE (°C) UNANNEALED @ 1.82 MPa	ASTM D-648	123-133	123-133	123-133
GLASS TRANSITION (°C)	DSC (10°C/MIN)	151-153	151-153	151-153
DIELECTRIC CONSTANT @ 10^5 Hz	ASTM D-150	2.95	2.95	2.95
OXYGEN INDEX (% O_2)	ASTM D-2863	36-41	36-41	25-27
UL STANDARD 94 VERTICAL BURN TEST				
1/16" RATING	UL 94	V-2	V-0	V-2
1/8" RATING		V-0	V-0/5V	V-2

Literature Cited

1. Fox, D. W.; Peters,E.N. In Applied Polymer Science, second edition ACS Symposium Series No. 285; 1985; pp 495-514.
2. Margolis, J. M., editor. Engineering Thermoplastics, Properties and Applications; Marcel Dekker: New York, 1985.
3. Arnold, C. J. of Polymer Science, Macromolecular Reviews. 1979, 14, 265-378.
4. Lanrock, A. H. Handbook of Plastics Flammability and Combustion Toxicology; Noyes Publications, Park Ridge, NJ, 1983; pp 57-63.
5. Johnston, N. W.; Joesten, B. L. J. of Fire and Flammability 1972, 3, 154-164.
6. Kourtides, D. A.; Parker, J. A. Polymer Engineering and Science, 1978, 18 (11), 855-860.
7. Hilado, C. Flammability Handbook of Plastics; 1982; pp 38-47.
8. Chang, I. Q. SAMPE Quarterly 1988 (July), 29-30.
9. Publication NMAB 318-2. Fire Safety Aspects of Polymeric Materials, Vol. 2 - Test Methods, Specifications and Standards; National Academy of Sciences: Washington, D. C., Technomic Publishing Co., 1979.
10(a). Babrauskas, V. New Technology to Reduce Fire Losses and Costs; Grayson, S. J., Smith, D. A., Eds.; Elsevier, 1986; pp 78-86.
10(b).Bassett, W. Proc. of the Fire Retardant Chem. Assoc., 1989, pp. 143-153.
11. Shue, R. S. Proc. of the 33rd Intl. SAMPE Symposium, 1988, pp. 626-633.
12. Theberge, J. E.; Crosby, J. M.; Talley, K. L. Plastics Engineering, 1988 (August), 47-52.
13. Norris, J. M.; et al. J. of Fire and Flammability-Combustion Toxicology 1974, 1, 52-77.
14. Mischutin, V. Am. Dyestuff Reporter 1977 (November), 51-56.
15. Brauman, S.; Chen, I. J. J. of Fire Retardant Chem. 1981, 8, 28-36.
16. Day, M.; et al. J. of Applied Polymer Science 1988, 35, 529-535.
17. Day, M.; Cooney, J. D.; Wiles, D. M. J. Thermal Analysis 1988, 34, 733-747.
18. Burleigh, P. H.; et al. J. of Fire Retardant Chem. 1980, 7, 47-57.
19(a).Burditt, N. A. Proc. Fire Retardant Chem. Assoc., 1987, pp 207-218.
19(b).U.S. Patent 4 352 909, Ferro Corp.
    U.S. Patent 4 200 703, Huels Corp.
    U.S. Patent 3 975 354, Ciba-Geigy Corp.
20. U.S. Patent 4 732 921, Hoechst Celanese, 1988.
21. Siegmann, A.; et al. Proc. RP/C40th Annual Conf. SPI Session 8E, 1985, pp 1-12.
22. Great Lakes Chem. Corp.-Product Information on Flame Retardant Chemicals, 1983.
23(a).U.S. Patent 4 764 550, Penn Walt Corp.
23(b).Braksmeyer, D. P.; Bohen, J. M.; Duffin, R. J. Proc. of the Fire Retardant Chemicals Assoc., 1989, pp 49-68.
24. Carnahan, J.; et al. Proc. 4th Intl. Conf. on Flammability and Fire Safety, 1979, pp 312-323.

25. Fenimore, C. P.; Martin, F. J. <u>Combustion and Flame</u> 1969, <u>13</u>, 495.
26. Mark, V. <u>Organic Coatings and Plastics ACS</u> 1980, <u>43</u>, 71-78.
27. Webb, J. L.; Cipullo, M. J.; Louie, M. L. <u>Organic Coatings and Plastics ACS</u> 1980, <u>43</u>, 79-83.
28. Ballisteri, A.; et al. <u>J. of Polymer Science Part A</u> 1988, <u>26</u>, 2113-27.
29. U.S. Patent 4 757 103, Anic SPA.
30. U.S. Patents (a) 4 254 015, 1981
                 (b) 4 335 038, 1982
                 (c) 4 366 283, 1982
                 (d) 4 486 560, 1984, Dow Chemical Co.
31. U.S. Patent 4 757 107, General Electric.
32. Thomas, L. S.; Ogoe, S. A. <u>Proc. of the Fire Retardant Chem. Assoc.</u>, 1985, pp 179-192.

RECEIVED November 20, 1989

Chapter 17

Brominated Phosphate Ester Flame Retardants for Engineering Thermoplastics

Joseph Green

FMC Corporation, P.O. Box 8, Princeton, NJ 08543

A brominated triaryl phosphate ester was shown to be a
highly efficient flame retardant additive for
engineering plastics such as polycarbonate,
polybutylene terephthalate, polyethylene terephthalate,
and various alloys including polycarbonate/ABS. In
some resins antimony oxide or sodium antimonate is not
required or desirable. The brominated phosphate has
excellent thermal stability and also does not discolor
at high processing temperature. Processing studies
show the flame retardant disperses readily and aids
processability. Flammability data, mechanical
properties, impact, and HDT are reported for various
resins and compared with resins containing commercial
bromine-containing flame retardants.

The combustion of gaseous fuels is believed to proceed via a free
radical mechanism ($\underline{1}$, $\underline{2}$). A number of propagating and chain-
branching reactions are critical for maintaining the combustion
process. Some of these reactions are illustrated below:

$$CH_4 + O_2 \longrightarrow CH_3 + H + O_2$$

$$H + O_2 \rightleftharpoons OH + O$$

$$CH_4 + OH \longrightarrow CH_3 + H_2O$$

$$CH_3 + O \longrightarrow CH_2O + H$$

$$CH_2O + OH \longrightarrow CHO + H_2O$$

$$HCO + O_2 \longrightarrow H + CO + O_2$$

$$CO + OH \longrightarrow CO_2 + H$$

Here H, OH and O radicals are chain carriers, and the reaction of H radical with O_2 is an example of chain branching in which the number of carriers is increased. The reaction of CO with OH radical converting CO to CO_2 is a particularly exothermic reaction.

The function of halogen-containing compounds as flame retardants has been explained by the radical trap theory. Liberated halogen acid (HX) competes in the above reactions for those radical species that are critical for flame propagation.

$$CH_4 + X \longrightarrow HX + CH_3$$

$$H + HX \longrightarrow H_2 + X$$

$$OH + HX \longrightarrow H_2O + X$$

$$O + HX \longrightarrow OH + X$$

The active chain carriers are replaced with the much less active halogen radical slowing the rate of energy production and helping flame extinguishment.

Antimony oxide is known as a flame retardant synergist when used in combination with halogen compounds. Volatile antimony oxyhalide (SbOX) and/or antimony trihalide (SbX_3) are formed in the condensed phase and transport the halogen into the gas phase (3). It has been suggested that antimony is also a highly active radical trap (4).

The flame retardant mechanism for phosphorus compounds varies with the phosphorus compound, the polymer and the combustion conditions (5). For example, some phosphorus compounds decompose to phosphoric acids and polyphosphates. A viscous surface glass forms and shields the polymer from the flame. If the phosphoric acid reacts with the polymer, e.g., to form a phosphate ester with subsequent decomposition, a dense surface char may form. These coatings serve as a physical barrier to heat transfer from the flame to the polymer and to diffusion of gases; in other words, fuel (the polymer) is isolated from heat and oxygen.

Triaryl phosphate esters are thermally stable, high-boiling (>350°C) materials. They can volatilize without significant decomposition into the flame zone, where they decompose. Flame inhibition reactions, similar to the halogen radical trap theory, have been proposed (6):

$$H_3PO_4 \longrightarrow HPO_2 + PO + etc.$$

$$H + PO \longrightarrow HPO$$

$$H + HPO \longrightarrow H_2 + PO$$

$$OH + PO \longrightarrow HPO + O$$

In modified polyphenylene oxide, the preponderance of evidence suggests that the phosphate ester flame retardant functions mainly in the gas phase (7) as a radical trap to help quench the flame.

The patent literature contains many claims for halogen/phosphorus synergy. A careful examination of the data does not support these

claims. Nevertheless, the additive effect with aliphatic chlorine
results in a flame retardant system of considerable commercial
significance for urethane polymers.

Commercially available flame retardants include chlorine- and
bromine-containing compounds, phosphate esters, and chloroalkyl
phosphates. Recent entry into the market place is a blend of an
aromatic bromine compound and a phosphate ester (DE-60F Special) for
use in flexible polyurethane foam (8). This paper describes the use
of a brominated aromatic phosphate ester, where the bromine and
phosphorus are in the same molecule, in high temperature
thermoplastic applications.

We previously reported that brominated aromatic phosphate esters
are highly effective flame retardants for polymers containing oxygen
such as polycarbonates and polyesters (9). Data were reported for
use of this phosphate ester in polycarbonates, polyesters and
blends. In some polymer systems, antimony oxide or sodium
antimonate could be deleted. This paper is a continuation of that
work and expands into polycarbonate alloys with polybutylene
terephthalate (PBT), polyethylene terephthalate (PET) and
acrylonitrile-butadiene-styrene (ABS).

Description of Flame Retardants

Three flame retardants were compared in this study, namely, a
brominated polycarbonate oligomer (58% bromine), a brominated
polystyrene (68% bromine), and a brominated triaryl phosphate ester
(60% bromine plus 4% phosphorus). These are described in Table I.
Figures 1 and 2 compare the thermal stability of the brominated
phosphate with commercial bromine-containing flame retardants by
thermogravimetric analysis (TGA) and by differential scanning
calorimetry (DSC). The brominated phosphate melts at 110°C and
shows a 1% weight loss at 300°C. Brominated polycarbonate and
brominated polystyrene are polymeric and are not as volatile at
elevated temperatures as the monomeric flame retardants.

Brominated phosphate heated in a glass tube in air at 300°C for
30 minutes remains a water-white liquid. This was compared with
commercial bromine-containing flame retardants which melt; they all
turn color. The excellent color stability of this brominated
phosphate ester makes it suitable for the high temperature
processing of engineering plastics.

The solubilities of the flame retardants in toluene are shown in
Table I. It is believed that the high solubility of the phosphate
in an aromatic solvent accounts in part for the ease of compounding
into various aromatic resins. This is discussed further in the
section on compounding.

Flame Retardant Polycarbonates

The brominated phosphate is an efficient flame retardant for
polycarbonate resin. UL-94 ratings of V-0 with oxygen index values
of greater than 40 are obtained. Polycarbonate resin containing
brominated phosphate processes with greater ease than resin
containing brominated polycarbonate as measured by injection molding
spiral flow measurements. The heat distortion temperature is reduced

and the high Gardner impacts are retained. The resultant products
are transparent and water-white (Table II).

Table I. Flame Retardant Description

Description	Trade Name	% Br	% P	M.P. °C	Toluene Solubility g/100g
Brominated polycarbonate oligomer	BC-58	58	-	230-260	7
Brominated polystyrene	68PB	68	-	215-225[1]	0.2
Brominated aromatic phosphate ester	PB-460	60	4	110	25

[1]Softening Point

Table II. Flame Retarding Polycarbonate Resin

Polycarbonate Resin	93	93
Brominated polycarbonate	7	-
Brominated phosphate	-	7
Oxygen Index	32.1	>39.6
UL-94, rating (1/16")	V-0	V-0
time, sec.	3.8	2.4
Heat Distortion Temp. @ 264 psi, °F	261	239
Gardner Impact, in. lbs.	>320	>320
Spiral Flow, Inj. Molding, in.	23.5	29.5
	transparent, water-white	

Flame retardant sulfonate salt polycarbonate resin gives a UL-94
rating of V-0 at 1/16 inch thickness. At 1/32 inch thickness,
however, the product drips to give a V-2 rating. The addition of 3%
brominated phosphate renders the product V-0 at 1/32 inch thick
(Table III).

Flame Retardant Polybutylene Terephthalate (PBT)

Mineral filled PBT polyester resin containing 12% brominated
phosphate and 4% antimony oxide yields a V-0 product with a 29.7
oxygen index. A product containing 16% brominated phosphate and

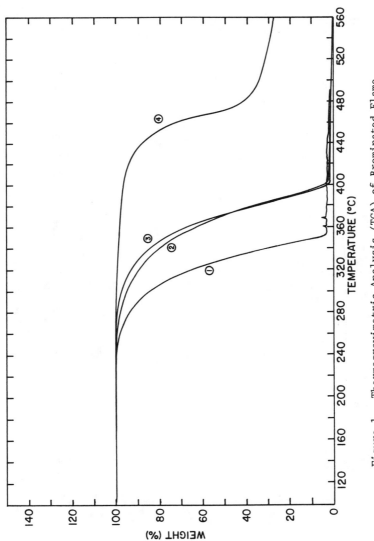

Figure 1. Thermogravimetric Analysis (TGA) of Brominated Flame Retardants - 10°C/min. under nitrogen
1. bis-(tribromophenoxy)ethane; 2. octabromodiphenyl oxide; 3. brominated aromatic phosphate ester; 4. brominated polycarbonate oligomer

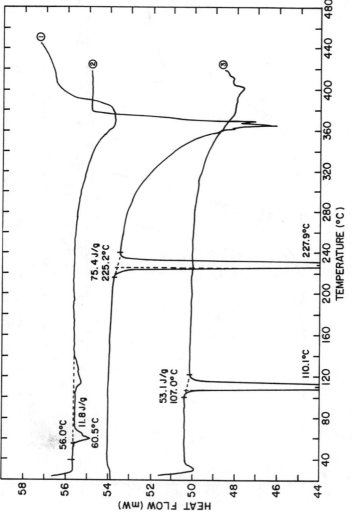

Figure 2. Differential Scanning Calorimetry (DSC) of Brominated
Flame Retardants - 10°C/min under nitrogen

1. octabromodiphenyl oxide; 2. bis-(tribromophenoxy)ethane;
3. brominated aromatic phosphate ester

no antimony oxide is also V-O with a 31.2 oxygen index. The use of
brominated polycarbonate requires the use of antimony oxide as a
synergist. This shows phosphorus to be highly effective as a flame
retardant in PBT (Table IV). The advantages of eliminating antimony
oxide are numerous. The lower oxygen index values for the systems
containing both phosphorus and halogen have been reported (9).

Table III. Flame Retardant Polycarbonate Resin

FR Polycarbonate (sulfonate salt)	100	99	99
Brominated phosphate	-	1	3
Oxygen Index	33.6	34.8	37.5
UL-94			
@ 1/16" rating	V-O	V-O	V-O
sec.	1.2	1.9	0.9
@ 1/32" rating	V-2	V-2	V-O
sec.	4.7	2.4	1.4
Melt Index, g/10 min. (250°C)	7.0	-	9.1

Table IV. Flame Retarding Mineral-Filled PBT

PBT Mineral Filled	84	84	84	84
Brominated polycarbonate	12	-	16	-
Brominated phosphate	-	12	-	16
Antimony Oxide	4	4	-	-
Oxygen Index	31.8	29.7	29.1	31.2
UL-94, Rating (1/16")	V-O	V-O	B	V-O
Time, sec.	0	3.7	-	3.1

With 30% glass filled PBT, brominated phosphate requires the use
of antimony oxide. A drip inhibitor was used in these studies and
as little as 0.3% Teflon 6C fibrous powder is adequate to inhibit
dripping. As little as 10% brominated phosphate will give a V-O
product (Table V).

The three flame retardants are compared in Table VI. Brominated
phosphate disperses readily in the resin presumably due to its high
solubility in aromatics. Resin containing brominated polycarbonate
is relatively difficult to process as measured by injection molding
spiral flow measurements.

Properties of the resins are similar with the brominated
phosphate containing resin showing a slightly lower heat distortion

temperature and a slightly higher Izod impact.   The flex modulus is
lowest for the brominated phosphate containing resin.

Flame Retardant Polyethylene Terephthalate (PET)

Table VII shows that sodium antimonate is antagonistic with the
phosphorus/bromine compound in 30% glass filled PET polyester resin.

Table V.   Flame Retarding Glass Filled PBT

PBT/30% Glass	86.5	82.5	86.5	82.5
Brominated Polystyrene	10	14	-	-
Brominated Phosphate	-	-	10	14
Antimony Oxide	3.5	3.5	3.5	3.5
Teflon 6C	0.5	0.5	0.5	0.5
Oxygen Index	28.2	30.6	27.9	30.0
UL-94, Rating (1/16")	V-O	V-O	V-O	V-O
Sec.	1.4	0.2	2.4	1.6

Table VI.   Flame Retardant PBT/30% Glass

PBT/30% Glass	82.5	82.5	82.5
Antimony Oxide	3.5	3.5	3.5
Teflon 6C	0.3	0.3	0.3
Brominated Phosphate	14	-	-
Brominated Polystyrene	-	14	-
Brominated Polycarbonate	-	-	14
UL-94, rating (1/16")	V-O	V-O	V-O
sec.	0.9	0	0.1
Oxygen Index	29.7	32.7	33.0
HDT, 264 psi, °C	197	204	200
Izod Impact (1/8")	1.3	1.0	1.1
Spiral Flow, in.*	47	46	37
Melt Index, g/10 min. (250°C)	13.6	13.0	12.5
Flex Strength, psi	24,600	22,800	26,500
Flex Modulus x 10^6 psi	0.83	1.13	1.15

*Control (no FR) gives 48 inches flow

V-O products with high oxygen index values are obtainable using brominated phosphate alone. PET fibers can be readily spun and the resulting fiber is white.

Flame Retardant Polycarbonate/PBT Polyester Blend

A 50/50 blend of polycarbonate resin and PBT polyester containing 13.5% brominated phosphate and no antimony oxide results in a product with a V-O rating and an oxygen index of 33. An equivalent product containing brominated polycarbonate has a low oxygen index and burns in the UL-94 test (Table VIII).

Various blend ratios of polycarbonate and PBT polyester were flame retarded with the three flame retardants. These data are shown graphically in Figure 3. Brominated phosphate is the most efficient and brominated polycarbonate the least efficient flame retardant. At a 50/50 ratio of polycarbonate/PBT, brominated phosphate is significantly more effective than brominated polystyrene.

Flame Retardant Polycarbonate/PET Polyester Alloy

The flame retardant performance of the three flame retardants in a commercial polycarbonate/PET polyester alloy were compared. Brominated phosphate is a very efficient flame retardant as measured by oxygen index and UL-94 (Table IX and Figure 4). The melt index of the resin does not change with the addition of brominated polycarbonate, doubles with brominated polystyrene, and doubles again with brominated phosphate (Table IX).

Brominated phosphate was also evaluated in a commercial glass filled polycarbonate/PET polyester alloy. A concentration of 10% gives a V-O rating with an oxygen index value of about 35 (Table X).

Flame Retardant Polycarbonate/ABS Alloy

The flame retardant performance of various flame retardant additives in a commercial polycarbonate/ABS alloy were compared. No antimony oxide was required. The data shows brominated phosphate to be a highly efficient flame retardant in this alloy (Table XI). An alloy composition containing 14% brominated phosphate and no antimony oxide gives a V-O rating (Table XII). The melt index of this alloy containing 12% brominated polystyrene was 7.6 g/10 min. (at 250°C); the equivalent resin containing brominated phosphate had a melt index of 13.3 g/10 min.

Compounding Characteristics

It was observed that brominated phosphate blends easily into various resins in a single or twin screw extruder. Compounding rates also are increased. It has been assumed that this is partly due to its high degree of solubility in aromatic solvent. This is in contrast with the polymeric flame retardants which are more difficult to incorporate or compound into various resins.

A study was conducted in a Brabender Plastic-Corder. Brominated phosphate was compared with the polymeric flame retardants brominated polycarbonate and brominated polystyrene. Various resins with and without glass were used and the temperature adjusted for the resin.

When the polymeric flame retardants are added in increments, the viscosity increases to a point and then decreases and plateaus as the original viscosity or at a higher viscosity. When brominated phosphate is added in increments, the viscosity decreases immediately and then plateaus, at a lower viscosity than the starting viscosity.

Table VII.　Flame Retardant PET/30% Glass Resin

PET/30% Glass	82	80	80
Brominated Phosphate	18	20	15
Sodium Antimonate	-	-	5
Teflon 6C	0.5	0.5	0.5
Oxygen Index	30.6	36.0	29.4
UL-94, rating (1/16")	V-2	V-0	V-2
sec.	1.2	0.1	5.1

Table VIII.　Flame Retarding Polycarbonate/PBT Blends
(No Antimony)

Polycarbonate	43	43
PBT Polyester	43	43
Teflon Powder	0.5	0.5
Brominated Polystyrene	13.5	-
Brominated Phosphate	-	13.5
Oxygen Index	24.8	33.0
UL-94, rating (1/16")	Burn	V-0
time, sec.		0.9

Table IX.　Flame Retarding Polycarbonate/PET Polyester Alloy

Polycarbonate/PET Alloy	90	86	90	86	90	86
Brominated Polycarbonate	10	12	-	-	-	-
Brominated Polystyrene	-	-	10	12	-	-
Brominated Phosphate	-	-	-	-	10	12
Teflon 6C	0.5	0.5	0.5	0.5	0.5	0.5
Oxygen Index	27.3	27.9	30.6	30.9	31.2	35.4
UL-94, Rating (1/16")	V-1	V-1	V-0	V-0	V-0	V-0
sec.	11.3	6.1	1.0	2.7	4.9	2.3
Melt Index, g/10 min.* (275°C)	-	20	-	37	-	71

*Virgin resin - 24 g/10 min.

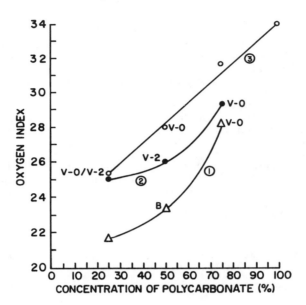

Figure 3. Flame Retarding Polycarbonate/PBT Polyester Blends (12% Flame Retardant)
1. brominated polycarbonate oligomer; 2. brominated polystyrene; 3. brominated aromatic phosphate ester

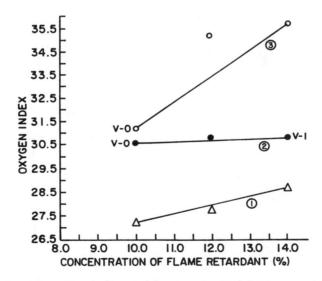

Figure 4. Flame Retarding Polycarbonate/PET Polyester Alloy
1. brominated polycarbonate oligomer; 2. brominated polystyrene; 3. brominated aromatic phosphate ester

Conclusions

A brominated triaryl phosphate ester was shown to have excellent
thermal stability and excellent retention of color after high
temperature processing. It is a highly efficient flame retardant in
engineering thermoplastics and alloys such as polycarbonates, PBT
polyester, PET polyester, and polycarbonate alloys with PBT, PET,
and ABS. With many polymers antimony oxide or sodium antimonate is
not needed presumably due to the presence of the highly effective
phosphorus. Brabender Plasti-Corder studies show the brominated
triaryl phosphate ester to blend easily into the aromatic resins,
presumably due to its high solubility in aromatic solvent, unlike
polymeric flame retardants which can be difficult to blend into the
engineering thermoplastics. This confirms the ease of compounding
observed in a single or twin screw extruder. Increased melt index
and greater spiral flow when brominated triaryl phosphate ester is
used confirms the ease of injection molding observed.

Table X.   Flame Retarding Polycarbonate/PET Polyester
Alloy - Glass Filled

Polycarbonate/PET Alloy (20% Glass)	90	86	82
Brominated Phosphate	10	14	18
Teflon 6C	0.5	0.5	0.5
Oxygen Index	34.8	36.6	>39.6
UL-94, rating (1/16")	V-O	V-O	V-O
sec.	1.9	2.5	0

44/44 PC/30% glass filled PET plus 12% Brominated Phosphate gives
V-O/0.9 sec.and 36.9 O.I.

Table XI.   Flame Retarding Polycarbonate/ABS Alloy

Polycarbonate/ABS Alloy	82.5	82.5	82.5	82.5
Teflon 6C	0.5	0.5	0.5	0.5
Brominated Polycarbonate	17.5	-	-	-
bis-(tribromophenoxy)ethane	-	17.5	-	-
Brominated Polystyrene	-	-	17.5	-
Brominated Phosphate	-	-	-	17.5
Oxygen Index	26.4	27.0	27.0	28.2
UL-94, Rating (1/16")	V-1	V-O	V-O	V-O
sec.	22	2.0	2.6	1.4

Table XII.   Flame Retarding Polycarbonate/ABS Alloy

Polycarbonate/ABS Alloy	86	86	86
Brominated Polystyrene	10	-	-
Brominated Phosphate	-	10	14
Antimony Oxide	4	4	-
Oxygen Index	27.0	26.1	25.8
UL-94, Rating (1/16") sec.	V-O 0.8	V-O 0.1	V-O 1.3

Acknowledgments

I wish to acknowledge the excellent contributions of Charles A.
Tennesen for compounding and testing, Ray Skok for flammability
testing, and John Jessup for mechanical property testing.

Literature Cited

1. Avondo, G.; Vorelle, C.; Debourgo, R. Combust. Flame, 1978; 31,
    7.
2. Williams, F. A. Encyclopedia of Physical Science and Tech.,
    1987, 3, 211.
3. Pitts, J. J.; Scott, P. H.; Powell, D. G. J. Cell Plastics,
    1970, 6, 35.
4. Brauman, S. K.; Brolly, A.S. J. Fire Retardation Chem., 1976,
    3, 66.
5. Green, J. Plastics Cpding, 1987, 10, No. 3, 57.
6. Hastie, J. W. J. Res. NBS, 1973, No. 6, 733.
7. Carnahan, J.; Haaf, W.; Nelson, G.; Lee, G.; Abolins, V.; Shank,
    R. Fourth Int. Conf. on Flammability and Safety, San Francisco,
    Jan. 1979.
8. Green, J. U.S. Patent 4 746 682, 1988.
9. Green, J. Proc. Fire Retardant Chemicals Assoc. Mtg., Grenelefe,
    Fl. March, 1988.

RECEIVED November 20, 1989

Chapter 18

Polymers with Improved Flammability Characteristics

W. T. Whang and E. M. Pearce

Polymer Research Institute, Polytechnic University, Brooklyn, NY 11201

The flame resistance of polymeric materials was
enhanced with the modification of chemical structure
and the incorporation of additives. The polymers
with more fused heterocyclic structures showed higher
thermal stability and more char yield, thus - polybenzoxazole > poly(2,4-difluoro-1,5-phenylene trimellitic amide-imide) > poly(2,4-difluoro-1,5-phenyl-
ene isophthalamide). The poly(amide-imide) showed
good solubility in N,N-dimethyl acetamide and N,N-
dimethyl formamide with better processability than the
polyamide. Among the investigated additives zinc
chloride was the best additive to improve the flame
resistance of nonsubstituted poly(1,3-phenylene iso-
phthalmide) (PMI). The material system increase 40%
of the char yield and 5 units of the oxygen index
when compared with a pure PMI.

Usually the thermal stability of polymeric materials is related
to the chemical bond strength.[1] Thermal stability can be
enhanced with stronger bond strength which can be achieved using
resonance stabilization in condensed ring structures. Although
many publications have discussed structure relationships and
thermal stability, few of them have presented systematic
correlations. We were interested in further correlating
structure, thermal stability, and flame resistance of a series
polymers, which exhibited a similarity in the mechanism of
thermal degradation. When poly(2,4-difluoro-1,5-phenylene
trimellitic amide-imide) and poly(2,4-difluoro-1,5-isophthal-
amide) were pyrolyzed at 400 - 500°C, it was found that
benzoxazole groups were generated on the backbone of these two
polymers.[2,3] It is also very interesting that the onset of the
second-step decomposition of these two polymers was close to that
of polybenzoxazole.

0097–6156/90/0425–0266$06.00/0
© 1990 American Chemical Society

Experimental

Monomer Preparation and Solvent Preparation

1,3-phenylene diamine

1,3-phenylene diamine was purified by vacuum sublimation to essentially white solids. (m.p 62.5-63°C).

2,4-difluoro-1,5-phenylene diamine

This diamine was prepared from 2,4-dinitro-1,5-difluoro benzene by a catalytic reduction over $SnCl_2$/HCl at 50-60oC for 3 hours.[9] The diamine was recrystallized from benzene to yield a white product (mp. 112.5-113°C).

2.4-diamino-1,5-benzenediol dihydrochloride[10,11]

m-Phenylene diacetate was dissolved slowly in 4 to 5 times its volume of fuming nitric acid while cooling with an ice water bath for 2 or 3 minutes. The solution was poured onto cracked ice, filtered immediately, then washed with boiling alcohol. The yellow solid was saponified by boiling it in 30% hydrochloric acid solution for 30 minutes. Recrystallization from water gave yellow needles of 2,4-dinitro-1,5-benzenediol with mp. 212-213°C. 2,4-Dinitro-1,5-benzenediol was then reduced with stannous chloride in concentrated hydrochloric acid at 50-60°C. At the beginning, the solution was clear, then some white crystals precipitated form the reaction solution. After cooling, the resultant precipitate was collected by filtration and washed with cold water to give 2,4-diamino-1,5-benzenediol dihydrochloride-no mp. before it decomposed above 200°C.

Acid Chlorides

Isophthaloyl chloride was recrystallized from n-hexane, which was distilled over sodium wire.
Trimellitic anhydride acid chloride was purified by reduced pressure distillation.

N,N-Dimethylacetamide

Commercially available N,N-dimethylacetamide was purified by vacuum distillation over calcium hydride to clear, colorless liquids and then kept dry in a desiccator.

Preparation of Poly(phosphoric acid)

Poly(phosphoric acid) was prepared by adding a 1.52/1 weight ratio of phosphorus pentoxide to 85% phosphoric acid in ice bath and then heating at 150°C for 6 hours, with stirring under nitrogen atmosphere.

Polymerization

Poly(2,4-difluoro-1.5-phenylene isophthalamide) and poly(1.3-phenylene isophthalamide

The diamine was dissolved in N,N-dimethyl-acetamide by stirring under a nitrogen stream. After the diamine was completely dissolved, equal moles of isophthaloyl chloride was added all at once. Stirring was continued for about three hours after which the reaction mixture was poured into hot water and magnetically stirred. The polymer was washed with hot water at least three times and then extracted with acetone to remove low molecular weight species. The polymer was dried in vacuum oven at about 70°C overnight.

Poly(2,4-difluoro-1,5-phenylene trimellitic amide-imide

Poly(2,4-difluoro-1,5,phenylene trimellitic amide-imide) was prepared by a two-step procedure.[12] At the first step, the polyamic acid was prepared by reacting 2,4-difluoro-1,5-phenylene diamine with trimellitic anhydride acid chloride (with the mole ratio of one to one) in anhydrous N,N-dimethylacetamide at room temperature under nitrogen. After reaction, the polymer was poured into water and precipitated. After filtration, the white solid was washed with distilled water and dried in a vacuum oven. The poly(amide-imide) was obtained from heating the polyamic acid at 220°C for 3 hours. The polyamic acid was dissolved in N,N-dimethyl acetamide or N,N-dimethyl formamide, cast on glass plates, and the solvent evaporated in a vacuum oven to form a polyamic acid film before heating at 220°C.

Poly[benzo(1,2-d:5,4-d']-1,3-phenylene

Equimolar quantities of 2,4-diamino-1,5-benzenediol dihydrochloride and isophthalic acid were mixed in fresh poly (phosphoric acid) using a high-shear stirrer under a slow stream of nitrogen gas. The system was heated at 40°C for 6 hours, at 60°C for 18 hours, at 120°C for 6 hours, at 160°C for 8 hours, and at 220°C for 24 hours. The resultant mixture was dark brown. The polymer was precipitated from water. After filtration and washing with water and methanol, the solid product was then dissolved in methane-sulfonic acid, filtered and precipitated by the addition of methanol. The solid was washed with concentrated ammonium hydroxide, water, methanol, methanol/benzene mixtures (with a volume ratio of 1/1), and finally benzene. The final product was dark brown.

Characterization

The polymers were characterized by IR spectra in KBr pellets. A Digilab FTS-IMX Infrared Spectrometer was used for this purpose. The oxygen indices of the polymers were measured by using the General Electric oxygen index equipment. The oxygen index (OI) is defined as the minimum concentration of oxygen in an oxygen-nitrogen atmosphere that is necessary to support a flame.

$$OI = \frac{\text{volume of } O_2}{\text{volume of } O^2 + \text{volume of } N_2} \times 100$$

A DuPont 910 differential scanning calorimeter (DSC) and a DuPont 951 thermogravimetric analyzer (TGA) connected to a DuPont 1090 thermal analyzer were used to study the transition data, thermal stability, and char yield, respectively, for all the polymers. The DSC was run under a nitrogen stream at a flow rate of 80 c.c./min. and at a heating rate of 20°C/min..

Results and Discussions

The chemical structures of these polymers were characterized using FT-IR. Poly(1,3-phenylene isophthalamide) (PMI) and poly (2,4-difluoro-1,5-phenylene isophthalamide) (2,4-DIF-PMI) showed N-H stretching bands at 3400-3200 cm^{-1} and C==O stretching bands(amide I) at 1630-1650 cm^{-1}. Poly(2,4-difluoro-1,3-phenylene trimellitic amide-imide) (2,4-DIF-PMTAI) showed additional bands at 1740 and 1796 cm^{-1} corresponding to imide C==O stretching band at 1625 cm^{-1} and C-O-C stretching bands at 1255 and 1050 cm^{-1}.

As shown in Figure 1, the onset of decomposition temperature was in this order: polybenzoxazole > 2,4-DIF-PMTAI > 2,4-DIF-PMI. The char yield of these polymers followed the same sequence. It showed that the introduction of more fused heterocyclic rings into the backbone of the polymers enhanced the thermal stability and flame resistance of polymeric materials.

It was interesting that 2,4-DIF-PMTAI and 2,4-DIF-PMI showed a two-step decomposition. The onset of decomposition at the second step of these two polymers were close to each other. More interesting, they were also close to the onset of decomposition of polybenzoxazole. We have confirmed that the decomposition of these two polymers at 400-550°C resulted in the formation of benzoxazole units on the backbone.[16]

The high thermal stability and flame resistance of polybenzoxazole were due to the resonance stabilization of the aromatic and the heterocyclic structures, were contributed substantially to high bond strength.

The solubilities of 2,4-DIF-PMI and 2,4-DIF-PMTAI were fairly good in N,N-dimethyl acetamide and N,N-dimethyl foramide. The polybenzoxazole could not be dissolved in any organic solvent and concentrated sulfuric acid could dissolve it.

Four Lewis acids and a radical trapper were used as additives in PMI in order to alternate the degradation mechanisms to

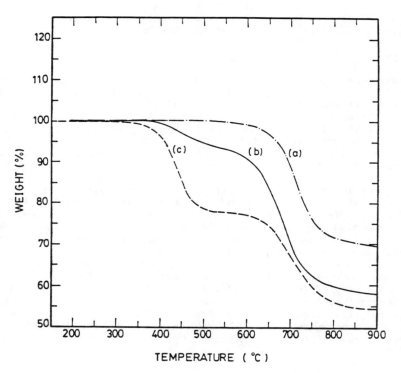

Figure 1.   TGA of (a) polybenzoxazole (— —·—·), (b) 2,4-DIF-
PMTAI (————) , and (c) 2,4-DIF-PMI(- - - --) at a heating rate of
20°C/min   under a nitrogen stream at 200 c.c./min.

favor the transformation of the polymer to char. The data on the thermal stability and flame resistance of PMI with and without 10% additives were shown in Table 1. The dynamic TGA thermograms of PMI with and without additives were shown in Figure 2.

Table 1. Effect of Additives in PMI on
Thermal Stability and Flammability

Additives 10%	IDT* (°C)	Residue of the pure additive at 900°C, (%)	Char Yield % 700°	800°	900°	900°C after Correction**	Oxygen Index
-	440	-	51.1	47.5	45.2	45.5	83
Clay	440	100	61.2	58.3	57.3	52.6	41
Charcoal	440	100	60.5	57.1	55.1	50.1	40
Al_2O_3	439	100	62.9	60.0	59.9	54.4	39
$ZnCl_2$	416	5	63.4	60.8	59.2	65.2	43
NiC12	422	16	52.6	50.0	48.2	52.4	39

* IDT Initial decomposition temperature.
** The char yield after correction for the inorganic additive.

The thermal stability of PMI with additives is not changed by the introduction of clay, charcoal, Al_2O_3. These additives were quite stable over the temperature range under study (up 900°C). The lower thermal stability observed for the cases of $ZnCl_2$ and NiC12 as additives may have resulted from a change in the degradation mechanism.

The char yield of 900°C, after correcting for the additives to give a corrected char yield, was calculated on the basis of the following equation:

$$\text{Char Yield (after correction)} = \frac{\text{Char Yield of PMI with Additives} - \text{Char Yield Contributed from Additives}}{\text{Weight Fraction of PMI before Degradation}}$$

The char yield contributed by the additives alone was evaluated by measuring the residue obtained from the pure additives by TGA.

Based on the char yield after correction, we found that all the additives improved the char yield of PMI. $ZnCl_2$ was 40% higher than that of pure PMI. The $ZnCl_2$ probably catalyzes the reaction of cyano groups with itself or with aromatic amide rings, but we have not experimentally studied this.

The other four additives gave 5-9% enhancement of char yields. Clay, NiC12 but were less efficient. The oxygen indices of PMI with these additives were higher than that of pure PMI, but differences were not significant.

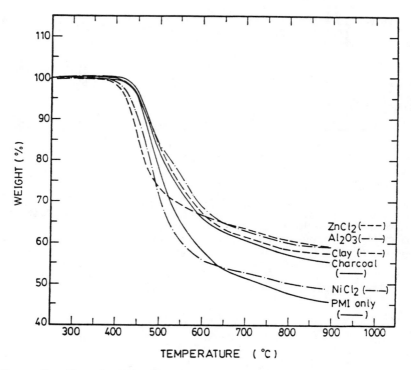

Figure 2. Dynamic TGA thermograms of 90% PMI with 10% additives:
1. ZnCl₂(– – –); 2. Al₂O₃(– • – • – •); 3. Clay(– – – –);
4. Charcoal(———); 5. NiCl2 (– • – •); 6. PMI(———) only at
a heating rate 20°C/min under a nitrogen stream at 200c.c./min.

Acknowledgment

The authors wish to express their thanks to the U.S. Army Research Office, Research Triangle Park, N.C. for sponsoring this work under the grant No. DAAG-29-79-C-0124.

Literature Cited

1. P. E. Cassidy, "Thermally Stable Polymers", Marcel Decker, N.Y. (1984.

2. W. T. Whang, M. Kapuscinska, and E. M. Pearce, J. Polym. Sci., Polym. Sym., 74, 109 (1986).

3. A. C. Karydas, W. T. Whang, and E. M. Pearce, J. Polym. Sci., Polym. Chem. Ed., 22, 847 (1984).

4. Y. P. Krasnov, V. I. Logunova, and S. B. Sokolov, Polym. Sci., USSR, 8, 2176 (1966).

5. G.F.L. Ehlers, K. P. Fisch, and W. R. Powell, J. Polym. Sci., A-1, 8, 3511 (1970).

6. H. L. Friedman, M. Goldstein, and G. A. Griffith, Thermal Anal., Proc. Int. Conf., 2nd, 1, 405 (1968).

7. D. A. Chatfield, I. N. Einhorn, R. W. Mickelson, and J. H. Futrell, J. Polym. Sci., Chem. Ed., 17, 1353 (1979).

8. D. A. Chatfield, I. N. Einhorn, R. W. Mickelson, and J. H. Futrell, J. Polym. Sci., Chem. Ed., 17, 1367 (1979).

9. B. H. Nicolet and W. J. Ray, J. Am. Chem. Soc., 49, 1801 (1927).

10. P. G. W. Typke, Ber. Dtsch. Chem. Ges., 16, 552 (1883).

11. J. F. Wolfe and F. E. Arnold, Macromolecules, 14, 909 (1981).

12. P. F. Frigerio, L. H. Tagle, and F. R. Diaz, Polymer, 22, 1571 (1981).

RECEIVED November 20, 1989

Chapter 19

Char Formation in Aromatic Engineering Polymers

A. Factor

Corporate Research and Development Center, General Electric Company, Schenectady, NY 12301

Since charring is an important route to flame retardancy, research has been done to better understand why different aromatic engineering polymers yield differing amounts of char upon combustion. Studies were made using thermogravimetric analysis, differential scanning calorimetry, electron spin resonance, infrared and Raman spectroscopies, x-ray scattering, solid state ^{13}C NMR and elemental analysis. These studies indicate that the polymers studied decomposed by a two step mechanism in which heating below ~ 550°C produces volatile tars and fuel gases and a primary char. Heating of this primary char above ~ 550°C produces a carbonaceous residue whose structure is virtually independent of the structure of the original polymer and is best described as a conglomerate of loosely linked small graphitic regions.

It is well known that char formation during polymer burning is an important mechanism by which polymers resist burning. For example Van Krevelen [1-2] has shown that a correlation exists between the oxygen index (OI) of a polymer containing no heteroelements and the amount of char it forms when pyrolyzed in the absence of air. As shown in Table I, a similar correlation is seen to hold for a number of common engineering thermoplastics.

However, while char formation is recognized as an important mode in achieving flame retardancy, little progress has been made in increasing the flame resistance of synthetic polymers by increasing char formation without significantly modifying the structure of the polymer. Towards this end, a careful study has been made of the char forming process for three aromatic

0097–6156/90/0425–0274$06.00/0

Table I. % Char vs. OI

	OI	% Char*
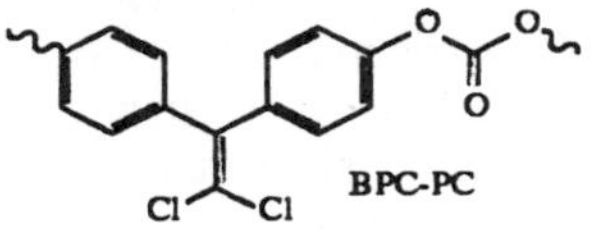 BPC-PC	56	54%
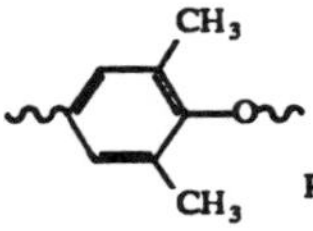 PPE	31	29%
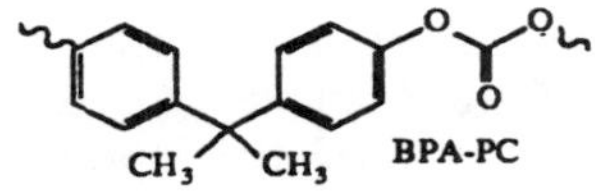 BPA-PC	27	26%
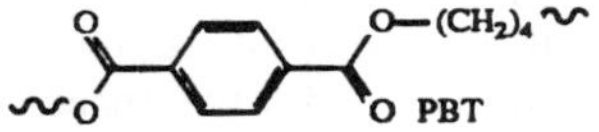 PBT	23	3%
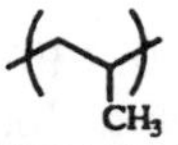 PP	17	0%

* % Residue in TGA in nitrogen at 700°C.

engineering polymers with varying degrees of charring tendencies (cf. Table I), namely, BPA polycarbonate (BPA-PC), the polycarbonate from 1,1-dichloro-2,2-bis(4-hydroxyphenyl)ethylene (BPC-PC) (3), and poly(2,6-dimethyl-phenylene oxide) (PPE).

Experimental

Materials. The following resin samples were utilized in this study: BPA-PC, Lexan® 140 resin (Lexan is a registered trademark of the General Electric Co.), $[\eta] = 0.50$ dl/g; BPC-PC, $[\eta] = 0.38$ dl/g; and PPE, $[\eta] = 0.50$ dl/g.

Char Preparation. Chars were prepared both by isothermal pyrolysis of 5 g samples of resin in a quartz boat heated in an atmosphere of flowing (0.5 SCFM) N_2 in a quartz tube oven (N_2 pyrolysis chars) and by open combustion of 1 g samples of resin exposed to 2.8 watts/cm² of radiant energy from an electric heating panel (4-5) (combustion or burn chars). All chars were finely ground with a glass mortar and pestle prior to analysis.

Analytical Methods. Photoacoustic (PA) FTIR spectra were run on a Nicolet 7199 system equipped with a EG&G Photoacoustic Cell to obtain high resolution spectra of dark chars (15). Raman IR spectra were obtained using SPEX Ramalog-10 Spectrometer. TGA's were run on a DuPont 9900 Thermal Analysis System scanning at 10°C/min in a N_2 atmosphere. DSC's were determined using a DuPont 912 Dual Cell Differential Scanning Calorimeter scanning at 20°C/min in a N_2 atmosphere. Solid state ^{13}C NMR spectra (CPMAS) were obtained using a GE GN-300 NMR spectrometer at 75.4 MHz using a spin rate ~ 3.5 kHz and a cross polarization contact time of 1 ms. ESR were determined with a JEOL MEX-1 X-band electron spin resonance spectrometer with 100 kHz field modulation using a Mn standard.

Results and Discussion

TGA and DSC Studies. The importance of oxidation in the production of fuel gases in polymer burning is currently a controversial question (6). In contrast, the effect of oxygen on char is better defined. Martel (7), for example, reports that while char formation for styrene polymers and aliphatic polyolefins is promoted by air, it had only a small effect on char formation in polymers containing aromatic rings in their main chain. Accordingly, in this study all thermoanalyses were run under an inert N_2 atmosphere. The TGA's of the three polymers of interest (Figure 1 and Table II) indicate that decomposition occurs between 400-550°C. The results of an analogous DSC study are shown in Table II and indicate that while the decomposition of BPC-PC and PPE are quite exothermic, that of BPA-PC is at best only mildly exothermic. Indeed the high exothermicity of BPC-PC explains a "zippering" effect observed during its burning with the radiant combustion apparatus used in this study in which ignition is observed to start on one side of a sample and move rapidly across the sample as a straight burn front across the sample's top surface.

Table II.   Thermo-Analyses

	TGA		ΔH
	$T_{10\%}$	% Char 700° Residue	DSC (cal/g)
BPC-PC	500°C	54	-250
PPE	460°C	29	-142
BPA-PC	550°C	26	~ 0

Key: $T_{10\%}$ - Temperature at 10% weight loss.

Elemental Analyses. The elemental analyses for both the N_2 pyrolysis chars and the combustion chars are reported in Table III and Figure 2. Plots of the % residue from the BPA-PC and BPC-PC N_2 pyrolyses studies are seen to

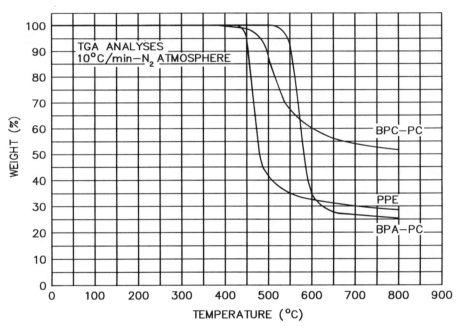

Figure 1. TGA Analysis - 10°C/min, N₂ Atmosphere.

track with the TGA results (compare Figures 1 and 2). Plots of the C/H ratios (Figure 3) versus the degradation temperature for BPA-PC and BPC-PC indicate a dramatic drop in the relative hydrogen content of the char with increasing pyrolysis temperature paralleling the observed weight loss. The low [Cl] in the BPC-PC burn char indicates that ~ 95% of the Cl was lost during combustion. Also of interest is the fact that the oxygen concentration in the BPA-PC burn char is ~ 3 times higher than in the 600°C char. This was confirmed by IR studies (vide infra) which show evidence for more oxygen containing groups in the burn char.

Table III. Elemental Analyses of Chars

BPA-PC[a]	% Residue	%C	%H	%O	C/H[b]
Theory	100	75.58	5.55	18.87	1.14
473°C	96.2	75.69	5.63	(18.69)[c]	1.12
510°C	68.7	79.90	5.67	(14.43)	1.18
528°C	32.1	90.33	5.04	(4.63)	1.50
548°C	28.6	91.35	4.95	(3.70)	1.55
591°C	26.5	93.64	4.21	(2.15)	1.87
Burn char	25.2	90.07	3.18	(6.76)	2.38

BPC-PC[a]	% Residue	%C	%H	Cl	C/H[b]
Theory	100	58.66	2.63	23.09	1.87
450°C	85.3	65.10	3.04	-	1.80
500°C	56.0	88.03	3.53	-	2.09
550°C	53.4	91.58	3.31	-	2.32
Burn char	55.2	87.72	2.96	1.21	2.49

a) The temperatures reported are the maximum temperatures reached during a pyrolysis.
b) C/H = g atoms C/g atoms H.
c) Bracketed values calculated by difference.

The above TGA and elemental analysis studies are consistent with Van Krevelen's two step model for polymer charring (2) in which a polymer first rapidly decomposes at ~ 500°C to fuel gases and a primary char residue characterized by modestly high hydrogen content. On further heating above ~ 550°C, this primary char is slowly converted in a second step to a nearly pure carbon residue by the loss of this hydrogen.

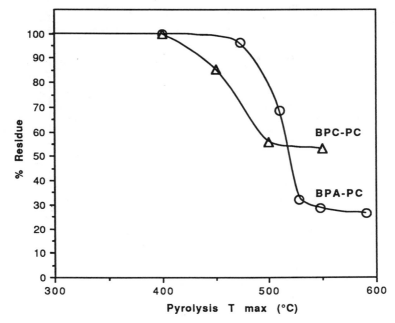

Figure 2.   % Char Residue vs Pyrolysis Temperature.

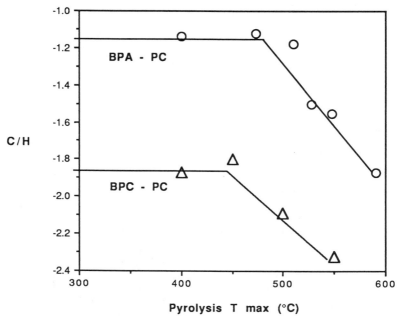

Figure 3.   C/H vs Pyrolysis Temperature.

Attempted Chemical Analyses. An attempt was made to apply the $CF_3CO_2H/H_2O_2/H_2SO_4$ reaction of Deno (8) to analyze for the aliphatic structures present in the char. This technique is reported to selectively oxidize only the aromatic portions of coal leaving the aliphatic groups as carboxylic acids. Indeed, treatment of the burn char from BPA-PC for 16 hrs at reflux produced traces of acetic acid indicating the presence of methyl groups. However, our studies indicated that in the 16 hrs reflux required to dissolve the burn chars in these studies, the dimethylmalonic acid initially formed by digestion of undegraded BPA-PC was readily oxidized to acetic acid with a half life of 1.5 hrs. Thus this approach was not pursued further.

ESR Studies. Jackson and Wynne-Jones (9) studied the ESR spectra from the chars of a number of polymers and found a correlation between d.c. resistivity and the free electron spin concentration but no correlation with C/H ratio. The g-values they measured were observed to be quite close to that of the free electron, i.e. 2.0026 for cellulosic char vs 2.0023 for the free electron. They concluded that it was not possible to determine whether the observed free spin was due to σ or π electrons.

In this work, only a brief study was made. The ESR spectra from BPA-PC and BPC-PC chars consisted of a single line and were quite similar to the signal reported for cellulosic char and to each other (Table IV), except that free spin concentration in the BPA-PC burn char was 15X greater than that from BPC-PC and ~ 2.5X greater than observed in the BPA-PC 600° N_2 pyrolysis char. It is too early in this study to make any conclusions about these results other than these chars contain significant concentrations of free radicals.

Table IV.   ESR Studies on Chars

	g-value	H 1/2 (gauss)[a]	Relative Normalized Intensity/mg
BPA-PC Burn Char	2.003	7.5	15
BPA-PC 600°C/N_2 Char	2.003	7.8	6
BPC-PC Burn Char	2.003	10.2	1

a) Peak width at half height.

PA-FTIR Studies. The IR spectra of the BPA-PC chars (Figure 4) shows progressive changes in the material showing loss of the carbonate group (1775 cm^{-1}), aliphatic C-H groups (2982 and 1385 cm^{-1}) and the growth or appearance of hydroxyl groups (3580 cm^{-1}), aromatic C-H groups (3012 and 3058 cm^{-1}), ester groups (1740 cm^{-1}), ether groups (1170 and 1260 cm^{-1}) and new aromatic groups (1900, 1610, 752, 820 and 880 cm^{-1}). Analogous progressive changes were seen in the IR spectra of pyrolyzed BPC-PC (Figure 5) as well as

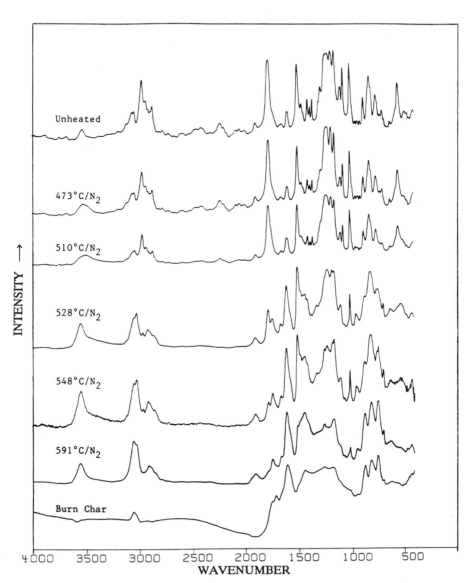

Figure 4.   PAS-FTIR Spectra of BPA-PC Chars.

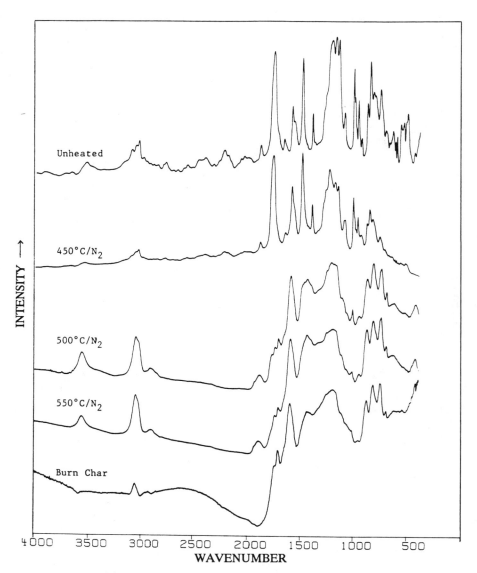

Figure 5.   PAS-FTIR Spectra of BPC-PC Chars.

those reported earlier for PPE (10). Comparison of the IR's of the burn chars with the highest temperature N_2 pyrolysis chars in Figures 4 and 5 indicate that pyrolysis and burning yield very similar products. As might be expected, the burn chars show more evidence for presence of carbonyl oxidation products ($\sim$ 1710 and 1750 cm^{-1}) while the N_2 pyrolysates showed the substantial presence of "OH" groups (3580 cm^{-1}) and more "C-H" groups (2900-2980 cm^{-1}). Comparison of the IR's from the burn chars from BPA-PC, BPC-PC and PPE (Figure 6) shows a remarkable similarity between them. The band in the 1600 cm^{-1} region has been previously observed in studies of carbonaceous materials (11-15) and has been assigned to the aromatic stretching mode which has been intensity enhanced by the presence of oxygen functionality. The 752, 820, and 880 cm^{-1} bands are likely due to out-of-plane aromatic C-H deformation vibrations corresponding to di-, tri-, tetra-, and penta-substituted aromatic rings. As previously suggested (12), these data indicate the presence of micro-graphitic regions loosely bonded by aromatic linkages and containing small amounts of various oxygen functionalities, e.g. Figure 7.

X-Ray Scattering and Raman Spectroscopy. X-ray scattering measurements were made on BPA-PC, BPC-PC and PPE burn chars in an attempt to detect the presence of graphite formation; however no evidence of crystallinity was detected. Nonetheless, previous Raman spectral studies of graphitic carbons (16-17) indicate the presence of two bands at 1340-1355 cm^{-1} and 1575-1590 cm^{-1}. The latter band is attributed to the E_{2g} species of infinite graphite crystals and the former band to the presence of small graphite crystals. In addition, it was found that ratio of the amplitude of the first line to the second line was inversely proportional to the graphite particle size (17). As reported in Table V, the Raman spectra of chars from BPC-PC, BPA-PC and PPE show similar bands with a range of intensities. Since there are other IR absorptions for carbon in the 1350 cm^{-1} region (17), the particle sizes calculated in Table V are no doubt undervalued. Nonetheless these observations confirm the presence of small graphitic regions in these chars.

Table V.   Raman Spectra of Chars

Sample	Relative Intensities		Calculated Particle Size[a]
	1354 cm^{-1}	1590 cm^{-1}	
BPC-PC Burn Char	133	198	64Å
PPE Burn Char	59	91	66Å
BPA-PC Burn Char	-	90[b]	-
BPA-PC 600°C Char	80	137	73Å

a) Particle size (Å) = $1000/(7.931 \times 10^{-2} + 23.252 (I_{1354}/I_{1590}))$.
b) Value of peak intensity in doubt due to weak spectrum.

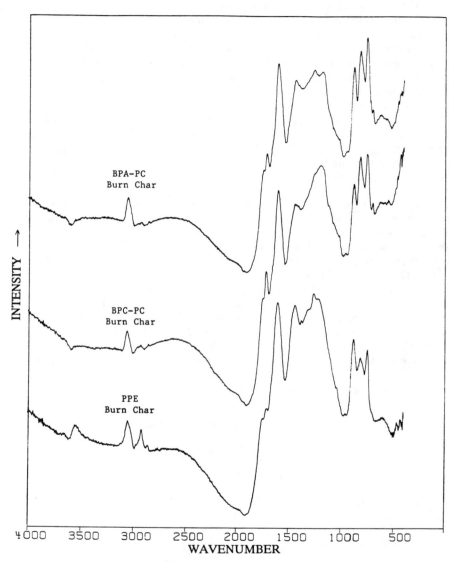

Figure 6.   PAS-FTIR Spectra of Burn Chars.

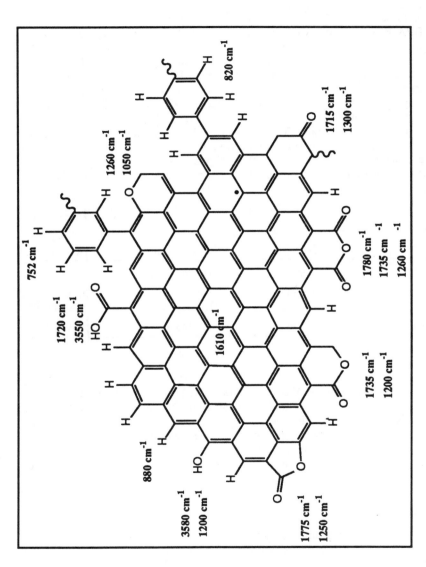

Figure 7.   Idealized Char Structure with IR Band Assignments.

^{13}C Solid State NMR. Earlier workers have used ^{13}C solid state NMR to monitor char formation during the pyrolysis of cellulose (18). They found the ^{13}C NMR of cellulose heated at progressively higher temperatures shows the loss of aliphatic carbons and the appearance of small broad new resonance peaks at 14 and 34 ppm (parafinic carbons) and a large broad peak at ~ 130 ppm (aromatic and/or olefin carbons). The ^{13}C NMR spectrum of BPC-PC, PPE and BPA-PC burn chars were measured (Table VI) and in each case showed the presence of a broad aromatic C peak at ~ 130 ppm ± 15 ppm. Generally the signals from fully substituted aromatic carbons appear from 130-142 ppm and those from protonated aromatic carbons at 117-130 ppm. The small peak due to aliphatic C's at 18 ppm in the spectrum from the PPE-burn char is consistent with the aliphatic C-H IR band found at 2920 cm^{-1} (cf. Figure 6). Thus the results in Table VI indicate the presence of both protonated and unprotonated aromatic carbons. These results are in agreement with the above elemental analysis and IR studies indicating that when BPC-PC, PPE or BPA-PC are heated, they lose most of their aliphatic groups resulting in predominantly protonated and unprotonated aromatic carbons in the residual char.

Table VI.   ^{13}C Solid State NMR Spectra of Burn Chars

| | Peak Position | | Peak |
	Peak Center	Peak Width	Assignment
BPC-PC	128 ppm	117-142 ppm	Aromatic C's
PPE	127 ppm	114-144 ppm	Aromatic C's
	18 ppm	(small peak)	Aliphatic C's
BPA-PC	127 ppm	115-145 ppm	Aromatic C's

Conclusions

Detailed spectroscopic studies on chars from several aromatic engineering polymers indicate that the polymers studied decomposed by a two step mechanism in which heating below ~ 550°C produces volatile tars and fuel gases and a primary char. Heating of this primary char above ~ 550°C produces a carbonaceous residue whose structure is virtually independent of the structure of the original polymer and is best described as a conglomerate of loosely linked small graphitic regions.

Acknowledgments

The author wishes to thank Nancy Marotta for elemental analyses and thermoanalyses, Dr. Elizabeth Williams for solid state ^{13}C NMR, Dr. Peter Codella for PAS-FTIR and Raman IR, and Paul Shields for X-ray diffraction, and to all of the above for helpful discussions and guidance.

Literature Cited

1. Van Krevelen, D. W. *Chimia* 1974, **28**, 504.
2. Van Krevelen, D. W. *Polymer* 1975, (16), 615.
3. Factor, A.; Orlando, C. M. *J. Polym. Sci., Chem. Ed.* 1980, **18**, 579.
4. Carnahan, J. C.; Haaf, W.; Nelson, G.; Lee, G.; Abolins, V.; Shank, P. Investigations into the Mechanism for Phosphorus Flame Retardancy in Engineering Plastics, Proceedings of the 4th International Conference on Flammability and Safety, San Francisco, 1979.
5. MacLaury, M. R. *J. Fire and Flammability* 1979, **10**, 175, Appendix A.
6. Kishore, A.; Nagovajan *J. Polymer Engineering* 1987, **7**, 319.
7. Martel, B. *J. Appl. Poly. Sci.* 1988, **35**, 1213.
8. Deno, N. C., et al *Fuel* 1978, **57**, 455; *Ibid* 1986, **65**, 611.
9. Jackson, C.; Wynne-Jones, W. F. K. *Carbon* 1964, **2**, 227.
10. Factor, A. *J. Polymer Science: Pt. A-1* 1969, **7**, 363.
11. Starsinic, M.; Taylor, R. L.; Walker, P. L., Jr.; Painter, P. C. *Carbon* 1983, **21**, 69.
12. Prest, W. M., Jr.; Mosher, R. A. Chapt. 11, p. 225 in M. Hair and M. D. Croucher Colloids and Surface Reprographic Technology, ACS Symposium Series 200, 1982.
13. Painter, P. C.; Snyder, R. W.; Starsinic, M.; Coleman, M. M.; Kuehn, D. W.; Davis A. *Applied Spectroscopy* 1981, **35**, 475.
14. O'Reilly, J. M.; Mosher, R.A *Carbon* 1983, **21**, 47.
15. Low, M. J. D.; Morterra, C. *Carbon* 1983, **21**, 275, 282.
16. Nathan, M. I.; Smith, J. E., Jr.; Tu, K. W. *J. Applied Physics* 1974, **45**, 2370.
17. Tuinstra, F.; Koenig, J. L. *J. Chem. Phys.* 1970, **53**, 1126.
18. Sekiguchi, Y.; Frye, J. S.; Shafizadeh, F. *J. Appl. Poly. Sci.* 1983, **28**, 3513.

RECEIVED November 20, 1989

Chapter 20

Effects of Coatings on the Fire Performance of Plastics

Gordon L. Nelson

College of Science and Liberal Arts, Florida Institute of Technology,
Melbourne, FL 32901–6988

Plastics used in electrical/electronic applications are frequently found painted in the final application. Data are reported from a study of the effects of EMI/RFI coatings on the fire test results of engineering plastics used for business machine housings. Tests included ignitability, flame spread, heat release, ease of extinction and smoke. Most EMI coatings decrease ignitability test results. Coatings tend to level diverse flame spread and ease of extinction performance. Coated samples have a tendency to show an increase in smoke formation under non-flaming conditions but are unremarkable under flaming conditions. Coatings can also positively interact with the substrate as shown with zinc on modified polyphyenylene oxide structural foam. 2 to 5 mil coatings, both metal and metal-filled organic, can significantly alter fire performance characteristics of engineering plastics.

Plastics find extensive use for business machine housing, applications which require significant fire retardancy. Previous work on plastic enclosures has shown excellent correlation between small scale and large scale fire test performance when a range of small scale tests are utilized (1-4). However, plastics used in a variety of applications and particularly for business machine housings are increasingly found painted in the final application. Both decorative and functional coatings are used. Interiors of business machine enclosures are routinely coated with an EMI (Electromagnetic Interference) coating, such coatings to provide shielding for critical electronic components. Little data have been published on effects of the presence of these coatings on the fire performance of the finished enclosure composite. This paper reports comprehensive data from a study of the effects of EMI coatings on fire test results of engineering plastics used for business machine housings (5-6).

Flammability Test Methods and Test Specimens

In previous papers, fire test results on plastics using the Ohio State University Heat Release Calorimeter (ASTM E-906) and the ASTM E-162 Radiant Panel Test have been discussed. Results were presented on test reproducibility and on the material factors affecting results, including thickness, moisture, aging, and pigments. Test results have been presented for a variety of engineering plastics (7-8). A comparative study was also presented on OSU Heat Release Test results and Radiant Panel Test results versus large scale room corner tests and box tests (9). This latter study showed that while neither OSU Heat Release Calorimetry nor Radiant Panel testing gave quantitative correlation with Room Corner Test or Box Test fire performance, Radiant Panel I_s values provided better qualitative correlation with Room Corner Test data and with Box Test data, and properly placed wood and plywood versus the plastic materials tested. Data also revealed a correlation of Maximum Rate of Heat Release (OSU) and Q (E-162), which suggested that flame spread test data in addition to heat release data are required to properly position the full scale fire performance of materials. Data also suggested that in evaluation of small scale test versus large scale or full scale performance of materials, it is essential that a significant number of materials in several forms be used and that tests representing several large scale scenarios be evaluated in order to insure the rigorousness and scope of correlation (9).

Given the above observations it was essential in the present study that multiple test methods be used, representing evaluation of the effects of coatings on ignitability, flame spread, heat release, ease of extinction, and smoke. Samples should be commercially prepared and representative of materials commonly used in business machine applications.

Commercial coated samples were obtained. Coating thickness was nominal 2 mils for organic-metal filled coatings (approximately 50% metal filler). Zinc metal coatings were zinc arc spray and were thicker, 5 mils, but normal for that process. Test results for each fire parameter are given as follows:

* Ignitability - (IEC 695) IEC Needle Flame and Glow Wire Tests
* Flame Spread - E162 Radiant Panel
* Heat Release - E162 Radiant Panel
* Ease of Extinction - D2863 Oxygen Index
* Smoke - E662 NBS Smoke Chamber (optical)
  - D4100 - Arapahoe Smoke Chamber (gravimetric)

All structural foam substrates were 1/4 inch in thickness except for RIM polyurethane, which was 1/2 inch in thickness. These comments hold for samples throughout the study. For solid plaques, samples were 3/32 inch thick.

Needle Flame Test

The Needle Flame Test is designed to simulate a small flame as might be encountered from a small electrical malfunction in an appliance or a piece of office equipment. The test apparatus consists of a hypodermic needle 35 mm in length with a bore of 0.5 mm and an outer diameter not exceeding 0.9 mm. The hypodermic needle has the tapered end cut off to avoid any interference with the flame. The gas used is butane. With the axis of the burner vertical, the gas supply is adjusted so that the length of the flame is 12 ± 1 mm. With the needle at 45° to a vertical test specimen, the sample is subjected to the flame for 30 seconds. The needle is kept 5 mm from the test sample. After the required time, the flame is removed and the after-burn time and the height of the after-burn flame recorded. Upon extinguishment the damaged area is recorded.

Results are presented in Table I. While several coatings show longer burn times than their uncoated counterparts, most coated samples showed significantly smaller areas of damage than the corresponding uncoated sample. Zinc showed smaller damage areas on all substrates. Other coatings which showed small damage areas included nickel/polyurethane on polycarbonate and on modified polyphenylene oxide structural foam substrates, and copper/epoxy on polyurethane and on modified-polyphenylene oxide structural foam.

Glow Wire Test

The Glow-wire test is designed to give qualitative information describing the response of a sample to a glowing wire such as might be encountered in the malfunction of the wiring in a household appliance or a piece of office equipment. A glow wire apparatus was constructed per IEC 695-2-1, using 4mm Nickel/Chromium (80/20) wire.

The apparatus used for the Glow-wire test, shown in Figure 1, consists of a carriage moving on a platform. The carriage holds the 3" x 3" polymer sample which is moved horizontally towards the glowing Nickel/Chromium wire which can be heated to a temperature of 660 C and 960°C. The polymer sample is subjected to the wire for thirty seconds with a force of 1.8 to 2.0 Newtons.

In order to heat the wire to a temperature of 660°C and 960°C, the apparatus had to be connected to a Tieg welder. Using the Tieg welder, very high amperage at low voltages can be obtained. The amperage output can be increased until the desired temperature of the wire is obtained. The calibration of the heated wire at 660°C is carried out by using a foil of 99.9 percent pure aluminum placed on the tip of the wire. For a temperature of 960°C 99.8 percent pure silver foil is used. The foil of 99.8 percent pure silver melts at 960°C.

After the desired temperature is obtained the sample is then placed on the carriage and is subjected to the glowing wire for thirty seconds. When the required thirty second time increment is completed, the sample is pulled away from the heat source. At this time the height of the after burn, the after-burn time, and the penetration are recorded. The average damage area is then determined by measuring the height and width of the badly damaged area. Each type of sample is tested three times.

Glow wire data are presented in Table II. Glow wire tests are particularly sensitive to coatings, with copper/epoxy and zinc coatings, for example, showing significant reduction in damage areas and after burn time for modified-polyphenylene oxide. Copper/acrylic on modified-polyphenylene oxide showed a large increase in both after-burn time and damage area in Glow Wire tests at 660°C.

Table III permits a cross comparison of Needle Flame and Glow Wire data for modified-polyphenylene oxide. Nickel/polyurethane, copper/epoxy and

Table I. Needle-Flame Tests on Engineering Structural Foam Samples with EMI Coatings

Substrate	Coating	After-Burn Time (s)	Damage (mm^2)
Polycarbonate (white)	uncoated	0	380
	Ni/polyurethane	2	240
	Cu/epoxy	0	310
	zinc	0	290
Polyurethane (RIM)	uncoated	1	500
	Ni/polyurethane	1	440
	Cu/epoxy	0	120
	zinc	1	230
Modified-Polyphenylene Oxide	uncoated	7	680
	Ni/polyurethane	14	110
	Cu/epoxy	1	290
	zinc	1	120
	nickel/acrylic	2	380
	graphite/acrylic	0	590
	copper/acrylic	18	710

Average of determinations by two workers.

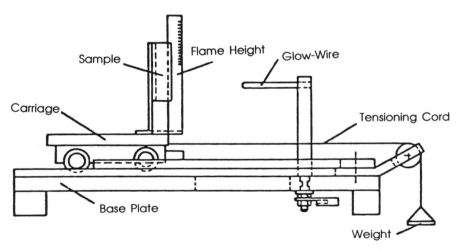

Figure 1.   IEC Glow Wire Test Apparatus

Table II. Glow Wire Tests on Engineering Structural Foam Samples with EMI Coatings
(30 second exposure)

	660°C		960°C	
	T_b (s)	Damage (mm^2)	T_b (s)	Damage (mm^2)
Poly-carbonate (white)				
Uncoated	0	170	0	1040
Ni/poly-urethane	0	120	0	980
Cu/epoxy	1	210	2	360
Zinc	0	70	0	280
Polyure-thane (RIM)				
Uncoated	7	610	3	910
Ni/poly-urethane	0	130	7	600
Zinc	1	160	3	480
Modified-PPO				
Uncoated	2	500	7	2070
Ni/poly-urethane	4	190	7	660
Cu/epoxy	0	70	4	260
Zinc	0	20	0	140
Ni/acrylic	0	300	--	----
Graphite/acrylic	0	100	8	1560
Cu/acrylic	13	1200	13	1100

zinc coatings show significantly smaller damage areas under all three ignitability conditions, while copper acrylic showed generally higher damage areas and after burn times.

In a third ignitability test, UL-94, all substrates were rated V-1/5V or better by UL-94.

Radiant Panel Test

The flammability of solids may be considered to be a function of the heat release rate and critical ignition energy of the material being studied. Flammability is an inverse function of the actual ignition energy of the material in question, and it is directly related to the rate of heat liberated after ignition of the sample.

The Radiant Panel Test method is designed to measure both of these properties in a single test (8). The test specimen faces the heat source but at a 30° angle to it such that the upper portions of the specimen are severely exposed (Figure 2). Since irradiance varies along the length of the specimen, the time progress of ignition down the specimen serves to measure critical ignition energy down the sample (this is the F_s, or Flame Spread Factor). The stack and the associated thermocouples placed above the specimen serve as a heat-flux meter for measuring the rate of heat release. This is the Heat Evolution Factor, Q (Q correlates with Maximum Rate of Heat Release from E906). Thus, measurements are made of the position of the flame front on the exposed surface of the specimen as a function of time, and the maximum temperature rise of the stack thermocouples. These two measurements are combined to give the flame spread index I_s (F_s x Q = I_s). In practice flammability behavior is influenced by size, geometry, orientation, and other sample parameters. Rather than a precise determination of physical fire parameters, an empirical index is chosen for the reporting of Radiant Panel data which employs asbestos hardboard with an index of 0 and the mathematics adjusted to give the I_s value for red oak at 100, this latter to provide some measure of agreement with ASTM E84. The instrument is calibrated in terms of heat flux measurements, however, rather than continual use of red oak as the reference material. Table IV shows Radiant Panel Test results.

The substrates tested alone have substantially different I_s values. Polycarbonate (1/4 inch) structural foam has an I_s of 27.5, modified-polyphenylene oxide (1/4 inch), 84.4, and RIM polyurethane (1/2 inch), 173.3. These I_s values compare with 164.4 for 1/4 inch hardboard and 139.1 for 1/4 inch plywood. A comparison of graphite, nickel, and copper/acrylic coatings on polycarbonate and modified-polyphenylene oxide substrates illustrate a dramatic result. Despite a factor of 3 difference in substrate performance, the I_s, Q and F_s values for the coated samples are very similar. The Q for the modified-polyphenylene oxide samples are 0.7 to 0.5 that of the uncoated sample. One would expect a similarity in F_s for the coated sample, but such a reduction in Q is dramatic. Both Q and F_s are determined by the 2 mil surface.

Two of the zinc surface samples show results that might intuitively be expected for metal coated surfaces. The polycarbonate and RIM-polyurethane substrates show much lower F_s values with the zinc coating as expected, but Q values are very similar to that of the uncoated substrates. For modified-polyphenylene oxide, however, a very low Q value is obtained suggesting flame

Table III. Needle-Flame vs Glow Wire Results Modified-PPO Structural Foam (Damage-mm^2)

Coating	Needle Flame	Glow-Wire 660°C	Glow-Wire 960°C
Uncoated	690	500	2070
Ni/Pu	110	190	660
Cu/Epoxy	290	70	260
Ni/Acrylic	380	300	---
Graphite/Acrylic	590	100	1560
Cu/Acrylic	710	1200	1100
Zinc	120	20	140

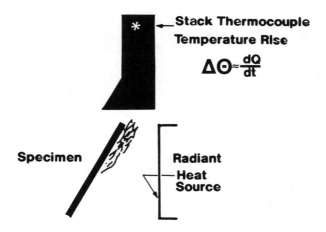

Figure 2. The Radiant Panel Test was designed to measure both critical ignition energy and rate of heat release. A sample is mounted facing a controlled heat flux but at a 30° angle to it such that the upper part of the specimen is more severely exposed. Since irradiance decreases down the specimen, the time progress of ignition down the specimen serves to measure central ignition energy. Thermocouples in the stack above the specimen serve as a measure of heat release rate.

Table IV. Radiant Panel Test Results for Structural Foam Samples

Materials*	Replicates	I_s	Q	F_s
				Average Values
P o l y -carbonate				
Uncoated (white)	4	27.5	14.8	1.87
Uncoated (grey)	1	31.5	16.8	1.88
graphite/ a c r y l i c (grey)	4	59.8	19.6	3.08
nickel/ a c r y l i c (grey)	5	48.0	15.9	3.33
copper/ a c r y l i c (grey)	1	85.1	26.0	3.27
zinc (white)	4	14.4	14.3	1.35
Modified-Polyphenyl-ene Oxide				
uncoated	6	84.4	30.6	2.77
graphite/ acrylic	4	64.3	15.6	3.99
nickel/ acrylic	4	68.1	20.0	3.36
copper/ acrylic	4	63.0	22.1	2.84
zinc	4	2.9	2.9	1.00
RIM-Poly-urethane				
uncoated	4	173.3	28.5	6.07
copper/ epoxy	3	78.3	28.3	2.76
nickel/ urethane	3	17.7	8.9	1.97
zinc	4	43.4	24.0	1.81
Hardboard	39	164.4	51.7	3.21
Plywood	3	139.1	33.8	4.12

Instrument Constant β, was 23.7 to 25.4, Average 24.7

*Samples conditioned to ambient room temperature and humidity.

retardant synergism of substrate and coating for this system. A low Q value is also obtained for the nickel/urethane coating on the RIM-polyurethane substrate.

Coatings can significantly alter the fire performance properties of plastics. A 2-mil coating can reduce the I_s value for a more flammable substrate by an order of magnitude while more flame retardant substrates can see a doubling of the I_s value.

Oxygen Index

Oxygen Index measures the ease of extinction of materials, the minimum percent of oxygen in an oxygen/nitrogen atmosphere that will just sustain combustion of a top ignited vertical test specimen (10). Three substrates were available--modified-polyphenylene oxide, polycarbonate and polystyrene. All materials were 1/4" by 1/2" by 5" coated on all sides. The oxygen indices of the base substrates varied from 23.6 to 32.1. Data are shown in Table V. Oxygen Index values are leveled for different substrates with the same coating. The oxygen indices of more flame retardant materials are reduced and that of less flame retardant materials elevated by the presence of a surface coating. Table VI presents data comparing Oxygen Index and Needle Flame test data. A qualitative correlation of OI and damage area is seen.

Smoke Chamber Results

Smoke evolution was measured using the National Bureau of Standards Smoke Chamber (ASTM E-662). In this test 3" by 3" samples are placed vertically in front of a radiant heat source (2.5 W/cm2). Non-flaming samples are tested using the heater alone. For flaming conditions, a six flamelet burner is positioned at the base of the sample and used in conjunction with the radiant heat source. Test results are obtained by measuring the percent transmission of a light beam which travels from the bottom of the chamber, through the accumulating smoke to the photomultiplier tube at the top of the chamber versus time. The resulting percent transmittance is then converted to specific optical density, D_s, which is calculated by using Equation 1, where D_s is the specific optical density,

$$D_s = V(\log(100/T))/LA \qquad (1)$$

V is the chamber volume, L is the light beam path length, A is the sample area, and T is the transmittance.

The EMI coatings that were tested were copper/acrylic, nickel/acrylic, graphite/acrylic, zinc, nickel/polyurethane and copper/epoxy. The coatings tested and their corresponding substrates are shown in Table VII. Uncoated samples were tested for references. All samples were conditioned in a humidity chamber for at least 24 hours at 70°C and 50% RH prior to testing and were tested in both flaming and non-flaming modes. Test averages were based on three replicates.

The tabulated results for all samples tested can be found in Tables VIII and IX, with Figures 3 and 4 depicting the results graphically. Non-flaming conditions produce the most dramatic results. For the polycarbonate substrates all coatings except zinc increase the amount of smoke produced. Zinc shows

Table V. Oxygen Index Data for Structural Foam Samples

| Coatings | Substrate | | |
	Modified PPO	Polycarbonate	Polystyrene
uncoated	26.5	32.1	23.6
nickel/acrylic	28.3	31.2	----
graphite/acrylic	23.9	32.4	----
copper/acrylic	24.6	27.5	----
nickel/urethane	29.0	30.5	27.7
copper/epoxy	30.6	31.0	29.3

Table VI. Comparison of Needle-Flame Test Results and Oxygen Indices for Modified-Polyphenylene Oxide Structural Foam Samples

Coating	Needle Flame Damage Area (mm^2)	OI
uncoated	684	26.5
Ni/polyurethane	108	29.0
Cu/epoxy	286	30.6
nickel/acrylic	379	28.3
graphite/acrylic	592	23.9
copper/acrylic	708	24.6
zinc	120	---

Table VII. NBS Smoke Chamber Tests EMI Coating and Corresponding Substrate

| Coating | Substrate | | | |
	Grey Polycarbonate	White Polycarbonate	Polyurethane	Modified Polyphenylene Oxide
Copper/acrylic	-	-	-	X
graphite/acrylic	X	-	-	X
nickel/acrylic	X	-	-	X
zinc	-	X	X	X
nickel/urethane	-	X	X	X
copper/epoxy	-	X	X	X
uncoated	X	X	X	X

Table VIII. NBS Smoke Chamber Data for EMI Coated Structural Foam Samples Average Non-Flaming Data*

Coating	D_s	D_s/wt	Sample Weight (g)	Weight burned (g)
		Polycarbonate		
uncoated (white)	63	1.80	34.5	11.1
nickel/ urethane	175	4.90	35.9	10.7
copper/ epoxy	132	3.70	35.4	10.7
zinc	67	1.86	36.1	9.3
uncoated (grey)	84	2.50	33.5	0.7
graphite/ acrylic	187	6.30	33.9	3.6
nickel/ acrylic	216	6.60	32.7	1.7
		RIM-Polyurethane		
uncoated	>660	--	45.8	10.7
nickel/ urethane	>660	--	55.5	13.3
copper/ epoxy	>660	--	46.5	6.5
zinc	440	9.9	44.2	9.2
		Modified-Polyphenylene Oxide		
uncoated	399	16.5	24.4	3.0
nickel/ urethane	323	11.0	29.2	4.2
copper/ epoxy	478	18.1	26.4	3.8
zinc	35	1.2	29.6	0.3
nickel/ acrylic	344	10.7	31.7	2.0
graphite/ acrylic	374	11.7	33.4	4.5
copper/ acrylic	>660	--	33.2	3.9

*Average of three replicates

Table IX. NBS Smoke Chamber Data EMI Coated Structural Foam Samples (Average Flaming Data*)

Coating	D_s	D_s/wt	Sample weight (g)	Weight burned (g)
Polycarbonate				
uncoated (white)	419	12.3	33.9	10.4
nickel/ urethane	455	12.9	35.4	16.3
copper/ epoxy	457	12.8	35.8	17.8
zinc	377	10.4	36.5	24.2
uncoated (grey)	306	9.2	33.4	19.0
graphite/ acrylic	240	7.2	33.3	19.1
nickel/ acrylic	384	11.5	33.6	19.4
RIM-Polyurethane				
uncoated	>660	---	45.1	6.8
nickel/ urethane	>660	---	52.2	6.6
copper/ epoxy	>660	---	49.5	5.8
zinc	>660	---	42.5	4.7
Modified-Polyphenylene Oxide				
uncoated	>660	---	26.9	5.3
nickel/ urethane	>660	---	28.6	4.2
copper/ epoxy	>660	---	27.0	3.7
zinc	>660	---	38.7	4.1
nickel/ acrylic	>660	---	31.6	5.3
copper/ acrylic	>660	---	35.2	5.7
graphite/ acrylic	>660	---	33.6	6.2

*Average of three replicates

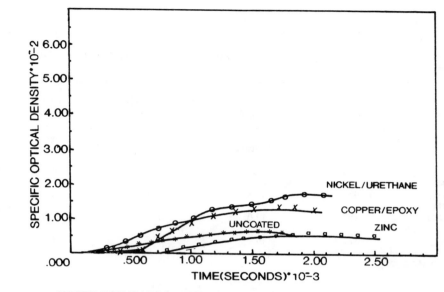

WHITE POLYCARBONATE (NON–FLAMING)

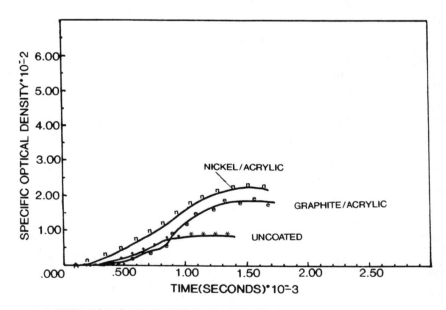

GREY POLYCARBONATE (NON–FLAMING)

Figure 3. NBS Smoke Chamber results (non-flaming) for two grades of polycarbonate structural foam. *Continued on next page.*

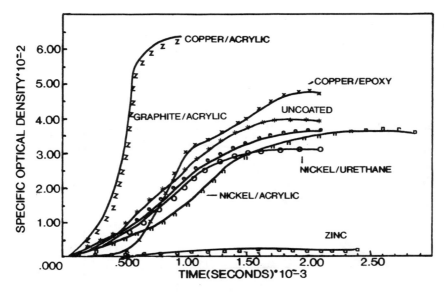

MODIFIED POLYPHENYLENE OXIDE (NON-FLAMING)

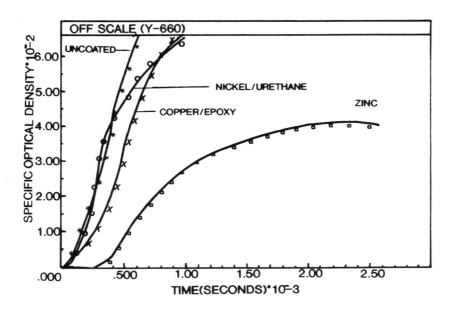

POLYURETHANE (NON-FLAMING)

Figure 3. Continued. NBS Smoke Chamber results (non-flaming) for modified-polyphenylene oxide and RIM polyurethane structural foam.

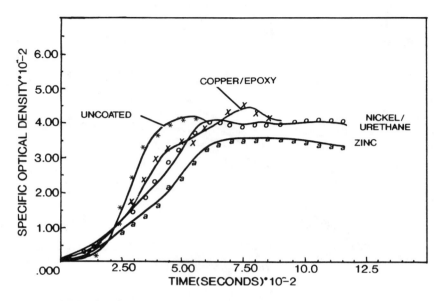

WHITE POLYCARBONATE(FLAMING)

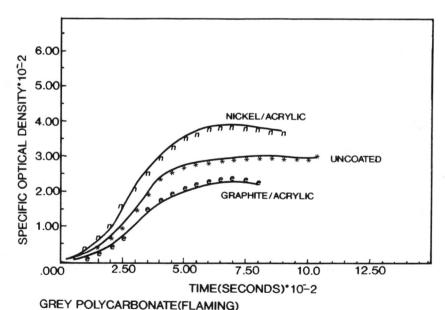

GREY POLYCARBONATE(FLAMING)

Figure 4.   NBS Smoke Chamber results (flaming) for two grades of polycarbonate structural foam. *Continued on next page.*

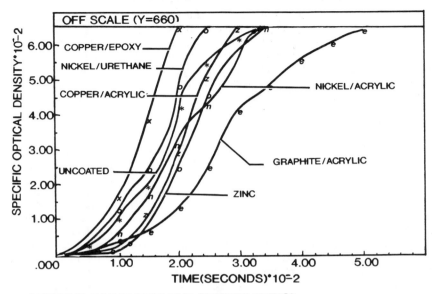

MODIFIED POLYPHENYLENE OXIDE(FLAMING)

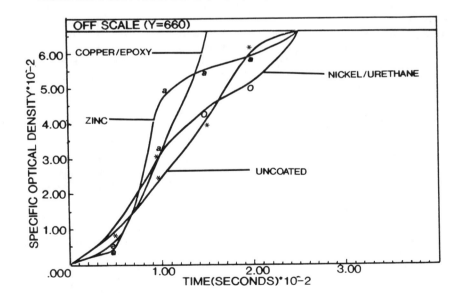

POLYURETHANE(FLAMING)

Figure 4. Continued. NBS Smoke Chamber results (flaming) for modified-polyphenylene oxide and RIM polyurethane structural foam.

a delay in smoke generation. For modified-polyphenylene oxide and for RIM-polyurethane substrates all coatings except zinc and copper/acrylic show little substantial effect on either the intensity of smoke produced or the time of production. A 10 fold reduction in D_s is observed for the modified-polyphenylene oxide sample with the zinc arc spray coating. This sample also showed the most significant reduction in I_s on Radiant Panel Testing. Significant flame retardant synergism with the zinc coating and modified-polyphenylene oxide substrate is present.

The results under flaming conditions show nothing as remarkable (Table IX). The most effective coating on reducing smoke (marginally) appears to be the graphite/acrylic coating on both modified polyphenylene oxide and on polycarbonate. The coating which appears to yield smoke earlier (marginally) is the copper/epoxy coating on both modified polyphenylene oxide and polyurethane structural foam.

In addition to structural foam samples, limited samples were available for EMI coatings on 3/32-inch thick solid plaques. The EMI coatings tested were from different manufacturers than those reported for the structural foam samples. Coatings were nominal 2 mils in thickness. Data are shown in Tables X and XI and Figure 5. For the modified-polyphenylene oxide substrate coated samples show higher smoke than the uncoated sample in non-flaming conditions. This result is reminiscent of the polycarbonate structural foam samples under non-flaming test conditions. Nickel/acrylic I shows the greatest increase in smoke. Under flaming conditions the four coated samples show a delay in smoke formation versus the uncoated substrates. Given the differing ability of solid and foam samples to flow under the influence of heat it is to be expected that solid and foam samples will show differing performance.

In Table XII are presented data for electroless nickel on injection molded polycarbonate. The coated sample shows slightly more smoke smoldering and somewhat less smoke flaming. A very thin metal coating ($< <1$ mm) has minimal effect on polycarbonate.

Arapahoe Smoke Chamber

Substrates were also tested using the Arapahoe Smoke Chamber. A weighed sample is placed in a special holder in a burn chamber. Smoke is drawn through weighed filter paper as the sample is burned in a butane flame. The sample is subjected to the flame for thirty seconds. The sample is removed and weighed. Char is removed from the sample in a sand mill. The example is weighed again. The filter is weighed to determine the amount of smoke produced. The percent of sample burned which goes to smoke and to char is reported in Table XIII. In Table XIV is shown a comparison of data from the NBS Smoke Chamber under flaming conditions with Arapahoe Smoke Chamber data. A clear qualitative correlation is present. For modified-polyphenylene oxide and for polycarbonate, structural foam (SF) samples show a higher percent char than non-foam samples. The presence of glass fiber (about 5 percent) as a foam nucleating agent and a foam structure might be expected to yield this result.

Table X. NBS Smoke Chamber Data for EMI Coated Solid Plaque Samples (3/32 inch thickness) Average Non-Flaming Data*

Coating	D_s	D_s/wt	Sample Weight (g)	Weight Burned (g)
		Modified-Polyphenylene	**Oxide**	
uncoated	106	6.3	16.8	0.4
copper/ acrylic I	162	9.3	17.5	0.8
nickel/ acrylic I	367	20	17.5	1.9
nickel/ acrylic II	279	16.4	17.0	1.5
		ABS		
uncoated	396	22.4	17.7	3.6
copper/ acrylic II	386	21.1	17.4	2.7

*Average of three replicates

Table XI. NBS Smoke Chamber Data for EMI Coated Solid Plaque Sample (3/32 inch thickness) Average Flaming Data*

Coating	D_s	D_s/wt	Sample Weight (g)	Weight Burned (g)
		Modified-Polyphenylene	**Oxide**	
uncoated	>660	---	15.7	10.8
copper/ acrylic I	>660	---	17.2	6.3
nickel/ acrylic I	>660	---	17.1	5.6
nickel/ acrylic II	>660	---	15.8	9.7
		ABS		
uncoated	>660	---	17.3	4.6
copper/ acrylic II	>660	---	17.4	4.8

*Average of three replicates

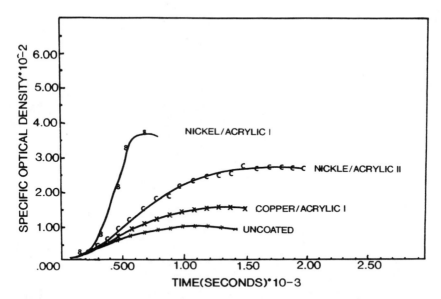

MODIFIED POLYPHENYLENE OXIDE (INJECTION MOLDED) NON-FLAMING

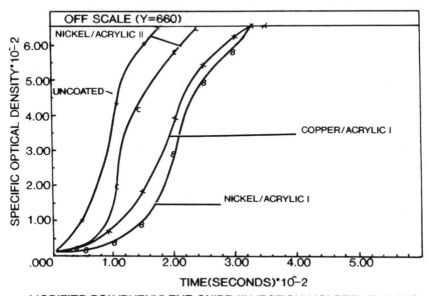

MODIFIED POLYPHENYLENE OXIDE (INJECTION MOLDED) FLAMING

Figure 5. NBS Smoke Chamber results for solid injection molded modified-polyphenylene oxide samples. *Continued on next page.*

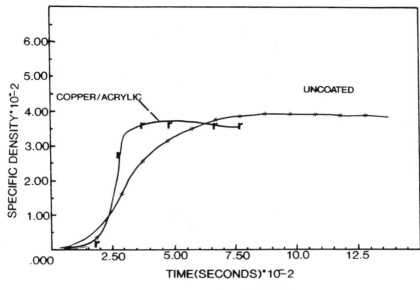

ABS (INJECTION MOLDED) NON-FLAMING

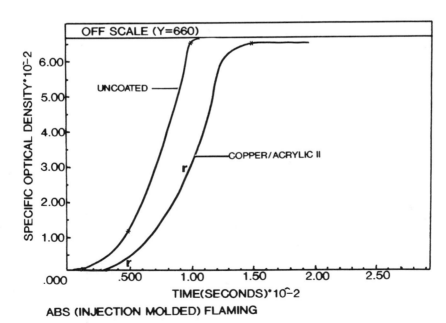

ABS (INJECTION MOLDED) FLAMING

Figure 5. Continued. NBS Smoke Chamber results for molded ABS samples.

Table XII.   NBS Smoke Chamber Data for Electroless Nickel on Injection Molded Polycarbonate (1/8 inch thickness)

	D_s	D_s/wt	Sample Weight (g)	Weight Burned (g)
		Non-Flaming		
uncoated	63	2.6	22.7	1.2
coated	70	3.1	22.9	1.3
		Flaming		
uncoated	245	11.4	21.7	16.1
coated	190	8.6	22.0	13.2

Table XIII. Arapahoe Smoke Chamber Tests % of weight burned

	Exposure(s)	% Smoke	% Char
Polystyrene (SF)	30	33.6	48.9
ABS (Inj)	30	30.8	52.6
ABS (Cycolac)	30	30.0	59.0
Mod-PPO (SF)	30	24.0	48.0
Mod-PPO (N190)	30	25.7	28.3
Mod-PPO (Inj)	60	25.9	28.0
Polyurethane (SF)	30	15.1	35.8
Polycarbonate (white-SF)	60	12.7	51.7
Polycarbonate (grey-SF)	60	13.0	47.4
Polycarbonate (sheet)	30	12.8	35.2
Polycarbonate (inj-FR)	60	9.5	43.5

Table XIV. Comparison of NBS Smoke Chamber Data (Flaming) with Arapahoe Smoke Chamber Data

Sample	% Smoke (Arapahoe)	D_s (NBS)
polycarbonate (white-SF)	12.7	419
polycarbonate (grey-SF)	13.0	306
polyurethane (SF)	15.1	>660 @ 2.4 x 10^2 seconds
mod-PPO (SF)	24.0	>660 @ 3.4 x 10^2 seconds
mod-PPO (inj)	25.9	>606 @ 2.0 x 10^2 seconds
ABS (inj)	30.8	>606 @ 1.0 x 10^2 seconds

Conclusion

These results show that thin coatings can significantly effect the fire performance of plastic substrates. Most EMI coatings decrease ignitability test results. Coatings tend to level diverse flame spread and ease of extinction performance. A 2-mil coating can reduce the I_s value for a more flammable substrate by an order of magnitude, while more flame retardant substrates can see a tripling of the I_s value.

Coatings can interact with the substrate as shown with zinc on modified-polyphenylene oxide structural foam. Examination of zinc on modified-polyphenylene oxide has been undertaken to ascertain the origin of the synergism and is reported separately. The presence of zinc shows an increase in char, and a decrease in low molecular weight volatiles at pyrolysis temperatures versus the uncoated substrate.

In general, for NBS Smoke Chamber data, coated samples have a tendency to show an increase in smoke formation under non-flaming conditions. Smoke results under flaming conditions are unremarkable and specific coating dependent.

Clearly 2 to 5 mil coatings can significantly alter fire performance. Proper evaluation and choice of coating for fire performance is an important, and often missed, opportunity.

Acknowledgments

The work of the following students is gratefully acknowledged:

Undergraduate Students	Graduate Students
W. Ronald Rose	J. Kent Newman
Kenneth L. Weaver	
Manual A. Bosarge	Post-doctoral Fellow
Anthony N. Morris	Eddie K.M. Chan

References

1. R.D. Bieniarz, Fire Experiments on Structural Foam Plastic Equipment Enclosures. Study for the Society of the Plastics Industry, Underwriters Laboratories, 1981.
2. Bernie Miller, Small-Scale Tests are Respectable-Foam Flammability, Plastics World, 1981 (9), 78-81.
3. G.L. Nelson, Combustibility of Structural Foam Plastics, J. Cellular Plastics, 18, 36 (1982).
4. G.L. Nelson, Flame Tests for Structural Foam Parts, Plastics Technology, 23 No. 12, 88-92 (1977).
5. G.L. Nelson, M.L. Bosarge, and K.M. Weaver, Proceedings of the Twelfth International Conference on Fire Safety, Part I, (Jan. 12-16, 1987) 12, 271-282 (1987).
6. G.L. Nelson, M.L. Bosarge, K.M. Weaver, Effect of EMI Coatings on the Fire Performance of Plastics, Part II, Proceedings of the Thirteenth International Conference on Fire Safety, January 11-15, 1988, 13, 367-378 (1988).

7. A.L. Bridgman and G.L. Nelson, Heat Release Rate Calorimetry and Engineering Plastics, J. Fire and Flammability, 13 114 (1982).

8. A.L. Bridgman and G.L. Nelson, Radiant Panel Tests on Plastics, Proceedings of the International Conference on Fire Safety, Jan. 17-20, 1983, 8, 191-226 (1983).

9. A.L. Bridgman and G.L. Nelson, Heat Release Calorimetry and Radiant Panel Testing: A Comparative Study, Proceedings of the International Conference on Fire Safety (Jan. 13-17, 1986), 11, 128-139 (1986).

10. G.L. Nelson and J.L. Webb, Oxygen Index, Flammability and Materials, in Advances in Fire Retardant Textiles, V.M. Bhatnagar, ed., (Technomic, Westport,CT) 271-370.

RECEIVED November 20, 1989

Chapter 21

Synergistic Fire Performance between a Zinc Coating and a Modified Poly(phenylene oxide) Substrate

Gordon L. Nelson and Eddie K. Chan

College of Science and Liberal Arts, Florida Institute of Technology, Melbourne, FL 32901–6988

EMI coatings affect the fire performance properties of engineering plastics. Zinc arc spray on modified-polyphenylene oxide is particularly effective. The enhanced fire retardancy has its origins in several factors assisted by the fact that zinc melts (420°C) just at the early stage of the decomposition of m-PPO, allowing intimate contact with the charring substrate. As in pure polystyrene, char formation is enhanced in air in m-PPO, and this is further enhanced by the presence of zinc. As seen in SEM, zinc oxide which is part of the thermal stabilization package of m-PPO is present at the charring surface enhanced by the zinc coating. It is observed that volatilization of small molecules is reduced for m-PPO with zinc present at temperatures under 700°C, with preference for volatilization of the triaryl phosphate flame retardant, styrene trimer, and PPO dimers. The flame retardant and larger entities formed preferentially lead to enhanced char formation.

Plastics find extensive use in business machine housings. Business machine housings are increasingly found painted in the final application. Both decorative and functional coatings are used. Interiors are routinely coated with an EMI (Electromagnetic Interference) coating, such coatings to provide shielding of critical electronic components. Comprehensive data have been provided in a separate report on the effects of EMI coatings on the fire performance of engineering plastics (1-3).

Test results show that thin coatings (2-5 mils) can significantly effect the fire performance of plastic substrates. Most EMI coatings decrease ignitability test results. Coatings tend to level diverse flame spread and ease of extinction performance. A 2-mil coating can reduce the I_s value in ASTM E162 Radiant

0097–6156/90/0425–0311$06.25/0

Panel tests for a more flammable substrate by an order of magnitude, while more flame retardant substrates can see a tripling of the I_s value.

For NBS Smoke Chamber data, coated samples have a tendency to show an increase in smoke formation under non-flaming conditions. Smoke results under flaming conditions are coating specific and unremarkable.

Of particular interest, coatings can interact with the substrate as shown with zinc arc spray on modified-polyphenylene oxide structural foam. While this synergism is evident in ignitability, flame spread, heat release, and smoke tests, the ASTM E162 Radiant Panel test results are of most interest. One would expect a metal coating (as opposed to an organic coating) to affect the F_s (flame spread) component of Radiant Panel data, but not the heat release portion. Since the substrate is the same, so should be the maximum rate of heat release. As shown in Table I, that is observed for both 5 mil zinc arc spray on RIM polyurethane and polycarbonate structural foam. On modified-polyphenylene oxide structural foam, however, a dramatic reduction of both heat release (Q) and flame spread (F_s) are seen resulting in an I_s 1/30th of the uncoated substrate. The purpose of this paper is to discuss the origin of the synergism of zinc on modified-polyphenylene oxide.

Modified-polyphenylene oxide (or ether) is a blend of high impact polystyrene (PS) and polyphenylene oxide (PPO), plus thermal stabilizers and a triarylphosphate flame retardant. Studies of the mechanism of the flame retardant in modified-polyphenylene oxide have shown some evidence for both solid phase and vapor phase inhibition (4). Indeed, one is always interested to know whether flame retardant action is on the solid or vapor phase.

The charring process plays an important part in the combustion of many polymers; certainly so for modified-polyphenylene-oxide. Martel has shown that for PPO/PS alloys, TGA experiments yield 33% char in air versus only 10-12% in nitrogen. PPO yields 40% char in nitrogen (5). It is known that polystyrene alone does not char under nitrogen, yielding monomer, benzene, toluene, and dimer and trimer (6). Polystyrene containing ammonium phosphate chars in TGA experiments carried out under oxygen but volatilizes without carbonization under nitrogen (7). When polystyrene is subjected to a small hydrogen flame a layer of carbon is formed on the polymer in contact with the flame (8). In detailed TGA/DTA studies of polystyrene, Martel (5) found a single endothermic peak corresponding to carbon-carbon scission at about 410°C. In air, two exothermic peaks were observed at about 410°C and 530°C, corresponding to the charring process. If the DTA experiment was interrupted at $400\text{-}420^\circ$C and cooled, the carbonaceous residue (10% of original) did not volatilize under nitrogen, but yielded a strong exothermic oxidation peak at 530°C. A DTA experiment of char obtained in an oxygen index experiment gives the same exothermic peak at $500\text{-}550^\circ$C. Therefore the peak at 530°C is characteristic of the oxidation of char formed in earlier oxidative degradation. The charring process was strongly dependent on the oxygen concentration in the atmosphere down to 10% oxygen. The oxygen concentration, however, had little effect on the DTA exothermic peak at 410°C. It was concluded that the charring process is dependent upon the competition between two reactions--the endothermic scission of carbon-carbon bonds with formation of monomer and volatilization, and an exothermic process which leads to the formation of a precursor to char. As the heating rate is reduced, volatilization is reduced in favor of the charring process. Char results from formation of and subsequent transformation of olefinic bonds in the polymer chain, first to crosslinked or polycyclic structures, then to char.

Results and Discussion

In the present study DSC and TGA data were run on DuPont 910 and DuPont 951 instruments, respectively. Arapahoe Smoke Chamber results were obtained on a commercial apparatus. Coated samples used were commercially prepared zinc arc spray samples on Noryl® FN 215 Structural Foam.

An initial experiment involved determination of Arapahoe Smoke Chamber results for samples with and without the zinc coating present. Data are presented in Table II. Depending upon orientation of the sample, an increase in char occurred for some samples with zinc present, while no change in smoke formation was seen. Initial pyrolysis GC/mass spectroscopy results at $900°C$ in helium showed no difference in volatiles formed with or without zinc. These results suggested enhanced char formation as the origin of the Radiant Panel results for zinc on modified-polyphenylene oxide (m-PPO). Zinc oxide is a known, effective thermal stabilizer in the alloy. The next work then focused on DSC/TGA studies.

DSC. DSC was used to study the thermal behavior of the decomposition of m-PPO with and without a zinc coating, when heated in an air or nitrogen atmosphere. Different heating rates were used and varied from $2.5°C/min$ to $20°C/min$. In an inert atmosphere (nitrogen), the DSC trace at different heating rates for m-PPO are shown in Figure 1. The decomposition temperature (Td) increases with higher heating rate. Figure 2 shows the results of Td both in nitrogen and in air. The Td in air is a few degrees higher than in nitrogen. In air, m-PPO is slightly more stable to decomposition than in nitrogen due to the formation of char, which will be discussed later. Figure 3 shows the decomposition behavior of m-PPO in air and in nitrogen. m-PPO resin, when heated in air will absorb energy (endothermic) from $380°C$ to $480°C$ and evolve a variety of decomposition products. The shoulder between $450°C$ to $480°C$ is char formation in the presence of oxygen. In air, above $480°C$, heat is evolved (exothermic) as char is decomposed. Figure 4 shows the heating curves of m-PPO with and a without zinc coating in nitrogen. As is shown, the Zn melted at about $420°C$ (melting point of zinc by DSC is $420.7°C$, during the early stages of m-PPO decomposition). The decomposition temperature of m-PPO is almost the same with or without the zinc coating. Figure 5a and 5b show the m-PPO coated with Zn heated under different rates in nitrogen and in air, respectively.

TGA. Thermogravimetric analysis measurements of m-PPO are shown in Figure 6. In air char and its decomposition is shown between 450 and $650°C$. Figure 7a shows the TGA of m-PPO under different heating rates in nitrogen. In all cases, the m-PPO resin started to lose weight at about $190°C$ and stopped at $\sim350°C$, the weight loss is approximately 5% which is due to the loss of the triaryl phosphate. Decomposition temperatures were measured at the point of maximum by the derivative of the weight loss curve. Td at $5°C/min$ is $448°C$, which increases to $471°C$ at a rate of $20°C/min$. A residue of $\sim17\%$ by weight is found at all three heating rates. In air, Figure 7b, the first part of the weight

Table I. Radiant Panel Test Results for Structural Foam Samples

Materials*	Replicates	I_s	Average Values Q	F_s
Polycarbonate				
Uncoated (white)	4	27.5	14.8	1.87
Uncoated (grey)	1	31.5	16.8	1.88
graphite/acrylic(grey)	4	59.8	19.6	3.08
nickel/acrylic (grey)	5	48.0	15.9	3.33
copper/acrylic (grey)	1	85.1	26.0	3.27
zinc (white)	4	14.4	14.3	1.35
M o d i f i e d - Polyphenylene Oxide				
uncoated	6	84.4	30.6	2.77
graphite/acrylic	4	64.3	15.6	3.99
nickel/acrylic	4	68.1	20.0	3.36
copper/acrylic	4	63.0	22.1	2.84
zinc	4	2.9	2.9	1.00
RIM-Polyurethane				
uncoated	4	173.3	28.5	6.07
copper/epoxy	3	78.3	28.3	2.76
nickel/urethane	3	17.7	8.9	1.97
zinc	4	43.4	24.0	1.81
Hardboard	39	164.4	51.7	3.21
Plywood	3	139.1	33.8	4.12

Instrument Constant β, was 23.7 to 25.4, Average 24.7
*Samples conditioned to ambient room temperature and humidity.

Table II. Arapahoe Smoke Chamber Data for Modified-PPO and Zn

Sample	% Smoke*	% Char*
Experiment 1		
Uncoated	2.6 ± 1.0	4.3 ± 1.9
w/Zn	2.7 ± 0.8	5.1 ± 3.1
Experiment 2		
Uncoated	3.5 ± 1.3	5.7 ± 2.4
w/Zn	4.6 ± 1.5	11.8 ± 3.6
Experiment 3		
Uncoated	3.1 ± 1.3	4.2 ± 1.2
w/Zn	2.7 ± 0.5	3.8 ± 1.7

* of initial sample weight

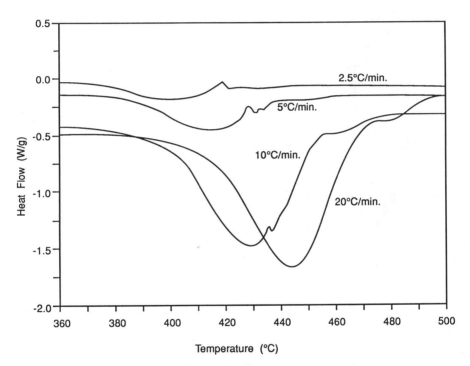

Figure 1.   DSC on m-PPO-Different Heating Rate, In N_2

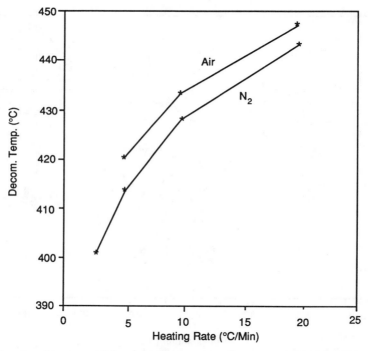

Figure 2.   Decomposition Temperature by DSC with Different Heating
Rate

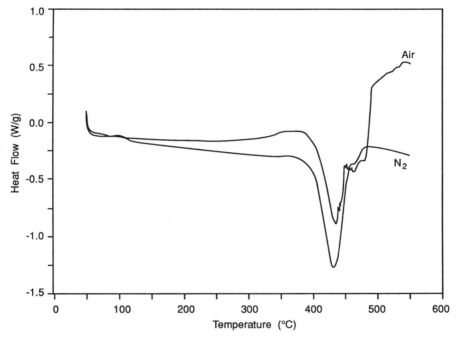

Figure 3.   DSC on m-PPO-In Air and In N_2, 10°C/min

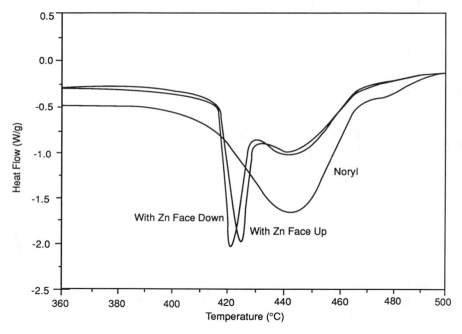

Figure 4.   DSC on m-PPO-With and Without Zn Coating, 20°C/min in N_2

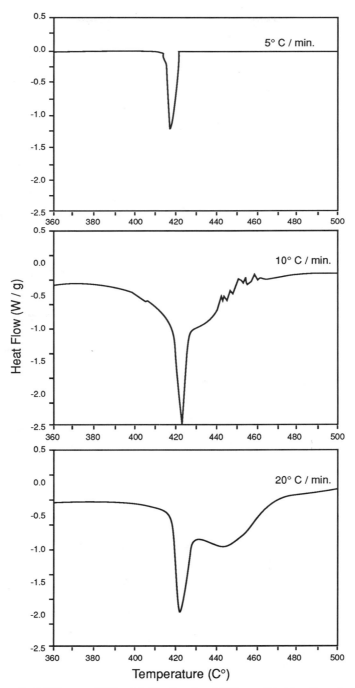

Figure 5a. DSC on m-PPO with Zn Coating, Different Heating Rate in N_2.

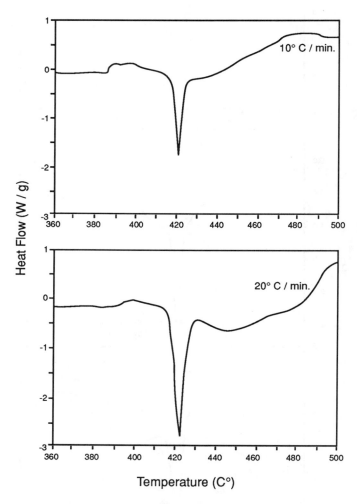

Figure 5b. DSC on m-PPO with Zn Coating, Different Heating Rate in Air.

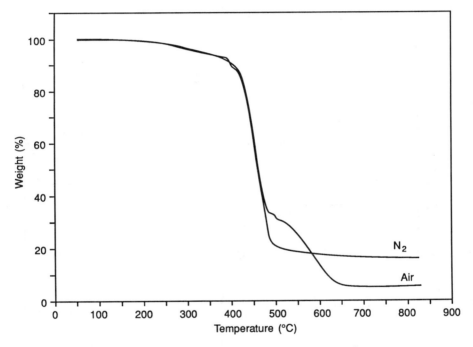

Figure 6.   TGA on m-PPO-In Air and In N_2, $10°$ C/min

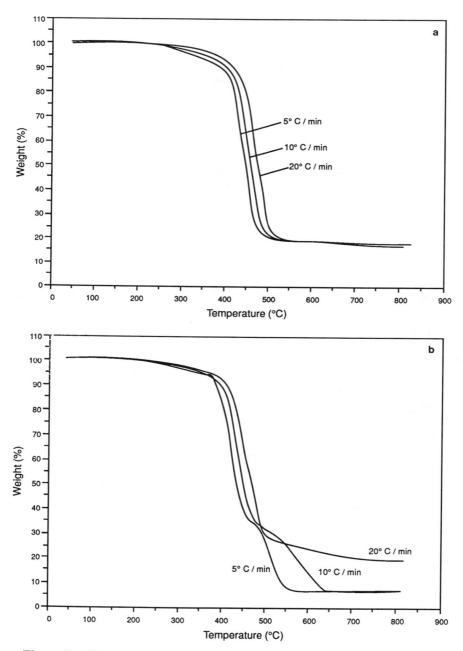

Figure 7.   TGA on m-PPO-Different Heating Rate (a) In N_2; (b) In Air

loss is due to triaryl phosphate as well. The Td at $5°C/min$ is $431°C$ and the second decomposition between $475°C$ to $570°C$ is the formation of char in the presence of oxygen as shown in the DSC. Char formation is also found at heating at $10°C/min$, and in both cases the residue is ~6%. At $20°C/min$, the Td is $460°C$ and char decomposition is less evident. The residue is ~18%. Figure 6 compares the TGA of m-PPO heated at $10°C/min$ in air and in nitrogen. The Td's of both are about the same, $455°C$. In air, the m-PPO coated with zinc yielded an additional weight loss feature between $500°C$ to $600°C$ which is the formation of additional char as shown in Figure 8. The zinc coating enhanced char formation and protected the substrate better. After $600°C$, oxidation of zinc to ZnO yields a gain of 3% by weight. The high weight percent remaining is due to a large amount of zinc in the tested sample. In Figure 9 are shown the decomposition curves of m-PPO and several coated m-PPO's. The m-PPO coated with zinc has the highest Td and highest residue. The zinc metal coating is 5 mils compared to the other commercial coated samples which were nominal 2 mils. The zinc coated sample shows added char, however, between 500-700°C versus the other two EMI coatings.

The decomposition temperature (Td) measured at maximum rate of heat loss from TGA curves was as follows:

Sample	Td(°C)
m-PPO	471
m-PPO/Zn	475
m-PPO/Ni-polyurethane	460
m-PPO/Cu-epoxy	463

modified-Polyphenylene oxide/zinc shows a marginally higher Td(°C). Clearly the metal filled organic coatings show lower Td values.

Isothermal TGA. m-PPO and several coated m-PPO's were isothermally held at $400°C$ for 90 min in nitrogen and air, Figure 10. In air, the weight loss is mainly in the first 20 min. Except for the copper/epoxy coating, the weight is almost the same after 30 min. In a nitrogen atmosphere, m-PPO without coating is the most stable one for the first 15 min. m-PPO with zinc coating is more stable in air than in nitrogen as char is formed in air. In nitrogen the metal filled organic coatings seem to promote degradation.

Pyrolysis-GC/MS. Pyrolysis-GC/MS were performed for m-PPO and m-PPO coated with zinc at $300°C$, $400°C$, $450°C$, $500°C$, $700°C$ and $900°C$ in helium as shown in Figure 11. The object of this study was to see the effect of the different pyrolysis temperatures. At $300°C$, m-PPO started to release volatiles such as phenol, trimethyl phenol, 1,3-diphenylpropane, triphenyl phosphate, styrene, benzene, ethylbenzene and bibenzyl. For m-PPO with zinc, lesser amounts of low molecular weight volatiles are formed at low pyrolysis temperatures. At $400°C$, the volatiles start to be released for m-PPO coated with zinc. After ~700°C, the amounts of volatiles are the same with or without the zinc coating.

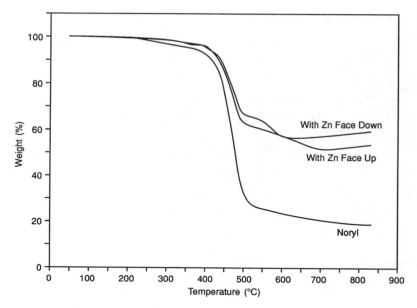

Figure 8. TGA on m-PPO-With and Without Zn Coating, 20°C/min, In Air

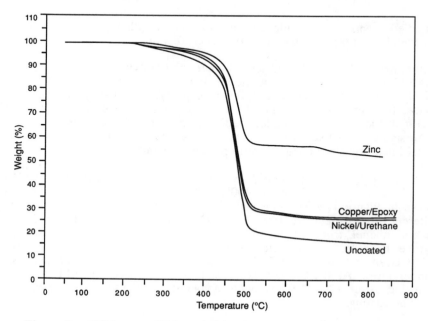

Figure 9. TGA on m-PPO and Coated m-PPO's, 20°C/min, In N₂

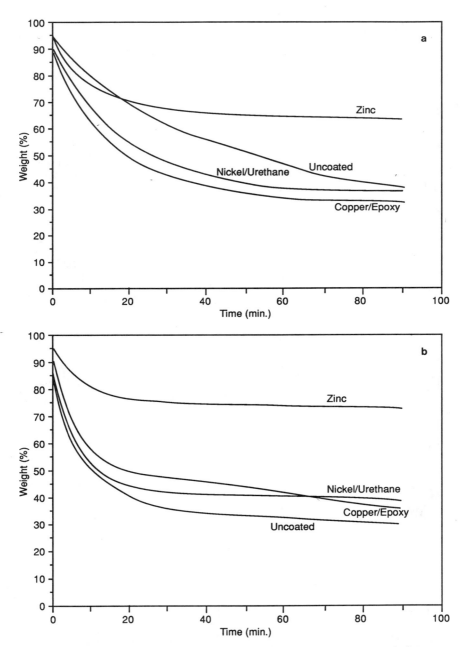

Figure 10. Isothermal TGA on m-PPO and Coated m-PPO's at 400°C (a) In N$_2$; (b) In Air

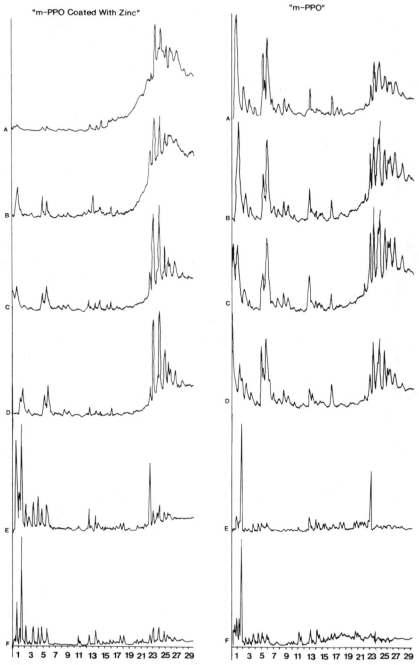

Figure 11. Pyrolysis of M-PPO and M-PPO coated with Zn at (A) 300 °C, (B) 400 °C, (C) 450 °C, (D) 500 °C, and (F) 900 °C.

Char Analysis. Analyses of char samples were performed on specimens prepared at $20°$C/minute and held at temperature for 30 minutes. Below $550°$C carbonaceous char is present. Above $550°$C in air and above $600°$C in nitrogen the residue consists of zinc, zinc oxide, glass and other inorganic species as shown in Table III.

Scanning Electron Microscopy. SEM photographs were taken of samples in air and nitrogen, with and without a zinc coating present. Figure 12 shows the pyrolysis product in air at 200x magnification. Figure 13 shows the pyrolysis products in air (a) and (b) nitrogen at 10000x magnification. Figure 14 shows the pyrolysis products of a zinc coated sample in air (a) and (b) in nitrogen. Zinc oxide nodules are clearly seen in Figure 13 and 14 for all samples, but substantially more so for the zinc coated samples.

Table III. Analysis of Carbon and Hydrogen Content of Char Samples for Modified Polyphenylene Oxide With and Without Zinc Coating

		% Residue on Ignition	% Carbon on Residue	% of Original Weight as Carbon	% Hydrogen on Residue
m-PPO/Zn	400°C	96.17	59.63	57.3	4.85
Nitrogen	500°C	37.82	42.25	38.2	2.07
	550°C	28.79	23.13	6.66	1.06
	600°C	31.17	7.19	2.24	0.35
	700°C	28.53	0.54	0.15	0.08
Air	400°C	96.33	56.24	54.2	4.72
	500°C	26.77	5.39	1.44	0.31
	550°C	20.37	4.56	0.93	0.41
	600°C	25.15	0.23	0.06	0.05
	700°C	22.57	0.25	0.06	0.06
m-PPO	400°C	93.62	78.87	73.8	6.27
Nitrogen	500°C	21.57	55.81	12.1	2.74
	550°C	17.09	42.90	7.3	2.10
	600°C	13.05	18.57	2.4	1.00
	700°C	9.57	1.16	0.11	0.13
Air	400°C	93.88	80.23	75.3	6.45
	500°C	10.09	30.30	3.1	1.12
	550°C	5.72	12.43	0.71	0.65
	600°C	5.16	0.66	0.03	0.17
	700°C	5.20	0.52	0.03	0.09

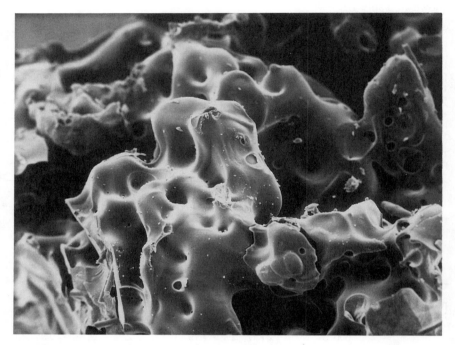

Figure 12. Scanning electron micrograph of m-PPO char pyrolysis in air (mag. 200x)

Conclusion

The presence of a zinc coating on modified polyphenylene oxide shows results different from that expected for a "simple" metal coating. The enhanced fire retardancy has its origins in several factors assisted by the fact that zinc melts (420°C) just at the early stage of the decomposition of m-PPO, allowing intimate contact with the charring substrate. As in pure polystyrene, char formation is enhanced in air in m-PPO, and this is further enhanced by the presence of zinc. As seen in SEM, zinc oxide which is part of the thermal stabilization package of m-PPO is present at the charring surface enhanced by the zinc coating. It is observed that volatilization of small molecules is reduced for m-PPO with zinc present at temperatures under 700°C, with preference for volatilization of the triaryl phosphate flame retardant, styrene trimer, and PPO dimers. The flame retardant and larger entities formed preferentially lead to enhanced char formation.

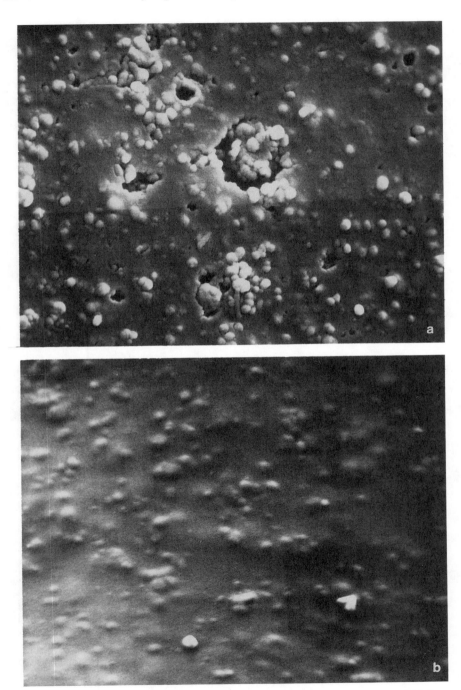

Figure 13. Scanning electron micrographs of m-PPO char (mag. 10,000x); (a) pyrolysis in N_2 and (b) pyrolysis in air

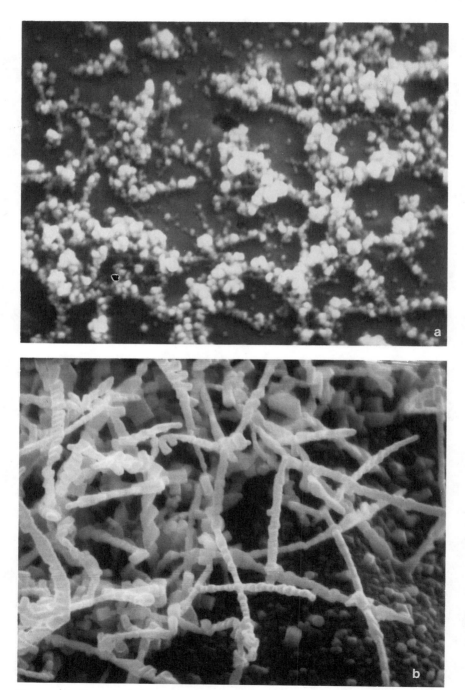

Figure 14. Scanning electron micrograph of m-PPO coated with Zn (mag. 10,000x) (a) pyrolysis in air, (b) pyrolysis in N_2.

Figure 14c. Scanning electron micrograph of m-PPO coated with Zn (mag. 10,000x) pyrolysis in N_2.

References

1. G.L. Nelson, M.L. Bosarge, and K.M. Weaver, Proceedings of the Twelfth International Conference on Fire Safety, (Jan. 12-16, 1987) 12, 271-282 (1987).
2. G.L. Nelson, M.L. Bosarge, K.M. Weaver, Proceedings of the Thirteenth International Conference on Fire Safety, (Jan. 11-15, 1988) 13, 367-378 (1988).
3. G.L. Nelson, Effects of Coatings on the Fire Performance of Plastics, this book.
4. J. Carnahan, W. Haaf, G. Nelson, G. Lee, V. Abolins and P. Shank, Proceedings of the 4th International Conference on Flammability and Safety, (Jan. 15-19, 1979) 4, 312-323 (1979).
5. B. Martel, Journal of Applied Polymer Science, 35, 1213 (1988).
6. H.A. Mackay, Carbon, 8, 517 (1970).
7. K. Kishore and K. Mohandas, Combust. Flame, 43 145 (1981).
8. F.R.S. Clark, J. Polym. Sci., Polym. Chem. Ed., 22 263 (1984).

RECEIVED November 20, 1989

FIRE AND CELLULOSICS

Fire and Cellulosics

The combustion of cellulosics impacts a broad range of interests ranging from uses of wood fuel, through building fires, manufactured and waste material fires, to forest fires. Building fires, especially in domestic buildings, very often involve smoldering combustion of "cellulosic" materials such as insulation and timber. Such materials should more properly be described as "lignocellulosics" because they contain the three major constituents of wood, viz. cellulose, hemicelluloses, and lignin. The same comment also applies to newsprint (the major original source of cellulosic insulation in buildings), which is made predominantly from mechanical wood pulp with very little chemical treatment beyond bleaching.

Lignocellulosics have a rather unique tendency to smoldering combustion in which active combustion occurs predominantly in a pyrolytic char phase, and this type of combustion precedes a portion of domestic building fires. It is also of major importance in propagation and resurgence of forest fires. Yet the whole field of research into ignition and combustion of such materials is remarkably sparsely populated, possibly because of the very complex and variable nature of the lignocellulosic fuels. The situation is dramatically illustrated by a reading of titles from any recent meeting of the Combustion Institute, which abounds with reports of research on combustion of synthetic polymers or liquid fuels, but only rarely includes any aspect of fire and cellulosics. The field of fire retardants, especially in cellulosic textiles, has been very extensively studied in the past, at both fundamental and applied levels, but there has been almost no serious research in this area during the past decade. The papers presented in this section provide representative coverage of current research on fire and cellulosics.

Two papers by Back and by Kubler are concerned with the complex question of pyrolytic and oxidative heat balance in real-life situations at relatively low temperatures (80–230 °C). This heat balance determines the catastrophic onset of "spontaneous" ignition of materials such as reconstituted board packs, haystacks, paper wastes, wood chip and sawdust heaps, etc. The review by Kubler represents a landmark in correlation of our understanding of such complex events.

The chapter by Tran describes measurements of heat release from full-scale wooden wall fires, and that by Hendrickson is concerned with optimization of water consumption in fighting forest fires by use of highly stabilized high expansion foams.

Finally, two laboratory-scale studies are reported. Schultz and coworkers describe an investigation of various approaches to data analysis of the pyrolysis phase in isothermal thermogravimetric analysis of cellulose with various fire retardant additives. Hshieh and Richards report on one of the very few chemical studies currently underway on fire and cellulosics, describing the often dramatic influences of natural and of added metal ions on ignition and combustion of wood.

The range of the fire situations encompassed in these few chapters is remarkable, covering as it does the spectrum from mountains of wood waste (Kubler) to milligrams of solid wood (Hshieh).

G. N. Richards
Wood Chemistry Laboratory
University of Montana
Missoula, MT 09812

Chapter 22

Effect of Crystallinity and Additives on the Thermal Degradation of Cellulose

Tor P. Schultz, Gary D. McGinnis, and Darrel D. Nicholas

Mississippi Forest Products Laboratory, Mississippi State University, Mississippi State, MS 39762

Thermogravimetric analysis (TGA) measures cellulose pyrolytic mass loss rates and activation parameters. The technique is relatively simple, straightforward and fast, but it does have disadvantages. One disadvantage is that determination of the kinetic rate constants from TGA data is dependent on the interpretation/analysis technique used. Another disadvantage of TGA is that the rate of mass loss is probably not equivalent to the cellulose pyrolysis rate.

In this study five cellulose samples of different crystallinities (10, 41, 63, 67, and 74%) were treated to 10% by weight with H_3PO_4, H_3BO_3, and $AlCl_3 \cdot 6H_2O$. These treated samples and untreated (control) samples were isothermally pyrolyzed under N_2 at selected temperatures and the TGA data analyzed by four methods (0-, 1st-, and 2nd-order; and Wilkinson's approximation) to obtain rates of mass loss. From these rates, activation energy (E_a), activation entropy ($\Delta S^{\ddagger}$) and enthalpy ($\Delta H^{\ddagger}$) values were obtained. E_a was also determined by the integral conversion method.

Both 1st- and 2nd-order rate expressions gave statistically good fits for the control samples, while the treated samples were statistically best analyzed by 2nd-order kinetics. The rate constants, 1st-order activation parameters, and char/residue yields for the untreated samples were related to cellulose crystallinity. In addition, $\Delta S^{\ddagger}$ values for the control samples suggested that the pyrolytic reaction proceeds through an ordered transition state. The mass loss rates and activation parameters for the phosphoric acid-treated samples implied that the mass loss mechanism was different from that for the control untreated samples. The higher rates of mass loss and

0097–6156/90/0425–0335$07.50/0

low E_a's of the phosphoric acid samples also suggest
that wood products treated with fire retardants which
contain ammonium phosphates and then exposed to
moderate temperatures over long times may undergo
extensive degradation. Under the conditions studied
boric acid appeared to be the best fire retardant.
This conclusion is based on a high char yield and
similar rates of mass loss at 300°C for untreated and
boric acid treated samples. Boric acid samples also
had much higher $\Delta H^{\ddagger}$'s and, consequently, higher E_a's.
Our results suggest that certain thermally-stable,
weak polybasic acids which can complex with
polysaccharides may provide fire-resistant properties
to lignocellulosics. The results and conclusions were
strongly influenced by the technique used to analyze
the TGA data.

In order to fireproof wood and cotton products and to thermally
convert biomass into chemicals, researchers must understand
cellulose pyrolysis. Extensive research has been conducted in this
area and several reviews are available (1-7).
    Thermogravimetric analysis (TGA) has often been used to
determine pyrolysis rates and activation energies (E_a). The
technique is relatively fast, simple and convenient, and many
experimental variables can be quickly examined. However for
cellulose, as with most polymers, the kinetics of mass loss can be
extremely complex (8) and isothermal experiments are often needed to
separate and identify temperature effects (9). Also, the rate of
mass loss should not be assumed to be related to the pyrolysis
kinetic rate (4) since multiple competing reactions which result in
different mass losses occur. Finally, kinetic rate values obtained
from TGA can be dependent on the technique used to analyze the data.
    Analysis of cellulose TGA data involves obtaining a corrected
mass (Mc), which is dimensionless, to account for residual char and
is calculated by:

$$Mc = \frac{M - Mr}{Mo - Mr} \qquad (1)$$

where M is the mass at time t; Mr is the char or residual material;
and Mo the initial mass (4,10). From the Mc values over time,
kinetic rates of mass loss based on 0-order (11), 1st-order
(2-4,7,8,10-16) and 2nd-order reactions (17) have been examined.
Other complex reactions reported include multiple competitive
(8,18-20) or consecutive reactions (18,21-23) and combinations of
these (4,6). Techniques are also available for determination of
rate, order and/or E_a from isothermal TGA data with no
interpretation or assumptions necessary (17,24,25).
    Cellulose crystallinity has been shown to affect pyrolysis
rates and E_a's (2,26,27). The initial low temperature decomposition
is reported to occur first in the amorphous region (5,26,27). Also,

levoglucosan is believed to be formed mainly from crystalline regions while dehydration to form char and furan derivatives takes place in the amorphous areas of cellulose (2,6,28).

Additives, such as fire retardants, can have a major effect on pyrolysis, and even trace amounts of ash have been shown to influence pyrolysis (6). Generally, fire retardants work by increasing the dehydration reaction rate to form more char and as a direct result give fewer flammable volatile compounds (1,3,7). Several papers have noted that phosphoric acid and its salts decrease the E_a (13,18,22,29), aluminum chloride has little effect (22) on E_a and boric acid increases the E_a (12,18). The reaction order for treated samples has been generally reported as 1st-order (12,13,18,29) which is also the most commonly used rate expression for analysis of TGA data of untreated cellulose.

The objective of this research was to examine the effect of crystallinity, additives and data analysis technique on isothermally pyrolyzed cellulose. The E_a, activation enthalpy ($\Delta H^{\ddagger}$) and activation entropy ($\Delta S^{\ddagger}$) were determined from the mass loss rates. This data was used to develop an understanding of how cellulose pyrolysis is affected by crystallinity and additives and how the results obtained are dependent on the data analysis technique.

Experimental

Cotton linter samples were decrystallized by ball milling for various times to give crystallinity indices (CI) of 10, 41, 63, 67 and 74% (27,30). These samples were then treated with aqueous solutions of H_3PO_4, H_3BO_3, or $AlCl_3 \cdot 6H_2O$ to provide retentions of 10% by weight of the solute, then vacuum dried and frozen prior to pyrolysis experiments. The effect of wetting on the CI was not determined. Ash was not removed from the samples because we believed that removal of the ash would further affect the molecular weight and because the crystalline and amorphous regions would be altered to different extents.

The TGA system was a Perkin-Elmer TGS-2 thermobalance with System 4 controller. Sample mass was 2 to 4 mgs with a N_2 flow of 30 cc/min. Samples were initially held at $110^{\circ}C$ for 10 minutes to remove moisture and residual air, then heated at a rate of $150^{\circ}C/min$ to the desired temperature set by the controller. TGA data from the initial four minutes once the target pyrolysis temperature was reached was not used to calculate rate constants in order to avoid temperature lag complications. Reaction temperature remained steady and was within $2^{\circ}C$ of the desired temperature. The actual observed pyrolysis temperature was used to calculate activation parameters. The dimensionless "weight/mass" Mc was calculated using Equation 1. Instead of calculating Mr by extrapolation of the isothermal plot to infinity, Mr was determined by heating each sample/additive to $550^{\circ}C$ under N_2. This method was used because cellulose TGA rates have been shown to follow Arrhenius plots (4,8,10-12,15,16,19,23,26,31). Thus, Mr at infinity should be the same regardless of the isothermal pyrolysis temperature. A few duplicate runs were made to insure that the results were reproducible and not affected by sample size and/or mass. The Mc values were calculated at 4-minute intervals to give 14 data points per run. These values were then used to

calculate 0-order ($\underline{11}$), 1st-order ($\underline{2-4,8,14,15}$) and 2nd-order ($\underline{17}$)
rate constants. Wilkinson's approximation method ($\underline{24,25}$) was also
used to obtain rates and reaction orders. Because of the
dimensionless value of Mc and the way in which all rate constants
were determined, all rates had units of sec^{-1} and can be directly
compared. E_a and the rate constants at $300^\circ C$ were determined by the
Arrhenius equation, and $\Delta H^{\ddagger}$ and $\Delta S^{\ddagger}$ were calculated by the Eyring
equation. E_a was also calculated by the integral conversion method
($\underline{17}$).

Results and Discussion

Tables I, III, V, and VII give the kinetic mass loss rate constants.
Tables II, IV, VI, and VIII present the activation parameters. In
addition to the activation parameters, the rates were normalized to
$300^\circ C$ by the Arrhenius equation in order to eliminate any
temperature effects. Table IX shows the char/residue (Mr), as
measured at $550^\circ C$ under N_2.

Control Samples. The 0-, 1st- and 2nd-order mass loss rate
constants (Table I) generally gave statistically good fits, with
most $r^2 > 99\%$. Thus, no kinetic expression seems "best" for
expressing mass loss. This is partially because only small mass
changes occurred at the lower temperatures, especially for the more
crystalline samples. With small mass losses, an essentially
straight and almost horizontal line is obtained regardless of the
method used. These include a normal plot for 0-order, log Mc for
1st-order, and inverse Mc for 2nd-order. The 1st- and 2nd-order
rate expressions both appeared to give slightly better fits than
0-order.
    Researchers in previous studies generally used 1st-order
kinetics to describe cellulose pyrolysis, but rarely have they
examined 2nd-order kinetics. Thus, discussion of our results for
untreated samples will concentrate on 1st-order rate constants so
that our results can be directly compared with results from prior
studies. A true reaction order of cellulose pyrolysis based on TGA
data is essentially meaningless, however, since mass loss involves
complex competing multiple reactions ($\underline{2,4,8}$). In addition, reaction
order was calculated on a dimensionless mass value rather than on
the correct but uncalculable molar concentration term.
    For all samples, including the treated celluloses, Wilkinson's
approximation generally gave unrealistic rate constants, reaction
orders, and especially E_a's. Although this method is easy and
requires no assumption of order, values obtained by this technique
should be viewed cautiously.
    The 1st-order rate constants, and also the 0- and 2nd-order
rates, appeared to be related to crystallinity, with the more
crystalline samples having smaller rate constants. The log of the
$300^\circ C$ 1st-order rate constants (Table II), as measured by the
Arrhenius equation to avoid slight temperature errors, gave a good
correlation ($r^2 = 0.94$) when regressed against crystallinity. As
mentioned earlier, crystallinity has been previously recognized to
affect pyrolysis ($\underline{2,6,26-28}$). The 1st-order E_a's (Table II) and the
Arrhenius plot of the 1st-order rates (Figure 1) also show a

crystalline effect, which is in agreement with previous studies
(5,26). Specifically, the amorphous sample had a significantly
lower E_a (13 kcal) than the other samples (33-40 kcal). The 0-order
but not the 2nd-order E_a's also indicate a crystalline influence
(Table II). Finally, the char yield (Table IX) is also related to
crystallinity which agrees with previous studies (2,6,28) that the
amorphous regions form more char and crystalline cellulose more
volatiles. We found an excellent fit (r^2 = 0.96) when the log of
the char yield was regressed against crystallinity.

The cellulose pyrolytic mechanism has been generally proposed
to be either a thermal homolytic (radical) bond cleavage or a
heterolytic (ionic) internal nucleophilic displacement of the
glucosidic group by an adjacent hydroxyl to give an anhydro-sugar
(4,6,7,21,31,32). The highly negative $\Delta S^{\ddagger}$ values which we obtained
suggest that the reaction may proceed through a very ordered
transition state (33). An ordered transition indicates a
nucleophilic displacement rather than a radical mechanism. This is
in agreement with recent work (6) which reports that radicals are
associated with char formation (minimal mass loss) and that
levoglucosan formation with a corresponding high mass loss proceeds
via the nucleophilic displacement pathway.

If the mechanism does proceed through an internal nucleophilic
pathway and if the rate of mass loss is equivalent to the pyrolysis
rate, the $\Delta H^{\ddagger}$ and $\Delta S^{\ddagger}$ values should be affected by crystallinity.
Specifically, the nucleophilic hydroxyl in an amorphous sample would
break relatively few hydrogen bonds (low $\Delta H^{\ddagger}$) (26) in order to
obtain the cylic transition state. However, the amorphous sample
would experience a largely negative $\Delta S^{\ddagger}$ value, since the transition
state is ordered while the amorphous "reactant" is relatively
disordered. Conversely, a hydroxyl group in a crystalline region
would require extensive hydrogen bond cleavage and consequently a
large $\Delta H^{\ddagger}$. In addition, the $\Delta S^{\ddagger}$ of crystalline cellulose should be
more positive than that for amorphous cellulose since the "reactant"
is more ordered. In Table II the 1st-order activation parameters
but not the 2nd-order activation parameters appear to follow this
trend.

E_a's were also determined by the integral conversion method
(17). This method does not require assumption of order or
determination of rate constants. The integral conversion method may
have limited usefulness since the values obtained did not always
agree with the E_a values obtained by the Arrhenius equation of the
0-, 1st- or 2nd-order constants.

Boric Acid. The statistical fits for the different rate expressions
(Table III) continued to give high r^2 values for the cellulose/H_3BO_3
samples. Unlike the control samples, the 2nd-order equation
generally gave the statistically best fit.

Wilkinson's approximation method for analyzing TGA data gave
high rates and reaction orders, and sometimes gave impossibly
negative E_a's and $\Delta H^{\ddagger}$'s (Table III, IV). E_a values obtained by
integral conversion were not always similar to the value obtained by
the Arrhenius equation of the 2nd-order rates.

Table I. Comparison of Cellulose Pyrolysis Rates (Rates of Weight Loss) Analyzed by Differential Methods for Control (Untreated) Samples. All Rates are in Units of 10^-4 / Sec

Crystallinity Index, %	Method (1)	Reaction Temperature (2), °C					
		300			310		
		Rate ± S.D.	(r^2)	n	Rate ± S.D.	(r^2)	n
10	A	1.17 ± 0.07	(0.958)	0	*1.34 ± 0.11	(0.920)	0
	B	2.08 ± 0.07	(0.987)	1	*2.94 ± 0.14	(0.973)	1
	C	3.78 ± 0.02	(0.999)	2	*6.78 ± 0.09	(0.999)	2
	D	8.06 ± 0.62	(0.987)	2.8	*11.47 ± 0.65	(0.995)	2.3
41	A	0.32 ± 0.01	(0.982)	0	*0.48 ± 0.01	(0.993)	0
	B	0.35 ± 0.01	(0.985)	1	*0.54 ± 0.00	(0.996)	1
	C	0.38 ± 0.01	(0.989)	2	*0.62 ± 0.01	(0.999)	2
	D	1.10 ± 0.07	(0.964)	10.3	*1.46 ± 0.11	(0.934)	6.8
63	A	0.22 ± 0.01	(0.986)	0	*0.34 ± 0.01	(0.994)	0
	B	0.23 ± 0.00	(0.988)	1	*0.37 ± 0.01	(0.996)	1
	C	0.24 ± 0.01	(0.990)	2	*0.41 ± 0.01	(0.998)	2
	D	0.31 ± 0.09	(0.825)	6.7	*0.93 ± 0.15	(0.927)	9.5
67	A	0.13 ± 0.01	(0.981)	0	*0.30 ± 0.00	(0.999)	0
	B	0.13 ± 0.01	(0.983)	1	*0.32 ± 0.00	(0.999)	1
	C	0.14 ± 0.01	(0.984)	2	0.35 ± 0.00	(0.999)	2
	D	0.59 ± 0.06	(0.958)	25.6	0.64 ± 0.05	(0.830)	9.6
74	A	0.12 ± 0.00	(0.989)	0	0.26 ± 0.00	(0.999)	0
	B	0.13 ± 0.00	(0.989)	1	0.28 ± 0.00	(0.999)	1
	C	0.13 ± 0.00	(0.990)	2	0.30 ± 0.00	(0.999)	2
	D	0.45 ± 0.04	(0.940)	26.7	0.54 ± 0.04	(0.807)	10.5

(1) A = zero-order; B = first-order; C = second-order; D = Wilkinson's Approximation, where slope is equal to n/2 and the intercept is the inverse of the rate.

(2) Indicates target temperature. The actual temperature differed ± 2 deg.C but remained fairly constant The actual temperature was used to calculate activation parameters.

* Indicates a sample which was run in duplicate.

Continued on next page

Table I. Continued

Crystallinity Index, %	Method (1)	Reaction Temperature (2), °C					
		320			330		
		Rate ± S.D.	(r^2)	n	Rate ± S.D.	(r^2)	n
10	A	1.26 ± 0.16	(0.884)	0	0.90 ± 0.14	(0.781)	0
	B	3.58 ± 0.27	(0.934)	1	3.63 ± 0.38	(0.884)	1
	C	11.40 ± 0.68	(0.959)	2	16.09 ± 0.98	(0.957)	2
	D	19.57 ± 0.71	(0.999)	2.0	41.10 ± 2.12	(0.999)	2.2
41	A	0.73 ± 0.01	(0.998)	0	1.35 ± 0.01	(0.999)	0
	B	0.88 ± 0.00	(0.999)	1	1.95 ± 0.03	(0.996)	1
	C	1.08 ± 0.02	(0.997)	2	2.88 ± 0.12	(0.979)	2
	D	1.75 ± 0.13	(0.885)	4.1	2.58 ± 0.17	(0.833)	2.0
63	A	0.53 ± 0.01	(0.996)	0	0.84 ± 0.01	(0.999)	0
	B	0.61 ± 0.01	(0.998)	1	1.05 ± 0.01	(0.999)	1
	C	0.70 ± 0.01	(0.999)	2	1.31 ± 0.03	(0.995)	2
	D	1.24 ± 0.08	(0.912)	5.5	1.73 ± 0.12	(0.854)	3.3
67	A	0.40 ± 0.01	(0.995)	0	* 0.57 ± 0.01	(0.999)	0
	B	0.44 ± 0.01	(0.997)	1	* 0.64 ± 0.00	(0.999)	1
	C	0.49 ± 0.01	(0.998)	2	* 0.74 ± 0.01	(0.998)	2
	D	0.96 ± 0.06	(0.919)	2.4	* 0.97 ± 0.05	(0.853)	4.1
74	A	0.34 ± 0.01	(0.992)	0	* 0.59 ± 0.00	(0.999)	0
	B	0.37 ± 0.01	(0.992)	1	* 0.68 ± 0.01	(0.999)	1
	C	0.40 ± 0.01	(0.992)	2	* 0.78 ± 0.01	(0.996)	2
	D	0.74 ± 0.11	(0.878)	8.5	* 1.06 ± 0.07	(0.799)	4.2

Table II. Activation Parameters for Pyrolysis (Rate of Weight Loss)
for Untreated Cellulose Samples Based on Data in Table I

Crystallinity Index, %	Method (1)	Activation Parameters
		Ea (kcal/Mole) ± S.D.
10	A	6 ± 4
	B	13 ± 3
	C	34 ± 2
	D	36 ± 5
	E	48 ± 4
41	A	34 ± 3
	B	40 ± 5
	C	47 ± 6
	D	19 ± 2
	E	42 ± 1
63	A	31 ± 2
	B	35 ± 2
	C	40 ± 2
	D	36 ± 9
	E	44 -
67	A	31 ± 5
	B	33 ± 5
	C	35 ± 5
	D	11 ± 3
	E	27 -
74	A	35 ± 3
	B	36 ± 3
	C	39 ± 4
	D	20 ± 2
	E	36 -

(1) A = zero-order; B = first-order; C = second-order;
     D = Wilkinson's Approximation; E = Integral Conversion
     Method.
(2) Determined using the Arrhenius equation.

Continued on next page

Table II. Continued

$k_{300}(10^{-4}/sec)$ (2)	Activation Parameters	
	$\Delta H^{\ddagger}$ (kcal/Mole) $\pm$ S.D.	$\Delta S^{\ddagger}$ (e.u.) $\pm$ S.D.
1.32	7 $\pm$ 4	-90 $\pm$ 7
2.28	12 $\pm$ 3	-56 $\pm$ 5
4.40	33 $\pm$ 2	-18 $\pm$ 4
7.37	35 $\pm$ 5	-13 $\pm$ 8
-	-	-
0.29	33 $\pm$ 3	-24 $\pm$ 6
0.31	39 $\pm$ 5	-12 $\pm$ 5
0.32	46 $\pm$ 6	0 $\pm$ 10
1.08	18 $\pm$ 2	-47 $\pm$ 3
-	-	-
0.21	30 $\pm$ 2	-28 $\pm$ 2
0.22	34 $\pm$ 2	-21 + 3
0.23	38 $\pm$ 2	-14 $\pm$ 4
0.45	35 $\pm$ 9	-19 $\pm$ 15
-	-	-
0.15	30 $\pm$ 5	-30 $\pm$ 8
0.16	32 $\pm$ 5	-26 $\pm$ 8
0.17	34 $\pm$ 5	-23 $\pm$ 8
0.59	10 $\pm$ 3	-62 $\pm$ 6
-	-	-
0.13	33 $\pm$ 3	-24 $\pm$ 5
0.14	35 $\pm$ 3	-22 $\pm$ 5
0.14	38 $\pm$ 3	-16 $\pm$ 6
0.43	19 $\pm$ 2	-47 $\pm$ 3
-	-	-

Table III. Comparison of Cellulose Pyrolysis Rates (Rates of Weight loss) Analyzed by Differential Methods for Cellulose Samples Treated with 10% Boric Acid. All Rates are in Units of 10^{-4} / Sec

Crystallinity Index, %	Method (1)	Reaction Temperature (2), °C					
		290			300		
		Rate ± S.D.	(r^2)	n	Rate ± S.D.	(r^2)	n
10	A	0.44 ± 0.01	(0.993)	0	0.64 ± 0.02	(0.993)	0
	B	0.53 ± 0.01	(0.996)	1	0.81 ± 0.01	(0.997)	1
	C	0.64 ± 0.01	(0.998)	2	1.04 ± 0.01	(0.999)	2
	D	2.91 ± 0.35	(0.962)	7.1	3.43 ± 0.39	(0.955)	5.2
41	A	0.15 ± 0.00	(0.996)	0	0.38 ± 0.00	(0.999)	0
	B	0.16 ± 0.00	(0.996)	1	0.43 ± 0.00	(0.999)	1
	C	0.17 ± 0.00	(0.996)	2	0.49 ± 0.00	(0.999)	2
	D	1.69 ± 0.29	(0.967)	18.4	1.52 ± 0.19	(0.907)	8.6
63	A*	0.14 ± 0.00	(0.995)	0	0.28 ± 0.00	(0.998)	0
	B*	0.15 ± 0.00	(0.994)	1	0.31 ± 0.00	(0.998)	1
	C*	0.16 ± 0.00	(0.999)	2	0.35 ± 0.01	(0.997)	2
	D*	1.48 ± 0.36	(0.960)	20.0	1.42 ± 0.20	(0.919)	11.5
67	A	0.09 ± 0.00	(0.987)	0	0.21 ± 0.00	(0.998)	0
	B	0.10 ± 0.00	(0.987)	1	0.24 ± 0.00	(0.999)	1
	C	0.11 ± 0.00	(0.987)	2	0.29 ± 0.00	(0.999)	2
	D	5.02 ± 0.91	(0.995)	17.4	1.68 ± 0.25	(0.957)	14.2
74	A	0.10 ± 0.00	(0.990)	0	0.20 ± 0.00	(0.999)	0
	B	0.10 ± 0.00	(0.990)	1	0.24 ± 0.00	(0.999)	1
	C	0.11 ± 0.00	(0.989)	2	0.25 ± 0.00	(0.999)	2
	D	3.14 ± 0.64	(0.988)	19.6	2.27 ± 0.38	(0.967)	13.7

(1) A = zero-order; B = first-order; C = second-order; D = Wilkinson's Approximation, where slope is equal to n/2 and the intercept is the inverse of the rate.

(2) Indicates target temperature. The actual temperature differed ± 2 deg.C but remained fairly constant (± 0.2 deg.C) during the run. The actual temperature was used to calculate activation parameters.

* Indicates a sample which was run in duplicate.

Continued on next page

Table III. Continued

Crystallinity Index, %	Method (1)	Reaction Temperature (2), °C					
		310			320		
		Rate ± S.D.	(r^2)	n	Rate ± S.D.	(r^2)	n
10	A	0.96 ± 0.06	(0.960)	0	1.04 ± 0.11	(0.876)	0
	B	1.41 ± 0.06	(0.980)	1	1.79 ± 0.15	(0.924)	1
	C	2.09 ± 0.05	(0.993)	2	3.15 ± 0.18	(0.961)	2
	D	5.22 ± 0.36	(0.984)	3.4	9.64 ± 0.25	(0.999)	3.0
41	A	0.77 ± 0.02	(0.991)	0	1.18 ± 0.08	(0.952)	0
	B	0.96 ± 0.02	(0.996)	1	1.72 ± 0.07	(0.979)	1
	C	1.2 ± 0.01	(0.999)	2	2.54 ± 0.05	(0.995)	2
	D	2.13 ± 0.17	(0.909)	4.1	3.62 ± 0.09	(0.993)	2.5
63	A	0.53 ± 0.01	(0.999)	0	1.01 ± 0.03	(0.991)	0
	B	0.63 ± 0.01	(0.999)	1	1.30 ± 0.02	(0.998)	1
	C	0.75 ± 0.01	(0.998)	2	1.70 ± 0.01	(0.999)	2
	D	2.13 ± 0.26	(0.910)	6.3	1.99 ± 0.09	(0.925)	2.6
67	A	0.37 ± 0.01	(0.997)	0	0.88 ± 0.01	(0.997)	0
	B	0.42 ± 0.01	(0.997)	1	1.11 ± 0.02	(0.998)	1
	C	0.47 ± 0.01	(0.995)	2	1.40 ± 0.03	(0.994)	2
	D	1.37 ± 0.18	(0.879)	8.8	1.77 ± 0.16	(0.765)	3.0
74	A	0.41 ± 0.00	(0.999)	0	0.93 ± 0.01	(0.998)	0
	B	0.46 ± 0.00	(0.999)	1	1.18 ± 0.01	(0.999)	1
	C	0.51 ± 0.01	(0.999)	2	1.50 ± 0.03	(0.996)	2
	D	1.05 ± 0.10	(0.842)	7.4	1.77 ± 0.12	(0.826)	2.8

Table IV. Activation Parameters for Pyrolysis (Rate of Weight Loss)
for Cellulose Samples Treated with 10% Boric Acid,
Based on Data in Table III

Crystallinity Index, %	Method (1)	Activation Parameters Ea (kcal/Mole) ± S.D.
10	A	20 ± 4
	B	28 ± 3
	C	36 ± 3
	D	26 ± 5
	E	58 ± 1
41	A	46 ± 5
	B	53 ± 4
	C	60 ± 3
	D	17 ± 7
	E	55 -
63	A	43 ± 2
	B	47 ± 2
	C	52 ± 3
	D	5 ± 5
	E	40 -
67	A	50 ± 2
	B	52 ± 3
	C	55 ± 5
	D	-23 ± 14
	E	40 -
74	A	51 ± 1
	B	56 ± 2
	C	59 ± 3
	D	-17 ± 13
	E	33 -

(1) A = zero-order; B = first-order; C = second-order;
D = Wilkinson's Approximation; E = Integral Conversion Method.

(2) Determined using the Arrhenius equation.

Continued on next page

Table IV. Continued

Activation Parameters

k_{300} (10^-4/sec) (2)	$\Delta H^{\ddagger}$ (kcal/Mole) ± S.D.	$\Delta S^{\ddagger}$ (e.u.) ± S.D.
0.60	19 ± 4	-47 ± 6
0.79	27 ± 3	-32 ± 6
1.04	35 ± 3	-17 ± 6
3.73	25 ± 5	-31 ± 9
-	-	-
0.31	45 ± 5	- 3 ± 8
0.35	51 ± 4	10 ± 6
0.40	59 ± 3	22 ± 4
1.79	16 ± 7	49 ± 12
-	-	-
0.25	42 ± 2	- 8 ± 4
0.28	46 ± 2	- 1 ± 4
0.32	50 ± 3	8 ± 4
1.69	4 ± 5	-70 ± 8
-	-	-
0.17	49 ± 2	3 ± 3
0.19	51 ± 3	8 ± 5
0.22	53 ± 5	12 ± 8
2.65	-24 ± 14	-118 ± 24
-	-	-
0.17	50 ± 1	6 ± 2
0.19	55 ± 2	14 ± 4
0.20	58 ± 3	20 ± 6
2.26	-18 ± 13	-108 ± 22
-	-	-

Table V. Comparison of Cellulose Pyrolysis Rates (Rates of Weight Loss) Analyzed by Differential Methods for Cellulose Treated with 10% Phosphoric Acid. All Rates are in Units of 10^-4 / Sec

Crystallinity Index, %	Method (1)	240 Rate ± S.D.	(r^2)	n	250 Rate ± S.D.	(r^2)	n	260 Rate ± S.D.	(r^2)	n
10	A	0.40 ± 0.03	(0.927)	0	0.55 ± 0.04	(0.946)	0	0.67 ± 0.05	(0.937)	0
	B	0.54 ± 0.04	(0.940)	1	0.83 ± 0.05	(0.962)	1	1.13 ± 0.07	(0.960)	1
	C	0.73 ± 0.05	(0.951)	2	1.26 ± 0.06	(0.976)	2	1.92 ± 0.08	(0.978)	2
	D	9.52 ± 0.97	(0.996)	5.8	11.03 ± 1.35	(0.994)	4.4	13.73 ± 1.43	(0.995)	3.7
41	A	0.59 ± 0.05	(0.933)	0	0.67 ± 0.05	(0.930)	0	0.77 ± 0.07	(0.900)	0
	B	0.92 ± 0.06	(0.955)	1	1.15 ± 0.07	(0.956)	1	1.67 ± 0.12	(0.941)	1
	C	1.44 ± 0.07	(0.972)	2	2.00 ± 0.09	(0.976)	2	3.69 ± 0.18	(0.973)	2
	D	11.79 + 1.24	(0.995)	4.2	14.63 ± 1.50	(0.996)	3.6	21.59 ± 1.84	(0.973)	2.9
63	A	0.48 ± 0.04	(0.928)	0	0.55 ± 0.05	(0.921)	0	0.60 ± 0.05	(0.920)	0
	B	0.68 ± 0.05	(0.945)	1	0.85 ± 0.06	(0.942)	1	1.05 ± 0.07	(0.946)	1
	C	0.96 ± 0.06	(0.959)	2	1.33 ± 0.08	(0.960)	2	1.85 ± 0.10	(0.967)	2
	D	10.11 ± 1.06	(0.995)	5.1	12.84 ± 1.27	(0.996)	4.3	17.04 ± 1.75	(0.997)	3.6
67	A	0.47 ± 0.03	(0.945)	0	0.63 ± 0.05	(0.941)	0	0.67 ± 0.04	(0.953)	0
	B	0.64 ± 0.04	(0.958)	1	0.96 ± 0.06	(0.961)	1	1.06 ± 0.07	(0.972)	1
	C	0.87 ± 0.05	(0.969)	2	1.49 ± 0.07	(0.977)	2	1.70 ± 0.06	(0.986)	2
	D	8.41 ± 0.90	(0.994)	5.5	10.62 ± 1.07	(0.994)	4.1	10.73 ± 1.20	(0.992)	3.9
74	A	0.46 ± 0.03	(0.941)	0	0.65 ± 0.05	(0.934)	0	0.74 ± 0.06	(0.916)	0
	B	0.61 ± 0.04	(0.954)	1	1.02 ± 0.06	(0.957)	1	1.35 ± 0.09	(0.948)	1
	C	0.81 ± 0.04	(0.966)	2	1.63 ± 0.07	(0.975)	2	2.50 ± 0.12	(0.973)	2
	D	7.51 ± 0.77	(0.994)	5.7	11.44 ± 1.47	(0.994)	4.0	15.79 ± 1.41	(0.997)	3.3

(1) A = zero-order; B = first-order; C = second-order; D = Wilkinson's Approximation.

(2) Indicates target temperature. The actual temperature differed ± 2 deg.C but remained fairly constant (± 0.2 deg.C) during the run. The actual temperature was used to calculate activation parameters.

* Indicates a sample which was run in duplicate.

Continued on next page

Table V. Continued

Crystallinity Index, %	Method (1)	Reaction Temperature (2), °C					
		270			280		
		Rate ± S.D.	(r^2)	n	Rate ± S.D.	(r^2)	n
10	A	* 0.73 ± 0.07	(0.901)	0	0.57 ± 0.08	(0.806)	0
	B	* 1.48 ± 0.11	(0.938)	1	1.48 ± 0.17	(0.861)	1
	C	* 3.05 ± 0.16	(0.967)	2	3.96 ± 0.36	(0.908)	2
	D	*20.26 ± 1.72	(0.998)	3.0	43.35 ± 2.75	(0.999)	2.8
41	A	0.68 ± 0.08	(0.860)	0	0.51 ± 0.07	(0.821)	0
	B	1.64 ± 0.15	(0.911)	1	1.57 ± 0.17	(0.879)	1
	C	4.05 ± 0.27	(0.950)	2	4.97 ± 0.41	(0.926)	2
	D	30.72 ± 2.30	(0.999)	2.8	56.45 ± 4.24	(0.999)	2.6
63	A	0.65 ± 0.06	(0.898)	0	0.60 ± 0.07	(0.874)	0
	B	1.42 ± 0.11	(0.937)	1	1.69 ± 0.14	(0.924)	1
	C	3.16 ± 0.17	(0.967)	2	4.84 ± 0.28	(0.961)	2
	D	26.33 ± 2.62	(0.998)	3.0	40.49 ± 3.69	(0.999)	2.6
67	A	0.71 ± 0.08	(0.878)	0	0.64 ± 0.08	(0.848)	0
	B	1.55 ± 0.13	(0.923)	1	1.76 ± 0.16	(0.907)	1
	C	3.45 ± 0.21	(0.957)	2	4.95 ± 0.32	(0.951)	2
	D	24.73 ± 1.95	(0.999)	2.9	38.08 ± 3.00	(0.999)	2.6
74	A	0.75 ± 0.08	(0.892)	0	0.61 ± 0.08	(0.814)	0
	B	1.59 ± 0.12	(0.935)	1	1.58 ± 0.17	(0.873)	1
	C	3.43 ± 0.18	(0.967)	2	4.21 ± 0.36	(0.921)	2
	D	21.53 ± 1.88	(0.998)	3.0	39.62 ± 2.47	(0.999)	2.7

Table VI. Activation Parameters for Pyrolysis (Rate of Weight Loss)
for Cellulose Samples Treated with 10% Phosphoric Acid,
Based on Data in Table I

Crystallinity Index, %	Method (1)	Ea (kcal/Mole) ± S.D.
10	A	6 ± 3
	B	15 ± 2
	C	25 ± 1
	D	21 ± 4
	E	64 ± 7
41	A	− 1 ± 3
	B	8 ± 3
	C	18 ± 3
	D	22 ± 4
	E	56 ± 4
63	A	4 ± 1
	B	13 ± 1
	C	23 ± 2
	D	20 ± 2
	E	65 ± 10
67	A	4 ± 2
	B	14 ± 2
	C	24 ± 2
	D	21 ± 4
	E	68 ± 9
74	A	4 ± 3
	B	13 ± 3
	C	23 ± 3
	D	22 ± 2
	E	47 ± 9

(1) A = zero-order; B = first-order; C = second-order;
D = Wilkinson's Approximation; E = Integral Conversion Method.

(2) Determined using the Arrhenius equation.

Continued on next page

Table VI. Continued

$k_{300}(10^{-4}/sec)$ (2)	$\Delta H^{\ddagger}$ (kcal/Mole) ± S.D.	$\Delta S^{\ddagger}$ (e.u.) ± S.D.
	Activation Parameters	
0.86	5 ± 3	-69 ± 6
2.70	14 ± 2	-51 ± 4
8.83	24 ± 1	-32 ± 2
59.34	20 ± 4	-36 ± 7
-	-	-
0.58	- 3 ± 3	-83 ± 6
2.25	7 ± 3	-64 ± 5
9.09	17 ± 3	-44 ± 5
90.71	21 ± 5	-33 ± 5
-	-	-
0.71	2 ± 1	-74 ± 4
2.46	12 ± 1	-55 ± 1
8.77	22 ± 2	-35 ± 3
64.85	19 ± 2	-37 + 3
-	-	-
0.80	3 ± 2	-73 ± 4
2.69	13 ± 1	-54 ± 3
9.31	23 ± 2	-33 ± 5
59.51	20 ± 4	-34 ± 8
-	-	-
0.81	3 ± 3	-73 ± 6
2.66	12 ± 3	-55 ± 5
9.01	22 ± 3	-36 ± 5
65.27	21 ± 2	-33 ± 3
-	-	-

Table VII. Comparison of Cellulose Pyrolysis Rates (Rates of Weight Loss) Analyzed by Differential Methods for Cellulose Treated with 10% Aluminum Chloride Hexahydrate. All Rates are in Units of 10^{-4} / Sec

Crystallinity Index, %	Method (1)	Reaction Temperature (2), °C					
		280			290		
		Rate ± S.D.	(r^2)	n	Rate ± S.D.	(r^2)	n
10	A	0.28 ± 0.02	(0.953)	0	0.37 ± 0.02	(0.959)	0
	B	0.43 ± 0.03	(0.961)	1	0.60 ± 0.03	(0.969)	1
	C	0.67 ± 0.03	(0.969)	2	0.99 ± 0.04	(0.978)	2
	D	22.80 ± 3.29	(0.998)	5.0	21.39 ± 2.99	(0.997)	4.3
41	A	0.40 ± 0.01	(0.988)	0	* 0.71 ± 0.03	(0.984)	0
	B	0.56 ± 0.01	(0.993)	1	* 1.14 ± 0.03	(0.994)	1
	C	0.80 ± 0.01	(0.996)	2	* 1.86 ± 0.02	(0.999)	2
	D	9.67 ± 1.69	(0.991)	5.3	*10.08 ± 1.34	(0.986)	3.7
63	A	0.34 ± 0.01	(0.985)	0	0.51 ± 0.02	(0.987)	0
	B	0.47 ± 0.01	(0.989)	1	0.76 ± 0.02	(0.993)	1
	C	0.66 ± 0.02	(0.993)	2	1.14 ± 0.02	(0.996)	2
	D	11.19 ± 1.96	(0.994)	5.8	10.56 ± 1.56	(0.989)	4.6
67	A	0.25 ± 0.01	(0.982)	0	0.39 ± 0.01	(0.984)	0
	B	0.33 ± 0.01	(0.985)	1	0.55 ± 0.02	(0.990)	1
	C	0.45 ± 0.01	(0.988)	2	0.78 ± 0.02	(0.994)	2
	D	12.73 ± 2.08	(0.996)	6.6	0.90 ± 1.53	(0.993)	5.4
74	A	0.24 ± 0.01	(0.982)	0	0.45 ± 0.01	(0.989)	0
	B	0.32 ± 0.01	(0.986)	1	0.66 ± 0.01	(0.994)	1
	C	0.43 ± 0.01	(0.989)	2	0.98 ± 0.01	(0.997)	2
	D	12.96 ± 2.04	(0.996)	6.7	11.43 ± 1.77	(0.991)	4.8

(1) A = zero-order; B = first-order; C = second-order; D = Wilkinson's Approximation.

(2) Indicates target temperature. The actual temperature differed ± 2 deg.C but remained fairly constant (± 0.2 deg.C) during the run. The actual temperature was used to calculate activation parameters.

* Indicates a sample which was run in duplicate.

Continued on next page

Table VII. Continued

Crystallinity Index, %	Method (1)	Reaction Temperature (2), °C					
		300			310		
		Rate ± S.D.	(r^2)	n	Rate ± S.D.	(r^2)	n
10	A	0.48 ± 0.03	(0.965)	0	0.59 ± 0.04	(0.959)	0
	B	0.89 ± 0.04	(0.979)	1	1.28 ± 0.05	(0.979)	1
	C	1.65 ± 0.05	(0.989)	2	2.81 ± 0.07	(0.992)	2
	D	22.86 ± 3.30	(0.996)	3.6	25.50 ± 3.39	(0.996)	3.1
41	A	0.90 ± 0.04	(0.974)	0	1.03 ± 0.06	(0.957)	0
	B	1.65 ± 0.04	(0.992)	1	2.22 ± 0.07	(0.988)	1
	C	3.07 ± 0.02	(0.999)	2	4.88 ± 0.02	(0.999)	2
	D	11.57 ± 1.36	(0.988)	3.1	14.11 ± 1.53	(0.991)	2.7
63	A	0.76 ± 0.03	(0.984)	0	1.02 ± 0.05	(0.974)	0
	B	1.26 ± 0.03	(0.995)	1	1.99 ± 0.04	(0.994)	1
	C	2.10 ± 0.02	(0.999)	2	3.98 ± 0.02	(0.999)	2
	D	10.24 ± 1.35	(0.985)	3.6	11.69 ± 1.34	(0.986)	2.8
67	A	0.64 ± 0.02	(0.984)	0	1.00 ± 0.04	(0.985)	0
	B	1.00 ± 0.02	(0.993)	1	1.79 ± 0.02	(0.998)	1
	C	1.58 ± 0.02	(0.998)	2	3.25 ± 0.03	(0.999)	2
	D	10.18 ± 1.40	(0.987)	4.0	9.33 ± 1.13	(0.979)	3.0
74	A	0.58 ± 0.02	(0.991)	0	0.86 ± 0.02	(0.991)	0
	B	0.89 ± 0.01	(0.997)	1	1.34 ± 0.01	(0.999)	1
	C	1.35 ± 0.01	(0.999)	2	2.12 ± 0.01	(0.999)	2
	D	9.76 ± 1.45	(0.986)	4.3	7.24 ± 0.92	(0.973)	3.5

Table VIII. Activation Parameters for Pyrolysis (Rate of Weight Loss)
for Cellulose Samples Treated with 10% Aluminum Chloride Hexahydrate,
Based on Data in Table VII

Crystallinity Index, %	Method (1)	Activation Parameters Ea (kcal/Mole) ± S.D.
10	A	16 ± 0
	B	24 ± 1
	C	31 ± 3
	D	3 ± 2
	E	79 ± 11
41	A	20 ± 4
	B	30 ± 3
	C	40 ± 3
	D	7 ± 3
	E	49 ± 9
63	A	24 ± 1
	B	32 ± 0
	C	40 ± 1
	D	1 ± 2
	E	42 ± 5
67	A	29 ± 0
	B	35 ± 1
	C	42 ± 2
	D	- 6 ± 1
	E	42 ± 1
74	A	27 ± 3
	B	30 ± 4
	C	33 ± 5
	D	-12 ± 2
	E	30 -

(1) A = zero-order; B = first-order; C = second-order;

   D = Wilkinson's Approximation; E = Integral Conversion Method.

(2) Determined using the Arrhenius equation.

Continued on next page

Table VIII. Continued

k_{300} $(10^{-4}/sec)$ (2)	$\Delta H^{\ddagger}$ (kcal/Mole) + S.D.	$\Delta S^{\ddagger}$ (e.u.) + S.D.
	Activation Parameters	
0.47	15 ± 0	-53 ± 0
0.88	23 ± 1	-39 ± 2
1.69	30 ± 3	-25 ± 4
23.54	1 ± 2	-69 ± 3
-	-	-
0.84	19 ± 4	-45 ± 6
1.56	29 ± 3	-27 ± 6
2.92	38 ± 3	- 9 ± 6
11.80	6 ± 3	-63 ± 6
-	-	-
0.73	23 ± 1	-38 ± 2
1.25	31 ± 0	-24 ± 0
2.17	38 ± 1	- 9 ± 2
10.96	- 1 ± 2	-74 ± 3
-	-	-
0.61	28 ± 0	-30 ± 0
0.96	34 ± 1	-18 ± 1
1.54	40 ± 2	- 7 ± 3
10.25	- 7 ± 1	-86 ± 1
-	-	-
0.58	25 ± 3	-35 ± 6
0.88	29 ± 4	-35 ± 7
1.34	32 ± 5	-21 ± 8
9.23	-13 ± 2	-97 ± 4
-	-	-

Table IX.   Char/Residue (Wr) Formed by Pyrolysis
Under N_2 at 550°C

%CI	% Char/Residue			
	Control	H_3BO_3	$AlCl_3 \cdot 6H_2O$	H_3PO_4
10	29.0	50.0	40.0	51.5
41	11.5	32.0	27.0	44.0
63	9.0	31.0	26.0	43.0
67	8.5	31.5	25.0	43.0
74	7.0	31.5	26.0	43.0

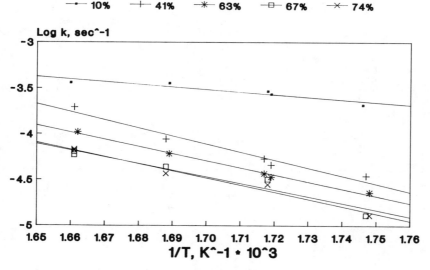

Figure 1.   Arrhenius plot of the 1st-order rates for the
untreated cellulose samples.

The mass loss rates for the boric acid samples were comparable
to the untreated samples, despite a higher char yield (Table IX)
for the treated samples. This was unexpected since the role of a
wood fire retardant is to increase the char by increasing the
dehydration reaction (1,3,7). Thus, a fire retardant treated sample
will actually pyrolyze at a lower temperature. Data from Table III
suggests that boric acid may form more char by suppressing formation
of flammable volatiles instead of by increasing the dehydration
rate.

Significant changes were observed in the activation parameters
(Table V) as compared to the control activation parameters (Table
II). Specifically, the treated samples had increased E_a's, $\Delta H^{\ddagger}$'s
and $\Delta S^{\ddagger}$s. An increase in E_a has been previously reported (12,18).
The increase in $\Delta H^{\ddagger}$ and $\Delta S^{\ddagger}$ and the ability of boric acid to complex
with sugars (34-37) suggest that boric acid complexed with cellulose
and increased the overall bonds cleaved in attaining the transition
state ($\Delta H^{\ddagger}$) and consequently increased E_a. If formation of
flammable volatiles does proceed through the hydroxyl nucleophilic
mechanism, then formation of a complex between boric acid and the
"reactant" hydroxyl group might inhibit the flaming reaction.
Unfortunately, boric acid has been reported to oxidize when heated
in air (7) and as a result would be expected to be much less
effective in air than the authors found under N_2.

Phosphoric Acid. The 2nd-order rate method for analyzing the TGA
data was statistically best (Table IV) for the cellulose/H_3PO_4
samples. This suggests that the conclusions from a prior study
which assumed a 1st-order reaction (29) may need to be reexamined.
While Wilkinson's approximation method gave high r^2 values, the rate
constant is determined by the intercept rather than the slope in
this method. Thus, the standard deviation of the rates determined
by Wilkinson's approximation method is still relatively high when
compared to the other methods. In addition, the reaction order as
determined by the Wilkinson approximation method was unrealistically
high, ranging from 2.6 to 5.8.

E_a values (Table VI) as determined by integral conversion are
2-5 times higher than the other E_a values, while the E_a values
obtained using the 0-order reaction rates were unrealistically low.

The 2nd-order 300°C rates determined by the Arrhenius equation
(Table VI) show that the rates are extremely high compared to the
control or boric acid treated samples. In addition, the rate of
mass loss appears to be unaffected by crystallinity. E_a values were
lowered relative to the untreated control samples, except for the
amorphous sample, and also appeared to be unaffected by
crystallinity. Phosphoric acid has been previously reported to lower
E_a (29). The $\Delta H^{\ddagger}$ values are also lower than the control values,
except for the amorphous sample. In addition, the $\Delta S^{\ddagger}$ values are
all large positive numbers, which suggests that the transition state
is relatively disordered. The increased rates, the lack of any
crystallinity effect, and the change in activation parameters imply
that the weight loss mechanism for the phosphoric acid samples is
different than that for the control samples. This mechanism is
probably the commonly accepted acid-catalyzed dehydration reaction
(1,7), which explains the high char yields (Table IX). An alternate

explanation is that the strong acid may have affected the cellulose
DP during the initial ten-minute 110°C heating period before any
weight loss occurred. However, a recent paper (38) suggested that
DP would not have an appreciable effect on the E_a for the first
pyrolysis stage, which occurs below about 300°C.

Phosphoric acid is a relatively strong acid and can cause
extensive damage to polysaccharides. Ammonium phosphate salts are
good fire retardants (1,7,13,18,22), are inexpensive, and should not
cause acid-catalyzed hydrolysis or dehydration. However, if
ammonium phosphate-treated wood is exposed to heat, ammonia will be
given off and phosphoric acid will be left (7,39-43). The critical
temperature at which monoammonium phosphate thermally disassociates
has been reported to be 166° (39), 170° (42), or 190°C (40).
However, the relatively low reported E_a of 8.14 kcal/mol (43)
suggests that even moderately warm temperatures over several years,
such as roof structures experience during the summer months, may be
sufficient to form phosphoric acid. Ammonium sulfate, which can
also thermally disassociate to form an acid (7,40), may cause even
more degradation to lumber than ammonium phosphates (40,44). This
thermal disassociation of ammonium salts may be the cause of the
premature degradation recently observed in some fire retardant
treated plywood roofing material.

Aluminum Trichloride Hexahydrate. The 2nd-order method for
determination of TGA rates of $AlCl_3 \cdot 6H_2O$-treated cellulose was again
statistically best (Table VII). This data contrasts with a dynamic
TGA study which suggested a 0-order followed by a 1st-order reaction
(22).

Wilkinson's approximation method again gave high rate constants
and orders and unrealistic E_a's (Table VIII). E_a values determined
by the integral conversion method were generally similar to the
values determined from the 2nd-order rates except for the amorphous
sample. The 2nd-order rates measured at 300° by the Arrhenius
equation (Table VIII) were dependent on crystallinity, but the 41%
CI sample, rather than the most amorphous sample, had the highest
rate constant. The rates were lower than the phosphoric
acid-treated samples but were significantly higher than for the
control and boric acid-treated samples. Aluminum chloride is
reported to provide fire resistance properties to wood by the same
dehydration mechanism as phosphoric acid (7).

Conclusions

Cellulose pyrolysis kinetics, as measured by isothermal TGA mass
loss, were statistically best fit using 1st- or 2nd-order for the
untreated (control) samples and 2nd-order for the cellulose samples
treated with three additives. Activation parameters obtained from
the TGA data of the untreated samples suggest that the reaction
mechanism proceeded through an ordered transition state. Sample
crystallinity affected the rate constants, activation parameters,
and char yields of the untreated cellulose samples. Various
additives had different effects on the mass loss. For example,
phosphoric acid and aluminum chloride probably increased the rate of
dehydration, while boric acid may have inhibited levoglucosan

formation. Both of these pathways would increase the char yield and provide some fire-retardant properties to wood. The results - and conclusions based on the results - are dependent on the method used to analyze the TGA data. Our results suggest that a weak polybasic acid which forms a strong thermally-stable complex with cellulose might provide fire-resistant properties to lignocellulosic products.

Analysis of TGA pyrolysis data of complex polymers can be quite difficult. Assumption of a particular reaction order without examining other possible reaction orders may make the results and conclusions of a study suspect. Also, kinetic analysis techniques which require no assumptions might give unrealistic results.

Acknowledgments

Funding for this project was provided by the USDA McIntire-Stennis Program and the State of Mississippi.

Literature Cited

1. LeVan, S. L. In The Chemistry of Solid Wood, Rowell, R. M., Ed.; Am. Chem. Soc., Washington, D.C., 1984, Chp. 14.
2. Nguyen, T., E. Zavarin, E. M. Barrall, II. J. Macromal. Sci. - Rev. Macromal. Chem. C, 1981, 20, 1.
3. Shafizadeh, F. In The Chemistry of Solid Wood, Rowell, R. M., Ed.; Am. Chem. Soc., Washington, D.C., 1984, Chp. 13.
4. Shafizadeh, F. In Cellulose Chemistry and Its Application, Nevell and S. H. Zeronian, Eds.; Harwood, England, 1985, Chp. 11.
5. Antal, M. J., Jr. Adv. in Solar Energy, Vol. 1, 1983, p. 61.
6. Antal, M. J., Jr. In Advances in Solar Energy. Vol 2, K. W. Boer and J. A. Duffie, Eds.; Plenum Press, NY, 1985, Chp. 4.
7. Browne, F. L. Theories of the Combustion of Wood and Its Control, FPL Report 2136, USDA Forest Service, 1963.
8. Lewellen, P. C., W. A. Peters, J. B. Howard, Symp. (Int.) Combustion (Proc.), 1976, 16, 1471.
9. Dickens, B., J. H. Flynn. In Polymer Characterization, Craver, C. D., Ed.; Am. Chem. Soc., Washington, D.C., 1983, Chp. 12.
10. Chatterjee, P. K. J. Appl. Poly. Sci., 1968, 12, 1859.
11. Kato, K., N. Takahashi Agr. Biol. Chem., 1967, 31, 519.
12. Hirata, T. Mokuzai Gakkaishi, 1981, 27, 737.
13. Akita, K., M. Kase J. Poly. Sci., Part A-1, 1967, 5, 833.
14. Bilbao, R., J. Arauzo, A. Millera Thermochim. Acta, 1987, 120, 120.
15. Ramiah, M. V. J. Appl. Poly. Sci., 1970, 14, 1323.
16. Fung, D. P. C. TAPPI, 1969, 52(2), 319.
17. Cardwell, R. D., P. Luner. Wood Sci. Tech., 1976, 10, 131.
18. Hirata, T., K. E. Werner. J. Appl. Poly. Sci., 1987, 33, 1533.
19. Bradbury, A. G. W., Y. Sakai, F. Shafizadeh. J. Appl. Poly. Sci., 1979, 23, 3271.
20. Agrawal, R. K. Cand. J. Chem. Eng., 1988, 66, 403.
21. Chatterjee, P. K., C. M. Conrad. Textile Res. J., 1966, 36, 487.

22. Tang, W. K. J. Poly Sci., Part C, 1964, 6, 65.
23. Lipska, A. E., W. J. Parker. J. Appl. Poly. Sci., 1966, 10, 1439.
24. Greenberg, A. R., I. Kamel. J. Poly. Sci., Poly Chem., 1977, 15, 2137.
25. Basan, S., O. Guven. Thermochim. Acta, 1986, 106, 169.
26. Basch, A., M. Lewin. J. Poly. Sci., Poly. Chem. Ed., 1973, 11, 3071.
27. Schultz, T. P., G. D. McGinnis, M. S. Bertran. J. Wood Chem. Tech., 1985, 5, 543.
28. Kato, K., H. Komorita. Agr. Biol. Chem., 1968, 32, 21.
29. Kumagai, Y., T. Ohuchi, C. Nagasawa, M. Ono. Mokuzai Gakkaishi, 1974, 20, 381.
30. Bertran, M. S., B. E. Dale. Biotechnol. Bioeng., 1985, 27, 177.
31. Arthur, J. C., Jr., O. Hinojosa. Text. Res. J., 1966, 36, 385.
32. Berkowitz-Mattuck, J. B., T. Noguchi. J. Appl. Poly. Sci., 1963, 7, 709.
33. Hirsch, J. A. Concepts in Theoretical Organic Chemistry, Allyn and Bacon, Boston, MA, 1974.
34. Brimacombe, J. S., J. M. Webber. In The Carbohydrates, Vol. 1A, Pigman, W., D. Horton, Eds.; Academic Press, NY, 1972, Chp. 14.
35. Dekker, C. A., L. Goodman. In The Carbohydrates, Vol. 11A, Pigman, W., D. Horton, Eds.; Academic Press, NY, 1970, Chp. 29.
36. Scott, J. E. In Methods in Carbohydrate Chemistry, Vol. V, Wistler, R. L., J. N. BeMiller, M. L. Wolfrom, Eds.; Academic Press, NY, 1965, Chp. 11.
37. MacDonald, D. L. In The Carbohydrates, Vol. 1A, Pigman, W., D. Horton, Eds.; Academic Press, NY, 1972, Chp. 8.
38. Calahorra, M. E., M. Cortazar, J. I. Equiazabal, G. M. Guzman. J. Appl. Poly. Sci., 1989, 37, 3305.
39. Tang, W. K., H. W. Eickner. Effect of Inorganic Salts on Pyrolysis of Wood, Cellulose, and Lignin Determined by Differential Thermal Analysis; FPL Research Paper 82, U.S.D.A., 1967.
40. George, C. W., R. A. Susott. Effects of Ammonium Phosphate and Sulfate on the Pyrolysis and Combustion of Cellulose; Forest Service Research Paper INT-90, U.S.D.A., 1971.
41. Toy, A. D. F. In Comprehensive Inorganic Chemistry, Vol. 2, Trotman-Dickenson, A. F., Exec. Ed.; Pergamon Press, NY, 1975, pg. 497.
42. Margulis, E. V., L. I. Beisekeeva, N. I. Kopylov, M. A. Fishman. Zh. Prikl. Khim., 1966, 39(10), 2364; Chem. Abstr. 66:7974h, 1967.
43. Menlibaev, A., D. Z. Serazetdinov, A. B. Bekturov. Izu. Akad. Nauk. Kaz. SSR, Ser. Khim., 1976, 26(5), 55; Chem. Abstr. 86:47631x, 1977.
44. Middleton, J. C., S. M. Draganov. For. Prod. J., 1965, 15, 463.

RECEIVED November 1, 1989

Chapter 23

Influences of Oxygen Chemisorption and of Metal Ions in Ignition and Combustion of Wood

G. N. Richards and F.-Y. Hshieh

Wood Chemistry Laboratory, University of Montana, Missoula, MT 59812

The influences of removal of metal ions, or of addition of metal ions, on the subsequent ignition and combustion of wood and wood chars can be dramatic. The major indigenous inorganic constituents of wood consist of inorganic salts and of metal ions bound to the uronic acids of hemicelluloses. Both can be completely removed by washing the wood with dilute acid at room temperature. Individual metal ions can then be "added-back" by ion exchange with the appropriate acetate solution. The complete removal of cations was most effective in raising ignition temperature or preventing ignition. The physical form of the wood samples (i.e., fine and course powders and solid discs) also had considerable influence on ignition, either in oxygen or in air. The changes in metal ions had small but significant influence on oxygen chemisorption behavior of the derived chars.

The ignition and combustion of lignocellulosics is a major source of destructive residential and forest fires. Much of the relevant research in the literature has been carried out on cellulose, whereas in most fire situations the combustible material is lignocellulose such as wood and wood-derived materials (e.g. newsprint). In addition to cellulose, wood contains two other major types of constituent, viz. hemicelluloses, and lignin, and also extractives and inorganic species. However, in considering the actual ignition and combustion of wood we are frequently dealing effectively with the ignition and combustion of <u>char</u>, which has been generated from wood by pyrolysis prior to ignition. This is especially true of smoldering combustion, which is a common event in connection with domestic or house fires, and in propagation of forest fires. In this connection we should note that the chemical and physical structures of the chars generated by pyrolysis of a wide range of biomass (originating from more than 10 different plant species and morphol-

ogies) appear remarkably similar in that their gasification rates
correlate very closely with total metal ion content (Kannan, M.P.;
Richards, G.N. unpublished results). As a result, the inorganic
content of cellulosics (specifically the identity and content of
metal ions) appears to be a dominant factor in ignition and
combustion of cellulosics such as wood. This is provided of course
that physical factors such as particle size, rate of heat input and
output, etc. are constant.
     In the ignition of wood char, the heat of chemisorption of
oxygen is one of the major thermodynamic events (about -100 Kcal.
g.$^{-1}$ per mole oxygen ($\underline{1}$)) and it may play a major role in ignition.
The influence of oxygen chemisorption on ignition will of course be
dependent on the temperature (and hence the rate at which
chemisorption occurs), and on the rate of heat removal by convection
and radiation. Chemisorption of oxygen by carbonaceous material has
been studied by many workers using various char materials and a
variety of experimental techniques ($\underline{2}$). Bradbury and Shafizadeh ($\underline{2}$)
reported that the chars prepared from cellulosic fabric at different
heat treatment temperatures (HTT) (450-800°C) have different
affinities for oxygen. Maximum chemisorption occurred on chars
prepared by pyrolysis for 1.5 minutes at 550°C and they concluded
that the maximum in chemisorption coincided with a maximum in free
radical content in the char. Later work by DeGroot and Shafizadeh
($\underline{3}$) however indicated that functional groups on cellulose char other
than free radicals could play a major role in determining reactivity
of char. Inorganic additives may enhance or retard gasification
rates, but they have little effect on chemisorption of oxygen on
cellulose chars.
     This paper reports the influences on oxygen chemisorption of
(a) temperature of sorption, (b) extent of charring, (c) metal ion
content in wood. The effect of metal ions in wood and sample
configuration on ignition of wood char have also been investigated.
Our approach has been to carry out a sequence involving charring of
cottonwood samples in nitrogen, oxygen chemisorption (e.g. at 140°C),
and subsequent heating to ignition. All of these sequential
processes are carried out within the thermal balance without
interruption. This procedure avoids some uncertainties which can
arise from separate char preparation and handling before
chemisorption determinations, such as surface oxide contamination
and necessity for "cleaning" of char surfaces.

Experimental Approach

Wood samples from cottonwood sapwood (Populus trichocarpa) were used
either as small discs (5.5 mm diameter, 2 mm thick), or as Wiley-
milled powder. Newsprint was "repulped" by maceration with water in
a blender and formed by filtration into a mat of 2-3 mm air-dry
thickness from which discs of 5.5 mm diameter were cut. Metal
analyses were carried out by inductively coupled argon plasma
spectrometry. Indigenous metal ions were removed by degassing under
0.025 M hydrochloric acid and eluting with this acid overnight at
room temperature before rigorously washing with conductivity water.
Individual metal ions were "added back" to the acid-washed wood by

repeating this process using 0.1 M solutions of the appropriate acetate salts and again rigorously washing with water.

The pyrolysis of wood, oxygen chemisorption and oxidation of wood chars were carried out in a computerized coupled TG-FTIR system containing Cahn-R-100 electric balance, DuPont Model 990 thermal analyzer and Nicolet MX-1 Fourier transform infrared spectrometer. All of these sequential processes are carried out within the thermal balance without interruption.

Influence of Thermal History on Oxygen Chemisorption of Chars. Chars of HTT 450°C from solid wood discs, heated from 25° to 450°C at 50°C/min and held for 10 minutes before cooling at 50°C/min to chemisorption temperature (CST) were used to investigate the effect of CST on the chemisorption of oxygen on cottonwood char. Studies using coupled thermogravimetry and Fourier transform infrared spectrometry of evolved gases showed that gasification products could not be detected at temperatures equal to or below 140°C; therefore 140°C was chosen for subsequent chemisorption studies.

The initial rate of oxygen chemisorption was measured from the average rate of oxygen uptake during the first five minutes. The approximate activation energy of initial oxygen chemisorption from 120-180°C was found to be 11.7 kcal/mole. This result is in agreement with that obtained by Bradbury (2) for pure cellulose char (HTT 550°C, 12.6 kcal/mole). These low activation energies indicate that the initial oxygen chemisorption is a diffusion-controlled process both in the relatively compact solid wood char and in the more fibrous cellulose char.

The effect of HTT on the chemisorption of oxygen was studied at CST 140°C. Table I and Figure 1 compare HTT and the weight loss of solid wood discs during the charring process (5°/min in nitrogen) with the initial rate of subsequent oxygen chemisorption. The total amount of oxygen uptake during the first 2 hours also has the same trend as the initial rate of chemisorption. In both cases (i.e., with and without 10 min holding at HTT), there is a pronounced maximum in chemisorption activity (CSA) in the HTT region 450-500°C. The relationship is evidently complex. The potential CSA of the char as it forms passes through two phases. In the first phase (400-500°C), after the initial rapid pyrolysis weight loss, there appears to be some change in the char, without great loss of weight, which increases the CSA twofold. Then in the second phase (HTT 500-550°C) the char loses half of its CSA, again with very little weight loss.

Chemisorption of oxygen on char has often been discussed previously in terms of free radical concentration in the char (1,5,6). For cellulose chars Bradbury and Shafizadeh (1) found that free spin concentration reached a sharp maximum at HTT 550°C, coinciding with maximum CSA and drew the obvious conclusion that the extent of CSA was at least partly related to free radical content of the char. However, in subsequent work on cellulose char, DeGroot and Shafizadeh (3) have found that unpaired spin concentration continues to increase up to HTT 700°C. The CSA of the char must therefore depend on factors other than free radical concentration.

At present we favor a tentative line of interpretation of CSA dependence on HTT of char as follows. The "protochar" is formed with rapid weight loss most importantly by elimination of water from

Table I.   Effect of Heat Treatment Temperature (HTT) on Chemisorption of Oxygen
on Cottonwood Chars[a] at 140°C

Heat Treatment Temperature (°C)	Weight Loss (% of Dry Wood)	Initial Rate of Oxygen Chemisorption x 10^3 (mmole g^{-1} min^{-1})	Oxygen Chemisorption[b] (mmole g^{-1})
370	75.34 (77.71)	26.6 (37.1)	0.57 (0.99)
400	78.26 (78.61)	46.3 (54.7)	1.16 (1.28)
450	79.63 (80.95)	83.1 (72.9)	1.34 (1.39)
500	80.71 (81.88)	89.4 (71.3)	1.49 (1.13)
550	82.71 (82.44)	52.1 (37.0)	1.11 (0.77)
600	82.86 (82.81)	45.8 (33.4)	1.06 (0.72)

[a]Charring process: pyrolysis of wood (in N_2) from 25°C to HTT at 5°C/min. Values in parentheses
are the data for the chars prepared from the same procedure as above but held at HTT for 10 min.
[b]Oxygen uptake during the first 2 h.

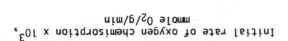

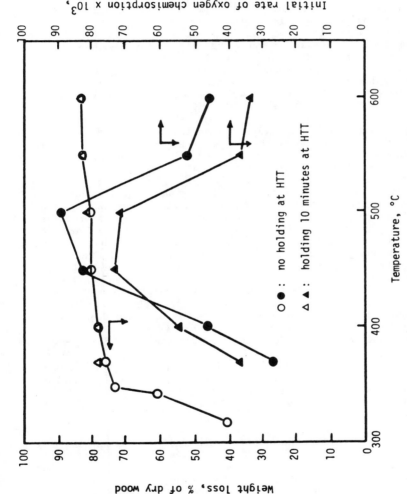

Figure 1. Initial chemisorption rates (140°C) and pyrolysis weight loss against maximum charring temperature. Pyrolysis at 5°/min in nitrogen. (Reproduced with permission from Ref. 19. Copyright 1989 Elsevier Scientific Publishing Company, Inc.)

polysaccharides and of methanol from lignin (5,6). The initial char undoubtedly contains many types of functional groups capable of interacting with oxygen, including aromatic ring systems. Above 400°C some chemical change must occur in such systems, without major weight change, which renders the chars initially more reactive and subsequently less reactive to oxygen. Earlier studies of the chemistry of cellulosic chars (7-10) using chemical and NMR methods indicated the following general conclusions as HTT increases above 400°C:

(a) a continuing increase in total aromatic content, not passing through any maximum which would correspond to the peak in CSA shown in Figure 1;

(b) a continuing decrease in non-aromatic functional groups;

(c) an increasing formation of polycyclic aromatic structures.

We are further investigating the phenomenon shown in Figure 1 to seek a chemical explanation, since it seems very unlikely that the changes in CSA between HTT 400°C and 500°C are due to any physical changes in the char. An FTIR investigation (Hshieh, F.-Y.; Richards, G.N., Combust. Flame, in press) of the chars shown in Figure 1 has confirmed the above generalizations regarding the effect of HTT on char chemistry, and suggests that oxygen chemisorption results in the introduction of carbonyl groups. Chars from newsprint showed a similar maximum in the relation of CSA to HTT.

At present we consider the most likely explanation of the peak in chemisorption when chars are prepared at about 500°C is associated with a facile reaction of oxygen with certain types of external regions of the polynuclear aromatic clusters in which reaction occurs more rapidly than with "normal" aromatic systems. As the polynuclear aromatic clusters grow with increasing char HTT, such "edge" regions will pass through a maximum in relation to total char weight. It has previously been suggested (11) that, on the basis of molecular orbital calculations, some regions (which have been called "K regions") in polynuclear aromatic hydrocarbons (PAH) have particularly high reactivity and hence have specific responsibility for carcinogenicity. In this context, a "K region" is defined as the external corner of a phenanthrene-type moiety in PAH.

The conditions used for char preparation in the present chemisorption studies (i.e., progressive slow charring of wood) are intended to be relevant to "real life" smoldering combustion situations. Most previous studies of chemisorption have used chars from cellulose (i.e., avoiding hemicellulose and lignin complications) and have normally used chars formed by "instantaneous" heating to a pre-set temperature.

Influence of Metal Ions on Oxygen Chemisorption and Ignition of Chars. We have carried out extensive studies of the influence of metal ions in wood on pyrolysis mechanisms (5,6) and this approach has now been extended to oxygen chemisorption of the chars. The metal ions occur in wood predominantly as the counterions of the uronic acid components of the hemicelluloses (12). We have shown that they can be almost completely removed by very mild acid treatment without any other major change in the chemistry of the wood. Table II shows that the major metal ions in cottonwood are Ca, K and Mg. The acid-washing process removed 98% of the metal ions in

Table II. Metal Ion Content (ppm) of Cottonwood and
Acid-washed Cottonwood

	Al	Ca	Co	Cu	Fe	Mg	P	K	Na	Zn	Total
Cottonwood	8	1500	4	4	2	200	140	1000	30	9	2897
Acid-washed cottonwood	2	13	0	0	3	2	10	15	15	3	63

the cottonwood discs. A single species of metal ion was then "added back" to the wood by normal ion exchange processes in order to isolate and identify influences of individual species of metal ions in the wood. Cobalt-exchange was included in this series for comparison, because it has been shown that this metal has rather unique properties in gasification of chars (13). Although the exact degree of ion-exchange for each metal ion cannot be determined from ash yield, the ash yield of the ion-exchanged wood discs (Table III) confirms that the same level of ion exchange was achieved as reported previously for wood powder (5,6).

Table III. Effect of Acid-washing and Ion-exchange on Ash Yield
of Cottonwood

Cottonwood	Ash Yield (% of Dry Wood)
Untreated	0.41
Acid-washed	0.01
Acid-washed, then Ca-exchanged	0.30
Acid-washed, then K-exchanged	0.40
Acid washed, then Co-exchanged	0.24

Figure 1 has shown that the maximum chemisorption of oxygen on chars from untreated wood occurs at HTT 450°-500°C. However, in order to understand better the effect of metal ions on the total process consisting of pyrolysis and subsequent chemisorption and oxidation of wood char, it was necessary to carry out pyrolysis, isothermal chemisorption and oxidation reactions in a single experiment. A typical overall pyrolysis, isothermal chemisorption (140°C) and oxidation curve is shown in Figure 2. The temperature program is: (1) heat from 25° to 500°C at 5°C/min, (2) cool at 50°C/min to 140°C, (3) hold for 2 hours, and (4) heat to 500°C at 3°/min. Oxygen was introduced at the time the temperature reached 140°C. The increase in temperature after the isothermal (140°C) region led to an increase in the rate of chemisorption, up to the temperature at which combustion (burn-off) becomes the dominant process resulting in rapid weight loss (ca. 270°C).

The influence of the metal ions on the three types of chemical

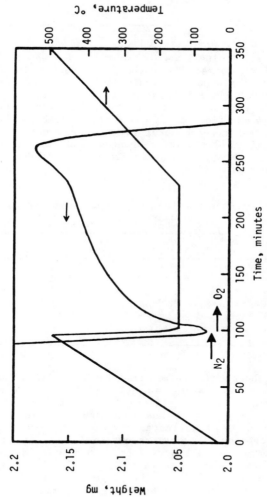

Figure 2. Pyrolysis - chemisorption - oxidation of cottonwood. (Reproduced with permission from Ref. 19. Copyright 1989 Elsevier Scientific Publishing Company, Inc.)

processes indicated in Figure 2 will be considered separately. The influence on the wood pyrolysis process has previously been studied in detail (5,6) and the same general effects were observed in this study. Thus the pyrolysis occurred rather more rapidly in the potassium exchanged wood than in the original wood and more slowly (i.e. at ca. 20°C higher temperature) in the calcium-exchanged wood, with acid-washed and cobalt-exchanged wood intermediate between the original and calcium-exchanged samples.

In the oxygen chemisorption region in Figure 2 the effects of metal ions were small, but significant. The CSA was lowest in the char from acid-washed wood and was increased by presence of all of the metal ions studied. Cobalt was outstanding in this respect and has also been shown to be unusually effective in catalysis of char gasification (13). The relative effect of cations on initial chemisorption rate was rather greater than the effect on total chemisorption in 2 hours. The most dramatic effects of metal ions however, occurred with ignition, which was most precisely measured by derivative thermogravimetry as shown in Figure 3. The peak temperatures (together with prior CSA values) are shown in Table IV. The sharp DTG peaks coincided with visual observation of ignition (emission of visible light) in the disc samples, except for the acid-washed sample which did not ignite. The ignition process therefore is defined as the onset of "run-away" combustion, implying the occurrence of a situation when the rate of heat generation in the sample begins to exceed the rate of heat removal, with consequent exponential increase in temperature and hence rate of combustion. The ignition appeared by eye to occur simultaneously over the whole disc and was accompanied by brilliant incandescence with no visible flame. The ease of ignition of the chars from woods differing only in cation content follows the order of the initial rate of oxygen chemisorption (Table IV), but presumably involves many other factors. All of the metal ions included in this study lowered the ignition temperature in comparison with the acid-washed wood. The dramatic influence of cobalt in lowering the ignition temperature is no doubt associated with its unusual effectiveness in catalysis of char gasification (13) and potassium appears to have a similar, though smaller influence. Calcium ions lower the ignition temperature much less effectively than potassium and in the original wood the ignition temperature is intermediate between that of potassium- and calcium-exchanged wood. This observation and also the fact that the char from acid-washed wood failed to ignite under the conditions of Figure 2 have obvious implications for further work on smoldering combustion of lignocellulosic materials.

Recently, Ahmed and Back (14) have proposed a mechanism for carbon-oxygen reaction. The simple mechanism was described as follows

$$C_F + O_2 \quad \text{--------> } C(O_2) \tag{1}$$
$$C_F + C(O_2) \quad \text{--------> } 2(CO)_c \tag{2}$$
$$(CO)_c \quad \text{--------> } CO_{(g)} + C_F \tag{3}$$
$$C_F + (CO)_c + O_2 \quad \text{--------> } CO_{2(g)} + (CO)_c + 2C_F \tag{4}$$
$$(CO)_c + C(O_2) \quad \text{--------> } CO_{2(g)} + (CO)_c + C_F \tag{5}$$

C_F refers to a free carbon site on the surface, $C(O_2)$ refers to

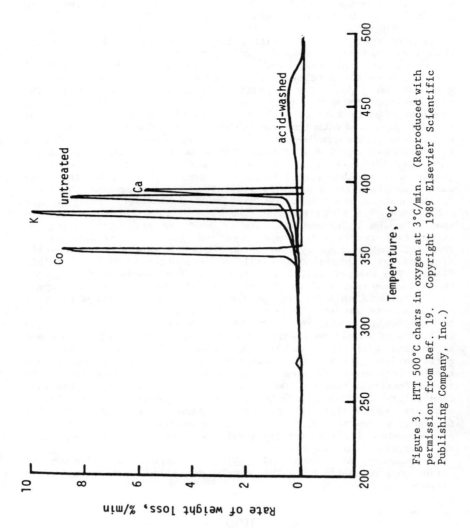

Figure 3. HTT 500°C chars in oxygen at 3°C/min. (Reproduced with permission from Ref. 19. Copyright 1989 Elsevier Scientific Publishing Company, Inc.)

Table IV

Effect of Acid-washing and Ion-exchange on Oxygen Chemisorption and Ignition of Cottonwood Chars

| Chars | Oxygen Chemisorption | | Initial Rate of Oxygen Chemisorption x 10^3 (mmole g^{-1} min^{-1}) | Ignition Temperature[c] (°C) |
	q^a (mmole g^{-1})	q_{max}^b (mmole g^{-1})		
Untreated	1.49	2.06	89.4	384
Acid-washed	1.35	2.33	64.1	no ignition
Acid-washed, then K-exchanged	1.41	1.75	92.1	377
Acid-washed, then Ca-exchanged	1.53	2.31	81.8	393
Acid-washed, then Co-exchanged	1.84	2.33	100.8	353

[a]Oxygen uptake during the first 2 h.
[b]Maximum oxygen uptake during isothermal chemisorption and oxidation process.
[c]Heating rate: 3°C/min.

an adsorbed oxygen molecule before formation of the surface oxide complex takes place, and $(CO)_c$ refers to the stable surface oxide complex. The first and second reactions represent formation of the surface oxide complex by the chemisorption process. The third reaction represents desorption of the surface oxide complex. The fourth and the fifth reactions represent gasification of the surface oxide complex.

With this type of concept in mind, a further series of experiments was carried out in which the char samples from untreated, acid-washed and ion-exchanged cottonwood discs were prepared in nitrogen at 20°C/min and held at 475°C for 10 minutes. Chars were then cooled to the desired temperature and immediately oxygen was introduced to the TG system. Iterative steps of 10°C were used for the temperature of switching to oxygen and the experiments were repeated until the minimum ignition temperature was obtained. The minimum ignition temperature described is therefore the minimum temperature at which fresh chars start to ignite. The results shown in Table V reveal the effect of metal ions in wood on the minimum ignition temperature of its derived wood char. These differences in ignition temperature are relatively smaller than those shown in Table IV, but are still significant in terms of fire situations.

Table V. Effect of Metal Ions on the Minimum Ignition
Temperature (in Oxygen) of Cottonwood Chars

Chars[a]	Minimum Ignition Temperature (°C)
Untreated	280
Acid-washed	310
Acid-washed, then K-exchanged	280
Acid-washed, then Ca-exchanged	280
Acid-washed, then Co-exchanged	280

[a]Charring process: pyrolysis of wood (in N_2) from 25°C to 475°C at 20°C/min and held at 475°C for 10 min.

There are three major heat or energy sources which are likely to influence ignition, i.e. from chemisorption, from combustion processes before ignition, and from the applied heat source. Bradbury and Shafizadeh ($\underline{1}$) have reported that preadsorption of oxygen on fresh cellulose char at temperatures below the minimum ignition temperature raised the subsequent ignition temperature of char (in oxygen) by 130°C. The data shown in Tables IV and V also reveal that preadsorption of oxygen on fresh wood char raised the subsequent ignition temperature of wood char by at least 73°C. This effect is undoubtedly due to the dispersion of the heat of chemisorption at temperatures too low for ignition and this energy is therefore no longer available to provide its "normal" dominant contribution to ignition. This effect has been applied to prevent spontaneous combustion of fresh charcoal by feeding air (preadsorption of oxygen) through fresh charcoal at temperatures below 150°C (15). The preadsorbed oxygen was said to "poison" some of the active sites on the charcoal surface and so raise the ignition temperature, but the type of explanation given above, now seems more likely.

Influence of Sample Configuration on Oxygen Chemisorption and Ignition of Chars. Ohlemiller (16) has shown that the configurations of both the heat source and the cellulosic material have major influence on ignition temperature. Thus the minimum heat source temperature required for ignition can be varied by at least 150°C as a result of variation in heat source geometry from concave to convex. Other parameters such as bulk density and even retardants may have much smaller influence. The layer thickness (e.g., of cellulosic insulation) can also have major effect on the temperature of smolder ignition (17). This temperature can range from 320-220°C for ignition of layers ranging from 3-30 cm thick. The lower ignition temperature has been related to decreased rate of heat loss in the thicker layers. On the other hand, more recent work by Davies and coworkers (18) on differential thermal analysis measurement of ignition temperature of small samples of cellulose powder showed that the ignition temperature was independent of sample mass and air flow rates over a wide range. In the latter work however, the heating rate had significant influence. This type of effect has now been studied with wood.

Three samples were used to investigate the influence of sample configuration on oxygen chemisorption and ignition. They were cottonwood discs (5.5 mm in diameter, 2 mm thick), coarse cottonwood powder (40 x 80 Tyler mesh) and fine cottonwood powder (-80 mesh). The minimum ignition temperature of the wood chars prepared from different samples can be seen in Table VI. Thus the char prepared from cottonwood discs ignited at much lower temperature than the char prepared from either of the cottonwood powders. The influence of sample configuration on the minimum ignition temperature of wood chars therefore is larger than that of metal ions in the wood from which the chars are prepared. The thermogravimetric analysis curves for pyrolysis-chemisorption-oxidation of the three different cottonwood samples were determined under conditions similar to Figure 2. In the pyrolysis step (25°-475°C at 20°C/min in nitrogen) there was no significant difference between the three samples. Similarly in the oxygen chemisorption phase (250°C for 15 min) there was no significant difference between the three samples. On further heating in oxygen at 10°C/min however, the disc sample ignited (shown by very rapid weight loss) at 392°C while the fine and coarse powders both ignited at 418°C. Thus the physical form of the char had a major influence on ignition temperature although the CSA of all chars was the same. The chars from the wood powders were sponge-like while the wood disc retained its shape during charring and the char had relatively high mechanical strength. It is concluded that the powder chars can probably dissipate heat of combustion and chemisorption by convection more rapidly than the disc samples. The above results certainly confirm the fact that the physical form and state of the sample plays an important role in ignition.

Table VI. Effect of Sample Configuration on the Minimum
Ignition Temperature (in Oxygen) of Cottonwood Chars

Cottonwood Chars[a]	Minimum Ignition Temperature (°C)
Char prepared from wood disc (5.5 mm in diameter, 2 mm thick)	280
Char prepared from coarse wood powder (40 x 80 Tyler mesh)	330
Char prepared from fine wood powder (-80 mesh)	340

[a]Charring process: pyrolysis of wood (in N_2) from 25°C to 475°C at 20°C/min and held at 475°C for 10 min.

Similar experiments to those just described were carried out on the three types of wood sample with the first charring step being carried out in air rather than nitrogen (i.e., oxidative pyrolysis) and with subsequent chemisorption and ignition also in air. In this case there were slight but significant differences in the oxidative pyrolysis stage, with the disc samples losing weight rather more slowly than the powders at low temperatures (250-300°C), but more rapidly at higher temperatures (ca. 320°C). Presumably these effects relate to the more ready access of oxygen to the powder samples and to the initial retention of early pyrolysis products in the discs. All samples however gave similar char yields (ca. 20% at 400°C) and the disc samples again ignited at lower temperature (439°C) than the powder samples (450°C). These temperatures are of course higher than the corresponding ignition temperatures in oxygen, but the differences between disc and powder samples are again significant and are assumed to be due to similar effects to those described above.

Experiments similar to those described above have been carried out with newsprint with generally similar results, as would be anticipated, since the newsprint is predominantly composed of mechanical wood pulp and therefore contains most of the original hemicelluloses and lignin (now "bleached") in addition to cellulose. The influences of some of the metal ions incorporated in both bleaching and printing can be detected in ignition processes. Thus, Table VII shows the high sodium content of newsprint due to bleach agents and the increase in aluminum content associated with colored inks. Acid washing of the newsprint led to removal of most of the metal ions and to significant increase in ignition temperatures (Table VIII). It is also notable that black ink appears to raise ignition temperatures of newsprint, while colored inks lower the ignition temperatures. We speculate that the former effect may be associated with quenching of free radicals at ignition by the carbon black or other components of black ink, while the ions such as aluminum, introduced with colored inks are likely to catalyze char gasification.

Table VII

Metal Ion Content of Newsprints (ppm)

Sample	Al	B	Ca	Cu	Fe	Mg	Mn	P	Si	Na	Ti	Zn	K	Total
Untreated
Without ink	246	8	817	4	20	192	56	35	193	1404	1	5	265	3246
Water-washed
Without ink	255	4	675	3	26	157	47	16	179	696	1	10	201	2270
With ink	187	3	702	9	50	108	61	21	91	384	4	4	137	1761
With color	474	2	1089	75	36	113	61	13	415	598	9	2	114	3001
Acid-washed
Without ink	42	1	0	0	9	31	0	14	65	31	0	1	25	219
With ink	4	3	16	2	8	2	0	6	27	57	0	2	43	170
With color	323	1	37	42	30	14	0	10	168	69	1	3	52	750

Table VIII.   Effect of Ink and Color on the Ignition Temperature
(in Oxygen) of Newsprint Chars[a]

Sample	Minimum Ignition Temperature (°C)	Ignition Temperature of Chemisorbed Chars[b] (°C)
Untreated		
Without ink	290	405
With ink	310	410
With color	270	386
Water-washed		
Without ink	290	397
With ink	310	403
With color	270	391
Acid-washed		
Without ink	320	476
With ink	330	478
With color	320	488

[a] Charring process: pyrolysis of newsprint (in N_2) from 25°C to 475°C at 20°C/min and held at 475°C for 10 min.
[b] Chemisorbed oxygen at 250°C for 15 min, subsequently heated at 10°/min.

Acknowledgement

This project was supported by a grant from the Center for Fire Research, National Institute of Standards and Technology.

References

1.   Bradbury, A.G.W.; Shafizadeh, F., Combust. Flame 1980, 37, 85-89.
2.   E.g., Bradbury, A.G.W.; Shafizadeh, F. Carbon 1980, 18, 109-115 and references therein.
3.   DeGroot, W.F.; Shafizadeh, F. Carbon 1983, 21, 61-67.
4.   Laine, N.R.; Vastola, F.J.; Walker, Jr., P.L. J. Phys. Chem. 1963, 67, 2030-2034.
5.   DeGroot, W.F.; Pan, W.-P.; Rahman, M.D.; Richards, G.N. J. Anal. and Appl. Pyrol. 1988, 13, 221-231.
6.   Pan, W.-P.; Richards, G.N. ibid 1989, 16, 117-126.
7.   Shafizadeh, F.; Sekiguchi, Y. Carbon 1983, 21(5), 511-516.
8.   Sekiguchi, Y.; Frye, J.S.; Shafizadeh, F. J. Appl. Polymer Sci. 1983, 28, 3513-3525.
9.   Sekiguchi, Y.; Shafizadeh, F. J. Appl. Polymer Sci. 1984, 29, 1267-1286.
10. Shafizadeh, F.; Sekiguchi, Y. Combust. Flame 1984, 55, 171-179.
11. Lowe, J.P.; Silverman, B.D. Accounts of Chem. Res. 1984, 17, 332-338.
12. DeGroot, W.F. Carbohydr. Res. 1985, 142, 172-178.
13. DeGroot, W.F.; Richards, G.N. Fuel, 1988, 67, 345-351.
14. Ahmed, S.; Back, M.H. Combust. Flame 1987, 70, 1-16.

15. Smith, R.H. U.S. Patent 4 170 456, 1979.
16. Ohlemiller, T.J. <u>Combustion Science and Tech.</u> 1981, <u>26</u>, 89-105.
17. Ohlemiller, T.J.; Rogers, F.E. <u>Combustion Science and Tech.</u> 1980, <u>24</u>, 139-152.
18. Davies, D.; Horrocks, A.R.; Greenhalgh, M. <u>Thermochemica Acta</u> 1983, <u>63</u>, 351-362.
19. Hshieh, F.-Y.; Richards,G.N. <u>Combustion and Flame</u> 1989, <u>76</u> 37-47.

RECEIVED January 11, 1990

Chapter 24

Rate of Isothermal Heat Evolution of Lignocellulosic Sheet Materials in an Air Stream

Ernst L. Back[1] and Frans Johanson[2]

STFI, Stockholm, Sweden

The rate of isothermal heat evolution in lignocellulosic sheet material was studied at temperatures between 150 and 230°C using a labyrinth air flow calorimeter and commercial hardboards, medium density boards and laboratory hardboards of holocellulose, bleached kraft and groundwood, the latter with and without fire retardants.

For all boards of mechanical or "thermomechanical pulp, the rate of isothermal heat evolution after an initiation period was maximum and then decreased with " reaction time. For boards of delignified pulp, i.e. of holocellulose and bleached kraft, this rate reached an intial plateau and then increased with time at temperatures above 200°C whereas it remained almost constant at lower temperatures. The following parameters are evaluated as a function of temperature:

1) the initial maximum or plateau rate of heat release
2) its subsequent change with time including corresponding rate constants
3) the overall heat release extrapolated to infinite time.

The initial rate of heat release around 200°C is 5 to 10 times higher for lignocellulosic boards than for the delignified boards. The activation energy of this initial rate varies from 15 kcal/mole for bleached kraft board to 30 kcal/mole for groundwood boards, with commercial boards of thermomechanical pulps in between.

For lignocellulosic boards the initial maximum rate of heat release declines with time – approximately according to first order kinetics. This decline is only moderately faster at higher temperature. The activation energy of the rate constant varied around 5 kcal/mole for hard and semi-hardboards, which

NOTE: This chapter is adapted with permission from ref. 10. Copyright 1989 De Gruyter.
[1]Current address: Feedback Consulting E & E Back KB, Vikbyvägen 42, 181 43 Lidingö, Sweden
[2]Current address: Pappersgruppen AB, Box 1004, 431 26 Mölndal, Sweden

indicates a reaction rate limited by diffusion, probably into the fiber or fibrous walls. Accordingly the total heat release extrapolate to infinite time, i.e. the total oxidation reaction increases very significantly with increasing temperature.

For delignified boards, a constant or at higher temperatures an increasing rate of heat release with time is interpreted as a radical initiated oxidation, maybe catalysed by some of the intermediate oxidation products - but retarded by radical scavengers formed in the degradation of lignin.

The corresponding chemical reactions are discussed. Oxidative formation of carboxylic and carbonylic groups is considered to be the main source of heat release.

In the production of wet process hardboards and semi-hardboards the press-dried boards are usually tempered or "cured" in hot air to increase their water resistance, dimensional stability, strenght and stiffness. A curing for 5 hours or more in hot air of 165°C is common. Higher curing temperatures reduce the period needed for each batch and thus increase the capacity of the heat treatment chamber, but they increase the auto-ignition risks.

During the heat treatment, oxidative auto-crosslinking of wood polymers and also some chain cleavage takes place in the lignocellulosic material (1-3). Wood extractives such as unsaturated fatty and resin acids prevalent in the hardboard are most accessible for such oxidation and take part in this auto-crosslinking. Among the wood polymers lignin is most reactive. The influence of temperature, time and added oxidation catalysts on the development of properties during these reactions has been described (1,2,4-6) The overall reaction is exothermic and unless this heat is dissipated rapidly enough from the material, self-heating and auto-ignition occurs. Such fires usually occur several times a year in the curing chambers and require constant attention. The cooling capacity required e.g. for the water showers in the recirculating air system to prevent such risks can be evalutated if the temperature dependence and time dependence of the heat evolution are known.

Such basic data are here presented from isothermal measurements on 5 by 90 cm lignocellulosic and cellulosic strips up to 13 mm thick. A suitable labyrinth air flow calorimeter operating at temperatures up to 250°C was built for these fairly large size samples. (7)

Experimental

The Labyrinth Air Flow Calorimeter.

The design and necessary calculations for the labyrinth calorimeter with turbulent air flow but low heat losses, have been presented earlier (7). See Figure 1. The advantage over micromethods such as DSC is that effects of sheet density and caliper of fiber entities

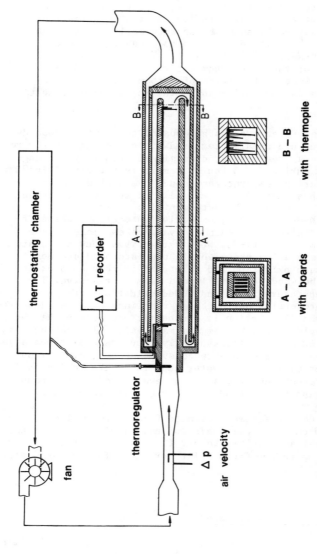

Figure 1. Labyrinth air flow calorimeter. Cross section A - A shows the measuring channel with four samples and the two outer insulating labyrinth channels. Cross section B - B shows one of the two piles of thermocouples which are placed at the channel inlet and the outlet to record the air temperature difference. (Reproduced with permission from ref. 10. Copyright 1989 De Gruyter.)

can be evaluated e.g. at high enough air speeds to keep temperature gradients in the material low.

Pre-thermostated air was circulated from a laboratory heating chamber by means of a centrifugal fan through the calorimeter and back to the heating chamber for re-thermostating by electrical elements, some of which were switched on continously and others on-off regulated by a thermistor probe placed at the inlet of the calorimeter. The air velocity was measured by means of a Pitot tube ahead of the calorimeter. The heat evolution channel had a cross section of 5.3 x 5.3 cm and a lenght of 100 cm. Depending on board caliper, 2 to 5 strips at a time were placed in the channel their length usually being 88 cm and their width 4.8 cm. The air flow in the channel along the samples was approximately 9 m/s and turbulent to ensure rapid heat and mass transfer. The velocity depended somewhat on the thickness and number of samples in the channel, as seen in Table I.

The increase in temperature of the air from the inlet to the outlet of the measuring channel was evaluated by means of a tenfold thermopile of chromel-constantan thermoelements. The calorimeter was insulated with mineral wool, the walls of the measuring and surrounding labyrinth channels being coated with aluminium foil. In the outer labyrinth channels the air flow was laminar. The heat loss through the walls between the inlet and the outlet of the measuring channel was measured with asbestos cement sheets, preheated at 250°C for 48 hours to remove organic material and moisture. In Figure 2 the temperature drop is plotted against the air temperature. Using this calibration correction, the rate of heat evolution in a board sample was calculated from the increase in the air temperature through the channel and the air velocity, both of which were recorded simultaneously.

It appeared relevant to relate the heat released to the mean temperature within the board sample. Neglecting the rather small temperature gradient over the thickness of the hardboard samples (7), it was decided to relate the data to the temperature in the surface layer of the board Ts according to the expression:

$$Ts - Ta = \Delta Q \cdot B/2 \ \alpha$$

where Ta is the mean temperature of the air, Δ Q the heat released per unit time and unit weight, B the grammage i.e. the "basis" weight per unit area of the board and α the heat transfer coefficient. The heat transfer coefficient was calculated to be 100kJ/m²°C in turbulent air of 8.5 m/s alongside the boards.

For a 12 mm insulating board with a thermal conductivity of about 0.38 kcal/m,h,°C (equal to 0.044 W/m°C) perpendicular to the sheet, the temperature difference between surface and thickness center under the conditions used was measured to be 10°C at a mean temperature of 190°C. For a hardboard with nearly three times this thermal conductivity it is negligible. For example, for a 3.2 mm hardboard it varies from 1 to 3°C over the 185° to 230°C range.

Procedure and Materials used.

With inert asbestos cement sheets in the measuring channel, the air-flow calorimeter was first equilibrated at an air temperature

Table I. Board Characteristics and Calorimeter Conditions

type of pulp	% lig-nin	% soluble in CH_2Cl_2	wood base	yield %	additives % on dry board	white water pH	board properties thickness mm	board properties density kg/m^3	calorimeter conditions samples in channel	calorimeter conditions air velocity m/s
holocellulose	3.9	0.1	Picea abies groundwood		none	5.4	2.8	890	4	7.4
bleached sulphate	0.1	0.4	Pinus silvestris		none	5.9	3.2	1100	4	8.0
groundwood	27.7	0.4	Picea abies	99	none	5.6	3.3	980	4	8.1
					10% diammonium phosphate	5.6	4.7	940	4	9.5
					10% borax/ boric acid	5.6	4.4	930	4	9.1
Asplund	28.5	0.5	80% Pinus 20% sawdust	94	0.7% $Al_2(SO_4)_3$ $18H_2O$	3.8	3.8	910	4	8.5
Masonite	32.1	1.8	45% Pinus 45% Picea 10% Betula	77	0.05% $Fe(SO_4)_2$ H_2O 0.1% $Al_2(SO_4)_3$ $18H_2O$	4.2	3.6	960	4	8.3
90% Asplund			75% softwood		0.4% $Al_2(SO_4)_3$$18H_2O$	4.2	13.1	310	2	11.4
10% groundwood			25% birch		40% asphalt					

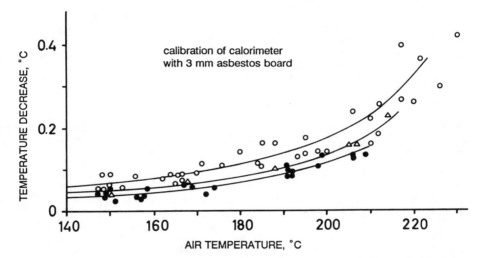

Figure 2. The temperature drop from the inlet to the outlet of the measuring channel due to heat losses versus the mean air temperature. Experiments with four 3 mm thick (open circles), three 6 mm thick (triangles), and two 10 mm thick asbestos sheets (filled dots). (Reproduced with permission from ref. 10. Copyright 1989 De Gruyter.)

5° to 10°C above the one to be used. The lignocellulosic board
samples were pre-heated in air of 150°C for 12 minutes in a separate chamber to avoid a large temperature drop in the calorimeter
when they were inserted. For a period of two minutes the fan was
stopped, the combined lid of the measuring and labyrinth channels
opened and the asbestos cement sheets replaced by the preheated
samples, the lid closed and the fan restarted. The temperature control of the heating chamber was switched off for another three
minutes until the air temperature had reached an equilibrium value.
The regulating temperature was then set to this temperature to
minimize the influence of the starting period. Both the temperature
difference over the channel and the air temperature at the inlet
were recorded. In some cases, the temperature at various points
over the board thickness and the temperature difference between the
board center and the surrounding air were also recorded.

Measurements covered periods ranging from 2 hours to 72 hours.
The total weight loss during the experiment was determined by
weighing the samples hot immediately before and after evaluation in
the calorimeter.

Some properties of the lignocellulosic and cellulosic boards
investigated are given in Table I. Figures 11,13,14,16 and 18 contains data for additional commercial boards of various thicknesses
and densities of pulps described in Table I, as indicated, in Table
II. Sheets of groundwood and of bleached kraft pulps were wet formed from commercial pulps dispersed for 30 minutes. The holocellulose pulp was produced by delignifying groundwood in sets of 550 g
dispersed at 60°C in two liters of water, to which was added 6
liters of a solution containing 400 g sodium chlorite and as a
buffer 136 g sodium acetate and 57 ml concentrated acetic acid at
65°C. Delignification took place at 60°C for 6 hours, followed by
dewatering and washing 5 times at 60°C. Thereafter delignification
was repeated once more using only 125 g sodium chlorite but a reaction time of 20 hours.

Wet formed sheets of groundwood, holocellulose pulp and
bleached kraft pulp were cold pressed for one minute at a 0.1 MPa
and hot pressed between two press wires at 150°C for one minute at
5.0 MPa, for another 4 minutes at 1.0 MPa and finally 12 minutes at
10.0 MPa.

Commercial Masonite and Asplund boards and the asphalt-impregnated insulation board were sampled before any commercial heat
treatment. The corresponding "thermomechanical" pulps had been
produced by pressurized refining of steamed chips. The higher pre-
steaming temperature used in the Masonite process resulted in a
yield of about 85% as compared to a fiber yield of about 94% of the
Asplund pulps, the remainder being dissolved.

Groundwood-based boards were impregnated with fire retardants
by adding a solution of either diammonium hydrogen phosphate or of
a 50/50 mixture of borax and boric acid to the moist sheets after
the cold pre-pressing. These impregnated sheets were then allowed
to dry at 20°C for 48 hours to achieve a solids content of about
60%, before being press-dried at 150°C as described. The amount of
fire retardants added to the boards were 10%, the amount retained
thereof being 80% for the borate and 100% for the phosphate.

Thickness values given in Table I include the indentations due
to the wire marks and the apparent density is based on this thickness and on the conditioned grammage at 65% RH and 20°C.

Table II. Constants of Arrhenius plots for the initial heat release rate and for the rate constant

Type of pulp (details in part I)	board properties thickness (mm)	density (kg/m³)	number of experiments	Equation 1 k_1 (kcal/kg h)	E_1 (kcal/mol)	Equation 3 k_3 (1/h)	E_3 (kcal/mol)	Equation 5 $\Sigma\ Q_T$ (kcal/kg) at 230°	at 160°
holocellulose	2.8	890	3	1.5×10^{10}	24	a	a		
bleached kraft	3.2	1 100	5	1.0×10^{8}	15	a	a		
groundwood	3.3	980	6	3.8×10^{15}	30	1.0×10^{4}	− 6	4 200	12
groundwood with 10% diammonium-phosphate	4.7	940	3	1.2×10^{13}	24	1.1×10^{2}	+ 7	2 000	140
groundwood with 8% borax/boric acid	4.4	930	3	3.5×10^{14}	27	7.6×10^{8}	+ 24	12 000	8 000
hardboards Masonite	2.2	950	5	1.4×10^{10}	18				
	2.7	910	3	1.0×10^{9}	16				300
	3.6	960	6	2.9×10^{9}	17			2 000	
hardboards Asplund	2.3	910	4	3.9×10^{10}	19				
	3.8	910	7	3.4×10^{9}	17				
	6.0	980	6	8.8×10^{9}	18	18.3	5.8		
semi-hardboards Asplund	12.6	500	3	3.2×10^{12}	23				
	9.3	760	5	1.1×10^{12}	22	5.8	4.4	5 800	260
	13.3	610	4	4.3×10^{13}	25				
asphalt impregnated insulation board	13.2	310	4	4.2×10^{11}	20	0.9	2.5	5 300	305

a) Hardboards of holocellulose and of bleached kraft could not be evalutated

Rates of Isothermal Heat Evolution

The rate of isothermal heat evolution at various air temperature
versus time for the groundwood-based hardboard without additives is
shown in Figure 3. Just after the samples achieve the set equilibrium temperature, the rate of heat evolution quickly rises to a
maximum and then decreases with time. The logarithm of the rate of
heat evolution decreases almost linearly with reaction time. Data
for the corresponding boards impregnated with 10% of a borax/boric
acid mixture are shown in Figure 4. Results for two commercial
hardboards, one of Asplund the other of Masonite pulp are given in
Figure 5.
    These figures show features characteristic for all hardboards,
semi-hardboards or insulating boards investigated, when they are
based on mechanical and "thermomechanical" pulps, and also for
those containing added asphalt or added tempering oils, i.e. a few
percent of unsaturated fatty acids and fatty acid esters. The absolute rates of theat evolution naturally vary batween these boards.
The addition of $Fe(SO_4)_2$ as catalyst in the Masonite boards
apparently did not change the approximately linear decrease of heat
release rate. The level of released heat is naturally highest for
the boards containing tempering oils or asphalt. More details are
given elsewhere (10).
    Corresponding hardboards based on delignified pulp behave
differently. For the hardboard of holocellulose pulp the results
are shown in Figure 6. The rates of isothermal heat evolution,
after the initiation period, remain constant with time at low
temperature, while they increase with time at the higher temperatures. The same is true for the hardboard of bleached kraft pulp as
shown in Figure 7. The logarithm of the rate of heat evolution here
increases with reaction time almost rectilinearly at high board
temperatures until the board ignites. The actual level of heat
release rate after the initiation period on the other hand is
slower for these delignified cellulose boards than for the non-
delignified ones of equal density.

Full Scale Measurements in a Curing Chamber for Hardboard

Some measurements were carried out in full scale chamber for batchwise heat treatment of hardboard by two procedures (8) as illustrated in Figure 8. One was to follow the total heat flux to the heat
exchanger of the chamber at equal air temperature first with an
empty carload and then with one fully charged measuring also heat
losses through the chimney. The heat exchanger used pressurized
water and the heat flux was evaluated continously. The chamber was
charged with 2.8 tons of hardboard made up of hundred single-laid
sheets at a few cm distance for the air to pass in between at 5.0
to 5.5 m/s. After the heating period the air flow through these
"daylights" serves mainly to transport the released heat from the
boards to the remaining part of the chamber, where heat is lost
partly by ventilation.
    In the second method, the temperature increase was measured
over a board length of about 5.5 meters after closing the sides of
one of these daylights to produce a closed air channel. A quadruple
thermo-chain was used in each of four such channels in a charge.

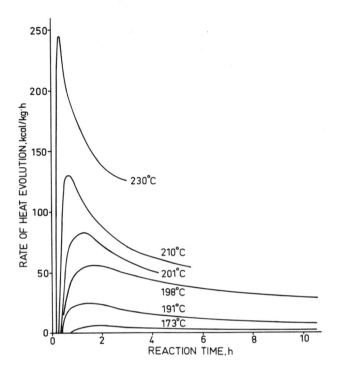

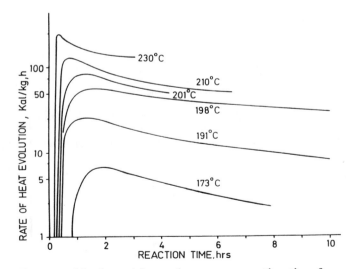

Figure 3. The rate of isothermal heat release versus reaction time for a 3.3 mm hardboard of groundwood. Above, on a linear scale; below, on a logarithmic scale. Released heat 1.0 kcal equal to 4.19 kJ. (Reproduced with permission from ref. 10. Copyright 1989 De Gruyter.)

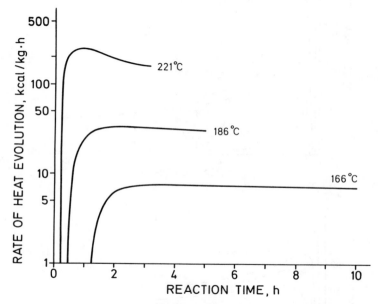

Figure 4. The rate of isothermal heat release versus reaction time for a 4.4 mm hardboard of groundwood containing 8% of a 50/50 mixture of borax and boric acid. (Reproduced with permission from ref. 10. Copyright 1989 De Gruyter.)

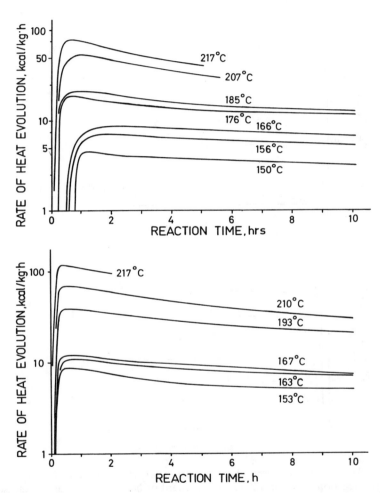

Figure 5. The rate of isothermal heat evolution on a logarithmic scale versus the reaction time. Top, for a 3.8 mm commercial hardboard of Asplund pulp; bottom, for a 3.6 mm commercial hardboard of Masonite pulp. Released heat 1.0 kcal equal to 4.19 kJ. (Reproduced with permission from ref. 10. Copyright 1989 De Gruyter.)

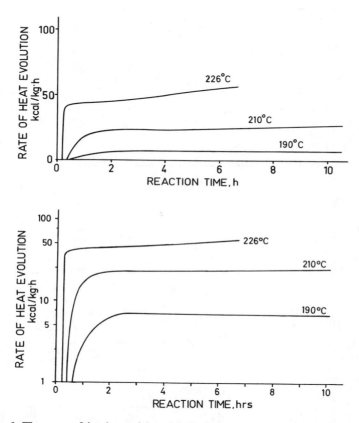

Figure 6. The rate of isothermal heat evolution on linear and logarithmic scales versus the reaction time for a 2.8 mm hardboard of holocellulose pulp. Released heat 1.0 kcal equal to 4.19 kJ. (Reproduced with permission from ref. 10. Copyright 1989 De Gruyter.)

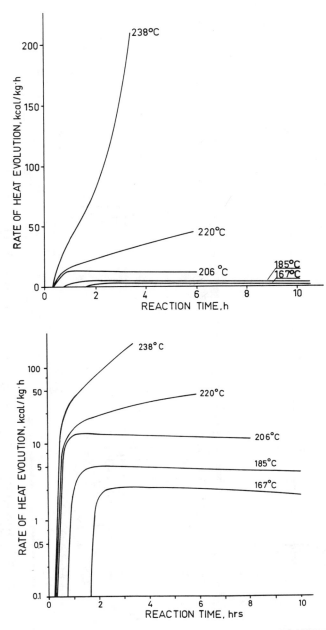

Figure 7. The rate of isothermal heat evolution on linear and logarithmic scales versus the reaction time for a 3.2 mm hardboard of bleached kraft pulp. Released heat 1.0 kcal equal to 4.19 kJ. (Reproduced with permission from ref. 10. Copyright 1989 De Gruyter.)

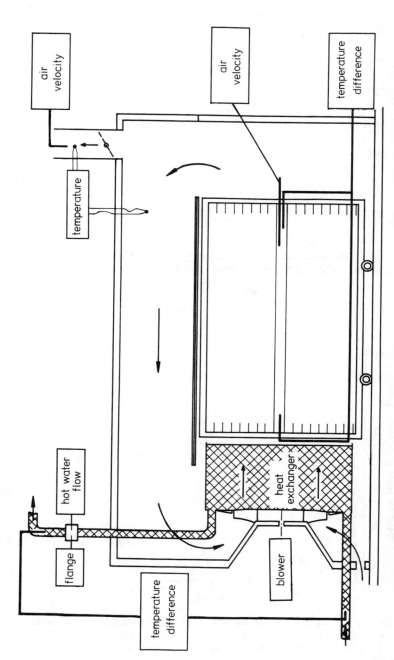

Figure 8. Measurements in a heat treatment chamber for hardboard. In a few channels of the carload of 100 single-laid boards, the air velocity and the temperature increase from ingoing air to outgoing air were measured as indicated. Also, measuring points used to determine the total heat balance in the chamber are indicated. (Reproduced with permission from ref. 10. Copyright 1989 De Gruyter.)

The thermoelements were placed just after a supporting steel beam crossing, where there is maximum air turbulence in this channel. The air velocity in there channels was also measured.

Figure 9 shows some results of these two procedures at a mean air temperature of $173^\circ C$. There are some differences between the four daylights due to differences in air temperature. Naturally there are also variations with time due to the regulating mechanism acting on the water flow through the heat exchanger to keep the chamber temperature at the pre-set level. These results refer to 3.2 mm commercial hardboard of Asplund pulp of density 950 kg/m^3. The mean isothermal heat release over the four hours' reaction time in this charge was 13 kcal/kg hardboard per hour or 52 kJ/kg·h. Other sets of measurements carried out with this procedure showed values up to 20 kcal/kg·h at an air temperature of $170^\circ C$ and 8-10 kcal/kg·h at $155^\circ C$. The initial maximum heat release at the full treatment temperature was quite pronounced in all these cases. These data are in reasonable agreement with those from the labyrinth flow calorimeter.

When, after a few hours of curing, the air temperature in the chamber was raised, another maximum heat relase occurred. Such measurements - carried out to evaluate the specific heat of the board - are reported elsewhere (9).

Temperature Gradients in the Samples During the Measurements

Within the boards the conditions are not fully isothermal, the temperature being somewhat higher in the center than in the outside layer. There is also a temperature gradient along the length of the boards from the channel inlet to the outlet. The magnitude of these temperature gradients was determined by means of thermocouples inside the samples. Figure 10 gives a typical gradient along the strip at an air temperature of $207^\circ C$ for a 6.0 mm Asplund hardboard of density 980 kg/m^3.

These temperature gradients increase with increasing rate of heat evolution, with increasing air temperature, with increasing board thickness, and with decreasing air velocity. An increase in board density, which increases the rate of heat evolution per unit volume, in some cases increased and in some cases decreased these temperature gradients, depending on the magnitude of the increase in thermal conductivity. In the experiments reported, the mean temperature of the board samples was kept at no more than $3^\circ C$ above that in the air flow, keeping the effect thereof within the experimental error. This was achieved by reducing the strip length of the board at the strips higher temperatures.

Released Heat Versus Weight Loss

Weight loss is due to the vaporization of oxidized and unoxidized degradation products of lignin, hemicelluloses and cellulose and loss of degraded or non-degraded pulp extractives. A general correlation between total weight loss and released heat is thus not expected. With the wide range if reaction times and with only a single end measurement of weight loss, a certain scattering of results is also to be accepted. Plots of these relations are presented in Figure 11.

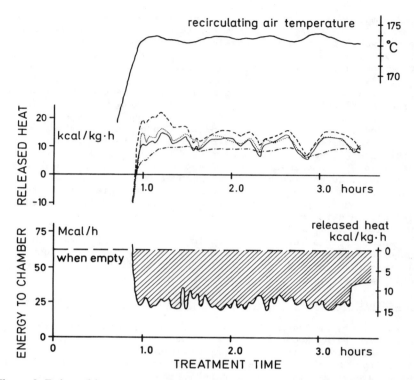

Figure 9. Released heat versus treatment time for a 2.8 ton carload of hardboard. Above, as evaluated on 4 separate channels each of this carload. Below, as evaluated by the total heat balance of the treatment chamber. Nearly one hour was necessary to bring the carload to full temperature. (Reproduced with permission from ref. 10. Copyright 1989 De Gruyter.)

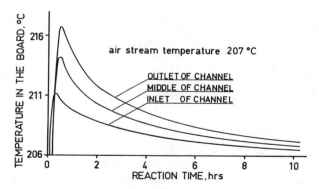

Figure 10. The temperature in the thickness center of the board measured at the inlet and at the outlet of the channel as well as in between versus the reaction time at an air temperature of 207 °C for a 6.0 mm hardboard of Asplund pulp. (Reproduced with permission from ref. 10. Copyright 1989 De Gruyter.)

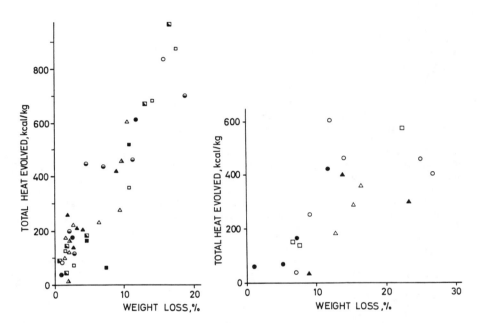

Figure 11. The total heat evolved versus the total weight loss.

Left for commercial hardboards of 2.3 mm (open circles), 3.8 mm (open squares), and 6.0 mm (open triangles), as well as for semi-hardboards of 9.3 mm and density 760 kg/m3 (filled squares), 12.6 mm and density 500 kg/m3 (filled circles) and 13.3 mm and density 610 kg/m3 (filled triangles), all of Asplund pulp. Also for commercial hardboards of Masonite pulp of 2.2 mm (semifilled circles), 2.7 mm (semifilled triangles) and 3.6 mm (semifilled squares).

Right for laboratory hardboards of groundwood, one 3.3 mm untreated (open circles), one 4.7 mm diammonium phosphate impregnated (open triangles) and one 4.4 mm borax/boric acid impregnated (open squares) as well as the 2.8 mm hardboard of holocellulose pulp (filled triangles), and the 3.2 mm hardboard of bleached kraft (filled circles). (Reproduced with permission from ref. 10. Copyright 1989 De Gruyter.)

For the commercial hardboards, the released heat per unit weight
lost is about 4400 kcal/kg. This means it is within the range of
the combustion value of softwood. For groundwood it is less, especially with increasing time. In the case of delignified pulp the
released heat per unit weight loss over longer periods is less than
this overall combustion value, maybe indicating volatilization also
of incompletely oxidized material.

In hardboards, some volatile material is already lost in the
pressurized refining process and in the subsequent press-drying at
$200^{\circ}C$ to $220^{\circ}C$ end temperature (3). The addition of fire retardants
did not appear to change released heat per unit weight loss at
these temperatures. The volatilized material has been analyzed in
commercial heat treatment and in laboratory scale by Nordenskjöld
and Östman (3). Carbon dioxide and carbon monoxide are the dominant
degradation products – when water is not measured. Remaining
products in order were total acids, such as acetic and formic etc.,
methanol and formaldehyde. Examples are given in Figure 12. Again,
there appears to be a maximum in CO_2 and CO release, after which
the evolution of these degradation products decreases with time.
The total emission in the commercial heat treatment of 5 to 8 hours
at 170 to $160^{\circ}C$ varied from 0.4 to 1.2% for CO_2 and 0.05 to 0.2%
for CO and 0.04 to 0.1% for total acids based on dry board. Some of
this emission might emanate from pyrolysis of higher molecular
weight material condensed and deposited on the walls of the heat
treatment chamber. The heat of formation of this CO_2 and CO is
about half the total heat release measured. Part of the oxidation
products might remain in the solid phase within the board material,
e.g. as bound carbonyl and carboxylic groups, partly followed by
heat consuming dehydration reaction.

Analysis of Results

The Initial Rate of Heat Evolution.

The "initial" rate or plateau rate of heat evolution, $(dq/dt)_i$
evident in Figures 3 to 7 has been plotted against the inverse
absolute temperature T, presenting so-called Arrhenius plots. For
lignocellulosic material such a plot appears to be relevant, but
for delignified cellulosic board it might appear less meaningful as
discussed below. Accordingly,

$$(dq/dt)_i = k_1 \cdot e \; -E_1/RT \qquad (1)$$

where E_1 is the activation energy, R the gas constant and k_1
the so-called frequency factor or preexponential factor. The
constants k_1 and E_1, calculated according to the least squares
method, are given for each board in Table II.

Figure 13 presents the Arrhenius plots for all boards of
density between 850 and 1 100 kg/m^3. Here the hardboard line is the
mean of both Asplund and Masonite type commercial hardboards,
presented in more detail in Figure 14. There was no significant
difference between hardboards from 2.3 to 6 mm thick made of the two
related types of coarse thermomechanical pulp. Semi-hardboards

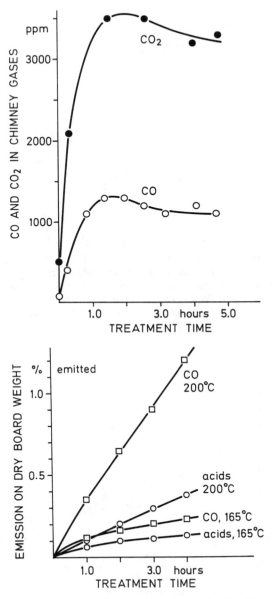

Figure 12. Top, carbon dioxide and carbon monoxide emitted in flue gases from batchwise commercial heat treatment of Asplund board at 165 °C versus time. In some plants the emission decreased more with time than here. Bottom, laboratory scale measurements at two temperatures. Data of emitted CO and total acids as weight % on dry hardboard. All data according to Nordenskjöld and Östman (3). (Reproduced with permission from ref. 10. Copyright 1989 De Gruyter.)

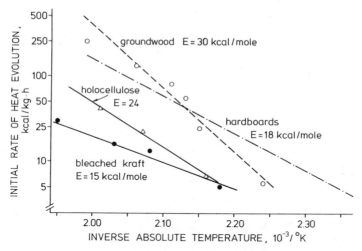

Figure 13. Arrhenius plots and activation energy E for the initial maximum or plateau rate of heat release, versus the inverse absolute temperature. All data refer to high density boards. The hardboard line represents a mean for the commercial Masonite and Asplund hardboards of Figure 14, with a caliper varying between 2.2 and 6 mm. (Reproduced with permission from ref. 10. Copyright 1989 De Gruyter.)

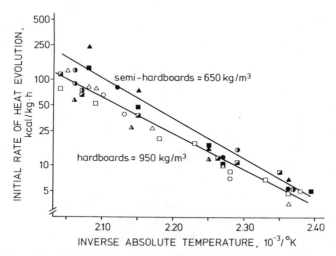

Figure 14. Arrhenius plots for the initial maximum rate of heat release versus the inverse absolute temperature for commercial boards. Data for hardboards of Asplund pulp with open symbols (circles 2.3 mm, squares 3.8 mm, and triangles 6.0 mm), and of Masonite pulp with semifilled symbols (circles 2.2 mm, triangles 2.7 mm, squares 3.6 mm). Also data for semi-hardboards of Asplund pulp from 500 to 750 kg/m3 density with filled symbols (squares 9.3 mm, circles 12.6 mm, and triangles 13.3 mm in calipers). (Reproduced with permission from ref. 10. Copyright 1989 De Gruyter.)

between 500 an 750 kg/m^3 density deviate slightly from this hardboard line, especially in the higher temperature range. At the
uppermost temperatures, part of this deviation may be due to
temperature gradients within these boards because of their lower
density. For an asphalt-impregnated insulating board of density
310 kg/m^3, the heat evolution rate is considerably higher, see
details in (10).

The temperature dependence of the initial rate of heat release
among the tested materials is greatest for the hardboard made from
groundwood showing an activation energy of 30 kcal per mole. It is
lowest for the hardboard made from bleached kraft, showing an
activation energy of 15 kcal per mole. Also, for delignified board
materials the initial plateau rate is only 10 to 20 % of that of
lignocellulosic boards. For commercial boards of thermomechanical
pulp, the activation energy increases slightly with decreasing board
density. The influence of board thickness, if any, is small. The
effect of the traces of the oxidation catalyst, ferro-sulphate,
added to the Masonite pulp was small compared with boards of the
catalyst-free Asplund boards.

The effect of two fire retardants added to groundwood is shown
in Figure 15. Both these fire retardants, the diammonium phosphate
and the 50/50 borax/boric acid mixture, increased the initial rate
of heat release over the temperature range investigated, particularly in the lowest range. This is in agreement with experimental
results obtained by other methods. On the other hand, the fire
retardants have sufficiently reduced this activation energy, so that
if extrapolated to higher temperatures such as above 230–250°C the
lines intersect. Above these temperatures the rate of heat release
will then be lower for boards with these fire retardants than for
boards without. In other words, the result could indicate that in
the presence of fire retardants the oxidative reactions at lower
temperatures take place with increased initial heat release rate
while above about 230 to 250°C the contrary is true. The data are in
agreement with other pyrolysis measurements in the presence of fire
retardants (11).

The Rate of Heat Release Versus Reaction Time.

In all the ligno-cellulosic boards, the logarithm of the rate of
isothermal heat evolution decreased approximately linearly with
reaction time. A formal representation according to a first order
reaction type is:

$$dq/dt = (dq/dt)_i \cdot e^{-k(t-t_i)}$$

or
$$dq/dt = (dq/dt)_i \cdot 10^{-r_{10}(t-t_i)} \tag{2}$$

where dq/dt is the rate of heat release at time t, and
$(dq/dt)_i$ is the initial maximum rate at time t_i, while r or
r_{10} is the temperature-dependent rate constant. Deviations from
this linearity of "Equation 2" occasionally can be seen at the

highest temperatures tested, e.g. above 220°C, especially just after
reaching the maximum rate, see Figures 3 to 7.

The rate constant r_{10} has been plotted against the inverse
absolute temperature for commercial boards in Figure 16 and for
groundwood based laboratory hardboard with and without fire retardants in Figure 17. In some experiments the reaction time was not
long enough for this evaluation.

The experimental data show a linear relationship between the
logarithm of the rate constant, r_{10}, and the inverse absolute
temperature, T,

$$r_{10} = k_3 \cdot e^{-E_3/RT} \tag{3}$$

where k_3 and E_3 are again material dependent constants, i.e. a
frequency factor and an activation energy in respect to reaction
rate. Mean values of k_3 and E_3 for groups of boards are included
in Table II.

It appears that the rate constant for most of the boards has a
smaller temperature dependence than the initial maximum rate, the
corresponding "activation energy" E_3 being around or less than
5 kcal/mol. An important conclusion is that the rate is diffusion
limited. This has to be compared to a mean activation energy around
20 kcal/mole for the initial maximum rate of heat release for the
commercial boards. As a consequence thereof the total heat release
extrapolated over infinite time does increase to a significant extent
with temperature from 150 to 230°C.

For the groundwood hardboards the treatment with fire retardants
resulted in an increased temperature dependence of the rate constant
r_{10}, while for the untreated board the rate constant actually
decreased slightly with increasing temperature (whereby a negative
"activation energy", E_3 is calculated). Thus, although in the lower
temperatures range the initial rate of heat release rate is larger
with the fire retardants, the rate decreases much more rapidly with
time than for untreated boards.

For the delignified boards of holocellulose or bleached kraft
this evaluation of the rate of heat release with time was not
possible.

The Total Heat Release and Its Temperature Dependence.

For lignocellulosic boards the total heat released over infinite time
under isothermal conditions based on equation (2) is

$$\Sigma Q_T = \int_0^{t_i} (dq/dt)\, dt + \int_{t_i}^{\infty} (dq/dt)_i\, 10^{+r_{10}(t_i - t)} \cdot dt =$$

$$= \text{approx } (dq/dt)_i / r_{10} \ln 10 \tag{4}$$

The first term covers the short initiation period until maximum heat
release rate occurs, which can be measured separately but is here
neglected in the approximation to the right of equation (4). Also
neglected are the existant deviations from first order kinetics.

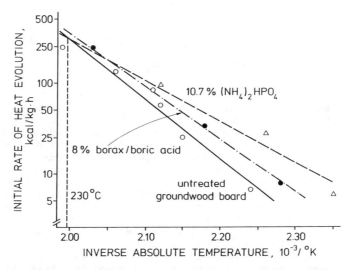

Figure 15. Arrhenius plots for the initial maximum rate of heat release versus the inverse absolute temperature for a laboratory hardboard of groundwood with and without added fire retardants, a 50/50 mixture of borax and boric acid respectively diammonium phosphate. (Reproduced with permission from ref. 10. Copyright 1989 De Gruyter.)

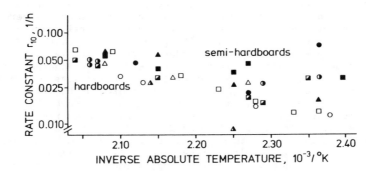

Figure 16. The rate constant for the decreasing heat release with time, r_{10} of "Equation 3" versus the inverse absolute temperature for the commercial boards of Figure 14. (Reproduced with permission from ref. 10. Copyright 1989 De Gruyter.)

By substitution from "Equation 1" and "Equation 3" the total heat
release as a function of temperature is given by

$$\Sigma Q_T \cong \frac{k_1 \; e^{-E_1/RT}}{k_3 \; e^{-E_3/RT} \cdot \ln 10} = \frac{k_1}{k_3 \cdot \ln 10} \cdot e^{(-E_1+E_3)/RT} \qquad (5)$$

Naturally such an extrapolation until the end of the reaction i.e. to
infinite time using "Equation 5" represents a rough estimate only.
Results for the mean of the 6 commercial hardboards, for the mean of
the 3 semihard boards as well as for the groundwood hard board are
given as a function of the inverse temperature in Figure 18. Accordingly, with increasing temperature there appears a very pronounced
increase in the overall extrapolated total isothermal heat release,
i.e. in the overall oxidation of the solid board. Considering the
general correlation between heat released and weight loss for each
lignocellulosic raw material shown in Figure 11, it might be concluded that oxidation at lower temperatures produces more residual
char or crosslinked material than at higher temperatures.

At higher temperatures e.g. 230°C, the hardboards at infinite
time have lost less than half their total combustion heat while
semi-hardboards will have lost all, if the extrapolation used is
valid. Table II includes the total calculated heat release at 230 and
160°C respectively.

Conclusions From the Analysis.

A few conclusions from this analysis of the heterogenous phase oxidation: For lignocellulosic boards of mechanical or thermomechanical
type, independent of additives such as the fire retardants here
tested, the heat evolution rate due to one or several, partly exothermic reaction has a maximum initial rate which rate then declines
according to approximately first order kinetics. Initially there thus
occur a number of reactive oxidation sites which are consumed or
blocked with time. This number of available active sites, "the active
mass", increases significantly with temperature, or the blockade of
oxidative heat release sites by side reaction is reduced correspon-
dingly with increasing temperature. The activation energy calculated
from this initial rate around 20 kcal/mole is in the lower range for
"chemical" activation of this maximum reactivity or maximum number of
reactive sites or as an alternative, indicates a chemical blocking of
heat releasing oxidation which blocking is reduced with increased
temperature.

On the other hand, the low temperature dependance of the rate
constants with activation energies around 5 kcal/mole indicates a
diffusion limited reaction rate which could refer to diffusion of
oxygene into the fibers of the board, i.e. into the fiberwalls. The
corresponding negative activation energy for the groundwood based
hardboard and the effect of fire retardants there upon are difficult
to understand.

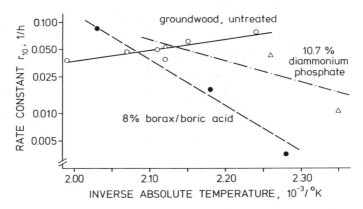

Figure 17. The rate constant for the decreasing heat release with time r_{10} versus the inverse absolute temperature for laboratory hardboards of groundwood with and without added fire retardants. While this rate constant is about equal in the range of 200 °C, it falls off more rapidly with fire retardants to lower temperatures, indicating a relatively larger remaining heat release rate there. (Reproduced with permission from ref. 10. Copyright 1989 De Gruyter.)

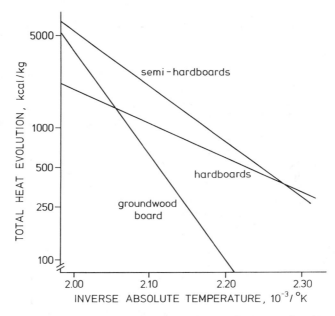

Figure 18. The total heat release to infinite time $\Sigma \, Q_T$ extrapolated according to "Equation 4" versus the inverse absolute temperature. Mean for the 6 commercial hardboards of Asplund and Masonite type and a mean for the 3 semi-hardboards of Figures 15 and 17, also data for the groundwood hardboard. (Reproduced with permission from ref. 10. Copyright 1989 De Gruyter.)

Between the lignin containing hardboards some differences occur. The groundwood hardboard with largely carbohydrate fiber surfaces has a much larger initial heat release, and chemical activation by temperature, than the commercial hardboards of coarse thermomechanical pulp which are made up of fiber bundles separated in the middle lamella, and surrounded by a rather smooth layer of partly condensed lignin.

For delignified boards made up of hemicellulose and cellulose, the initial rate of heat evolution is much lower than for boards of the lignocellulosic material. Thus at 200°C it is only about 15 % thereof. Besides for delignified boards this rate keeps constant below 200°C while above 200°C it increases very significantly with time to reach similar or higher values than those for lignocellulosic material. One interpretation can be an oxidation mechanism which is catalyzed by radicals or by intermadiate products formed, their amount increasing with time. Since lignin only makes up about 30 % of the thermomechanical pulp boards, and the last mentioned effect of cellulose-hemicellulose, does not show up there, an interaction with lignin or its degradation products seem to block this – with time increasing – oxidation rate of cellulose-hemicellulose material.

The initial time period until a plateau or maximum level of heat release occurs might be related to time required for the formation of a level of polymer radicals, while a minor part of this period also is required to reach the temperature equlibrium set.

Naturally for a complete understanding much more data would have been useful, e.g. separate data for boards of lignin, of hemicellulose and of cellulose and given combinations thereof.

Comparison With Ohter Heat Release Data

No similar direct measurements of the rate of heat release seem to have been published. One related set of data at 180°C only by Topf (12,13) is based on measuring the weight loss versus time of powdered material together with the combustion value of the solid residuals. The heat released is then taken as the difference in combustion value to that of the original sample.

The results were presented up to 1200 hrs and agree with those presented here. For α –cellulose the rate of heat release rate is constant for 300 hours at 180°C and similar to our corresponding value of about 4 kcal/kg h for bleached pulp. For Sipo sawdust from "Entandophragma utile", i.e. African mahony, the heat release at 180°C is initially about 15 kcal/kg h, similar to our data for hardboards, and then decreases with time, to e.g. 5 kcal/kg h after 200 hrs. In the range of 160°C to 180°C Topf also showed the rate for cellulose weight loss to increase with time over a few hundred hours before becoming constant and then decreasing, while the rate of weight loss rate for wood and particularly for lignin decreased with time after an initial maximum.

Other data on the rate of heat release of lignocellulosic materials are indirect. They are, for example, based on a rapid preheating of a mass of material to a given temperature, then following the development of temperature and of temperature gradients in this mass versus time, occasionally until autoignition occurred. The derivation of released heat per unit weight and unit time then requires assumptions concerning various material parameters such as the specific

heat and the heat conductivity. Data are then obtained over a range
of temperatures and occasionally over time. Activation energies for
such heat release rates have alsso been calculated and reported. They
usually refer to the initial rate of heat release.

Such activation energies of abiotic heat release for lignocellulosic material have recently been summarized and and compared in
detail by Kubler (14). These activation energies in the range from
$120^{\circ}C$ to $220^{\circ}C$, when air is present, usually are between 60 and 115
kJ/mole, i.e. 15 to 30 kcal/mole (14). Activation energies for the
delignified cellulose-hemicellulose material presented here are at
the lower end of this range.

For cotton cellulose such indirect measurements indicated first
a decreasing rate of heat release with time to a minimum reached
after 8 hours at $165^{\circ}C$, followed by an increasing rate with time
(15, 16) . In this case, the minimum rate was 0.7 kcal/kg h, which is
about half our initial rate for bleached kraft at that temperature
here. Also, this minimum rate was found to be greater for viscose
cellulose (17). In oxygene the rate was double that in air and still
showed the pronounced minimum level.

The greater rate of oxidation of lignin than of cellulose in the
temperature range of $150^{\circ}C$ to $250^{\circ}C$ with its dominating effect on
wood has also been pointed out in various papers (18,19).

The rate of heat release for freshly activated carbon similar to
that found here for lignocellulose, has an initial maximum (e.g. 1.2
kcal/kg h at $100^{\circ}C$), then decreases with time according to an
approximately first order kinetics (20). This behaviour was explained
as a consumption of reactive sites for oxidation.

In a more general sense pyrolysis rates and corresponding activation energies are often based on measurements of weight loss either
by dynamic differential gravimetry or at constant temperature, partly
carried out in nitrogen. Corresponding activation energies are usually in the same range as those for heat release while rate data
often fit first order or second order kinetics at constant temperature. These weight loss rates and the amount of char produced were
found to be larger, the lower the relative fraction of crystalline
material (21).

Various fire retardants affect the pyrolysis in specific ways.
Those which are or which on heating form strong acids, e.g. ammonium
salts of phosphoric acid, act partly by promoting auto-crosslinking
in the wood polymers, so that the relative amount of char is
increased while the total released heat is reduced. At lower temperatures, the total weight loss and total heat release are increased in
this case (11,22) (23). The activation energy based on weight loss
rates was found to be reduced by these fire retardants (21). These
results are in general agreement with our data on the rate of heat
released in the presence of such fire retardants.

Corresponding Chemical Reactions

Auto-Crosslinking Reactions.

The collected data were aimed for the commercial heat treatment of
hardboards (24) and for curing processes proposed for some paper
grades (4,6,25). They thus refer to quite a mild pyrolysis in air,
for a shorter treatment period than some of the data presented here

and in a rapid air stream to guarantee a high heat transfer. The
weight loss and the production of volatile material are only a few
per cent in these operations and the ignition of gases and board must
be prevented. An effort has been made earlier to optimize and understand these curing processes (1,2,24,26).

Figure 19 indicates some main reaction principles, with a radical
initiated cross-linking and chain-splitting with or without oxidation
(27), the details of which for wood polymers are not available.
Carbonylic groups as crosslinking intermediates as well as carboxylic
end groups are formed during this reaction in air, probably in all
the wood polymers (28). Radical formation to initiate such oxidative
reactions in cellulose has been shown by ESR measurements (29).

The auto-crosslinking produces the "desirable" moisture resistance in the commercial boards, such as the increased wet strength
and reduced swellability in water as well as an increased modulus of
elasticity. Since the cellulosic polymers and fibres are preferentially oriented in the sheet directions, the z-direction strength
perpendicular hereto is considerably improved, usually doubled by
such a commercial heat treatment (26). Chain-cleavage dominates at
the very end of the commercial treatment. Especially at higher temperatures it reduces strength properties. The brittleness of the material is increased both by auto-crosslinking and chain cleavage.
Arrhenius plots can be given for the rate of development of various
such material properties (1), e.g. for crosslinking evalutated by wet
strength and for chain cleavage evalutad as by molecular weight i.e.
water-extractable material. The pH of this water extract and of a
water suspension of the residual solid fell off with time as well.

Oleophilic material occurring, such as resin and fatty acids,
especially those which have conjugated double bonds, are most easily
oxidized and can also contribute to crosslinking in the system. Such
oleophilic material also redistributes, e.g. by sublimation, to cover
all hydrophilic surfaces in the material (1) (30).

The auto-crosslinking is assumed to take place in lignin and
hemicelluloses, while crystalline cellulose should be little affected
under the mild conditions here described. This also is indicated by
the greater relative amount of char formed when heating lignin and
hemicelluloses as compared to cellulose or when heating amorphous
cellulose as compared to crystalline cellulose, both in the absence
and presence of oxygen. Besides, these crosslinking reactions are
generally catalysed by acidic conditions and by metal ions with high
redox potential (1). Acidic groups formed within the residual material during the heat treatment might act similar to strongly acidic
fire retardands to promote such crosslinking and formation of char.
Measurements of the declining degree of polymerization of cellulose
during pyrolysis (28) overestimate the chain cleavage, since cross-
linked cellulosic material looses its solubility in cellulose
solvents (1).

The crosslinking reaction per se is exothermic, while a corresponding evaporation of water formed consumes heat.

Oxidation and Degradation Reactions.

The oxidative formation of carboxylic and carbonylic groups in the
wood polymers and in its extractives is probably the most important
heat releasing reaction in this lower temperature range (31,32) and

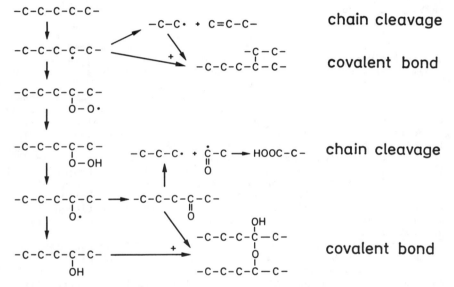

Figure 19. Example of reactions in wood polymers after activation such as by heat leading either to auto-crosslinking or to chain cleavage (27). (Reproduced with permission from ref. 10. Copyright 1989 De Gruyter.)

apparently the reactive mass herefore does increase significantly
with temperature. It might to a minor extent also take place in
nitrogen due to oxygene previously absorbed in and on the solid
phase. This oxidation would reduce the heat of combustion of the
solid residual. This indeed was found to be the case at 180°C for
wood and to a greater extent for lignin (12,13). Dehydration reactions, on the other hand, would instead increase the combustion value
of the residual solids as was found to be true for α-cellulose over
the first 400 hours at 180°C. These are the types of reactions for
auto-crosslinking in wood polymers by oxidative routes and/or by
activated dehydration. Crosslinking may be promoted by lower temperatures, leading to a lower total heat release over time. Carbon
dioxide was the predominant volatile substance released in the heat
treatment. The background is the heat consuming decarboxylation of
carboxylic groups at polymer ends or side chains, mainly of newly
formed such groups due to oxidation.

Uronic acids and pectic material have partly or mainly been
dissolved and removed from commercial boards on pressurized refining.
On heating wood they will contribute considerably to CO_2 release
(33). Organic acids such as formic and acetic acid are also split off
from side chains of wood polymers.

The significant release of carbon monoxide as well as of methanol can partly be referred to the degradation of lignin (33). Besides
the lignin already in the range below 200°C tends to split of phenolic compounds as well as methanol and carbon monoxide (34) (35). The
interaction of phenolic degradation products from lignin with cellulose or hemicellulose radicals can prevent grafting of cellulose
(36). It thus has a radical scavenger effect on carbohydrorates which
can explain the significant difference in heat release rate with
time found here when lignin is present or not. It might also affect
the composition of degradation products. As already mentioned, the
oxidation follows different routes in the different wood polymers
probably with some interaction.

In the range up to 250°C, various dynamic, DTA and DSC measurements on lignin without a significant peak show a weak exothermic
degradation in inert and especially in oxygen atmosphere. Similar
behaviour has been reported for various hemicelluloses, especially
xylan with peaks in the range of up to 250°C (37). For cellulose in
the presence of oxygen there is a more pronounced exothermic peak
somewhere around 300°C, depending on the rate of heating, just when
the rate of weight loss has a maximum (11) (22) (37). There are
probably differences between the crystalline and the amorphous cellulose in this respect. For example above the glass transition temperature of 220°C the reactivity of the amorphous cellulose might abrupt-
ly increase as indicated by data for sugars (38).

Specific surface area and total mass and density among others
can affect temperature gradients and gas diffusion rates and thus can
affect reaction rates and results (39). In fiber building boards,
especially oxygen and other gas diffusion into and out of single
fibers, appears to be most important.

The heat release data presented with their diffences between various cellulosic and lignocellulosic material in a general way thus
are in no disagreement with data in the literature. They can hopeful-
ly contribute to a better understanding of the oxidative pyrolysis.

Acknowledgments

The authors are indebted to Åke Isaksson – now Sunds Defibrator for
skillful technical assistance, to the Swedish Council for Applied
Research and to the Swedish Fire Research Board for support of this
work.

Literature Cited

1. Back, E. L. Pulp Pap. Mag. Can. 1987, 68, 247–258.
2. Stenberg, L. E. Svensk Papperstidn. 1978, 81, 49–54.
3. Nordenskjöld, M.; Östman, B. STFI-medd. 1981, A 669.
4. Back, E. L.; Stenberg, L. E. Pulp Pap. Mag. Can. 1976, 77,
    T264–270.
5. Back, E. L.; Stenberg, L. E. Pulp Pap. Mag. Can. 1977, 78,
    T271–275.
6. Anderson, R. G.; Back, E. L. Tappi 1975, 58(6), 88–91 and 58(8),
    156–159.
7. Eriksson, S. Å. I.; Back, E. L. Svensk Papperstidn. 1966, 69,
    300–304.
8. Gustavsson, N. U.; Isacsson, Å.; Back, E. L. Medd. Wallboardind.
    Centrallab. 1966, B 37.
9. Olsson, A. M.; Back, E. L. Nord. Pulp. Pap. Res. J. 1989, 4, in
    press.
10. Back, E. L.; Johanson, F. Holzforschung. 1989, 43, in press,
    part I and II.
11. Tang, W. K.; Eickner, H. US Forest Service Res. Paper. 1968, FPL
    82.
12. Topf, P. Holz Roh-Werkstoff. 1971, 29(7), 269–275.
13. Topf, P. Holz Roh-Werkstoff. 1971, 29(8), 295–300.
14. Kubler, H. For. Prod. Abstracts. 1987, 10(11), 299–327.
15. Walker, I. K.; Harrison, W. J. New Zealand J. Sci. 1977, 20(2),
    200.
16. Walker, I. K.; Harrison, W. J.; Jackson, F. H. New Zealand J.
    Sci. 1978, 21(2), 329–334.
17. Walker, I. K.; Harrison, W. J.; Jackson, F. H. New Zealand J.
    Sci. 1970, 13(4), 623–640.
18. Roberts, A. F. Combust. Flame. 1970, 14(2), 261–272.
19. Roberts, A. F. Combust. Flame. 1971, 17(1), 79–86.
20. Bowes, P. C. Self-heating, evaluation and controlling the
    hazard. Ed.; Elsevier Amsterdam, 1984.
21. Schultz, T. P.; McGinnis, G. D.; Nicholas, D. D. Paper presented
    at the 197th Am. Chem. Soc. Natl. meeting, Dallas, April
    9th–14th, 1989.
22. Tang, W. K. Research Paper, Forest Products Laboratory, USDA
    Forest Serivce. 1967, No. FPL-71:16 pp.
23. LeVan, S. L. In The Chemistry of Solid Wood. Advances in
    Chemistry Series, 207. American Chemical Society. 1984.
24. Back, E. L.; Klinga, L. Svensk Papperstidn. 1963, 66, 745–753.
25. Back, E. L.; Olsson, A. M. Tappi. 1989, 72,(10), in press.
26. Back, E. L. Holzforschung. 1987, 41(4), 247–258.
27. Back, E. L.; Gellerstedt, G. STFI-medd. 1983, D184.

28. Nevell, T. P.; Zeronian, S. H. Cellulose Chemistry and its
     Applications. Ed.; Ellis Horwood Ltd, Chichester, England. 1985;
     Chapter 10 and 11 by T. P. Nevell and F. Shafizadeh.
29. Arthur, J. C. Jr.; Hinojosa, O. J. Polymer Sci. Part C,36,
     53-71.
30. Swanson, J.; Cordingly, S. Tappi. 1959, 42(10), 812.
31. Shafizadeh, F.; Badbury, G. J. Appl. Polymer Sci. 1979, 23,
     1431-1442.
32. Philipp, B.; Baudisch, J.; Stöhr, W. Cellulose Chem. Technol.
     1972, 6, 379-392.
33. DeGroot, W. F.; Pan, W.-P.; Rahman, M.; Richards, G. N. J. Anal.
     App. Pyrolysis. 1988, 13, 221-231.
34. Gardner, D. J.; Schultz, T. P.; MacGinnis, G. D. Wood Chemistry
     and Techn. 1985, 5, 85-110.
35. Masuku, C. P.; Vuori, A.; Bredenberg, J. B. Holzforschung. 1988,
     42(6), 361-368.
36. Kobayashi, A.; Phillips, R. B.; Brown, W.; Stannet, V. T. Tappi.
     1971, 54(2), 215-221.
37. Beall, F. C. Wood Sci. Technol. 1971, 5, 159-175.
38. Richards, G. N. Int. Sugar J. 1986, 88(N.1052), 145-149.
39. Haw, J. F.; Schultz, T. P. Holzforschung. 1985, 39, 1289-296.

RECEIVED November 1, 1989

Chapter 25

Heat Release from Wood Wall Assemblies Using Oxygen Consumption Method

Hao C. Tran and Robert H. White

Forest Products Laboratory, U.S. Department of Agriculture Forest Service, One Gifford Pinchot Drive, Madison, WI 53705–2398

The concept of heat release rate is gaining acceptance in the evaluation of fire performance of materials and assemblies. However, this concept has not been incorporated into fire endurance testing such as the ASTM E–119 test method. Heat release rate of assemblies can be useful in determining the time at which the assemblies start to contribute to the controlled fire and the magnitude of heat contribution. Twelve wood wall assemblies were tested in an ASTM E–119 fire endurance furnace at the USDA Forest Service, Forest Products Laboratory. Heat release measurements using the oxygen consumption method were a part of this program. The data demonstrate the usefulness of the oxygen consumption calorimetric technique for calculating heat release rate. The tests reconfirmed previous work that showed heat release from assemblies protected with gypsum board is negligible during the first 30 min and is significantly less than that of unprotected assemblies for the duration of the test. With proper specifications for the fire endurance furnace, such as air supply and furnace pressure, the method can be incorporated into the ASTM E–119 test standard. Accuracy of the method is discussed.

The measurement of heat release rate using the oxygen consumption method has been developed and perfected for a number of applications. The major applications include heat release rate calorimeters, such as the Cone Calorimeter developed by the National Institute of Standards and Technology (NIST) (formerly the National Bureau of Standards) (1), furniture calorimeters (2), room fire tests (3), and the ASTM E–84 tunnel test (4). Because heat release rate of assemblies undergoing fire endurance testing are useful in assessing the overall fire performance of the assemblies, there has been some interest in incorporating this methodology into standard fire endurance tests such as the ASTM E–119 test method (5). The standard ASTM E–119 test method is used to evaluate the ability of an assembly to resist a severe fire exposure and maintain structural integrity for a specified period. Heat release measurements indicate the contribution of the assembly to the intensity of the fire.

At the USDA Forest Service, Forest Products Laboratory (FPL), Brenden and Chamberlain (6) examined the feasibility of measuring heat release rate from an ASTM E–119 furnace. Three methods of measuring heat release were considered: the substitution method, oxygen consumption method, and weight of material/heat of combustion method. The oxygen consumption method was shown to be the most advantageous way to measure heat release. However, data were limited to a few assemblies. Chamberlain and King *(7)* used the substitution method to evaluate similar assemblies at another test facility.

Available analytical methods for determining the fire resistance (or fire endurance) of wood members have been reviewed by White *(8)*. A finite-element heat transfer model for wood-frame walls was developed by the University of California–Berkeley *(9)* with funding from the FPL.

As a follow-up to previous work, we measured heat release rate for a series of 12 wall assemblies of four different constructions, using an improved gas sample analysis system and computer data acquisition.

Background

The heat release rate (amount of heat released per unit time) measures the potential of a material to contribute to a fire and has been the subject of many studies. The combustible volatiles produced when wood undergoes thermal degradation are responsible for flaming combustion and heat release. Browne and Brenden *(10)* calculated the heat of combustion of the volatile pyrolysis products of untreated and fire-retardant-treated ponderosa pine wood using an oxygen bomb calorimeter. By determining the heat of combustion of whole wood and partially pyrolyzed wood (char residue), they found heat of combustion of the volatile pyrolysis products varied with the degree of volatilization (pyrolysis). In general, heats of combustion of the volatile products were less than that of the original wood. Heat of combustion of the residual char was higher than that of the original wood.

Because heat of combustion of the volatile product is not the same as that of whole wood, one cannot estimate heat release rate based on mass loss rate as can be done with "ideal" fuels such as gases, liquids, and some noncharring solid materials. Thus, measuring heat release rate rather than mass loss rate is appropriate for wood and charring materials. Several bench-scale calorimeters have been developed to measure heat release rate of materials *(1,11,12,13)*.

Some early calorimeters use "thermal" methods based on principles of heat and mass balance *(12)* and temperature rise of a constant flow of air through the combustion chamber *(13)*. These calorimeters suffer from many drawbacks associated with their design. Heat and mass balance requires numerous measurements to account for all heat and mass flows. In most cases, thermal lag and losses in the equipment occur, which are not easily calculated.

Because of the problems associated with thermal methods, a gas analysis method, namely the oxygen consumption method, was devised. The landmark work for this method was done by Huggett *(14)*. He documented that the heat release per unit oxygen consumed is approximately the same for a wide range of organic materials. For example, cellulose and cotton have a net heat of combustion of 13.6 MJ/kg O_2 consumed. For maple wood, the value is 12.5 MJ/kg O_2. The average is 13.1 MJ/kg O_2 for most materials; the general variability is within 5 percent of the mean or better. Thus, the rate of heat release can be calculated with confidence when the rate of oxygen consumed is known. Huggett *(14)* also showed that for incomplete combustion, when part of the combustion products are carbon monoxide, aldehydes, or carboxylic acids, the heat of combustion per unit oxygen consumed is not much different from the average heat of combustion, if concentrations of these products are small. If the products are present in large amounts, proper corrections can be applied. The oxygen consumption method measures the chemical heat release rate, which is the sum of two major components: convective and radiative *(15)*.

Brenden and Chamberlain *(6)* measured heat release rate from wall assemblies having fire-retardant-treated studs and gypsum board as interior finish in the FPL fire endurance furnace using three methods: (a) the substitution method, by which the amount of fuel required to maintain the ASTM E–119 time-temperature curve for a noncombustible wall assembly is compared with that required for the assembly being tested (thus, two tests must be conducted, one with a noncombustible assembly and the other with the test assembly); (b) the oxygen consumption method, by which the rate of heat release is related to the rate of oxygen consumed by the combustion of pyrolysis products from the assembly; and (c) the weight of material/heat of combustion method, by which the heat of combustion of the material at the beginning of the test is compared to the residual heat of combustion of the material at the end of the test (this method does not provide rate data; instead, the total heat release is obtained). Evidently, the oxygen consumption method is more advantageous because rate of heat release is obtained with one single test. Because of some difficulties in the gas sampling system, Brenden and Chamberlain made a few assumptions for the data reduction of the oxygen consumption method.

Chamberlain and King *(7)* used the substitution method to measure heat release from assemblies similar to those used by Brenden and Chamberlain *(6)*. Steel stud walls of similar construction were used as reference. The heat release from the wood wall assemblies made of fire-retardant-treated framing and protected with type-X gypsum board was low enough to be within experimental error. The oxygen consumption method was attempted but it was not successful because of the complex geometry of the stack at the test facility, making measurement of flow of exhaust gases difficult.

The ASTM E–119 furnace available at the FPL is amenable to the implementation of the oxygen consumption method because (a) a long vertical exhaust duct is above the furnace, which allows combustion products to flow through without losses, and (b) the ratio of length to effective diameter of the exhaust is large enough to allow complete mixing of the exhaust gases. With the availability of modern equipment to monitor fire gases, we were able to refine heat release rate measurements.

Experimental Methods

The FPL vertical wall furnace used in our study was described in some detail by Brenden and Chamberlain *(6)*. This furnace is normally used to evaluate the fire endurance of wall assemblies. The basic guidelines for the furnace test method are given in the ASTM E–119 standard (5). The method was designed to evaluate the ability of a structure to withstand a standard fire exposure that simulates a fully developed fire. The furnace is gas fired, and its temperature is controlled to follow a standard time–temperature curve. A load may be applied to the assembly. The failure criterion can be taken as time at burnthrough, structural failure, or a specified temperature rise on the unexposed side of the wall—whichever comes first. The construction of the furnace is not specified in the ASTM E–119 standard.

Twelve wood-frame walls were tested as part of a fire endurance research project to develop and verify models to predict the fire endurance of wood wall assemblies. The assemblies consisted of three groups, A, B, and C, using different construction types (Table I). Eight tests (Groups A and C) were based on a 2^3 factorial design involving the type of interior finish (plywood or gypsum board), presence of a fiberglass insulation in the cavity, and two load levels, 11,500 and 22,400 lb (51 and 100 kN). The four tests of Group B were the same as the tests in Groups A and C, except no load was applied and no hardboard siding was used. The wall tests of Group B were added to obtain the third set of replicate temperature profile data. The walls of Group B also provided information on heat release rates for a longer test duration.

Calibration Test.   Before the wall tests were carried out, a calibration test was conducted to evaluate burner performance and to check for agreement between fuel gas and heat release calculations. For this test, the furnace was closed with a masonry wall lined with a layer of ceramic blanket material. The ASTM E–119 time–temperature curve was followed for 60 min.

Table I. Construction and Load Conditions of Wall Assemblies

Test no.	Siding	Load (lb)	Load (kN)	Interior finish[a]	Insulation
A-1	Hardboard	22,400	100	Plywood	No
A-2					Yes
A-3				Gypsum	No
A-4					Yes
B-1	None	0	0	Plywood	No
B-2					Yes
B-3				Gypsum	No
B-4					Yes
C-1	Hardboard	11,500	51	Plywood	No
C-2					Yes
C-3				Gypsum	No
C-4					Yes

[a]Interior finish for tests 1 and 2 of each group was plywood; finish for tests 3 and 4 was gypsum board.

Wall Assembly Construction. Because the materials and the design of the assemblies were specified in English units of measurement (for example 2 by 4 lumber, 24 in. on center), dimensions are expressed in English units with SI equivalents. By contrast, our equipment is calibrated in SI units, and consequently SI units are used in the remaining sections of this paper.

The 8 by 10 ft (2.44 by 3.05 m) walls for fire endurance tests were constructed of nominal 2 by 4 in. (38 by 89 mm) Douglas-fir or larch wood studs 24 in. (0.61 m) on center. The moisture contents of the studs ranged from 8.1 to 13.2 percent and their densities from 430 to 710 kg/m^3. The exterior sheathing was 5/8-in.- (16-mm-) thick C–D plywood with exterior glue. A double nominal 2 by 4 in. (38 by 89 mm) plate was at the top and a single plate was at the bottom. The plywood interior finish was 5/32-in.- (4-mm-) thick, simulated wood grain finish plywood paneling. The gypsum board interior finish was 5/8-in.- (16-mm-) thick, type-X (fire-rated) gypsum board. The insulation was 3.5-in.- (90-mm-) thick, unfaced, glass-fiber insulation (friction-fit). Walls with insulation also had a 4-mil clear polyethylene vapor barrier on their interior side. All wood and gypsum materials were conditioned at 23°C and 50 percent relative humidity, resulting in moisture content of 9 to 10 percent in wood materials. Figure 1 is a diagram of the wall frame construction. The hardboard siding was on the exterior side of the 5/8-in. (16-mm) plywood (not shown).

Furnace Operating Conditions. The furnace pressure was normally slightly positive at the top and negative at the bottom. In an effort to increase air supply, an additional blower was added for runs B–2, B–3, and B–4. We found no significant increase in air flow out of the stack. However, the furnace pressure increased significantly, resulting in significant leaks into the testing environment. Thus, we decided to use the original air supply system. To increase air available for complete combustion of pyrolysis products from the wall, the gas burners were adjusted to have a wider range. The lower limit was changed from 500 kW to 250 kW in series B and C. This helped provide more air to oxidize wood materials, especially when wood contribution was greatest. The operating conditions are summarized in Table II.

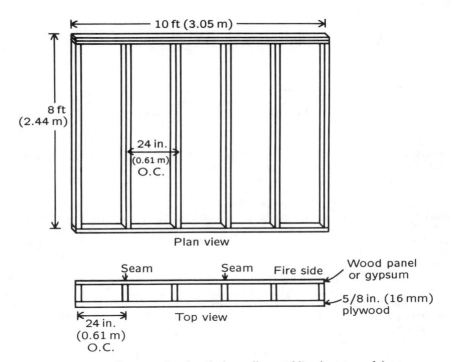

Figure 1. Construction details for wall assemblies (not to scale).

Table II. Furnace Operating Conditions

Test no.	Burners (kW)	Pressure Bottom	Top	Approximate test termination time[a] (min)
A–1	500–850	–	+	8
A–2		–	+	12
A–3		–	+	33
A–4		–	+	37
B–1	250–850	–	+	14
B–2		+	+	20
B–3		+	+	52
B–4		+	+	60
C–1	250–850	–	+	13
C–2		–	+	17
C–3		–	+	46
C–4		–	+	52

[a]For series A and C, these times correspond to structural failure. For series B, they correspond to burnthrough.

Instrumentation. To calculate rate of heat release on the basis of oxygen consumption, the flow rate of air into the system and the species concentrations in the stack must be measured. The furnace is operated by forced air via a blower. The amount of air coming into the system is split into two streams, called primary and secondary air. The pressure of the primary air is controlled to regulate the fuel-air mixture. The primary air pressure controls the amount of gas input to the burners. The secondary air is the bulk of the incoming air, which acts as a diluent to lower the temperature of the exhaust gases and supplies oxygen to the whole system. The incoming flow was not measured because of the complexity of the piping system; instead, flow rate out of the stack was continuously monitored.

We measured flow rate out of the system with a bidirectional probe placed at the center of the exhaust stack, which has a rectangular cross section of 12 by 20 in. (0.3 by 0.5 m). The equivalent diameter of the exhaust stack is 15 in. (0.38 m). The probe was positioned near the top of the 12-ft- (3.66-m-) long stack. The ratio of length to diameter of the duct (L/D) was close to 10:1. This distance ensured that the stack gas was well mixed prior to sampling. Two thermocouples were placed in the same cross section of the duct, one attached to the bidirectional probe and the other between the probe and the wall of the stack. A sample of the exhaust gases was pumped to the sampling system. The tube between the sample probe and the sampling system was heat traced to prevent water condensation. The location of the bidirectional probe, thermocouple, and gas sample probe in the duct is shown in Figure 2.

The sample handling system is diagrammed in Figure 3. As shown, the concentrations of oxygen, carbon monoxide, and carbon dioxide were on a dry basis because the water was removed. The water vapor analyzer did not function properly because condensation occurred in the instrument when water vapor was too high, often exceeding 20 percent. By using some approximations in the calculations, we did not need the exact concentration of water.

The flow rate of fuel gas into the furnace was measured using a volumetric flow meter. Data were collected every 30 s using a personal computer.

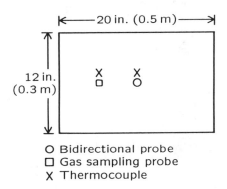

O Bidirectional probe
□ Gas sampling probe
X Thermocouple

Figure 2. Location of probes and thermocouples in stack (cross section).

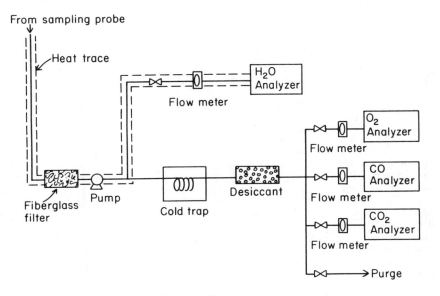

Figure 3. Sampling handling system for gas analysis.

Data Reduction and Calculations

Calculations of heat release rate by the oxygen consumption method as described by Parker (16) for various applications will be used here. Basically, heat release by the wall is heat release rate as measured in the furnace stack using the oxygen consumption method minus the heat release rate from the fuel gas. The comprehensive equation for heat release rate of the materials ($\dot{Q}_{wall}$) as given in Parker's paper (16) is

$$\dot{Q}_{wall} = \left[1 - \frac{X_{O_2}}{X_{O_2}^0}\left(\frac{1 - X_{O_2}^0 - X_{CO_2}^0}{1 - X_{O_2} - X_{CO} - X_{CO_2}}\right)\left(1 + \frac{E_{CO}' - E' }{2E'}\frac{X_{CO}}{X_{O_2}}\right)\right]E'X_{O_2}^0\dot{V}_0 - E'r'\dot{V}_g \quad (1)$$

where

$X_{O_2}, X_{CO}, X_{CO_2}$ are volume (or molar) fractions of gas species as obtained in analyzers,

$X_{O_2}^0, X_{CO_2}^0$ are ambient concentrations of gas species (assuming concentrations of O_2 and CO_2 of 21 percent and zero, respectively),

E' is heat release per volume of oxygen consumed, 17.2 MJ/m^3 at 25°C or 17.4 MJ/m^3 at 20°C (values based on average of 13.1 MJ/kg of oxygen consumed),

E_{CO}' heat produced per unit volume of oxygen consumed in burning of CO to CO_2, 23.1 MJ/m^3 O_2 at 25°C or 23.4 MJ/m^3 at 20°C,

$\dot{V}_0$ volume flow rate of air into system, corrected to a standard temperature (20°C),

r' volume of O_2 consumed in burning a unit volume of fuel, and

$\dot{V}_g$ volume flow rate of gas into the furnace, corrected to 20°C.

We used volumetric flows of air and fuel gas because they were directly measured. Using volumetric flow rates is awkward because the temperature and pressure of the gases in question have to be specified. In our case, the pressure of the fuel gas supply was about 1.5 kPa and that of the furnace 25 Pa above atmospheric pressure in extreme cases. These gauge pressures are small compared to absolute atmospheric pressure of >100 kPa. Thus, pressure corrections are insignificant. Temperature is the main factor that needs to be specified as the reference or basis for volumetric flow calculation. One alternative is the substitution of volumetric flow by mass flow. The conversion from volumetric flow to mass flow is straightforward using proper density values of the gases in question. The average heat of combustion of 13.1 MJ/kg of O_2 consumed is then applicable. However, for the purposes of this paper, either method will yield the same result.

Heat Release Rate From Furnace. Heat release rate from the furnace ($\dot{Q}_{furnace}$) is calculated from the information obtained in the stack: flow rate and species concentrations. Heat release rate is based on the first term in Equation 1.

$$\dot{Q}_{furnace} = \left[1 - \frac{X_{O_2}}{X_{O_2}^0}\left(\frac{1 - X_{O_2}^0 - X_{CO_2}^0}{1 - X_{O_2} - X_{CO} - X_{CO_2}}\right)\left(1 + \frac{E_{CO} - E'}{2E'}\frac{X_{CO}}{X_{O_2}}\right)\right]E'X_{O_2}^0\dot{V}_0 \quad (2)$$

When there is complete combustion, X_{CO} is negligible. Equation 2 is simplified to

$$\dot{Q}_{furnace} = \left[1 - \frac{X_{O_2}}{X_{O_2}^0}\left(\frac{1 - X_{O_2}^0 - X_{CO_2}^0}{1 - X_{O_2} - X_{CO_2}}\right)\right] E' X_{O_2}^0 \dot{V}_0 \qquad (3)$$

In our tests, the furnace was occasionally deficient of oxygen. When oxygen level went to zero, a significant amount of CO was produced as well as visible and sooty smoke. These are indications of incomplete combustion. The only product of incomplete combustion monitored was CO. Incomplete combustion products other than CO were not accounted for.

When X_{O_2} approaches zero, the term in the square brackets of Equation 2 approaches 1 or results in errors in the data reduction program because of X_{O_2} in the numerator and/or denominator. Equation 2 needs some modification to be sensitive to CO in these cases.

When $X_{O_2} \rightarrow 0$ and $(X_{CO}/X_{O_2}) >> 1$, Equation 2 can be approximated by

$$\dot{Q}_{furnace} = \left[1 - \frac{X_{O_2}}{X_{O_2}^0}\left(\frac{1 - X_{O_2}^0 - X_{CO_2}^0}{1 - X_{CO} - X_{CO_2}}\right)\left(\frac{E'_{CO} - E'}{2E'}\frac{X_{CO}}{X_{O_2}}\right)\right] E' X_{O_2}^0 \dot{V}_0 \qquad (4)$$

which can be rearranged to

$$\dot{Q}_{furnace} = \left[1 - \frac{X_{CO}}{X_{O_2}^0}\left(\frac{1 - X_{O_2}^0 - X_{CO_2}^0}{1 - X_{CO} - X_{CO_2}}\right)\left(\frac{E'_{CO} - E'}{2E'}\right)\right] E' X_{O_2}^0 \dot{V}_0 \qquad (5)$$

Air Flow Rate Into System.   The flow rate into the system is calculated based on the flow rate out of the system and an assumed expansion factor (α). The flow rate of exhaust air coming out of the system corrected to 20°C ($\dot{V}_s$) is based on the differential pressure across the bidirectional probe and the average temperature in the stack:

$$\dot{V}_s = 0.926kA\left(\frac{2\Delta P}{\rho_0}\frac{T_0}{T}\right)^{1/2} \qquad (6)$$

where

0.926 is a suitable calibration factor for bidirectional probe,

k the calibration factor for velocity profile in the stack (found to be 0.85 by Brenden and Chamberlain (6)),

ΔP differential pressure in pascals,

A cross-sectional area of stack, 0.155 m²,

T_0 reference temperature, 293 K,

ρ_0 density of air at reference temperature, and

T average temperature in stack in kelvins.

The chemical expansion α is defined as

$$\alpha = 1 + X_{O_2}^0 (\beta - 1) \qquad (7)$$

where β is the ratio of the moles of combustion products formed per mole of oxygen consumed. Parker (16) documented that for most materials, α ranges from 1.00 for carbon to 1.21 for hydrogen, with most organic fuels about 1.1. We assumed $\alpha = 1.084$ as recommended for most fuels in another application (3). A more accurate assumption can be made if the stoichiometry of combustion is known.

The flow rate into the system is related to the flow rate out as follows:

$$\dot{V}_s = (1 - \phi)\dot{V}_0 + \alpha\phi\dot{V}_0 \tag{8}$$

rearranged to give

$$\dot{V}_0 = \frac{\dot{V}_s}{1 + (\alpha - 1)\phi} \tag{9}$$

where ϕ is the depletion fraction, the fraction of the incoming air that has been totally depleted of its oxygen, which can be calculated based on the oxygen to nitrogen ratios in the incoming air and the stack *(16)* as

$$\phi = 1 - \frac{Z}{Z_0} \tag{10}$$

where Z and Z_0 are oxygen to nitrogen ratios in the stack and ambient air, respectively. Equation 10 can be expanded to

$$\phi = 1 - \frac{(X_{O_2}/X_{O_2}^0 - X_{O_2})}{(1 - X_{O_2} - X_{CO} - X_{CO_2})} \tag{11}$$

This equation allows us to calculate ϕ based on the species fractions.

Heat Release Rate From Fuel Gas. The fuel gas used in these tests was a mixture of natural gas supplied by the local gas company. This gas mixture contains approximately 90 percent methane and small fractions of ethane, propane, butane, CO_2, and nitrogen, as analyzed by Brenden and Chamberlain *(6)*. Although composition of the gas changes with time, the changes were small in our case. A statistical sample of gross heat of combustion of fuel gas over several months showed a coefficient of variation of 0.7 percent. Also, the gross heat of combustion of natural gas reported by the gas company on the day of the test did not vary significantly from test to test. Thus, we assumed that the net heat of combustion was constant.

Net heat of combustion was calculated based on the gross heat of combustion and the stoichiometry of combustion of the fuel mixture and a typical composition of the fuel mixture. The rate of heat release from burning fuel gas $(\dot{Q}_{fuel})$ is

$$\dot{Q}_{fuel} = \Delta H_g \dot{V}_g \tag{12}$$

where

ΔH_g is net heat of combustion of fuel gas, 33.9 MJ/m³ at 20°C, and

$\dot{V}_g$ volumetric flow rate of fuel gas in cubic meters per second at 20°C.

Heat Release Rate From Walls. The heat release rate from the wall assemblies should be the difference between $\dot{Q}_{furnace}$ and $\dot{Q}_{fuel}$. However, a direct subtraction is not correct because of the assumption that 13.1 MJ/kg oxygen consumed is released for all fuels. In fact, the heat release per unit oxygen consumed is 12.54 MJ/kg for methane as given by Huggett *(14)* and 12.51 MJ/kg for the fuel mixture as determined by Brenden and Chamberlain *(6)*. Thus, the heat release from the wall $\dot{Q}_{wall}$ is determined using a correction factor of 13.1/12.51 = 1.048:

$$\dot{Q}_{wall} = \dot{Q}_{furnace} - 1.048\dot{Q}_{fuel} \tag{13}$$

The term $1.048\dot{Q}_{fuel}$ is essentially the same as the second term in Equation 1 because $E'r' = (E'/E_g')\Delta H_g$, where E_g' is the correct E' value for fuel gas.

Results and Discussion

Heat release rates from the calibration run and the walls were made using Equations 1 to 13. The fire endurance results will be discussed in another paper. The tests were terminated shortly after structural failure or burnthrough of the wall assembly. The test termination times are given in Table II.

Calibration Test. Heat release rate is shown in Figure 4. Note that the heat release rate of the fuel calculated based on Equation 6 is slightly less than that of the total heat release. The heat release from the noncombustible wall calculated by Equation 13, using the proper correction of 1.048, is zero. Except for some perturbations in the beginning, which can be attributable to approximated response times of the analyzers, there is excellent agreement between the heat input by burning fuel gas and the total heat release rate calculated. We encountered a problem synchronizing the heat release rate calculated and the fuel gas rate because of the discrete data sampling time. The imperfectly synchronized rates resulted in "noises" in their differences, especially at the beginning of the test. No effort was made to interpolate data to match the two rate curves.

Heat Release Rate of Wall Assemblies. As examples, heat release rates are shown for series B because these tests lasted for longer times than the tests in series A and C (Figures 5–8). In walls B–3 and B–4, which had gypsum as the interior finish, combustion was complete throughout the entire runs. The concentration of CO was negligible.

In many cases when combustion was incomplete because of lack of oxygen, a significant amount of CO was formed. This amount of CO may further combust to CO_2 if extra air is added. Therefore, we can estimate the additional energy to be released if CO was allowed to burn to CO_2. The flow rate of CO can be calculated as follows:

$$\dot{V}_{CO} = X_{CO}\dot{V}_0 \frac{X_{N_2}^0}{X_{N_2}} \tag{14}$$

where the concentration of N_2 is on a dry basis. Equation 14 is expanded to

$$\dot{V}_{CO} = 0.79\dot{V}_0 \frac{X_{CO}}{(1 - X_{O_2} - X_{CO} - X_{CO_2})} \tag{15}$$

The flow rate of CO is multiplied by the heat of combustion of CO (11.7 MJ/m^3 of CO at 20°C) to obtain the heat release rate from CO if CO burns to CO_2. This heat release rate plus heat release rate from the wall is the potential heat release rate if CO is further oxidized to CO_2.

For walls B–1 and B–2, which had plywood as the interior finish, the heat release rate from the furnace (burners plus wall), heat release rate from the wall, and potential heat release rate are shown in Figures 5 and 6. The difference between the potential heat release rate and the wall contribution is the possible contribution by CO.

For those walls protected by gypsum (walls B–3 and B–4), furnace heat release rate and heat release rates from the fuel gas (Eq. (12)) and the wall (Eq. (13)) are shown in Figures 7 and 8. Similar to the noncombustible wall, the gypsum provided very little heat, except for a small peak in the beginning, which could have been caused by the gypsum paper. Heat release rates from walls B–3 and B–4 rose slowly as a result of the gypsum opening up at the seams or cracks.

Total Heat Release From Wall Assemblies. We examined the total heat release from the wall assemblies as total heat contribution. The total heat release is obtained by integrating the area under the heat release rate curve with time and it is expressed in megajoules (MJ). The total heat release data from ignition to different times are shown in Table III for the wall assemblies.

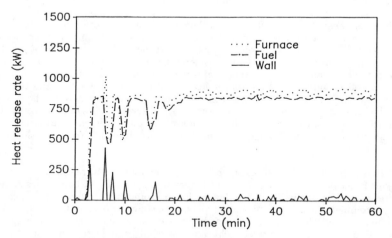

Figure 4. Heat release rate from calibration test with a blank wall.

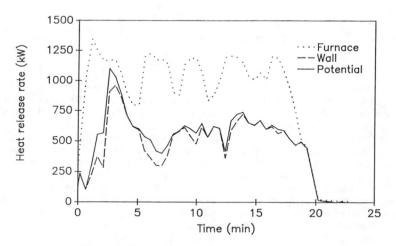

Figure 5. Rate of heat release for wall test B–1.

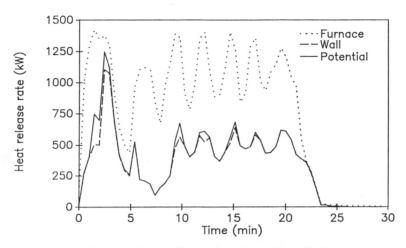

Figure 6. Rate of heat release for wall test B–2.

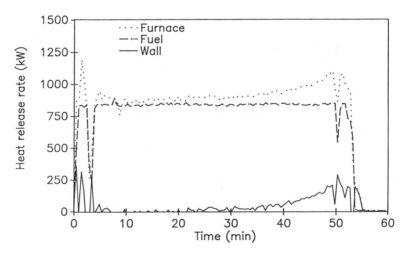

Figure 7. Rate of heat release for wall test B–3.

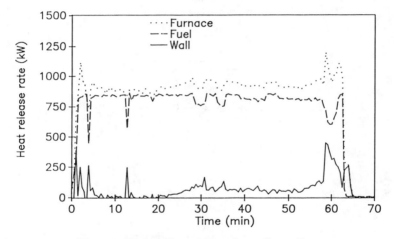

Figure 8. Rate of heat release for wall test B–4.

Table III. Total Heat Release From the Wall Assemblies

Test no.	Total Heat Release (MJ)							
	5 min	10 min	15 min	20 min	25 min	30 min	35 min	40 min
A–1	120	285	342	—	—	—	—	—
A–2	140	183	299	—	—	—	—	—
A–3	34	36	31	28	25	27	39	52
A–4	−11	−10	−16	−22	−27	−15	−5	17
B–1	196	350	533	687	—	—	—	—
B–2	181	280	434	586	650	—	—	—
B–3	28	23	20	20	21	28	36	49
B–4	30	33	40	40	48	70	96	117
C–1	200	390	517	518	—	—	—	—
C–2	195	265	424	588	—	—	—	—
C–3	29	38	41	48	59	72	89	117
C–4	40	51	56	61	77	94	115	138

In general, walls A–1, B–1, and C–1 had the highest heat release. Walls A–2, B–2, and C–2 had consistently less heat contribution because of the insulation. Walls A–3, B–3, and C–3 and A–4, B–4, and C–4 had no significant heat contribution. The values in these tests vary because of initial errors associated with the perturbations at the beginning of the tests. However, total heat release did not grow, indicating zero heat release rate when gypsum was present.

As we have noted, eight tests (series A and C) represented a 2^3 factorial design involving type of interior finish, presence of a fiberglass insulation in the cavity, and load level (Table I). Series B was similar to A and C except for the absence of load and hardboard siding. In most cases, the four tests of series B could also be included in the analysis.

The factorial approach to the design of experiments allows all the tests involving several factors to be combined in the calculation of the main effects and their interactions. For a 2^3 design, there are 3 main effects, 3 two-factor interactions, and 1 three-factor interaction. Yates' algorithm can be used to determine the main effects and their interactions *(17)*. The data can also be represented as a multiple linear regression model

$$y = B_0 + B_1 x_1 + B_2 x_2 + B_3 x_3 + B_4 x_1 x_2 + B_5 x_1 x_3 + B_6 x_2 x_3 + B_7 x_1 x_2 x_3 \qquad (16)$$

where y is the response variable, x_i is the factor variable for factor i, and B_j are parameters.

By eliminating insignificant variables, Equation 16 can be reduced to fewer terms. In addition to the three variables of the factorial design, the burner level was also included in the analysis.

Because of the large difference in the behavior of the thin plywood and the gypsum board, the type of interior finish was the dominant factor in the statistical analysis of the total heat release data (Table III). Linear regression of the data sets for 5, 10, and 15 min resulted in squares of the correlation coefficients $R^2 = 0.88$ to 0.91 with the type of interior finish as the sole variable. For the plywood, the average total heat release was 172, 292, and 425 MJ at 5, 10, and 15 min, respectively. For the gypsum board, the average total heat release was 25, 27, and 29 MJ at 5, 10, and 15 min, respectively.

For 5 min, the type of burner (lower limit of 250 kW (series B and C) compared to 500 kW (series A)) also had a significant effect. For a linear regression model with interior finish and the burner level as the variables, the $R^2 = 0.96$. Because the burner level primarily affected the total heat release from the plywood, the cross-product of burner level and type of interior finish is also a significant factor. At 10 min, the presence of insulation was more significant than the burner level. For a model with interior finish, insulation, and burner level as variables, the $R^2 = 0.95$ for 10-min data. At 15 min, insulation was no longer a significant factor. This is consistent with the visual observations that the insulations in the plywood tests were gone after approximately 10 min. A model of interior finish and burner level had an $R^2 = 0.96$ for the 15-min data. At 20 min, only the type of interior finish was a significant main factor (the cross-product of interior finish and burner was also significant). By 20 min, nearly all the plywood tests had been terminated.

Except for the load level, which should not have an effect on the total heat release, series B and C were identical. Comparison of the results for 5, 10, and 15 min indicates very good agreement between the two sets of data.

Comparison of Study to Previous Work.   The only work that can be compared to ours is that of Brenden and Chamberlain (6) using the same furnace and Chamberlain and King (7) using a different furnace. The assemblies used in these studies were of similar construction. The wall assemblies used contained fire-retardant-treated studs as the only combustibles. The interior and exterior sides were 5/8-in. (16-mm) type-X gypsum. Thus, heat release rates were understandably small. The onset of heat release started at 23 min after ignition. After that, a slow increase occurred, culminating to approximately 100 Btu/min · ft^2 (19 kW/m^2) at 60 min. For an exposed area of 8 by 10 ft (2.44 by 3.05 mm), this would be 140 kW.

Our results are comparable to the results of Brenden and Chamberlain (6) and Chamberlain and King (7). In our most comparable assembly having gypsum on the fire-exposed side and no insulation (B–3) (Fig. 7), contribution from the walls started approximately at 30 min and rose to about 200 kW at 50 min. Similar results were obtained despite differences in assembly constructions. The exterior sheathing in our tests was plywood whereas that of Brenden and Chamberlain (6) was gypsum. Moreover, we used untreated wood studs in our assemblies.

Accuracy of Oxygen Consumption Method.   We have demonstrated that the oxygen consumption method can be used to quantify the amount of heat contributed to the fire environment by the walls. The heat release rate can be used as a diagnostic tool to evaluate performance of assemblies in question.

However, many possible sources of error other than the accuracy of the measurements deserve attention. These sources of error are inherent in the assumptions that we made about average E', chemical expansion factor, and effect of dilution.

Average E'—As stated earlier, the oxygen consumption method is based on the finding that most materials have the same E', with accuracy within 5 percent of the mean.

Chemical expansion factor—The values of α and β vary from material to material as shown by Parker (16). The values of α vary from 1.00 for carbon to 1.21 for hydrogen. Thus, extreme errors in α can be 10 percent from the mean, especially when $\phi = 1.0$. An erroneous α results in wrong estimates of air flow rate into the system, which may lead to a maximum of 10 percent error in heat release rate.

Effect of dilution—In addition to the water present in the ambient air, the materials contained a certain amount of water. The water evaporated from the materials "diluted" the exhaust gases. This water is not accounted for in the chemical expansion factor. The magnitude of the error depends on the evaporation rate of water and the total flow rate out the exhaust stack. Because we did not have an accurate measurement of water vapor in the exhaust stream, the effect of dilution was not calculated. This error can be large at the start of the test when water is boiled out of the system and should become negligible thereafter.

Of all these sources for error, the first two are the most significant. The oxygen consumption method has been reported to be accurate within 15 percent, which is the additive effect of average E' and the chemical expansion factor. This extreme error can happen only in the case of complete oxygen depletion. The error is substantially reduced for most cases when the oxygen level in the exhaust is larger than zero. As shown in the calibration test, very good agreement between $\dot{Q}_{furnace}$ and $\dot{Q}_{fuel}$ was obtained. The method is accurate at least for fuel gas alone.

Conclusion

Several conclusions can be drawn from the analysis of heat release rate in this study. In terms of heat contribution from the walls, the gypsum wallboard provided complete protection to the assembly for 30 min. Even after 30 min, heat release rate from the assembly was below 200 kW up to 60 min of exposure compared to the 600 kW from the assemblies without gypsum. The insulation also helped protect the assemblies, although not as dramatically as gypsum; it reduced heat release rate between 5 and 10 min.

The versatility and accuracy of the oxygen consumption method in heat release measurement was demonstrated. The critical measurements include flow rates and species concentrations. Some assumptions need to be invoked about (a) heat release per unit oxygen consumed and (b) chemical expansion factor, when flow rate into the system is not known. Errors in these assumptions are acceptable. As shown, the oxygen consumption method can be applied successfully in a fire endurance test to obtain heat release rates. Heat release rates can be useful for evaluating the performance of assemblies and can provide measures of heat contribution by the assemblies. The implementation of the heat release rate measurement in fire endurance testing depends on the design of the furnace. If the furnace has a stack or duct system in which gas flow and species concentrations can be measured, the calorimetry method is feasible. The information obtained can be useful in understanding the fire environment in which assemblies are tested.

One critical factor that affects the heat release rate is the availability of air. The furnace has to be designed so that many requirements can be met simultaneously: (a) time–temperature curve of ASTM E–119, (b) adequate air supply, and (c) pressure requirement inside the furnace. To incorporate the heat release rate measurement into the ASTM E–119 standard, specifications must be made to address these three criteria. If these criteria can be agreed upon, the heat release rate measurement should be made a part of the existing test standard.

Literature Cited

1. Babrauskas, Vytenis. *Development of the Cone Calorimeter—A Bench-scale Heat Release Rate Apparatus Based on Oxygen Consumption.* U.S. Department of Commerce NBSIR 82–2611, 1982.
2. Babrauskas, Vytenis. *J. Fire Sci.* 1983, *1*, 9–32.
3. ASTM Annual Book of Standards. Proposed method for room fire test of wall and ceiling materials and assemblies. American Society for Testing and Materials: Philadelphia, PA, 1983; Vol. 04.07, pp. 958–978.
4. Parker, William J. *An Investigation of the Fire Environment in the ASTM E–84 Tunnel Test.* NBS Tech. Note 945. National Bureau of Standards: Washington, DC, 1977.

5. ASTM Annual Book of Standards. Standard methods of building construction and materials. Method E 119–83. American Society for Testing and Materials: Philadelphia, PA, 1985.
6. Brenden, John J.; Chamberlain, David L. *Heat Release Rates From Wall Assemblies: Oxygen Consumption and Other Methods Compared.* Res. Pap. FPL–RP–476. U.S. Department of Agriculture, Forest Service, Forest Products Laboratory: Madison, WI, 1986.
7. Chamberlain, David L.; King, Edward G., Jr. *Heat Release Rates of Construction Assemblies by the Substitution Method.* National Forest Products Association. Technical Report No. 9, 1987.
8. White, Robert H. In *SFPE Handbook of Fire Protection Engineering.* National Fire Protection Association, 3–130–3–142: Quincy, MA, 1988; Chapter 8.
9. Gammon, Barry W. Ph.D. Thesis, University of California, Berkeley, 1987.
10. Browne, F.L.; Brenden, J.J. *Heat of Combustion of the Volatile Pyrolysis Products of Fire-retardant-treated Ponderosa Pine.* Res. Pap. FPL 19. U.S. Department of Agriculture, Forest Service, Forest Products Laboratory: Madison, WI, 1964.
11. Tewarson, Archibald. In *Flame Retardant Polymeric Materials*; Lewin, M., Atlas, S.M., and Pearch, E.M., Eds.; Plenum Press: New York, 1982; Vol. 3, Ch. 3.
12. Brenden, John J. *J. Fire Flammability*, 1975, *6*, 274–293.
13. ASTM Annual Book of Standards. Standard test method for heat and visible smoke release rates for materials and products. Test Method E 906–83. American Society for Testing and Materials: Philadelphia, PA, 1985.
14. Huggett, Clayton. *Fire Materials* 1980, *4*, 61–65.
15. Tewarson, Archibald. In *The SFPE Handbook of Fire Protection Engineering.* Di-Nenno, P.J., Ed.; The National Fire Protection Association Press: Quincy, MA, 1988; Ch. 13, Sec. I.
16. Parker, William J. *J. Fire Sci.* 1984, *2*, 380–395.
17. Box, George E.P.; Hunter, William G.; Hunter, Stuart J. *Statistics for Experimenters.* John Wiley & Sons: New York, 1978.

RECEIVED November 1, 1989

Chapter 26

Self-Heating of Lignocellulosic Materials

Hans Kubler

**Department of Forestry, University of Wisconsin—Madison,
Madison, WI 53706**

Processes which generate heat in organic
materials are reviewed. At ordinary
temperatures, respiration of living cells and
particularly the metabolism of microorganisms
may cause self-heating, while at elevated
temperatures pyrolysis, abiotic oxidation,
and adsorption of various gases by charred
materials drive temperatures up whenever the
released heat is unable to dissipate out of
the material. The crucial rate of pyrolytic
heat release depends on exothermicity and
rates of the pyrolysis process.

In self-heating materials, the temperature rises without
input of energy from the surroundings, and without any
evident heat-generating process such as visible combustion.
Some self-heatings start at ambient conditions, while
others begin after temperature of the material has been
raised by outside energy. The amount of heat and the power
per unit mass of the material are generally small, but in
large piles or packs and under other conditions where
little of the generated heat dissipates, temperatures tend
to rise nevertheless.
 Self-heating is quite common, yet it receives
attention only when it causes economic losses or leads to
fires. Fresh grass clippings in garbage cans and plastic
bags heat within an hour or so. Piles of manure and
compost always seem to self-heat, as the *steam* rising from
opened piles indicates. Farmers have cause to be alarmed
when temperatures rise in hay, silage, straw, cotton,

0097–6156/90/0425–0429$06.25/0
© 1990 American Chemical Society

grain, and other plant materials; they experience self-heating in exposed piles, bales and other free storages as well as in silos.

Forest products industries know that temperature increases in piles of sawdust and bark. In pulp and paper mills, self-heating develops in amassed tree *chips*. Paper rolls stacked hot tend to self-heat, as occasionally do stored bales of waste paper. The wood-base panel products particleboard, hardboard, and fiberboard self-heat after being stacked too hot in the factory. Where in structures the framing lumber, wood-base panels, and lignocellulosic insulation is heated by items such as steam pipes, temperatures tend to rise above that of the heat source.

In landfills, organic materials such as grass clippings, brush, kitchen waste, and dumped forest products raise the temperature. The manufacture of charcoal from lignocellulosics in pits, retorts, and kilns involves self-heating, when after an initial period of combustion the air supply is shut off and the charred material releases heat without oxidation. Similar reactions may be the cause of underground *fires* in coal seams and dry peat bogs, which *burn* for months and sometimes years although air has no access.

All self-heatings consume material and partially convert it into substances such as water vapor and carbon dioxide. This consumption is desirable in landfills, but in nearly all other cases useful material is devalued or lost. Self-heating causes greatest damage when it leads to smoldering and flaming combustion.

The literature has been reviewed earlier in great detail, with emphasis on forest products (1). Here the review is updated and drastically condensed, and includes lignocellulosics other than wood. Of the former 194 references, only those used for illustrations, in support of crucial statements, and for new conclusions are cited again. Some cited early publications were not available in 1987.

Self-heating is caused by a series of processes, each of which has its particular temperature range, and raises the temperature up to the level or into the ranges of ensuing processes. The processes are abstracted one by one in the sequence of increasing temperatures, although the processes overlap, and most cases of self-heating involve several processes. Some earlier publications shall be evaluated from a new angle.

<u>Respiration</u>

Green leaves of plants photosynthesize, trapping light energy for the conversion of carbon dioxide and water into sugars. As a process which does not generate heat photosynthesis has no place in this review, but photosynthesizing cells are alive, and living cells must respire to maintain life. The living cells convert

chemical energy into heat, even though the cells of green
leaves at the same time absorb light energy and accumulate
chemical energy. The necessity of respiration is better
known in the cells of animals and people: when our brain
is not supplied with oxygen by the blood stream, brain
cells succumb within minutes.

Cutting grass, or more precisely separating leaf tips
from lower parts of the grass plant, is not immediately
lethal, and even chopping up these tips kills only a small
percentage of the cells. The leaves continue to respire;
they oxidize stored carbohydrates and other foods into
carbon dioxide and water, or into intermediate compounds.

In photosynthesizing leaves of growing plants the heat
of respiration can dissipate while in piles of harvested
plants it may be trapped and may accumulate, examples being
heaps of alfalfa, clover, and corn stalks. Garbage cans
and plastic bags with grass clippings have little space for
air. The clippings consume nearly all available oxygen,
then external oxygen diffuses into the containers to
sustain respiration somewhat. Seeds, living cells by
definition, consist mainly of starch and other stored food
which oxidizes extremely slowly, so that these food
reserves last a long time and seeds can survive for years,
consuming very little oxygen and releasing very little
heat. This seems to be why grain self-heats less than
other life crops, in spite of the huge volumes in grain
elevators.

Wood and bark of living trees are essentially dead
tissues. Heartwood in the center of mature trees and old
bark at the very tree stem surface contain no living cells,
whatsoever, whereas in sapwood and in young bark some cells
remain alive. In the cambial layer between wood and bark,
in which trees grow new wood and bark, practically all
cells live. Respiration of the living tree cells causes
self-heating in piles of sawdust, of bark, and of *chips* for
pulp; the higher the percentage of living cells, that is of
cambial cells and cells near the cambium, the more the
piles self-heat. *Whole-tree chips*, obtained by chopping up
entire trees, self-heat greatly especially when chopped
from young trees, because the foliage included consists
mainly of living cells (2).

Plant cells cease living and generating heat when they
have exhausted their food reserves after a few days, weeks,
or months. Most plant cells die earlier, due to lack of
moisture and oxygen, or from heat. Living cells of
lignocellulosics perish when the moisture content of the
embedding lignified cells drops below fiber saturation near
30% moisture content, based on ovendry weight. Seeds
survive down to much lower moisture contents.

The upper temperatures which plants can tolerate vary
from plant species to species. Tree cells survive up to
about 50°C. Heat generation may continue even after cell
death, when enzymes produced by the living cells continue
catalyzing the oxidation process post mortem. At the

other, low temperature extreme, many plants survive
freezing in the state of dormancy. The temperature
optimum, at which plants metabolize the fastest lies for
many species around 25°C; it depends on relative humidity
of the atmosphere, since high degrees of heat are usually
associated with dry air and the stress of moisture loss.

Metabolism of Microorganisms

Dead as well as live plants and animals serve as food for
fungi and bacteria. These microorganisms do not
photosynthesize, and like animals they feed on the organic
materials. The term *microorganism* indicates that the
individuals can be seen only under the microscope; many
fungi however form large, conspicuous fruiting bodies known
as mushrooms and toadstools, but their filaments in the
host substrate remain by far too thin to be visible to the
unaided eye. The microorganisms break their substrate down
to convert it into carbon dioxide and water; in this
respect their metabolism resembles respiration and
combustion. It is another heat-generating process.
    Bacteria and spores of fungi abound in the air as well
as at surfaces of almost everything, multiplying and
growing rapidly under suitable conditions. Most plants
fend off these microorganisms, while some live with the
parasites and others are killed by them. Separating higher
plants from their roots and chopping them up into fragments
during harvesting operations weakens their defenses and
causes early death; it is then only a question of time
until microorganisms under suitable conditions consume the
harvested material.
    How long plant tissues resist decomposition and how
quickly the decomposition releases heat, that depends on
the plant constituents. Starch and sugars in fruits,
roots, and sapwood is ready food, so that fungi feeding on
these low-molecular weight carbohydrates may spread
throughout large piles of some materials within a few days.
Cellulose, the most common plant constituent, is much more
resistant than starch and sugars, but still may last only a
few weeks as we know from the decomposition of leaves and
entire herbaceous (not lignified) plants. The
decomposition of cellulose nevertheless releases heat at
relatively fast rates, because cellulose makes up the mass
of the particular plant and occurs in great quantities, as
in hay for example. Lignin, next to cellulose the main
constituent of *woody* lignified tissue, adds months and
years to the resistance of lignocellulosics, which
therefore release heat at relatively low rates, but huge
masses may compensate for the low rate. The heartwood of
many tree species and other plant tissues contain
substances which poison the tissue as food for
microorganisms, and are the reason why such tissues resist
decomposition for years and even decades, depending on
temperature as well as the oxygen and moisture available.

In early stages of storage of harvested plants, microorganisms quasi supplement respiration, but after heat has killed all substrate cells around 50° or 60°C, the microorganisms make a great difference by raising temperatures considerably higher. Actually some microorganisms tolerate no more heat than substrate cells, but some heat resistant *thermophilic* microorganisms survive and metabolize up to about 80°C.

Microorganisms depend on moisture like all plants. They secrete enzymes in aqueous solution to break the food down into water soluble compounds, then absorb the compounds in solution. Hence fungi and bacteria can metabolize in lignocellulosic substrates only as long as the substrate contains at least 20% moisture. Drying and freezing force microorganisms into dormancy, but after remoistening and thawing they resume metabolizing. Thus piles of wood chips start self-heating also in winter, although slowly at first (2). Optimal temperatures of the microorganisms vary with the species; most metabolize fastest between 25° and 30°C. Waterlogged material self-heats relatively little, because it lacks oxygen which most microorganisms need: they are aerobic. Only a few belong to the anaerobic category. Some compact, dense masses self-heat little in this stage, due to lack of oxygen and also to substantial conduction of heat to the surface (3).

Microorganisms thrive in landfills. In the common *sanitary* type the relatively small, flexible, variable pieces of household garbage lend themselves to compaction by the heavy bulldozers and trucks driving on the landfill. The compaction and interspersed capping layers of clay-type dirt create an essentially air free mess in which mainly anaerobic bacteria decompose the organic materials into methane. This gas builds up pressure to seep, flow, and diffuse out into the air and into surrounding ground. Methane as an energy rich gas carries much chemical energy away, leaving little for exothermic aerobic decomposition and self-heating.

The situation appears to be different in *demolition* landfills used for dumping wrecked buildings and other structures in metropolitan areas. There bulldozers on studded caterpillar tracks crush and compact the pieces of lumber, wood-base panels, and other bulky items. Nevertheless the dump holds more air than sanitary landfills do, so that mainly aerobic microorganisms decompose the organic, largely lignocellulosic materials into carbon oxides and water vapor while releasing much heat. Some air, pulled in from the landfill hillside where the covering layer of dirt has been washed away, may sustain the aerobic microorganisms after they have consumed the innate oxygen. One landfill with a total volume between two and three million cubic meters self-heated to the level of pyrolysis. The resulting events will be discussed at the end of the following, *pyrolysis* part.

Heat of Pyrolysis

Lignocellulosics reach temperatures around 80°C in many ways besides. For example, planer shavings and peat insulation around hot pipes or in walls of dry kilns easily attain 100°C, and forest product industries stack still hot wood-base panels at temperatures around 80°C. In many cases the temperatures of the hot materials later on rose above the initial 80° or 100°C, first to levels of smoldering combustion, and finally to those of open flames. In air-exposed and ventilated materials oxidation could cause the heating above 80° or 100°C, but inside tight packs of panels pyrolysis must have been the heat source.

Up to about 70°C, plant tissues are thermally stable, as they must be in nature to avoid damage from prolonged direct exposure to the sun. Pyrolysis, the chemical decomposition by heat, starts in dry lignocellulosics around 100°C, in moist ones below 80°C. It accelerates as temperature rises, peaking in many organic materials between 275° and 300°C, at which point cellulose disintegrates.

The exothermicity of pyrolysis had been reported in the literature as early as 1910 (4). Nevertheless later most authors saw thermal disintegration as an endothermic reaction. Recent theoretical (5) and experimental work (6) confirmed the exothermicity for slow pyrolysis, in which the evolving volatiles contain large percentages of carbon dioxide and water, that is to say of energy-free substances, so that the sum of heats of combustion of the pyrolysis products remains below the heat of combustion of the original material. Consequently according to the first law of thermodynamics, slow pyrolysis must release heat. Fast, high-temperature pyrolysis in contrast, leads to plenty of high-energy volatiles such as hydrogen, indicating endothermicity; the vigorous molecular motion of high temperatures literally tears the compounds apart into high-energy fragments, with little opportunity for orderly arrangement. In this way expended heat energy winds up as chemical energy in the volatiles.

In trials with wood since 1910, several researchers did notice pyrolytic heat release, but others found the reaction endothermic. The contradictions can be explained with different sizes of the samples. It is believed that *primary* pyrolysis volatiles interact in *secondary*, exothermic reactions catalyzed by the solid residue. Long residence times of the volatiles in the disintegrating material favor secondary reactions, of course. Residence times are indeed long in large and in slowly disintegrating samples, in which the volatiles have a long path to the surface and migrate out slowly.

Many authors explored pyrolysis in *Differential Thermal Analysis (DTA)* and in *Differential Scanning Calorimetry (DSC)* trials, in which tiny samples weighing only one milligram up to a few grams were heated rapidly at rates from 0.5°C/min up to nearly 200°C/min. The exhibited heat balances changed as temperatures rose and were overall on the endothermic side, since in the tiny, rapidly heated samples the volatiles had little chance for secondary, exothermic reactions. DTA and DSC trials are therefore not representative of real situations.

Amounts of heat released in pyrolysis trials were small. Even heats measured at around 130°C, far above the 80°C margin, could adiabatically raise the temperature by less than one centigrade in an hour (6). With so little heat, inaccuracies of measurements could cause qualitative mistakes. Additionally, temperatures were usually determined with thermocouples, whose metallic wires conducted heat up to one thousand times faster than the pyrolzed material, and could again involve substantial errors, especially with small samples and steep temperature gradients.

Pyrolytic self-heating has caused quite a few fires. When new wood-base panels were stacked too hot in packs for storage or for transport, heat of pyrolysis drove temperatures deep inside the packs higher and higher, pyrolyzing the center into darkening brittle materials, and finally into black charcoal (7). Due to contraction from the charring and volatilization, the material cracked and crumbled and left cavities inside the packs. The char zones and the developing cavities were rarely exactly in the pack's center but always relatively far from the pack surface, so that only a fraction of the generated pyrolytic heat was conducted to the distant surfaces out of the pack.

Some of the pyrolysis volatiles effused out, and could have been perceptible symptoms of the deterioration and an imminent disaster, but in open yards, unattended warehouses, or on trucks nobody was present to smell the odors. The cavities grew in all directions, approaching a pack surface one or two weeks after stacking. At that point air gained access to the hot charred material and triggered ignition. The combustion spread rapidly along the pack surface, possibly also into the opened cavity as air rushed in.

To prevent the self-heating, factories cool fresh fiberboard to temperatures of 80° or lower before stacking. The United States Government as buyer of the panels requires the much safer temperature limit of 54.5°C (130° F) for wood fiberboard, and 60°C (140° F) for board made from sugar cane (8). Experiments with unrepresentative, small samples can be very misleading regarding safe stacking temperatures, since in small samples the self-heating may develop or become noticeable only above 150°C, while oxidative heating and resulting ignition appear only around 200°C (7).

In the case of the demolition landfill mentioned above, pyrolysis supplied heat after the microorganisms reached their temperature limit. Eighteen months from the time of landfilling, obnoxious pyrolytic volatiles effused as dense smoke out of the seemingly burning dump. Landfill operators fought the *fire* for six months by excavating smoking zones deep under the surface, and by injecting water. The degree of charring of excavated wood indicated that the temperature had reached about 200°C. Some zones which smoked only slightly and were not excavated continued to *steam* for several additional months, as 80°C-hot, moisture-laden gases (with the typical acid odor of steamed veneer logs) and high percentages of carbon dioxide and methane still streamed out of crevices in the clay-dirt capping.

The rate of pyrolytic heat release depends on rate and exothermicity of pyrolysis. According to the theoretical evaluation of these factors in the APPENDIX, the most rapid pyrolytic self-heating can be expected between 120° and 170°C.

Abiotic Oxidation

All kinds of oxidation release large amounts of heat, in the order of 15 kJ per gram of oxygen consumed in the case of lignocellulosics. Microorganisms and other living cells achieve *biotic* oxidation at ambient temperature by means of catalyzing enzymes. *Direct chemical* or *abiotic* oxidation, also known as *atmospheric* oxidation, generally occurs only at elevated temperatures; its threshold of detectable heat generation appears around 80°C, and somewhat below 80°C in the presence of catalyzing moisture. Oxidation shows itself in oxygen consumption, in evolution of carbon dioxide and water, as weight loss, and of course as heat release. Oxidation and oxidative heat generation intensify with increasing temperature in a very progressive manner, so that in the presence of air above 130°C at least, lignocellulosics release more oxidative than pyrolytic heat. In the landfill *fire* described in the preceding paragraph, excavated pieces of charred, at most 200°C hot wood, started glowing when sudden exposure to air and oxidation boosted their temperature to above 475°C.

Some extraneous constituents of plant tissues, such as alkali salts in many lignocellulosics and particularly resin in wood, oxidize at lower temperatures and faster than the regular constituents. Therefore some fires from self-heating have been attributed to extraneous constituents. Extrinsic substances added after harvesting may accelerate self-heating even more; the additives and contaminants either catalyze oxidation and pyrolysis, or the extrinsics themselves oxidize, with fats and oils being examples (9).

Metallic compositions serving as *siccatives* catalyze the well-known oxidation and polymerization of oil in paints and other finishes. Likewise, ferrous and other metallic objects boost self-heating in piles of lignocellulosics. Among the known catalyzing substances are iron sulphides and iron oxides from combustion gases of direct-fired driers, powdered metals, metal oxides, some acids, zinc chloride, sodium carbonate, and inorganic salts which modify pyrolysis and are used as fire retardants.

The influence of oxygen on hot lignocellulosics shows in many ways. Mitchell (10) observed property changes in wood samples heated at 150°C for 1 to 16 hours. In ovendry samples, the properties changed in oxygen more than in nitrogen atmospheres. Moisture amplified the effect of oxygen through hydrolysis, which can be considered a kind of pyrolysis and is similarly an exothermic reaction (5).

The classic example, and probably the most frequent cause of fires from self-heating of any substance, are oils rags discarded by painters. In the crinkled, crumpled rags oxygen has access to large surfaces of oil, while only some of the generated oxidative heat dissipates out of the insulating mess of air and rags, which tends to ignite within a few days or even after several hours.

Whether pyrolysis or oxidation generates heat, this is what determines how the self-heating should be controlled. Air blown into or sucked out of piles works against pyrolytic self-heating, whereas it promotes oxidation where much of the innate oxygen has been consumed. Air which flows through and cools piles is helpful up to at least 100°C. Dense, compacted zones may self-heat relatively much (11) due to the large mass of heat-generating material, while little of the generated heat is carried away by air convection.

In hay with its high proportion of air space and the large surfaces of combustible pieces, oxidation is at elevated temperatures hard to stop. Therefore in self-heating hay stacks, even temperatures around 60°C are already dangerous and call for fire fighters, who generally blow or suck out the heat rather than try to exclude air (11).

Some factories *temper* or *cure* fresh hardboard to increase water resistance, dimensional stability, and strength. "A curing for five hours or more at 165°C is common" (12). To find ways to prevent ignition during tempering, Back and Johanson (13) measured the heat evolving from air-exposed strips of the boards between 150 and 240°C. "In lignin-containing boards, the heat evolution rate passed rapidly through a maximum and then decreased with time, while for lignin-free boards the rate of heat evolution after reaching an initial plateau increased significantly with time above 200°C or stayed about constant below 200°C." The rapid initial heat evolution may be related to available oxidation sites in the boards, but overall the reactions must have been very complex.

Among lignocellulosic panel products, fiberboard (called *fiber insulation board* in earlier decades) seems to have caused more fires than the denser products hardboard, particleboard, and plywood. Fiberboard self-heats more because it conducts less of the generated heat out of the pack, but oxidation of the relatively porous material may be another factor. Self-heating and resulting ignition of fiberboard have been extensively analyzed (14).

Adsorptive Heat

Lignocellulosics as hygroscopic substances release *adsorptive* heat when adsorbing water, especially in the form of water vapor with its large latent theat. The adsorptive heat diminishes as the moisture content of the adsorbent increases. At rising temperatures, hygroscopic materials usually lose moisture and dry, whereby the evaporation process consumes heat and retards self-heating (15, 16). Hot hygroscopic substances adsorb moisture only under rare circumstances, for example when old steam pipes crack and steam escapes into the insulating sawdust. Wood in a hot atmosphere with increasing relative humidity, and panels in an unevenly dry pack adsorb little, and adsorptions of this kind rarely play a significant role.

Gas adsorption by charred lignocellulosics is another matter, and may contribute crucial heat in late stages of slow self-heating. Charred lignocellulosics, charcoal, and the lignocellulosics from which most commercial charcoal is pyrolyzed are obviously very different substances. The pyrolysis process gradually converts the raw material, so that one cannot define a certain point or pyrolysis condition beyond which the solid residue has become charcoal. The higher the temperature of pyrolysis and the longer it lasts, the more pronounced are the charcoal characteristics. Charcoal contains percentagewise more carbon than lignocellulosics; it is also more porous, has a larger internal surface, and is chemically more reactive. In the pyrolysis process lignocellulosics approach all of these charcoal properties.

Charcoal readily adsorbs gases, including oxygen out of the surrounding atmosphere. Unlike oxidation, oxygen adsorption does not lead to effusing carbon dioxide and water vapor, rather the solid gains weight; moreover, the heat released per unit mass of adsorbed oxygen diminishes with increasing amounts of the adsorbate, whereas in case of oxidation the heat of combustion remains constant. Charcoal adsorbs also carbon monoxide and carbon dioxide. Such adsorptions are by no means restricted to ambient temperatures, occurring in hot charcoal too although less than in cool charcoal, and furthermore not all types of hot charcoal adsorb.

These adsorptions appear to be inconsistent with the evolution of carbon dioxide and other volatiles out of the charring solid in the pyrolysis process. The adsorptive properties develop as pyrolysis frees sites for adsorption; debris escaping from thermally decomposing lignocellulosics leaves the char residue with a highly reactive, eagerly adsorbing inner surface.

Adsorptions are exothermic, releasing more than the latent heat of the adsorbate gas, which quasi liquefies and solidifies besides being attracted. The first traces of adsorbed oxygen release even more heat than the combustion of carbon to carbon dioxide; but after the first adsorption sites have been occupied, the adsorptive heat -- based on unit mass of adsorbed oxygen -- gradually drops to the level of condensation heat, far below the heat of combustion.

It has been claimed but not generally accepted that long-lasting heat between 100 and 200°C converts cellulose and lignocellulosics into *pyrophoric* carbon, which finally spontaneously ignites. It may be that during lengthy slow charring the residual solid adsorbs oxygen and carbon oxides, whereby the released adsorptive heat raises the temperature and triggers ignition. The charcoal reactions are rather complex (17) because properties of charcoal vary much, depending on the original material, conditions of pyrolysis, size of the pieces, and access for oxygen and moisture during storage, not to speak of storage temperature and time.

APPENDIX: Rate of Pyrolytic Heat Release

Heat release expressed in energy per unit mass of self-heating material per unit time, such as Joules (J) per gram (g) and hour (h), depends on the rate and on the exothermicity of the pyrolysis process. Both vary with the size of the material sample, as explained above in the *pyrolysis* part. The following estimates and calculations concern principally relatively large samples of one kilogram or so.

Figure 1 applies to pyrolysis in which the wood temperature is raised from 100°C in about 10 h to various final temperatures. The weight losses, depicted as *volatiles*, and *solid residues* have been determined after the trials, and originate mainly from Klason, v. Heidenstam, and Norlin (4), under consideration of data published by Goos (18) and Stamm (19). Measurements on small samples by Beall and Eickner (20), LeVan and Schaffer (21), and Elder (22) have been compared. Up to about 275 or 300°C -- the temperature range in which cellulose rapidly disintegrates -- increased final trial temperatures cause increased increments of volatiles; beyond 300°C the increments steadily diminish.

The slope of the *volatile* = f(final temperature) curve in Figure 1 amounts to volatile increments at particular temperature steps, and is a measure of the pyrolysis rate. The slopes have been obtained by differentiating the *volatile* curve of Figure 1 for drafting Figure 2. For example, volatiles at 200°C equal 4.6%, and 7.7% at 220°C; that is a 3.1% increment in the 20°C step and amounts to a

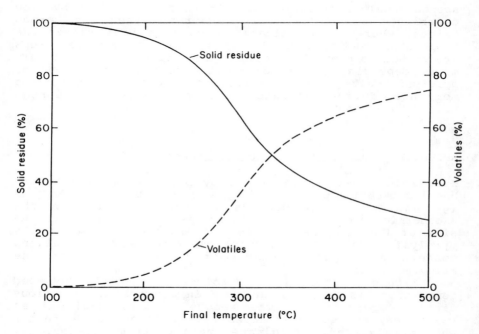

Figure 1. Pyrolysis of wood heated from 100°C to the
final temperature in ten hours.

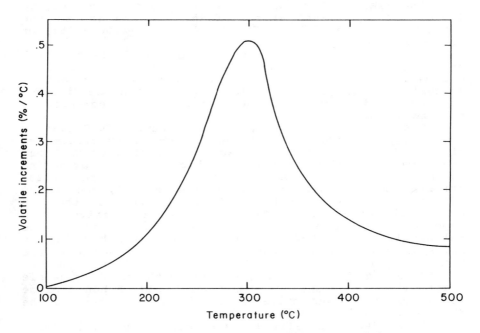

Figure 2. Pyrolysis increments in each centigrade step, expressed in mass of volatiles per mass of original wood.

secant slope of 0.16%/°C. Differentiation of the curve in
Figure 1 at the intermediate temperature 210°C renders the
nearly identical tangent slope 0.15%/°C.

Figure 2 gives rates of pyrolysis in trials with
gradually increasing temperatures. In the case of wood
heated to 500°C for example, the volatile increments
increase from temperature step to temperature step up to
300°C, dropping off in further steps. The sum of the
increments reaches 75% at 500°C, as they should according
to Figure 1. The area under the curve corresponds to that
sum. One can estimate the sum for each final temperature
from the areas. For example, in the case of heating to
250°C, the increments average 0.1%, and the sum of the
increments becomes 0.1 x (250 - 100) = 0.1 x 150 = 15%, or
0.15 g/g (compare Figure 1).

Some publications give amounts of pyrolysis heat (H)
per mass of volatiles; others include numbers for
estimating H. The heats appear to vary greatly from
author(s) to author(s) for trials under similar conditions,
but one trend is obvious: heat of pyrolysis decreases as
temperature increases.

Near the marginal pyrolysis temperature 100°C, mainly
the energy-free gases carbon dioxide and water effuse out
of the slowly disintegrating wood, as mentioned earlier.
Therefore the heat of pyrolysis is high, possibly exceeding
the heat of combustion of wood (20 kJ/g). In Klason's (23)
trials, some of which are included in Table I, amounts of
effusing water vapor and carbon dioxide were in the 5 h
trial relatively small (16.6% and 6%, respectively), as
opposed to 26.1% and 12.6% in the 336 h trial). The
relatively large, 800 g sample in the 336 h trial at
atmospheric pressure, versus the 100 or 250 g sample in the
5 h trial in vacuum, contributed to the different amounts
of effusing gases. In the 8 h trial with an 800 g sample
(Table I), the amounts of water vapor and carbon dioxide
(20.5 and 10.2%, respectively) were in between those of the
336 h and of the 5 h trials.

John (24) rationalized his excessive pyrolysis heats
(Table I) with catalytic combustion of primary pyrolysis
hydrogen. The heat of combustion of hydrogen (142 kJ/g) is
indeed very high, but the rationalization requires release
of hydrogen at low pyrolysis temperature, for which no
direct evidence exists. Besides that, oxygen for the
catalytic combustion would have to be torn out of
disintegrating wood under consumption of improbably small
energy. John's (24) explicit values H are not consistent
with other data of his experiments, and with H-values
reported by other researchers. The 60 kJ/g calculated on
basis of Kubler, Wang, and Barkalow's work (6) deserve not
much confidence either, since the trials were not designed
for exact measurements of volatiles.

Heats H of pyrolysis must be very high at low
pyrolysis temperatures (t °C), but values above 10 kJ/g may

Table I. Heat (H) of pyrolysis per gram volatiles in
trials with wood by various authors

Temperature ($^\circ$C)	Sample Mass (g)	Duration (h)	Solid Residue (%)	Heat of Pyrolysis (J/g)	Source
80-130[a]	2250	2400	96[i]	60000	(6)
120	21	650	99.5[i]	146000[b]	(24)
140	21	650	98.0[i]	55300[b]	(24)
160	21	650	93.7[i]	44500[b]	(24)
175	~3	50	--	335[c]	(25)
275	22.9	1	78.6	1170	(26)
305	27.8	0.6	52.5	340	(26)
325	24.2	0.6	39.7	290	(26)
375	30.1	0.5	30.9	310	(26)
435	25.0	0.4	28	-200[d]	(26)
100-400[e]	800	336	39.4	1160	(23)
100-400[e]	800	8	30.9	580	(23)
100-400[ef]	100-250	5	19.5	-250	(23)

[i] "incomplete", trial terminated long before residues reached constant weight.

[a] Series of trials at different temperatures. In some trials air had access and contributed to the weight loss, but the heat of pyrolysis value concerns only trials without air.

[b] The results published about this trial are not consistent; some results imply much lower heats of pyrolysis.

[c] Cubes of fiberboard were exposed to hot air, which could cause exothermic reactions in addition to that of pyrolysis; a weight loss of 2% is assumed.

[d] Calculated by Kung and Kalelkar (27).

[e] Temperature gradually raised to 400°C at the end of the trial.

[f] Trial in vacuum.

be unrealistic. Considering this and the various H-values
of Table I, a relation of the kind

$$\log H = 5 - t/100 \tag{1}$$

$$H = \text{antilog} \, (5 - t/100) \tag{2}$$

is possible. Accordingly, H would equal 10 kJ/g at 100°C,
1 kJ/g at 200°C, and 0.1 kJ/g at 300°C (Figure 3). Above
300°C or already above 275°C, pyrolysis is probably an
endothermic process, as indicated by the negative H of the
435°C trial in Table I. The rapid decrease of H with
increasing pyrolysis temperature appears to be one reason
why *pyrolysis* lines in the Arrhenius plot of Figure 5 are
less steep than the lines of *oxidation*.

Figures 2 and 3 facilitate estimating the heat release
per unit mass of original wood for each temperature by
multiplying

Pyrolysis Increment x Heat of Pyrolysis = Heat Release
g volatiles/g wood x J/g volatiles = J/g wood

Figure 4 shows the results. Above 200°C, the dashed line
reflects heats of pyrolysis according to Equation 2 and
Figure 3, while the more realistic, full line is based on
heats which approach zero at 300°C. According to the
figure, heat release peaks between 130 and 230°C. Since
heat dissipates out of self-heating materials in proportion
to temperatures above ambient, and at 230°C the dissipation
should exceed the dissipation at 130°C by a factor of two,
the most rapid pyrolytic self-heating can be expected
between 120 and 170°C.

Figure 4 tells how much heat evolves per unit mass of
material; it can be used for the comparison of heat
generation at different temperatures. Considering how the
depicted heat releases were compounded, the figure reveals
especially the role of pyrolysis exothermicity versus
pyrolysis rates. Figure 4 does not show the power
generated by the material, that is energy released per unit
time or units of Watt (W). The power can be estimated
roughly on the basis of times during which the quantities
of volatiles depicted in Figure 1 effused. The trials
lasted about 10 h, but the particular steps and volatile
increments took only about one hundredth of the 10 h or 0.1
h. Assuming the heats of Figure 4 were actually released
in 0.1 h, one arrives at the temperature range from 130 to
230°C with rates of heat release in the order of 3 W/kg.
This order of magnitude is consistent with extrapolated,
actually measured powers in Figure 5. The agreement speaks
for realistic values in Figure 4, but the wide, highly
temperature dependent range from 0.1 to 4 W/kg in Figure 4
shows also that Figure 4 is not suited for exact
determinations of the power.

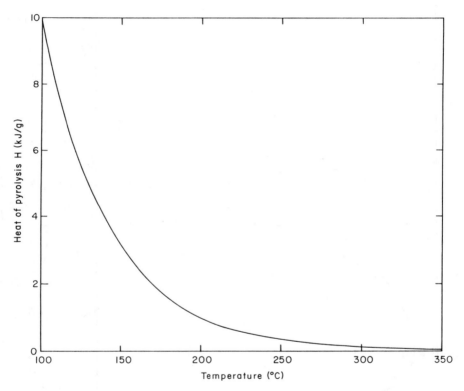

Figure 3. Heat of pyrolysis H per mass of volatiles at various pyrolysis temperatures, calculated with Equation 2.

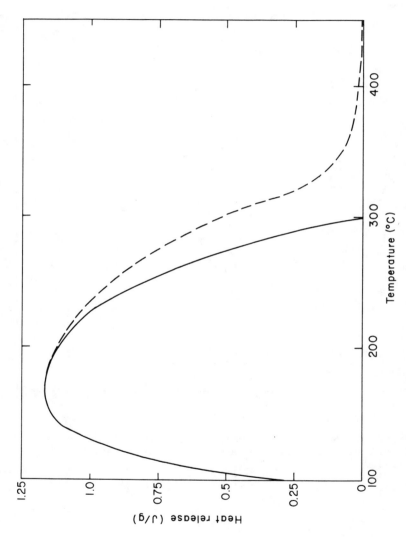

Figure 4. Pyrolytic heat release per gram of original wood in relation to the pyrolysis temperature.

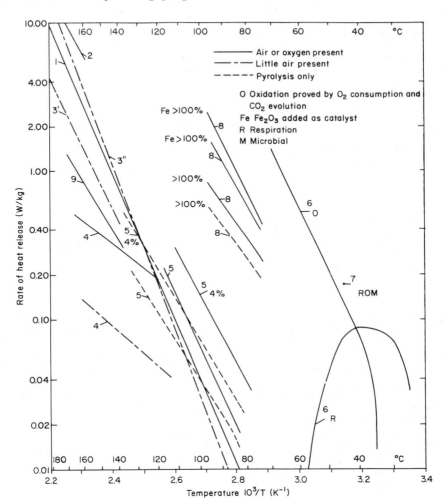

Figure 5. Rate of heat evolution in wood and cotton.
Experiments $\gtrsim$ 70°C included heat of pyrolysis; $\lesssim$ 80°C
the material was moist, $\gtrsim$ 130°C it was ovendry;
between 80° and 130°C, moisture contents are
mentioned in %, unless the material was ovendry. 1 --
Bowes (28); wood raspings; heat values extrapolated
from measurements at 190-335°C. 2 -- Eriksson and
Back (29); hardboard. 3 -- Gross and Robertson (30);
3' -- solid pine. 3" -- fiberboard. 4 -- John (24);
pine sawdust. 5 -- Kubler, Wang, and Barkalow (6);
hardwood sawdust. 6 -- Springer, Hajny, and Feist
(31); chips of aspen and Douglas fir. 7 -- Springer
and Zoch (32); aspen chips. 8 -- Walker and Harrison
(33); pine sawdust. 9 -- Walker, Harrison, and
Jackson (34); cotton. [Reproduced in adapted form
with permission from Ref. (1). Copyright 1987 C.A.B.]

LITERATURE CITED

1. Kubler, H. For. Prod. Abstr. 1987, 10, 299-327.
2. Sampson, G. R.; McBeath, J. H. For. Prod. J. 1989, 39
    (2), 53-7.
3. Steklenski, P. G.; Schmidt, E. L.; Haygreen, J. G.
    For. Prod. J. 1989. 39(2), 8-13.
4. Klason, P.; v. Heidenstam, G.; Norlin, E. Z. Angew.
    Chem. 1910, 23, 1252-7.
5. Kubler, H. Wood Fiber 1982, 14, 166-77.
6. Kubler, H.; Wang, Y.-R.; Barkalow, D. Holzforschung
    1985, 39, 85-9.
7. Amy, M. L.; Plicot, M. Chim. Ind. (Paris) 1960, 83,
    411-8.
8. Insulation Board, Federal Spec. LLL-I-555, U.S. GPO:
    Washington, DC, 1960.
9. Virtala, V.; Frilund, S. O. O. F. On Spontaneous
    Ignition and its Occurrence, State Institute Technical
    Research, Julkaisu 14: Helsingfors, Finland, 1949.
10. Mitchell, P. H. Wood Fiber Sci. 1988, 20, 320-35.
11. Blauss, H. Brandhilfe 1980, 34, 212-5.
12. Back, E. L.; Johanson, F. Holzforschung 1989, 43, in
    press.
13. Back, E. L.; Johanson, F. Holzforschung 1990, 44, in
    press.
14. Thomas, P. H. Self-Heating of Fibre Insulating Board,
    Fire Research Stn, Note 306: Boreham Wood, Herts., UK,
    1957.
15. Thomas, P. H.; Simms, D. L.; Law, M. Ignition of Wood
    by Radiation, Fire Research Stn, Note 280: Boreham
    Wood, Herts., UK, 1956.
16. Fosberg, M. A. Heat and Water Vapor Flux in Conifer
    Forest Litter and Duff, U.S. Forest Service, Research
    Pap. RM-152: Fort Collins, Co, 1975.
17. Shafizadeh, F. In Chemistry of Solid Wood; Rowell, R.,
    Ed.; ACS: Washington, DC, 1984, Chapter 13.
18. Goos, A. W. In Wood Chemistry; Wise, L. E.; Jahn, E.
    C., Eds.; Reinhold: New York, 1952; pp 826-51.
19. Stamm, A. J. Wood and Cellulose Science; Ronald: New
    York, 1964; p 307.
20. Beall, F. C.; Eickner, H. W. Thermal Degradation of
    Wood Components, U.S. Forest Prod. Lab., Research Pap.
    FPL 130: Madison, WI, 1970.
21. LeVan, S.; Schaffer, E. L. Therm. Insul. 1982, 5, 229-
    44.
22. Elder, T. Wood Fiber Sci. 1984, 16, 169-79.
23. Klason, P. J. Prakt. Chem. 1914, B90, 413-47.
24. John, R. Investigation on Spontaneous Ignition of
    Self-Heating Materials, Universitat, Arbeitsgem.
    Feuerschutz, Forschungsber. 24: Karlsruhe, Fed. Rep.
    Ger., 1972.

25. Thomas, P. H.; Bowes, P. C. Br. J. Appl. Phys. 1961, 12, 222-9.
26. Roberts, A. F.; Clough, G. 9. Int. Symp. Combust., Academic: New York, 1963; pp 158-166.
27. Kung, H. C.; Kalelkar, A. S. Combust. Flame 1973, 20, 91-103.
28. Bowes, P. C. Estimates of the Rate of Heat Evolution, Fire Research Stn, Note 266: Boreham Wood, Herts., UK, 1956.
29. Eriksson, S. A. I.; Back, E. L. Sven. Papperstidning 1966, 69, 300-4.
30. Gross, D.; Robertson, A. F. J. R. N. B. Stand. U. 1958, 61, 413-7.
31. Springer, E. L.; Hajny, G. J.; Feist, W. C. Tappi 1971, 54, 589-91.
32. Springer, E. L.; Zoch, L. L. Tappi 1970, 53, 116-7.
33. Walker, I. K.; Harrison, W. J. N. Z. J. Sci. 1977, 20, 191-200.
34. Walker, I. K.; Harrison, W. J.; Jackson, F. H. N. Z. J. Sci. 1970, 13, 623-40.

RECEIVED November 20, 1989

Chapter 27

Use of Highly Stabilized High-Expansion Foams in Fighting Forest Fires

Connie M. Hendrickson

AR'KON Consultants, 2915 LBJ Freeway, Suite 161, Dallas, TX 75234

Foams for woodland and grass fire-fighting have been developed with low drainage (water loss from foam) and high expansion (volume of foam per volume of starting liquid), two parameters traditionally considered to be inversely proportional. The stabilizing additive is poly (methyl vinyl ether/maleic anhydride), thought to cross-react with fatty alcohol components of the foam base via the anhydride function.

Foam use in regular fire-fighting, particularly in industrial situations, has been common for a number of years now. However, foam is a relative newcomer as a tool in combating forest and grassland fires. Wetting agents have been a standard in the field, used to "stretch" the available water supplies in situations where water is often at a premium.

It is indisputable that foam gives a fire-fighter more power for the water available, stretching supplies even further than use of a wetting agent. A high expansion foam delivers 400 to 800 times the volume of foam for the starting volume of concentrate and water (called the premix). A volume of concentrate, diluted 2% in water, then yields 20,000 to 40,000 times its original volume (Table I).

Table I. Advantages of Foam Use

High Expansion: Efficient use of materials
Expansion: 400-800: 1 (foam:liquid volume)

1 gal concentrate
↓
50 gal premix
↓
20,000 gal foam

0097–6156/90/0425–0450$06.00/0

However, there are a number of reasons for foam underuse in this particular segment of the industry:

1. Foam concentrate is more expensive than water or wetting agents.
2. Shelf life is often less than wetting agents.
3. Foam use requires generatons and more sophisticated equipment than wetting agents, and thus
4. Additional training of personnel is also required.
5. Many foams on the market contain ingredients which are not highly susceptible to biodegradation, or which may be harmful to the environment on a long-term basis.
6. High expansion foam, necessary for the coverage required in a woodland or grassland situation, usually loses its water content very rapidly (rate of water loss is called drainage). The remaining foam is then extremely dry and fragile, and susceptible to wind, water from other hoses, etc. (Table II).

Table II. Disadvantages of Using Foam

1. Foam is more expensive than water or wetting agents
2. Generators: more equipment needed and training
3. High expansion traditionally means
   a. fast drainage
   b. high fragility

Typically, high expansion foams have characteristics similar to those in Table III. As expansion increases, drainage also increases. A foam with an expansion of 400:1 loses 35% of its water in the first 15 minutes after formation; however, a foam with an expansion of 1000:1 loses 85% of its water in its initial fifteen minutes.

Table III. Typical Parameters of High Expansion Foams

Expansion	Drainage (% lost in 15 min)*	
400:1	35	
800:1	55	Fragility
1000:1	85	

*percentage of the total water lost in 15 min.

Foam used was Macrofoam B (Rockwood Systems):
2% premix, on a UL laboratory generator

There are a number of ways to produce high expansion while retaining a reasonable drainage, or stabilization of the foam (1).

In the case of foams formulated on the basis of alkyl and ethoxylated alkyl sulfates, the stabilizing material is generally the corresponding fatty alcohol (e.g., lauryl sulfate, lauryl alcohol). The relatively non-polar hydroxyl group is thought to insert between the charged sulfate "heads" in the micelle and then in the bubble wall; stabilization is a result of a more comfortable distance between the sulfate groups than would otherwise be the case. It is a passive process in that no chemical reaction takes place and no bonds are formed or broken. A complete discussion of these stabilization methods is included in Rosen ($\underline{1}$). In a typical formulation, however, as the percentage of fatty alcohol is increased and drainage decreases, the viscosity and pour point are altered. Viscosity increases slightly at a normal use temperature (65° F or above) and increases drastically as temperatures decrease (40° or below). The pour point (temperature at which the concentrate is easily poured or educted) is thus raised to levels which may render the now "stabilized" foam concentrate unusable under relatively moderate conditions. Addition of glycol or glycol ethers as anti-freeze ameliorates the pour point situation somewhat but is limited by the fact that these materials act as foam breakers in high concentrations.

The general feeling in formulations of this type has been that increased viscosity in the concentrate is a necessity for drainage control ($\underline{2}$). High viscosities and cold susceptibility simply rendered the product unusable in outdoor situations where such conditions might be encountered.

A number of polymers used for foam stabilization affect the viscosity in much the same manner but may contribute a higher degree of stabilization for the percentage of the additive used than the low molecular weight fatty alcohols. High molecular weight additives of this type, listed in Table IV, are also thought to be passive participants in that no chemical reaction takes place and no bond changes occur.

Table IV. Polymeric Additives for Foam Stabilization

Polyvinyl alcohol (cold-water soluble) ($\underline{3}$)
Polysaccharides (Kelzan) ($\underline{3}$)
Lignin sulfonates ($\underline{3}$)
Protein (animal hydrolysates, zein, albumin) ($\underline{3}$)
Methyl cellulose derivatives (3)
Polyacrylic acids (Carbopols) ($\overline{\underline{4}}$)

These polymeric additives generally introduce a whole new set of considerations, listed in Table V, which may limit or otherwise affect foam production or use.

Table V. Problems with Polymeric Additives

Dispersion during addition
Use of alcohol solvents
Stability and shelf life (gel and
    precipitate formation)
High viscosity at low temperatures
    (non-educting)

We have recently developed and obtained a patent for an ultra-stabilized high expansion foam (5) with expansion and drainage characteristics as shown in Table VI, using a polymeric additive for stabilization without harming expansion.

Table VI. Ultrastabilized High Expansion Foams

Expansion*	Drainage (% lost in 15 min)**
400:1 (fan)	3
750:1 (screen)	12-18
1000:1 (screen)	23

*dependent on generator
fan=water-driven fan
screen=nozzle/screen stationary
**percentage of the total water lost in 15 min

Foam used was Macrofoam (Rockwood) in a 2% premix. Gantrez AN139 (GAF) was used at a 3% concentration.

Although expansion and drainage are still increasing together, it can be easily seen that the polymer-stabilized foam has a drainage far below that of the typical foam (each with the same basic formulation) shown in Table III. The foam base is an alkoxy/linear alkyl sulfate, fatty alcohol, and glycol/glycol ether mixture, as outlined in Table VII.

Table VII. Foam Ingredients

A.	Alkyl and alkoxy sulfonates	Foamers
	Fatty alcohols and amides	Secondary stabilizers
	Glycols and glycol ethers	Cosolvent/antifreeze

B. Poly (methyl vinyl ether/maleic anhydride)

The drainage-stabilizing additive is a polymer: poly (methyl vinyl ether/maleic anhydride (PMVEMA) (Gantrez AN, GAF Corp.). The PMVEMA is not a passive additive but, via the anhydride function, a reactive species which will undergo hydrolysis and bond formation in the presence of an alcohol or water (Figure 1). Percentages added to the foam concentrate giving satisfactory performance range from 1-3%. The polymer must be added prior to the glycol antifreeze component in order to avoid extensive cross-linkages and gel formation.

We have visualized the mechanism of stabilization in the following steps:

In the concentrate:

1. The polymer is added to the other components of the concentrate prior to micelle formation. Effective stabilization does not occur if the polymer is added after the concentrate is allowed to age for several hours.
2. The anhydride functions react with the fatty alcohol molecules present in the mixture (as well as with water). Glycols are not added until after the anhydride is completely hydrolyzed (24 hours or more) (Figure 1).
3. The fatty alcohol hydrophobic "tails" are incorporated into the micelles as the concentrate is aged (Figure 1), shown by the increasing viscosity as the foam ages but after the anhydride hydrolysis reaction is complete.

In the premix (2% concentrate in water):

4. As the concentrate is diluted into premix and foam is generated, the hydrophobic "tails" become part of the bubble wall (Figure 2).
5. The width of the bubble wall decreases as the inner water layer drains downward under the influence of gravity (Figure 2).
6. The polymer strands prevent the narrowing of the bubble wall and cause slowing of the water drainage, thus decreasing drainage time and stabilizing the foam (see Bikerman (6) for a discussion of the mechanism of drainage).

The foam has been successfully used in a grassland fire by Rockwood Systems and in a forest fire situation (Bureau of Indian Affairs) (Latham, A., Rockwood Systems, personal communication, 1988). The layout used to combat the fire is shown in Figure 3. The foam chemicals were supplied as a premix (2.5% concentrate/-water) from a tanker, and the foam was produced from two different portable generators. The water-driven fan generator (Macrogen 50, Rockwood Systems) was used to lay a "barrier wall" of foam five feet from the fan; the screened nozzle type (Superjet X, Rockwood) was then used to throw a large quantity of foam over that barrier (15-20 ft. range). The barrier wall created by the water-driven fan acted to retain the looser foam produced by the screened nozzle and create a three-dimensional situation to fully utilize the power of the foam against the oncoming fire. Figure 4 is a schematic for the set-up both for a premix tanker, and for a water tanker utilizing concentrate and an eductor system.

The formulation is non-toxic, non-hazardous, and biodegradable (5, 7, 8) so that it may be safely used in situations where an adverse contaminant might have a large environmental impact. It has also been successfully used in three-dimensional fires, such as warehouse protection and airplane hanger systems, and as a spill fume suppressant on hydrocarbons and on oleum (9).

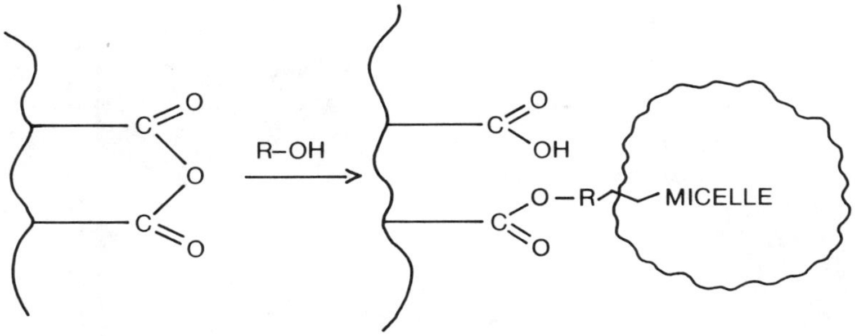

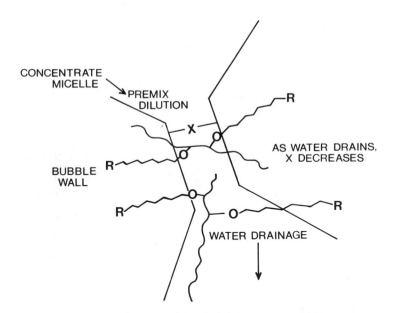

Figure 1.   Interaction of polymer and foam: an alcohol group in the foam reacts with an anhydride function on the polymer chain and is thus linked to the polymer chain.   The hydrophobic R group is subsequently incorporated into the micellar structure.

CONCENTRATE
MICELLE

PREMIX
DILUTION

R

X

AS WATER DRAINS,
X DECREASES

BUBBLE
WALL

R

R

O

WATER DRAINAGE

Figure 2.   Foam drainage and water loss from the bubble wall: concentrate is diluted into a premix, and the foam is generated from the premix. When foam is generated, hydrophobic R groups become part of the bubble wall (X = width of bubble wall).   X decreases as water drains downward under the influence of gravity.   The polymer strands prevent narrowing of bubble wall, decreasing drainage time and stabilizing the foam.

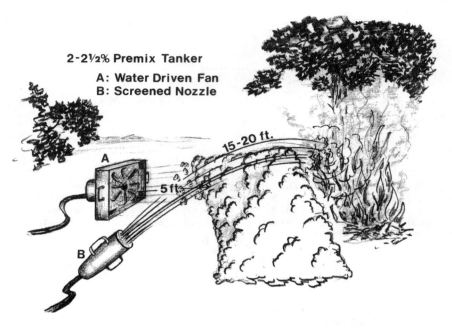

Figure 3.   Foam generators and placement for a forest, grassland, or other outdoor scenario.

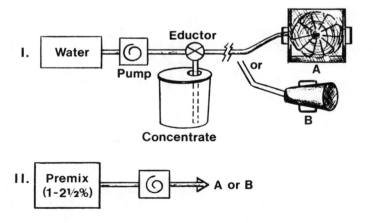

Figure 4.   Schematic of foam production: Tanker-stored water
using an eduction system (I) or a premix stored in a tanker
(II) may be used for foam generation using a water-driven fan
(A) or a screened nozzle (B).

The development of ultrastabilized high expansion foams must continue in order for foam use in forest and grassland scenarios to gain wider acceptance and popularity. The long-term potential of foams of this type is an exciting challenge for the surfactant chemist as well as the fire-fighting professional.

Acknowledgments

This work was done for Rockwood Systems Corp., Lancaster, Texas.

Literature Cited

1. Rosen, M. Surfactants and Interfacial Phenomena; Wiley and Sons: New York, 1989; pp 297-299.
2. Bikerman, J. J. Foams; Springer-Verlag: New York, 1973; pp 173-174.
3. Perri, J. M. "Fire-Fighting Foams" in Foams: Theory and Industrial Use, Bikerman, J. J., Ed.; Rheinhold: New York, 1953; pp 189-242.
4. Rand, P. U.S. Patent No. 4 442 018, 1984.
5. Hendrickson, C. M. U.S. Patent No. 4 836 939, 1989.
6. Bikerman, J. J. Foams; Springer-Verlag: New York, 1973; pp 159-183.
7. Longman, G. F. The Analysis of Detergents and Detergent Products; Wiley and Sons: New York, 1978; pp 494-539.
8. Rosen, M. Surfactants and Interfacial Phenomena; Wiley and Sons: New York, 1989; pp 7-16.
9. Hendrickson, C. M. Proc. 42nd Southwest Regional Meeting of the ACS, 1986, p 106.

RECEIVED November 1, 1989

FIRE PERFORMANCE, TESTING, AND RISK

Fire Performance, Testing, and Risk

The choice of materials makes a difference in fire safety. The appropriate choice of materials can prevent or retard ignition and flame spread. One assembly can withstand a given fire while a similar assembly of different materials cannot.

Many discussions of fire safety would lead one to believe that a number of separate fire parameters must be evaluated and incorporated into assessment of fire risk, and that these parameters must be traded off or averaged in the comparison of one material or system versus another. Implicit in such a discussion is the risk equation, where R is risk:

$$R = f(\text{ignition}) + f(\text{growth}) + f(\text{smoke}) + f(\text{toxicity}) + ...$$

Yet fire is a sequence of events. First, an object must be ignited; second, fire must spread from the point of ignition; third, as fire grows, smoke and toxic gases are generated in significant quantities; and fourth, fire penetrates barriers.

One must have ignition and spread of flame to create smoke and toxic gases. If ignition is prevented, spread of flame minimized, and smoke controlled, toxic gases are likewise affected. If the risk equation is viewed at its limits, if the probability of ignition is zero, then the risk function must go to zero. This suggests that the risk function is not an additive function, $fA + fB + fC$, but is a product function, $fA \times fB \times fC$; that is,

$$R = Af(\text{ignition})^a \times Bf(\text{growth})^b \times Cf(\text{smoke})^c \times Df(\text{toxicity})^d \times ...$$

If $f(\text{ignition})$ is very low, risk R is low even if $f(\text{toxicity})$ is high. If $f(\text{ignition})$ and $f(\text{growth})$ are large, then R is large even if $f(\text{toxicity})$ is in a "normal" range. Such an equation gives the expected response of external forces, such as sprinklers, on R. Sprinklers reduce growth and R is reduced. Implicit in risk are the probability and the severity of the event.

The exponent b in the growth term is not 1; in fact, for flame spread from a point, b is 2. It is not surprising, then that ignition and growth terms overwhelm toxic potency differences for differences less than one order of magnitude between materials.

Viewed in terms of toxic hazard, risk has the following form:

$$R(\text{toxic}) = Df(\text{toxicity})^d = Df[(\text{amount}) \times (\text{potency})]^d$$

where amount is proportional to $f(\text{growth})^b$. Clearly, toxic hazard itself is influenced by the other fire properties and is not an independent variable.

One does not want to sacrifice ignition and growth performance for toxic potency performance. Ignition and growth performance determine the probability of a fire occurring and the magnitude of the fire if it occurs. All fire atmospheres are highly toxic. The smoke plume exiting a room from a fire of moderate size (i.e., extinguishable with a garden hose) has a CO concentration of 5000 to 50,000 ppm, a lethal atmosphere. A single object weighing in pounds can produce sufficient CO to be life-threatening in a home-sized room if it burns rapidly. A handful of charcoal in a hibachi can cause death if brought into a room in a house.

Fire is an irreversible process. Different paths give different events, which makes the use of small-scale tests so difficult in a regulatory context.

Since 80% of fire deaths are the result of toxic gas exposure, regulators would like to regulate toxic hazard directly. Yet current regulations work because ignition and growth terms dominate the risk equation. Ignition and growth are readily evaluated and controlled in regulation by relative material performance specifications and engineering design requirements, plus use of detection and suppression devices.

Data over the past decade show a decrease in the number of residential fire deaths and a reduction in the number of fires. The CPSC believes that this decline is a direct result of fire prevention/detection activities: flammability standards for mattresses, the voluntary standard for upholstered furniture, and promotion of smoke detectors.

Fire risk assessment techniques, particularly Delphi techniques, also continue to evolve in the fire protection community.

Several general risk assessment formulae have been advanced:

Risk = Probability × Consequence (Hazard)
Risk = Probability of Occurrence × Probability of
       Exposure × Potential for Harm

In the second equation, the last term is intrinsic to the product and is the result given ignition. This is somewhat akin to the Government Accounting Office (GAO) scheme introduced for rating buildings a decade ago - given ignition, what is the probability of a fire reaching the walls or barriers or going beyond the barriers? It is interesting how closely different experts rate buildings using the GAO Delphi approach.

The final section of this book begins with a discussion by Hirschler of the general principles of fire hazard. Flisi discusses harmonization of fire testing in Europe, as Europe prepares for 1992. He describes the planned interim approach. Paul discusses the use of combustion modified polyurethane foam and its potential impact upon fires in the United Kingdom. Hirschler discusses the use of heat release equipment to measure smoke, equipment which provides data more useable in fire models.

Macaione and Tewarson present data on a variety of test methods to evaluate fiber reinforced composites for vehicles. Karlsson and Magnusson use bench scale tests to predict results in 1/3 scale and full scale for combustible wall lining materials. Hirschler, in the final chapter, uses fire model calculations to assess effects of wire coatings in plenums.

Chapter 28

General Principles of Fire Hazard and the Role of Smoke Toxicity

Marcelo M. Hirschler

BFGoodrich Technical Center, P.O. Box 122, Avon Lake, OH 44012

Fire hazard is a combination of several properties, including ignitability, flammability, flame spread, amount of heat released, rate of heat release, smoke obscuration and smoke toxicity.

A large number of procedures are now available for measuring fire properties, but many of them are of little interest since they represent outdated technologies. Thus, in order to obtain a realistic estimate of fire hazard for a scenario it is essential to measure relevant fire properties. Furthermore, the appropriate instruments have to be used, viz. those yielding results known to correlate with full scale fire test results.

True fire hazard can be determined only in a specific scenario. Therefore, it is necessary to determine which fire properties are most relevant to the scenario in question. These fire properties will then have to be measured and combined in order to obtain an overall index of fire hazard. As a general rule, it is clear that the most important individual property that governs levels of fire hazard is the rate of heat release: the peak rate of heat release is proportional to the maximum intensity a fire will reach.

A large number of small-scale tests have been designed to measure the toxic potency of the smoke of materials. These tests differ in many respects; the consequence of this is that the relative toxic potencies of smoke resulting from these various tests are different. The tests are not useful,

0097–6156/90/0425–0462$06.00/0
© 1990 American Chemical Society

therefore, to rank materials in terms of
their toxicity. The tests are useful,
however, in selecting those, very few,
materials with a much higher toxic potency
than the common materials in everyday use.

Fire safety can be improved by
decreasing fire hazard, but is unlikely to be
affected by small changes in toxic potency of
smoke, since the toxic potency of most
materials is very similar.

In order to understand the various concepts associated with
fire safety it is essential for all major terms to be
defined adequately.

Fire hazard can be defined as the potential for harm
associated with fire: it addresses threats to people,
property, or operations, resulting from a particular fire
scenario. A fire scenario involves those conditions
relevant to the initiation, development, or harm caused by
a fire. Fire risk is a combination of three elements: (a)
fire hazard, (b) probability of fire occurrence in the
scenario in question and (c) probability of the material
or product in question being present in the fire scenario.
Toxic potency (of smoke) is a quantitative expression
relating concentration and exposure time to a certain
adverse effect, on exposure of a test animal; the effect
is usually lethality. It is necessary to stress that the
toxic potency of smoke is also heavily dependent on the
conditions under which the smoke has been generated, since
the mode of generation will affect both the quality and the
quantity of smoke. Smoke is interpreted here as the sum
total of the gaseous, liquid and solid airborne products
of combustion. Exposure dose is an integration of the
toxic insult, as calculated from the smoke concentration
vs. time curve. If the insult results from an exposure to
a single toxicant, and its concentration is constant, the
exposure dose is simply the product of concentration and
time of exposure. Time to effect can be very different
from time of exposure, since many toxicants act with a
delayed effect, so that the test animal (or the victim) may
die long after the exposure.

Stages of a Fire and Fire Hazard

A major fire follows several stages:

1. Ignition (onset of fire)
2. Development of fire within original compartment
3. Involvement of other products
4. Full room involvement (or flashover)

5. Transport of fire to other compartments
6. Decay

The intensity of the fire will determine which stages the fire will traverse on its way from ignition to decay. The National Fire Protection Association (NFPA) stores statistical data collected from the fire marshalls' reports. It classifies [1] fires as:

(i) restricted to the object of origin
(ii) restricted to the area of origin
(iii) restricted to the room of origin and
(iv) extending beyond the room of origin.

It is clear that fires of type (i) will not go through stages 3-5, fires of type (ii) will skip stages 4 and 5, and fires of type (iii) will not reach stage 5, before decaying. These considerations are important because they will be an essential tool in deciding the properties to be measured for estimating fire risk or fire hazard.

As far as fire hazard to humans is concerned, the main aspects to be considered are:

* Heat effects
* Toxicity of smoke
* Lack of visibility

These phenomena all depend both on time and location. Thus, it is important to consider the following two aspects for all of them:

* Transport of smoke
* Decay of smoke components

The fire properties most relevant to each stage of a fire are:

1.* Ease of ignition of product first ignited
   * Ease of extinction of product first ignited
2.* Rate of heat release of product first ignited
   * Amount of heat released by product first ignited
   * Flame spread characteristics of product first ignited
   * Mass loss rate of product first ignited
   * Smoke factor of product first ignited
   * Rate of fire growth
   * Presence of fire suppression devices (e.g. sprinklers)
   * Toxicity of smoke
3-4.* Ease of ignition of other products
   * Rate of heat release of other products
   * Amount of heat released by other products
   * Flame spread characteristics of other products
   * Mass loss rate of other products

   * Smoke factor of other products
   * Rate of fire growth
   * Presence of fire suppression devices
   * Toxicity of smoke
5-6.* Fire performance of products in original
   compartment
   * Fire endurance of structural components of
   original compartment
   * Ease of ignition of products in other compartments
   * Same issues as in earlier stages, for new
   compartments
   * Overall fuel and oxygen supply
   * Geometric scenario considerations
   * Transport and decay of smoke
   * Fire protection measures:
   - Compartmentalisation
   - Sprinklers
   - Smoke detectors
   - Extinction capabilities
   * Effects of conditions on fire fighters
   - Visibility
   - Heat
   - Toxicity of smoke

     Some of the fire properties mentioned have been
measured and well understood for a long time, but others
are relatively new concepts. The most important fire
properties and proposed measurement methods will thus be
discussed in the following sections.

Rate of Heat Release and Associated Fire Properties

It has now become clear that the single property which most
clearly defines the magnitude of a fire is the maximum rate
of heat release [2, 3]. The peak rate of heat release is
an indication of the peak intensity of a fire. The rate
of heat release (RHR) can, thus, be used as a small scale
substitute for the burning rate of the full scale fire.
This property (RHR) governs not only the burning rate (and
mass loss rate) of the product being consumed but also the
amounts of other items which will be burnt. The rate of
heat release will also therefore govern the overall amount
of smoke and combustion products being generated in the
fire, since other products will be ignited only if enough
heat reaches them at sufficient speed. The rate of fire
growth can be represented by the rate of rise of the heat
release rate.
     Much research has been done to identify the
relationship between the properties of materials as
measured in small scale tests and the performance of
products made from them under real fire conditions [4, 5].
The best approach is to estimate the rate of growth of a
real fire (or perhaps the time available before flashover)
based on measuring, in a small scale test, the peak rate

of heat release for those materials used to manufacture the burning product. The rate of burning of a real fire can be expressed in terms of the rate of mass loss. Yield factors can be measured in small scale tests to give the amounts of heat, smoke and toxic gases generated per unit mass burnt. They can then be coupled with the burning rate of the real fire to estimate the potential build up of heat, smoke and toxic gases in the real fire [6]. Rate of heat release can be measured in instruments called rate of heat release (or RHR) calorimeters [7-10]. The data measured from one of these instruments (the cone calorimeter, developed at the National Institute for Standards and Technology, NIST) has been shown, repeatedly, to correlate well with those found in full scale fires [11-13]. The data from another RHR calorimeter (the Ohio State University instrument) has been shown to correlate with those from the cone calorimeter [14] and from full scale aircraft tests [15]. It is already being used to regulate aircraft interior materials [16]. A third RHR calorimeter (the Factory Mutual instrument) is being used to assign insurance risk to cables in non-combustible environments [17].

Rate of heat calorimeters can be used to measure a number of the most important fire hazard parameters, including the peak rate of heat release, the total heat release, the time to ignition and smoke factor (a smoke hazard measure combining the total smoke released and the peak RHR [14, 18-20]). The smoke factor will give an indication of the total amount of smoke emitted in a full scale fire.

In summary, thus, if RHR calorimeters are fitted with the appropriate instrumentation they can be used to measure:

        Rate of heat release
        Total heat released
        Ease of ignition
        Mass loss rate
        Smoke factor

Other Fire Properties Useful for Aspects of Fire Hazard

Some of the other properties of interest for fire hazard assessment cannot be measured with RHR calorimeters. They include flame spread, limiting oxygen index (LOI, or simply oxygen index, OI: both names have been used, but the author's preferred nomenclature is the one used here) and fire endurance.

It is outside the realm of this paper to discuss the instruments used for these tests in any detail. It is only worth mentioning a few general principles.

If a material does not ignite, it will not endanger lives or contribute to fire hazard. Most organic materials do, however, ignite; the hazard will, thus, be greater the lower the ignition temperature or the shorter the time to ignition.   Some of the most common ignitability tests, other than RHR instruments, are:

* ISO 5657:	Measures sample ignitability, with a conical combustion module. Normal sample orientation: horizontal.
* IEC 695-2:	This contains two ignitability tests: one uses a glowing wire and one a needle flame.
* ASTM D1929:	Setchkin ignition apparatus. Measures flash ignition and spontaneous ignition temperatures. Normal sample orientation: horizontal.

Once a material has ignited, the hazard associated with it will increase if its flammability is greater; one of the most reliable quantitative small scale flammability tests is the limiting oxygen index test (ASTM D2863). This test measures the limiting oxygen concentration in the atmosphere necessary for sustained combustion. It is not a good predictor of full scale fire performance, but can give an indication of ease of burning or ease of extinction. Tables of typical results have been published (e.g. refs. 21-23).

The LOI test cannot be used to predict full-scale fire performance.   However, if a material has, as a rule of thumb, an LOI value above 25-27 it will, generally, only burn under extreme conditions (high applied heat). It has been shown that the LOI does not, in fact, correlate well with other fire tests, not even a small-scale flammability test such as UL 94 [24].   It has, further, been suggested that there may be some advantage in using a modification that uses bottom ignition [25].   It is important to keep in perspective the utility of this test for ease of extinction: it can (a) give a first approximation to suggest whether a material is very flammable or not; (b) show whether changes in a base formulation have improved flammability characteristics and (c) be used as a quality control tool.

The next property of interest is flame spread, which can be measured by a variety of standard test methods, depending on the angle at which the flame impinges on the material. The two most widely used tests are ASTM E162 and ASTM E84.   In ASTM E162 a radiant panel ignites a 15 cm by 45 cm sample at an angle of $30°$ to the right of the vertical.   A variant of this test is the IMO (or Lateral Ignition and Flame Spread Test, LIFT) apparatus.   In ASTM E84 (Steiner tunnel test) a pair of gas burners ignite a horizontal 7.5 m long sample from below.   A variety of other flame spread tests exist, but they are generally associated with specific applications or scenarios (e.g. cable tests, floor covering tests, etc.).

Fire endurance properties are always measured directly on finished products, and are specific for a particular application.

Smoke Obscuration

Another important property of materials is their tendency to decrease visibility. The most common method for measuring this property is the NBS smoke chamber in the vertical configuration at 25 kW/m^2 (ASTM E662). This instrument has now been shown to be associated with a variety of deficiencies, the most important of which is its lack of correlation with the results of full scale fires [26-28]. A variety of other devices are also used for measuring smoke obscuration, and details are beyond the scope of this paper. Suffice it to mention that, in order to obtain results meaningful for fire hazard assessment it is necessary either to avoid full sample consumption or to compensate for it in some way, for those materials or products which do not burn up completely in a fire. Furthermore, when samples are exposed vertically to flame they may melt or drip and, thus, avoid being consumed by letting the material artificially escape the action of the flame. The best methods for assessing smoke obscuration are those that combine smoke and heat release measurements.

Toxic Potency of Smoke

During the 1970's and early 1980's a large number of test methods were developed to measure the toxic potency of the smoke produced from burning materials. The ones most widely used are in refs. 29-32. These tests differ in several respects: the conditions under which the material is burnt, the characteristics of the air flow (i.e. static or dynamic), the type of method used to evaluate smoke toxicity (i.e. analytical or bioassay), the animal model used for bioassay tests, and the end point determined. As a consequence of all these differences the tests result in a tremendous variation of ranking for the smoke of various materials. A case in point was made in a study of the toxic potency of 14 materials by two methods [33]. It showed (Table I) that the material ranked most toxic by one of the protocols used was ranked least toxic by the other protocol! Although neither of these protocols is in common use in the late 1980's, it illustrates some of the shortcomings associated with small scale toxic potency of smoke tests.

Smoke is not a uniform substance and its composition depends on the exact conditions under which it was generated. Therefore, the composition of the smoke generated from the same material in different tests can vary broadly and, consequently, so will its toxicity.

Table I. Comparative Mortality Data of Combustion Products of Polymers

Toxicity Ranking (Toxicity Increases Upwards)

Ranking	STATIC CHAMBER Sample	STATIC LC$_{50}$ g	DYNAMIC CHAMBER Sample	DYNAMIC LC$_{50}$ g
1	Red Oak	9	Wool	0.4
2	Cotton	10	Polypropylene	0.9
3	ABS (FR)	21	Polypropylene (FR)	1.2
4	SAN	23	Polyurethane foam (FR)	1.3
5	Polypropylene (FR)	25	Poly(vinyl Chloride)	1.4
6	Polypropylene	28	Polyurethane foam	1.7
7	Polystyrene	31	SAN	2.0
8	ABS	33	ABS	2.2
9	Nylon 6,6	37	ABS (FR)	2.3
10	Nylon 6,6 (FR)	37	Nylon 6,6	2.7
11	Polyurethane foam (FR)	47	Cotton	2.7
12	Polyurethane foam	50	Nylon 6,6 (FR)	3.2
13	Poly(vinyl Chloride)	50	Red Oak	3.6
14	Wool	60	Polystyrene	6.0

The toxic potency of the smoke of most common materials (natural or synthetic) is very similar (see Figure 1). In fact, the difference between the toxic potency of almost all combustible materials is less than one order of magnitude. Therefore, the relative rankings of materials are heavily dependent on the exact composition of the smoke being tested, i.e. on the combustion procedure being used.

The fact that the main direct cause of death in fires has always been the toxicity of combustion products was already discussed in the National Fire Protection Association (NFPA) Quarterly in 1933 [34]. Smoke contains mainly two types of toxic gases: asphyxiants and irritants, but the individual toxic gas associated with the largest fire hazard is carbon monoxide (CO).

Toxic Gases in Fires

CO is present in all fires, because it is a combustion product of any organic material, and it causes the formation of carboxyhemoglobin (COHb) in blood. Although the exact lethal level of COHb is heavily dependent on the individual affected, any value above 20% can lead to death [35]. Even if a very conservative estimate is taken of the lethal level of COHb (viz. 50%), it alone accounts for 60% of all fire deaths, while over 91% of fire victims have levels above 20% COHb [36]. Other factors can lead to a lower tolerance towards CO, even in the absence of other toxic gases: typically heart disease, blood alcohol level, burns and age [35]. However, it is interesting to note the great similarity found between the blood COHb level distributions in two studies, one involving the notorious 1980 MGM Grand Hotel fire [37] and the other one involving deaths from CO evolution due to malfunctioning gas heaters [38]. This indicates that deaths in fires correlate well with deaths from carbon monoxide poisoning. A recent statistical analysis of a data base of of over 2,000 fatalities involving carbon monoxide from fire and non-fire sources [35] has shown that, once the controlling factors of age, disease and blood alcohol level have been accounted for, the COHb distribution from fire and non-fire fatalities are very similar [39].

A variety of other gases are also given off by burning materials: In two studies fire fighters went to address actual buildings on fire, equipped with combustion product monitors [40, 41]. Both studies had the same conclusions: the overwhelming hazardous toxicant in a fire is carbon monoxide.

These studies also pointed out that a potentially very dangerous gas in fires is acrolein, because the ratio of its concentration, as measured in the atmosphere of real fires, to its lethal exposure dose (LED) is higher than for many other common fire gases. The ratios of concentrations

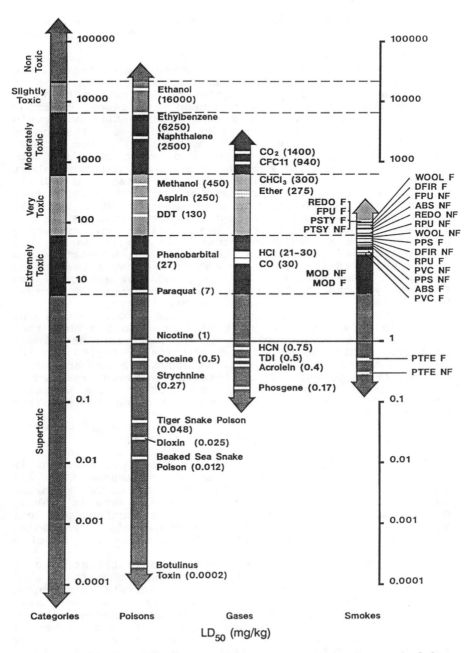

Figure 1. Categories of toxicity and lethal doses of various poisons and of the smoke from polymeric materials according to the NBS Cup Furnace Smoke Toxicity Protocol.

to LED found for two other common toxic combustion products, hydrogen chloride (HCl) and hydrogen cyanide (HCN), were much lower.

This discussion does not address the mechanism of action of these toxicants, i.e. whether the toxicant is an asphyxiant (as CO or HCN) or an irritant (as acrolein or HCl). Table II shows the highest concentration found in these studies for the 4 most common fire gases, together with their lethal levels and their odor detection levels [42-47]. The peak concentrations found were 7,450 ppm of CO, 100 ppm of acrolein, 280 ppm of HCl and 10 ppm of HCN. As regards HCl, it is relevant to point out that its airborne concentration remains at a peak value for a short period only before decaying [48]. The decay of HCl is particularly fast in the presence of sorptive surfaces such as most ordinary construction materials [49, 50]. The rate of HCl decay can be fast enough so that within 30 min its airborne concentration may have fallen to virtually nothing. The majority of other common fire gases (CO, carbon dioxide, hydrocarbons) are virtually unreactive and do not decay [51]. The gases that can decay either do so at a much lower rate than HCl (e.g. HCN) or are much less frequently present in fires (e.g. hydrogen fluoride, because less fluorinated polymers are in use).

A number of studies have been made of combinations of individual toxic gases. Most of these studies show that the effects of these combinations of toxic gases are simply additive. This has been found empirically for CO and HCl [52] and for CO and HCN [44], although the mechanisms of action are different. These results can be interpreted as each toxicant taking its toll and acting on a weakened system. The CO-carbon dioxide combination has been claimed to be synergistic [53].

Fire Hazard and Smoke Toxicity

It has already been stated that the principal toxicant in a fire scenario is carbon monoxide, generated when all carbonaceous materials burn. Moreover, the carbon monoxide concentration in full scale fire scenarios depends heavily on fire load (i.e. how much material is burning, per unit volume) and on geometrical arrangements, including ventilation, while the dependence on materials is of a lower order.

This secondary effect of materials is illustrated by the difficulties encountered, in a recent study [54], when attempts were made to correlate CO concentrations measured in small scale and full scale fire tests. The same small scale equipment (typically the cone calorimeter rate of heat release test) could predict adequately a number of very important full scale fire properties, including ignitability, rate of heat release, amount of heat release and smoke obscuration. It could not, however, be used to

Table II. Toxicity of Combustion Gases

	Lethal exposure dose ppm min	Animal	ODL ppm	Peak in Fire ppm
Carbon monoxide[a]	138,000	rats	–	7,450
Carbon monoxide[b]	90,000	mice	–	7,450
Acrolein[c]	2,500 – 5,000	baboons	0.16	98
Hydrogen cyanide[d]	4,800	rats	0.58	9
Hydrogen chloride[e]	112,000 – 169,000	rats	0.77	280
Hydrogen chloride[f]	ca.150,000	baboons	0.77	280
Hydrogen chloride[g]	38,250	mice	0.77	280

ODL: Odor detection level. Ref. 38.
[a]: Ref. 39, 30 min exposure, within exposure deaths.
[b]: Ref. 40, 30 min exposure, within exposure deaths.
[c]: Ref. 41, 5 min exposure, post-exposure deaths.
[d]: Ref. 40, 30 min exposure, within exposure deaths.
[e]: Ref. 42, 30-60 min exposures, post-exposure deaths.
[f]: Ref. 41, 5-15 min exposures, no deaths.
[g]: Ref. 43, 15 min exposure, incapacitation and death.

predict full-scale CO concentrations. The latter concentrations were controlled by the geometry of the full-scale fire, the ventilation/oxygen content and the mass loading, and were little affected by the chemical composition of the burning materials.

Thus, smoke toxicity is often very closely associated simply with the mass loss rate, since the toxicity in a fire scenario will be primarily a function of the mass ofsmoke per unit volume and per unit time being emitted into the ambient atmosphere.

This discussion indicates that toxic potency measurements are a small portion of the overall toxicity picture. They may serve a useful purpose only in identifying those materials (or products) with a toxic potency outside that of the majority of other products. Such materials (or products) may well have to be looked at somewhat more closely.

One method for quick screening of the toxic hazard of materials is to calculate the ratio of their toxic potency and their mass loss rate parameter. The mass loss rate parameter is the ratio of the average mass loss rate and the time to ignition [55], and thus represents the product of mass loss rate and flame spread rate. If the mass loss rate parameter differs by more than an order of magnitude from that of ordinary materials the material in question should be investigated more thoroughly [55, 56]. The choice of a factor of 10 is typical of the difference, in classical toxicology, between toxicity categories [57].

Fire Hazard Assessment

Probably the best way of assessing fire hazard is by calculations via mathematical fire growth and transport models, such as HAZARD I [58], FAST [59], HARVARD [60] or OSU [61]. These models predict times to reach untenable situations. They are often combined with fire escape models and will, then, yield times to escape.

It is possible, however, to estimate effects on fire hazard in a particular scenario by simpler means. In some cases, an adequate choice of fire properties can be made. Then, the combination of test results into a matrix form, or into a single parameter, can indicate, even if only semi-quantitatively, the effect of varying a particular material or fire protection measure on fire hazard.

Full scale tests are particularly valuable to obtain information on fire hazard. They can be used to validate small scale tests, and to validate mathematical fire models. The most important additional dimension full scale tests add are effects, e.g. radiation from the fire itself, which are difficult to simulate in a smaller scale. Full scale tests are very expensive and time consuming. It is essential, thus, to design them in such a way as to (a) make them most relevant (b) minimize their number and (c)

help replace them by appropriate small scale tests and fire models.
Fire risk assessment is made in order to determine the overall value of decreasing fire hazard in a particular scenario. The level of fire risk that is acceptable for a situation is, normally, a societal, and not a technical, decision. Therefore, fire hazard assessments are generally more common than fire risk assessments. The NFPA Research Foundation has undertaken a project to develop a methodology for fire risk assessment. It has done this by studying four cases in detail: upholstered furniture in residential environments, wire and cable in concealed spaces in hotels and motels, floor coverings in offices and wall coverings in restaurants.

Conclusions

Fire safety in a particular scenario is improved by decreasing the corresponding level of fire risk or of fire hazard. Technical studies will, more commonly, address fire hazard assessment. Fire hazard is the result of a combination of several fire properties, including ignitability, flammability, flame spread, amount of heat released, rate of heat release, smoke obscuration and smoke toxicity.
The most important fire property associated with fire hazard is the rate of heat release: the peak rate of heat release is an indication of the maximum intensity of a fire.
Categories of toxicity are classically distinguished by differences in orders of magnitude. The toxic potency of the smoke of most common materials is very similar, and thus, the toxicity of smoke is usually governed simply by the amount of material burnt per unit time.
Toxic potency of smoke data can be used as one of the inputs in fire hazard assessment. In particular, they can be combined with average mass loss rates and times to ignition to obtain a quick estimate of toxic fire hazard.
In order to improve fire safety for each scenario, the most relevant fire properties for that scenario have to be measured, with the appropriate instruments.

References

1.   NFPA 901 Standard, <u>Uniform Coding for Fire Protection</u>; 1976 Edn.
2.   Thomas P. H., Int. Conf. <u>FIRE: control the Head – Reduce the Hazard</u>; Fire Research Station, October 24-25, 1988, London, UK, paper 1.
3.   Babrauskas, V., <u>Int.Conf. FIRE: control the Heat – Reduce the Hazard</u> Fire Research Station, October 24-25, 1988, London, UK, paper 4.

4.   Babrauskas, V., Fire Mats 8, 81 (1984).
5.   Babrauskas, V., Bench-Scale Methods for Prediction of
     Full-Scale Fire Behaviour of Furnishings and Wall
     Linings, SFPE Technology Report 84-10, Soc. Fire Prot.
     Engineers, Boston, MA, 1984.
6.   Tewarson, A., in Handbook Society Fire Prevention
     Engineers (Ed. P. di Nenno), Chapter 1/13, p. 1-179,
     NFPA, Quincy, MA, 1988.
7.   Smith   E. E., in Ignition, Heat Release and
     Noncombustibility of Materials, A.S.T.M. STP 502, (Ed.
     A.F. Robertson), Am. Soc. Test. Mats, Philadelphia,
     PA, 1972, p. 119.
8.   Parker, W. J. and Long, M. E., in Ignition, Heat
     Release and Noncombustibility of Materials, A.S.T.M.
     STP 502, (Ed. A.F. Robertson), Am. Soc. Test. Mats,
     Philadelphia, PA, 1972, p. 135.
9.   Tewarson, A. and Pion, R. F., Combust. Flame, 26, 85
     (1976).
10. Babrauskas, V., Development of the Cone Calorimeter.
     A Bench-Scale Heat Release Rate Apparatus Based on
     Oxygen Consumption, Nat. Bur. Stands, NBSIR 82-2611
     (1982).
11. Babrauskas, V., J. Fire Sci. 2, 5 (1984).
12. Babrauskas, V. and Krasny, J., Fire Safety, Science
     and Engineering, A.S.T.M. STP 882, (Ed. T.Z.
     Harmathy), ASTM, Philadelphia, PA, p. 268 (1985).
13. Ostman, B., Int. Conf. FIRE: control the Heat - Reduce
     the Hazard, Fire Research Station, October 24-25,
     1988, London, UK, paper 8.
14. Hirschler, M. M., this volume.
15. Hill, R. G., Eklund, T. I., Sarkos, C. P., Aircraft
     Interior Panel Test Criteria Derived from Full-Scale
     Fire Tests, DOT/FAA/CT-85/23, September 1985.
16. Dept.   of   Transportation,   Federal   Aviation
     Administration, Improved Flammability Standards for
     Materials Used in the Interiors of Transport Category
     Airplane Cabins; Final Rule, 14 CFR Parts 25 and 121,
     Federal Register, 53 (165) pp. 32564-81, Aug. 25,
     1988.
17. Factory Mutual Research Corporation, Specification
     Testing Standard for Less Flammable, Class Number
     3972, Norwood, MA, May 1989.
18. Hirschler, M. M. and Smith, G. F., in Proc. Fire
     Safety Progress in Regulations, Technology and New
     Products, Fire Retardant Chemicals Assoc. Fall
     Conference, Oct. 18-21, Monterey, CA, p. 133 (1987).
19. Coaker, A. W. and Hirschler, M. M., Proc. 13th. Int.
     Conf. Fire Safety (Ed. C.J. Hilado), Product Safety,
     p. 397, San Francisco, CA, (1988).
20. Hirschler, M. M. and Smith, G. F., Proc. 14th. Int.
     Conf. Fire Safety (Ed. C.J. Hilado), Product Safety,
     p. 68, San Francisco, CA, (1989).

21. Cullis, C. F. and Hirschler, M. M., <u>The Combustion of Organic Polymers</u>, Oxford University Press, Oxford, 1981.
22. Hilado, C. J., <u>Flammability Handbook of Plastics</u>, 3rd. Edn, Technomic, Lancaster, 1982.
23. Nelson, G. L. and Webb, J. L., in <u>Adv. Fire Retardant Textiles, Pt. 3, Progress in Fire Retardancy, Vol. 5</u> (Ed. V.J. Bhatnagar), Alena Enterprises of Canada, Technomic, Westport, CT, 1975, p. 271.
24. Wharton, R. K., <u>Fire Mats</u> <u>5</u> 93 (1981).
25. Stuetz, D. E., A.H. Di Edwardo, F. Zitomer and B.P. Barnes, <u>J. Polym. Sci., Polym. Chem. Edn</u> <u>18</u>, 987 (1980).
26. Babrauskas, V., <u>J. Fire Flammability</u> <u>12</u>, 51 (1981).
27. Quintiere, J.G., <u>Fire Mats</u> <u>6</u>, 145 (1982).
28. Smith, G. F. and Dickens, E. D., <u>Proc. 8th. Int. Conf. Fire Safety</u> (Ed. C.J. Hilado), Product Safety, p. 227, San Francisco, CA, (1983).
29. Levin, B. C., Fowell, A. J., Birky, M. M., Paabo, M., Stolte, A. and Malek, D., <u>Further Development of a Test Method for the Assessment of the Acute Inhalation Toxicity of Combustion Products</u>, Nat. Bur. Stands, NBSIR 82-2532 (1982).
30. Alarie, Y. and Anderson, R. C., <u>Toxicol. Appl. Pharmacol.</u> <u>51</u>, 341 (1979).
31. Alexeef , G. V. and Packham, S. C., <u>J. Fire Sci.</u> <u>2</u>, 306 (1984).
32. Kimmerle, G., <u>J. Combust Toxicol.</u> <u>1</u>, 4 (1974).
33. Cornish, H. H., Hahn, K. J. and Barth, M. L., <u>Env. Hlth Perspect.</u> <u>11</u>, 191 (1975).
34. Ferguson, G. E., <u>NFPA Quarterly</u>, <u>27</u>(2), 110 (1933).
35. Nelson, G. L., Canfield, D. V. and Larsen, J. B., <u>Proc. 11th. Int. Conf. Fire Safety</u> (Ed. C. J. Hilado), p. 93, Product Safety, San Francisco, 1986.
36. Birky, M. M., Halpin, B. M., Caplan, Y. H., Fisher, R. Sl, McAllister, J. M. and Dixon, A. M., <u>Fire Mats</u>, <u>3</u>, 211 (1979).
37. Birky, M. M., Malek, D., and Paabo, M., <u>J. Anal. Toxicol.</u> <u>7</u>, 265 (1983).
38. Consumer Product Safety Commission, <u>Fed. Reg.</u> <u>45</u> (182), 61880, Sept. 17, 1980.
39. Debanne, S., Case Western Reserve University, private communication, 1989.
40. Burgess, W. A., Treitman, R. D. and Gold, A., <u>Air Contaminants in Structural Firefighting</u>, N.F.P.C.A. Project 7X008, Harvard School Public Health, 1979.
41. Grand, A. F., Kaplan, H. L. and Lee, G. H., <u>Investigation of Combustion Atmospheres in Real Fires</u>, U.S.F.A. Project 80027, Southwest Research Institute, 1981.
42. Amoore, J. E. and Hautala, E., <u>J. Appl. Toxicol.</u> <u>3</u>, 272 (1983).
43. Babrauskas, V., Levin, B.C. and Gann, R.G., <u>Fire Journal</u>, <u>81</u>(2), 22 (1987).

44. Esposito, F.M. and Alarie, Y. R., *J. Fire Sci.* 6, 195 (1989).
45. Kaplan, H. L., Grand, A. F., Switzer, W. G., Mitchell, D. S., Rogers, W. R. and Hartzell, G. E., *J. Fire Sci.* 3, 228 (1985).
46. Hartzell, G. E., Packham, S. C., Grand, A. F. and Switzer, W. G., *J. Fire Sci.* 3, 195 (1985).
47. Hinderer, R. K. and Hirschler, M. M., *J. Vinyl Technology* 11(2) 50 (1989).
48. Galloway, F. M. and Hirschler, M. M., *Mathematical Modeling of Fires A.S.T.M. STP 983*, (Ed. J. R. Mehaffey), Amer. Soc. Test. Mater., Philadelphia, PA, p.35 (1988).
49. Bertelo, C. A., Carroll, Jr., W. F., Hirschler, M. M. and Smith, G. F., *Proc. 11th. Int. Conf. Fire Safety* (Ed. C. J. Hilado), p. 192, Product Safety, San Francisco, 1986.
50. Galloway, F. M. and Hirschler, M. M., *Fire Safety J.*, 14, 251 (1989).
51. Beitel, J. J., Bertelo, C. A., Carroll, Jr., W. F., Gardner, R. A., Grand, A. F., Hirschler, M. M., and Smith, G. F., *J. Fire Sci.* 4, 15 (1986).
52. Hartzell, G. E., Grand, A. F. and Switzer, W.E., *J. Fire Sci.* 5, 368 (1987).
53. Leviin, B. C., Paabo, M., Gurman, J.L. and Harris, S. E., *Fund. Appl. Toxicol.* 9, 236 (1987).
54. Babrauskas, V., Harris, R. H., Gann, R. G., Levin, B. C., Lee, B. T., Peacock, R. D., Paabo, M., Twilley, W., Yoklavich, M. F. and Clark, H. M., *Fire Hazard Comparison of Fire-Retarded and Non-Fire-Retarded Products*, NBS Special Publ. 749, July 1988, National Bureau of Standards, Gaithersburg, MD.
55. Hirschler, M. M., *J. Fire Sci.* 5, 289 (1987).
56. Babrauskas, V., *Int. Conf. FIRE: control the Heat - Reduce the Hazard*, Fire Research Station, October 24-25, 1988, London, UK, paper 7.
57. Casarett, L. J., in *Toxicology - The Basic Science of Poisons*, (Eds. L. J. Casarett and J. Doull), Macmillan, New York, p. 24, 1975.
58. Bukowski, R. W., Jones, W. W., Levin, B. M., Forney, C. L., Stiefel, S. W., Babrauskas, V., Braun, E. and Fowell, A. J., *Hazard I, Volume I: Fire Hazard Assessment Method*, NBSIR 87-3602, Natl Bur. Stands (July), 1987.
59. Jones, W. W., *A Model for the Transport of Fire, Smoke and Toxic Gases (FAST)*, NBSIR 84-2934, Natl Bur. Stands (September), 1984.
60. Mitler, H. E. and Emmons, H. W., *Documentation for CFC V, the Fifth Harvard Computer Fire Code*, Harvard University Report to National Bureau of Standards, NBS-GCR-81-344 (1981).
61 Smith, E. E. and Satija, S., *J. Heat Transfer*, 105, 281 (1983).

RECEIVED October 3, 1989

Chapter 29

Harmonization of Fire Testing in the European Community

Umberto Flisi

CSI (Montedison Group), Viale Lombardia, 20, 20021 Bollate (MI), Italy

A description is given of the initiatives carried out within the European Community for the harmonization of fire testing. The technical and economic reasons are explained for such initiatives, which are taken in order to remove barriers to trade from the European internal market. Of the various fire aspects, only fire reaction testing is taken into consideration here, because it appears as a major technical obstacle to the free circulation of construction materials. All possible approaches are considered for the attainment of such a harmonization and one, the so called interim solution, is fully described. The proposed interim solution, is based on the adoption of three fundamental test methods, i.e. the British "Surface Spread of Flame", the French "Epiradiateur" and the German "Brandschacht", and on the use of a rather complicated "transposition document", which should allow to derive most of the national classifications from the three test package.

The Single European Act /1/, effective from 1 July 1987, commits all the twelve countries of the European Economic Community to remove all barriers to trade by 31 December 1992.

The different fire safety requirements within the Member States of the Community certainly present an obstacle to this objective, because a manufacturer who wants to market his product all over Europe, at present, has to perform a lot of different fire tests,

0097–6156/90/0425–0479$06.00/0
© 1990 American Chemical Society

following different procedures. He may even have to repeat the same
tests in different countries, for instance in France, Spain and
Portugal or in Belgium, United Kingdom and Ireland, since in general
there is no mutual acceptance of test data between countries.

In addition, national classification systems and, in some
countries, quality control procedures are required such that the same
production process may have to be supervised by several national
inspection procedures.

All these different requirements not only add extra costs to the
goods, ultimately paid by the consumer, but they also distort
production patterns, increase stock holding costs, discourage
business cooperation and fundamentally frustrate the creation of a
common market.

The "technical barriers to trade" formed by "reaction to fire"
requirements are represented by:
- differences in national building regulations,
- differences in national classification systems,
- different national test specifications,
- different national quality systems.

Resolution of the problem, therefore, requires collaboration
among people of:
- the regulatory institutions,
- the standardization bodies,
- the testing laboratories,
- the certification bodies.

However, the differences between the national test methods are
considered to be the major barrier to trade. Whilst the harmonization
of test and classification systems is insufficient on its own to
provide for a free market, it is undoubtedly a necessary condition to
it. Without a common method of evaluating the fire behaviour, there
is no basis for a common regulatory specification.

A common regulatory system, prescribing identical fire
classifications based on unified test specifications, for materials
in the same usage, cannot be obtained by the end of 1992. It is a
much longer term objective.

However, in the interim it is possible to remove some major
components of the obstacle, i.e. the classification system, the tests
and the quality control requirements.

This paper outlines some proposals and actions, which may give a
partial solution to the problem prior to 1992.

Possible Approaches to Fire Testing Harmonization.

During the past years, several approaches have been suggested to the
Commission of the European Communities, in order to reduce the
problem created by the large number of reaction to fire tests.

ISO Tests. The evident solution in a situation like this would be to adopt the appropriate international standards, which the ISO Technical Committee TC 92 is charged with developing. Since 1961, this committee has been developing reaction to fire tests for building materials, with the aim of determining such parameters as:
- non combustibility
- ignitability
- surface spread of flame
- rate of heat release
- smoke density.

So far, only the non-combustibility (ISO 1182) and the ignitability (ISO 5657) tests are available as international standards. For the other parameters, the test methods are still at a development stage. A solution to the 1992 objective is not yet possible, even if the necessary tests can be finalised prior to that date, because many practical difficulties still remain to be solved.

Firstly, the industry has little or no experience with most of these tests, especially those which have not reached their finalised format yet. Whilst the industry ideal may be to produce materials that are "fire safe", in reality it has to satisfy certain targets in the form of "classifications", based on performance in fire tests. The industry, therefore, needs time to familiarize itself with the new tests and to reformulate their products, if necessary, to meet the new requirements.

Secondly, the question arises how to use the new toolkit of tests for building regulatory purposes. The various national regulators are equally as lacking as the industry in an understanding of material behaviour in these new tests. Since there is a very intimate inter-relationship between the regulations and the test methodology at each national level, adoption of the new toolkit may not only require amendment to regulations, but may generate a whole new range of performance classifications, if these tests are adopted in advance of an agreed European Classification system, with guidance to national regulators on the use of the system.

It is clear that the ultimate solution rests with the ISO tests used as a part of a European Classification System adopted as a part of a common or model building code for the European Community.

National Tests. It is clearly possible to remove the technical barrier to trade represented by different national fire test procedures, by providing a facility for a manufacturer to conduct the relevant tests once and within his home country, with a guarantee of acceptability of the results by all Member States. This requires laboratories to equip themselves with all the necessary equipment and for an extensive interlaboratory collaboration and calibration procedure to be introduced, which would ensure mutual acceptance of test results.

This represents an easily achievable part solution to the problem, since the current level of technical co-operation between laboratories is very high within Europe. However, the burden on the manufacturer remains high, if he wishes to market in all Member States, since he will still be required to conduct all the tests used throughout the Community.

Translation of National Test Data. A less cumbersome solution could be found if there was the possibility of satisfying a regulatory requirement of one Member State by test results obtained from a fire test procedure of another Member State.

The European Commission wanted to verify if there was such a possibility and a few years ago contracted Prof. Blachere and others to carry out such an investigation. The results of this investigation /2/, indicated some interesting prospects based on a theoretical evaluation of the individual tests.

In accordance with this proposal, the manufacturer would need only to perform those tests necessary at his national level and, by reference to a "translation document" plus some additional complementary tests, he would be able to obtain the national classification of another Member State.

The main advantage of the proposal is that it needs no change of the national regulations and the manufacturer continues to evaluate his material performance by a test method he knows and understands.

However this way appears impracticable for several reasons.
- A lot of practical work needs to be done to demonstrate that the theoretical analysis works in practice. This involves detailed analysis of all test methods versus all legal requirements.
- The system is not necessarily equitable, since for some Member States it is possible to interpret its results and satisfy a large number of other Member States requirements,whereas for other Member States no interpretation is possible.
- Commercial competition between industry may lead them to seek out the most favourable rooting to suit their material.
- The system requires a level of confidence by the various regulators in the individual tests used in all the other countries of the Community, which in the short time is unlikely to be achieved. National authorities cannot be expected to accept materials on the basis of tests they do not know, nor understand, and which are conducted in foreingn laboratories. Currently they do not accept tests to their own national Standards from other countries.

The Interim Solution /3/. One of the major difficulties in the prospect of satisfying any one regulatory classification requirement from any single or collection of national tests is the large number of different tests used in the various Member States. The question addressed by Deakin, Klingelhoefer and Vandevelde in 1986 was: "What is the minimum number of national tests which will satisfy the maximum number of regulatory requirements ?".

It was the consideration of this question, together with the prospects outlined by Prof. Blachere in his report which have led to the proposal "interim solution".

What is proposed is that material fire properties are determined by a carefully chosen set of national tests, appropriate to the material usage, which will cover the main field of interests of most regulators directly. Those which cannot be satisfied directly have to be satisfied by a translation or "transposition document" derived from the Blachere Report.

The set of national tests must fulfill the following requirements:
- minimum number
- greatest experience
- widest range of information
- widest acceptance.

The Proposed Interim Test Package.

The test methods chosen for the interim solution are listed in Table 1 against the proposed use of the materials. It can be seen that the test methods are grouped in four packages according to the type and level of performance.

Highest Level of Performance. For the highest level of performance, i.e. for non combustible materials, two test methods are proposed irrespective of the end use:
- the non-combustibility test ISO 1182, shown in Figure 1, which satisfies the national requirements of Belgium, Denmark, Germany (not completely), Greece, Ireland, Italy, The Netherlands and United Kingdom;
- the calorific potential test ISO 1716, which satisfies the national requirements of France, Germany, Portugal and Spain.

Materials passing these tests will not be required to be subjected to any other reaction to fire test.

Other Tests. For the other fire reaction tests, distinction is made according to the material application, and three groupe of tests are considered.

Wall and Ceiling Linings. Three national test procedures dominate the European market-place for wall and ceiling linings:
- French "Epiradiateur"   NF P92-501 (see Figure 2)
- British "Surface Spread of Flame" BS 476: Part. 7 (Figure 3)
- German " Brandschacht"   DIN 4102, Teil 1 (Figure 4)

Table 1. Test Methods Proposed for the Interim Solution
in the European Community

HIGHEST LEVEL OF PERFORMANCE: ISO 1182 "Non-combustibility"
                                                    +
                                          ISO 1716 "Calorific Potential"

OTHER TESTS

WALLS + CEILINGS	FLOORINGS	ROOFS
NFP 92 - 501 + BS 476 Part 7 + DIN 4102 - Teil 1	ISO DTR 9239 (radiant flooring panel) + BS 476 - Part 7	Radiant Panel Test (French arreté 10.09.70 or ISO test) + DIN 4102 - Teil 7 = NEN 3882

SIMPLE IGNITION TEST

SMOKE AND TOXICITY

Smoke: ISO DTR 5924 if necessary
Toxicity: no test method

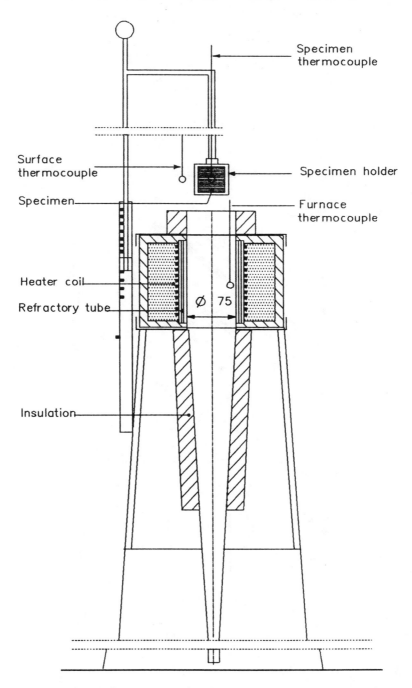

Figure 1. General arrangement of the ISO 1182 test apparatus.

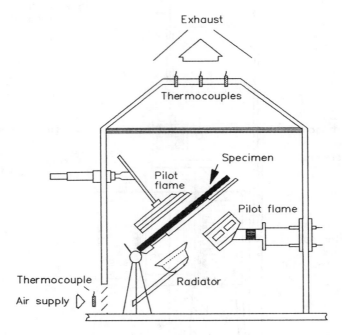

Figure 2. Scheme of the epiradiateur.

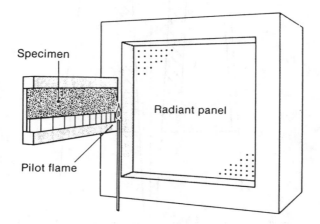

Figure 3. Scheme of the British Radiant Panel.

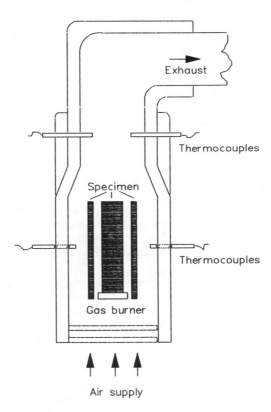

Figure 4. Scheme of the Brandschacht.

Table 2 shows that many of the Member States requirements may be satisfied in whole or in part by the three methods, and it is possible to extend the coverage to all countries but Denmark, via a "transposition document", which translates the national classification performance requirements to results obtained from one of the three tests.

Floor Coverings. Two tests have been proposed, i.e.
- the ISO "Flooring Test", ISO/DTR 9239 derived from NBSIR 75-950, shown in Figure 5;
- the British Surface Spread of Flame, BS 476: Part 7.
Probably the ISO method alone will be accepted by all Member States.

Roofs. Two tests are proposed:
- the German "Basket Test", DIN 4102, Teil 7;
- a radiant panel test.
The choice of the radiant panel only depends upon the rate of development of the international test procedure. If this is not available in sufficient time, the French method will be taken.

Simple Ignition Test.   The proposed tests for combustible materials may need to be supplemented by a simple ignition test, involving contact of a small flame in the absence of any impressed irradiance. Some of these tests are already used in the Member States, in particular Germany and Italy (see Figure 6). Such a test could be used as an additional test or a screening test.

Smoke.   A few Member States have a regulatory requirement for smoke production of materials. These requirements are based on national procedures which are considered inferior to the current ISO development detailed in ISO/DTR 5924, and it is recommended that if a real requirement exists, and has to be perpetuated, then the national requirements should be translated into performance in the ISO/DTR.

Toxicity.   There is only one Member State (Germany) having a test which is used to assess toxic hazards of combustion gases. The test is used mainly to evaluate non-combustible materials and is based on bio-assay techniques. The philosophies of other countries consider non-combustible materials as presenting no, or negligible toxic hazard.

A satisfactory test for toxicity of combustion gases is not yet available at national or international level. The ISO/TC92/SC3 working groups are actively working on all aspects of toxic hazards in fire and it is proposed that the outcome of this work is awaited and not pre-empted by an interim solution.

Table 2. Proposed Fire Reaction Test Methods for the European Community

WALLS AND CEILINGS

Proposed E.E.C. test methods 1. NF P 92 - 501 + 2. BS 476 - Part 7 + 3. DIN 4102 - Teil 1	National requirements based on	
	Main national test method(s)	Additional national test method(s)
Belgium	NF P 92 - 501 or BS 476 - Part 7	
Denmark	DS 1058-1 DS 1058-3 now DS/INSTA 410 DS/INSTA 412	DS 1058-3
Germany	DIN 4102 - Teil 1 Brandschacht	DIN 4102 - Teil 1 Kleinbrenner
France Spain Portugal	NF P 92 - 501 Epiradiateur or NF P -503 Bruleur electrique	1) Tests for fusible materials 2) Flame propagation test
Italy	CSE - RF 1/75/A or CSE - RF 2/75/A CSE - RF 3/77	
The Netherlands	NEN 3883 = Flame propagation + Flash over + Smoke	
U. K. Ireland Greece	BS 476 - Part 7 + BS 476 - Part 6	BS 2782: 4 additional tests for plastic materials

⊘⊘⊘⊘⊘ : The proposed methods cover the national needs
⊠⊠⊠⊠⊠ : Transposition is possible to national needs
▒▒▒▒▒ : Covered if simple ignition test is included

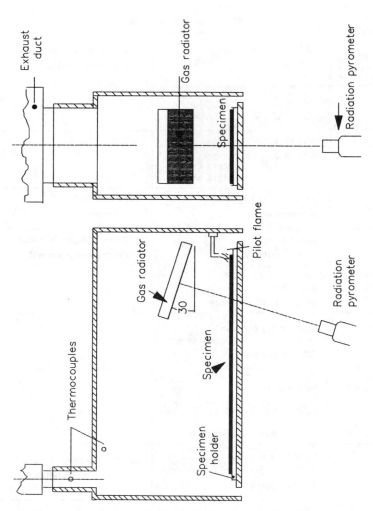

Figure 5. Scheme of the ISO/DTR 9239 test method. Left: longitudinal section; right: cross section.

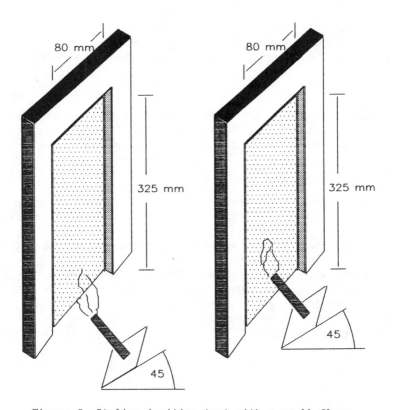

Figure 6. Italian ignition test with a small flame.
Left: CSE-RF1 flame applied to specimen edge.
Right:CSE-RF2 flame applied to specimen surface.

The Classification System.

It is envisaged that the interim solution will operate in parallel
with the various national test and classification procedures until
such a time that the final unified European system is available.
     A manufacturer will, therefore, have the option of conducting
the existing national tests required by a Member State, to which he
wants to export, most probably still in a laboratory of that Member
State, as is the case now. Alternatively, he can opt for conducting
the "Interim Test Package" in any official laboratory, within the
European Community, and, provided that he satisfies the necessary
requirements, he cannot be asked to undertake any further testing.

CEN/TC127

The European Committee for Standardization (CEN) has recently formed
a new technical committee, TC 127, with the following scope /4/:
"To develop standards utilizing relevant existing work, where
available e.g. in ISO, CEC and EFTA, for assessing the fire
behaviour of building products, components and elements of
construction.
To develop standards for classification of products, components and
elements of construction, appropriate to the fire risks related to
their application.
To develop standards for assessing fire hazard and for providing
fire safety in buildings".
     The Commission of the European Communities has asked /5/ CEN TC
127 to produce the necessary standards concerning fire safety in
support of the "Construction Products Directive" CEN/TC127 met first
in London on 18-20 May 1988 and took some resolutions which may
favour a rapid development of the interim solution mentioned in
chapter 3.
     The most important, and debated resolution was the sixth one,
which concern the three tests package for wall and ceiling linings.
In this context the committee accepted the CEC mandate to prepare a
standard based on the three national norms:
- NF P 92 501
- BS 476: Part 7
- DIN 4102 Parts 15 and 16.
     Several working groups were created with the task of:
- considering the question of criteria and harmonizing the field of
  application of the two standards (ISO 1182 and ISO 1716) for
  non-combustibility (Ad Hoc 1);
- considering the various test methods for ignitability by direct
  flame impingement under zero impressed irradiance (Ad Hoc 2);
- preparing the harmonized test package for the interim solution and

the transposition document from the interim test package to national classifications (Working Group 1);
- preparing a method of test standard for floor coverings, using ISO/DIS 9239 (Ad Hoc 3);
- preparing a method of test standard for the external exposure of roofs to fire (Ad Hoc 4);
- proposing a European reaction-to-fire classification system, using the tests included in ISO/TR 3814 (Working Group 2).

In this way all the work has been set up for a rapid development of the interim solution and for building a solid basis of the future final norms.

CEN/TC 127 also deals with fire resistance, but in this case the harmonization problems are less difficult to solve, because all countries at least refer to the same test method, i.e. ISO 834.

EGOLF

Some years ago a group of official testing laboratories of different countries of the European Community met together to form an international association called EGOLF (European Group of Official Laboratories for Fire-testing) with the aim to promote interlaboratory collaboration and acceptance of test data.

EGOLF gives a strong support and assistance to the European Commission for the implementation of its harmonization programme by:
- providing a forum for discussion of problems related to standard fire tests;
- promoting research and development within the field of fire testing;
- defining the quality level for equipment and expertise to be maintained by the fire test laboratories;
- organising seminars on fire subjects of topical interest.

EGOLF includes about 20 laboratories and is considered as the precursor of the future European system of fire testing and certification.

SPRINT RA 25 Programme.

SPRINT RA 25 is a transnational collaborative project among seven laboratories, each from a different Member State of the EEC. These laboratories are:

- University of Gent	Gent	Belgium
- CSTB	Champs sur Marne (Paris)	France
- Materialprufungsamt NW	Erwitte (Dortmund)	Germany
- CSI	Bollate (Milano)	Italy
- TNO	Delft	The Netherlands
- LGAI	Barcelona	Spain

- Warrington Fire Res. Centre Warrington   United Kingdom
   The major aims of the project have been to:
- ensure the repeatability and reproducibility of the three test
  methods (Brandschacht, Epiradiateur and Spread of Flame) of the
  interim solution;
- disseminate the information required to test and evaluate materials
  according to these tests;
- provide the basis for the mutual acceptance of test data.

   Each of the seven laboratories has received a thorough training
by a pilot laboratory in the use of the three apparatuses and in the
assessment of materials and their classifications according to the
three national standards, i.e. in France NF P 92.501, in Germany DIN
4102 Parts 15 and 16, and in the U.K. BS 476 Part 7.

   A range of 11 different materials, representative of building
products commonly used in Europe, has been distributed and will be
tested in each of the laboratories, on each test method, and the
results will be assessed for repeatability and reproducibility.

   The RA 25 group have begun the task given by CEN/TC127
(resolution 6 /6/) of preparing the text of a reference document on a
three part method of test standard for wall and ceiling linings.

   The RA 25 contract has been extended by CEC and renumbered RA
100 in order to continue this exercise with the ISO 1182 apparatus.
However, the major part of the work of the RA 100 group will be the
preparation of unified test reports for each of the test methods,
which will ensure that all the necessary details are recorded to
enable a product to be classified in any of the EEC Member States.
Two new laboratories, i.e. the French Laboratoire National d'Essai
and the Danish Dantest, joined the previous seven.

The Final Solution

CEN/TC 127 has created, as it has been said in chapter 5, a Working
Group (WG 2) charged with the responsibility of deriving a European
classification system based on the ISO tests. WG 2 is composed of 21
members from 14 countries and is convened by Sweden. It has very
recently formulated /7/ its philosophy, based on the following
statements:
- The European classification system of reaction to fire must reflect
  the behaviour of tested products in real fires. Test methods should
  be designed so that the results can be used as important part of
  risk assessment.
- The classification of products should be achieved by small scale
  tests which have been validated by large scale tests /8,9/. In
  special cases, when relevant information cannot be obtained in
  small scale, a standardized large scale test, i.e. the Room/Corner

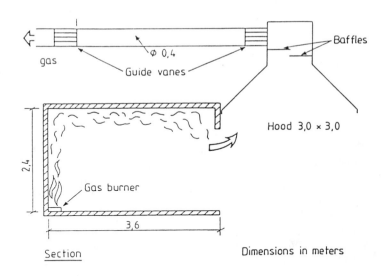

Section   Dimensions in meters

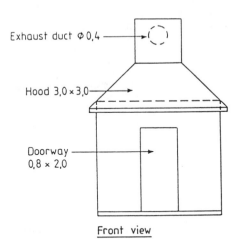

Front view

Figure 7. Schematic view of the Room/Corner Test Apparatus.

Test (see Figure 7), will be necessary to achieve directly the in-
formation needed for a classification.
. The number of small scale test methods, used for classification
purposes, should be limited and based on ISO tests, presumably the
Cone Calorimeter /10/ (see Fig. 8) and possibly the ISO Surface
Spread of Flame test /11/.

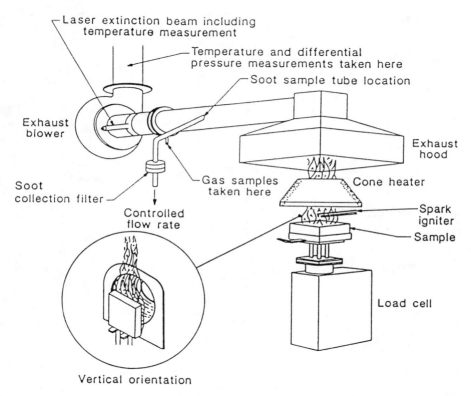

Figure 8. The Cone Calorimeter according to ISO/DP 5660.

The group also agreed to work with the Swedish proposal /12/ employing the Cone Calorimeter and the Room/Corner Test and explore whether it proves to be adequate in terms of technical relevance, costs, etc.

In addition the room-corridor scenario will also be investigated in full scale and attempts made to seek correlations with the Cone Calorimeter and the ISO spread of flame test.

An informal survey revealed that in the near future as many as 22 Cone Calorimeters will be installed in 10 European countries and 15 Romm/Corner Test facilities will be available in 8 countries.

Literature Cited.

/1/ Commission of the European Communities - European Single Act. Bulletin of the European Communities. Supplement 2/86 (1986).

/2/ BLACHERE, G.; TEPHANY, H.; TROTTEIN, Y.; MARTIN, J. - A Propos des Essais de Reaction au Feu dans la CEE. CEC Document III/3197/88 (June 1988).

/3/ BLACHERE, G.; DEAKIN, A.G.; KLINGELHOEFER, H.G.; VANDEVELDE, P.; TEPHANY, H.; - Proposal for the Harmonization to Fire Tests in Europe. CEN/TC 127 Document N. 79.

/4/ CEN/TC 127 "Fire Safety in Buildings" - Resolution N. 3 Document CEN/TC 127 N. 42 (1988).

/5/ CEN/TC 127 Document N. 33.

/6/ CEN/TC 127 Document N. 42.

/7/ CEN/TC 127 Document N. 103.

/8/ SUNDSTROM, B.; - The New ISO Full Scale Test Procedure for surface Linings. ISO/TC92/SC1/WG7 Document N. 50.

/9/ FLISI, U.; - Polymer Degradation and Stability 1989, 23;385.

/10/ BABRAUSKAS, V.; - Development of the Cone Calorimeter. NBSIR 82-2611, Nat. Bur. Stand. (1982).

/11/ ISO/DP 5658 - Document ISO/TC92/SC1/WG3 N. 95.

/12/ CEN/TC 127 - Documents N. 16 and 32.

RECEIVED January 26, 1990

Chapter 30

British Regulations for Upholstered Furniture

Combustion-Modified Polyurethane Foam

K. T. Paul

Rapra Technology Limited, Shawbury, Shrewsbury, Shropshire, United Kingdom

In 1988, the UK introduced Safety Regulations which ultimately will require that domestic upholstered furniture and mattresses will meet and resist a cigarette test and that fabrics and fillings will resist specified flaming ignition sources. Polyurethane foams must also be of the specified as combustion modified types which in the UK include foams containing melamine and exfoliated graphite.

It is nearly 20 years ago that the unacceptable fire behaviour of "modern" upholstered furniture became highlighted in the UK by Fire Brigade reports of domestic fires. This poor performance was blamed on the use of flexible polyurethane (PU) foam upholstery and demands were made to ban PU foam or at least to insist on the use of flame retarded PU foam.

The Home Office summary of UK fire statistics for 1987(1) shows that although only about 18% of fires occur in occupied dwellings, these result in typically three quantities of fatal and non-fatal injuries. The greatest single cause of fires in dwellings is the unintentional misuse of cooking equipment (43%) but these result in relatively few fatalities (10%) but a larger proportion of non-fatal injuries (37%). In contrast, fires initiated by smokers materials and matches together cause 18% of fires in dwellings but result in 51% of fatalities and 28% of non-fatal injuries. Fire brigade reports indicate that a major proportion of fires initiated by smokers materials and matches involves upholstered furniture, bedding or mattresses either at the ignition stage or are directly involved in the rapid growth stages of the fires Tables I and II.

The response to this situation in the early 1970's was to set-up three major research programmes to investigate the fire performance of furniture and furnishings. These programmes, (2), (3), (4), not surprisingly, produced similar conclusions that the adverse fire behaviour of upholstered furniture and made-up beds

0097–6156/90/0425–0498$06.50/0

Table I. Location of Fires in UK

Location of Fire	1977	1982	1987
Occupied Buildings	93.4	96.2	104.1
Dwellings	51	56.4	63.2
Industrial	14	11.3	10.8
Public Area*	8.9	9.2	10.9
Hospitals	2.1	1.8	1.7
Schools	1.8	4.3	4.3
Others	15.5	15.3	16.2
Total No. of Fires	326.8	357.9	354.0

* Hotels, hostels, boarding houses, clubs, public houses, restaurants, places of public entertainment and shops.

Table II. Causes of Fires and Casualties in Dwellings

Cause	No. of Fires 1000's		No. of Fatalities		No. of Non-fatal Injuries	
	1978	1988	1978	1988	1978	1988
Cooking	19.4	23.6	38	63	1545	3123
Matches 1	4.1	4.0	49	77	526	685
Smokers Materials	4.5	6.2	198	255	859	1788
Heating	5.7	4.0	192	122	895	976
Other and Unknown	17.0	16.6	215	138	1292	1815
Total	50.7	55.5	692	653	5117	8387

arose because of easy ignition by small everyday sources
(cigarettes and matches) and rapid burning producing high rates
of heat, smoke and toxic fire gas generation. Major contributory
factors were the use of smoulderable and easily ignitable
fabrics, melting fabrics and flexible PU foams. Recommendations
were to use materials which did not smoulder, to use fabrics
which were not easily ignited by small flames, and to restrict
the amount of flammable materials used. The use of interliners
and barrier fabrics and foams was demonstrated and also the
adverse effects of using certain types of flame retarded PU foams
(previously demanded) with flammable fabrics.

The work (3) confirmed the correctness of the decision not
to enforce the use of flame retarded PU foam and showed that
although high resilience PU foams and flame retarded PU foams
themselves were more difficult to ignite, this advantage was
often lost when they were used in furniture as fabric covered
composites. Early flame retarded PU foams were more likely to
produce greater smoke and in some circumstances could burn more
rapidly than composites containing standard PU foam.

Various UK government authorities and especially The Crown
Suppliers were already "fire conscious" and further extended the
use of fabrics of reduced flammability, barrier fabrics and
barrier foams, with high resilience PU foams in their purchase
specifications for upholstered furniture and bed assemblies for
use in the Crown Estate, public buildings, hospitals etc.
Typically, products were required to meet cigarette and No. 5
wooden crib sources see Table I when tested to Crown Suppliers
Tests (5), (6).

As British Standard tests (7, 8, 9) for upholstered
furniture ignitability became available, these were progressively
used both by The Crown Estate and by government agency
specifications. Home Office Codes of Practice (10, 11) were
gradually extended to additional public areas and advised similar
ignition performance to that originally used for public areas.
i.e. a minimum ignition resistance of cigarette and crib No.5 of
BS 5852 Parts 1 and 2. Additional codes have been added as they
are prepared. The nett result is the upholstered furniture and
bed assemblies in the UK public and government buildings are
controlled by a complex combination of Acts, Codes and
Specifications. In essence these require that upholstered
furniture resists at least the cigarette and small gas flame
(No.1) in low risk areas, the cigarette and No.5 wood crib in
medium risk areas and the cigarette and No.7 wood crib in high
risk areas. See Table III. These requirements have recently
been published as British Standards (12) (13). In the UK, the
public sector has, since the early 1970's progressively
introduced and formalised ignition resistance performance
specifications for upholstered furniture, mattresses and bed
assemblies either on the product itself or on the upholstery
composite. Early tests in the mid 1970's for rate of fire growth
were based on room-corridor test rigs in which the product was
burned to completion and fire temperatures, smoke density and
volume and carbon monoxide concentration were monitored

Table III.   Characteristics of Ignition Sources

Ignition sources of BS 5852 Parts 1 and 2	Theoretical heat of combustion approx. (KJ)	Flame height (mm)	Flame temp. (°C)	Local heat flux. (kW/m²)	Rate of burning	Duration of flaming
Cigarette	16	-	-	-	-	approx 20 mins.
Gas flame 1	2	35-40	880	30-40	45 ml/min.	20 secs.
Gas flame 2	12	140-150	890	20-40	160 ml/min.	40 secs.
Wood crib 5	285	510	725	20	10g/min.	approx 3 mins.
Woodcrib 7	2100	660	725	49	36g/min.	approx 7 mins.

SOURCE: Adapted from ref. 34.

continuously. Product specifications were not issued but the tests were used for comparative development and type approval tests for critical installations (5).

The UK public area situation contrasted markedly with the UK domestic situation. In spite of the recommendations of the early research programmes, full scale tests with domestic furniture in the late 1970's showed, regrettably, that not only had domestic upholstered furniture design and construction not followed the recommendations of early research, but that the fire performance of such furniture had arguably deteriorated.

In 1978, the Home Office report (14) concerning the fire behaviour of new (i.e. post 1945) materials specifically criticised the fire performance of upholstered domestic furniture and recommended manufacturers to improve ignition resistance, to reduce flammability, and to reduce the rates of smoke and toxic gas generation.

In 1980 the Consumer Protection Act (15) required manufacturers to take reasonable steps to ensure that fabric - soft infill composites of domestic furniture resisted the cigarette test of BS5852: Part 1. These regulations did not apply to rouchings, trims to seat platforms etc. but only to the primary upholstery composites. The Consumer Protection Amendment of 1983 required specified children's furniture containing cellular materials had to resist Source 5 of BS 5852 Part 2. (16)

The UK National Bedding Federation introduced a voluntary code which required its members only to manufacture mattresses which resisted a smouldering cigarette on its own and when covered by an insulating, non-combustible fibre pad (17). This test is now defined within BS 6807(8) while BS 7175(18) refers to bedding and pillows.

The 1988 Consumer Safety Regulations relate to both upholstered furniture and mattresses(19).

It is not proposed to review the US situation but in some ways the UK situation is similar to that in the US with controls for domestic and public area furniture but whereas the requirements apply throughout the whole of the UK, in the US they are rather fragmented. The UFAC scheme and its derivatives such as California TB116 require upholstery composites to resist smouldering cigarettes while other cigarette tests may be applied to upholstered furniture and mattresses. California originated small flame ignition requirements for PU foams for furniture (TB117) and also more severe and comprehensive tests for high risk and public occupancies (TB133). The latter involve ignition and simplified smoke and carbon monoxide assessments. Boston also originated fire tests for upholstery for some public buildings including prisons, hotels etc. A proposal which is under consideration by various legislatives is the International Association of Fire Fighters proposal which is based on the Californian TB133 test. The Underwriters Laboratories test UL1056 is based on the NBS furniture calorimeter and as such is a much more comprehensive test procedure. Other tests are applied in special environments e.g.

prison mattresses. US flame ignition and flammability tests were received in detail by Babraukas in reference (20)(21).

DEVELOPMENT OF FLAME RETARDED POLYURETHANE FOAMS(22)(23)(24)

Early flexible polyurethane foams were of the so called standard type and were readily ignited by a small flame. Once ignited they produced considerable amounts of heat and smoke. Flame retarded flexible PU foams became available in 1954-55, i.e. within a few years of flexible PU foams becoming available in commercial quantities(22). These FR PU foams contained trichloroethyl phosphate or brominated phosphate esters and resisted ignition from small flame sources. Unfortunately they may burn when subjected to a larger ignition source or when covered by a flammable fabric and may then produce as much heat and more smoke than the standard grade of PU foam(3). This was identified by UK room tests in the early 1970's and has been confirmed more recently by furniture calorimeter tests at the NBS(21).

"Cold cure" or high resilience PU foams tended to liquify before igniting were developed and gave good results in standard tests such as BS 4735(25) (similar to the discontinued ASTM D1692) but could still burn when used with flammable fabrics in furniture (Table IV). However they have been used most successfully in the UK in combination with flame retarded cotton interliners and fabrics of low flammability e.g. wool, nylon, FR cotton etc. and formed the basis of public area furniture used in the UK since the early 1970's (Table V).

A different approach was to impregnate standard PU foam with resin bonded hydrated alumina. This material was originally developed as a fire barrier material and when used with high resilience PU foams and fabrics of low flammability gave upholstered seating and mattresses which resisted the highest ignition source of BS 5852 Part 2, the No.7 126 gram wood crib. The impregnated PU foam was subsequently developed for used as a full depth mattress for prison and psychiatric hospital use where vandalism and arson can be a problem but this foam is unsuitable for use in upholstered chains etc. The use of barrier fabrics and foams were unacceptable to the UK domestic furniture industry which required single foam soft upholstery materials.

The flammability properties of the hydrophilic PU foam developed in USA in the late 1970's were outstanding even under severe fire conditions but it's physical properties and high densities restricted its use for other than highly specialised high risk areas.

The next major improvement was the development of combustion modified PU foam. The original CMHR polyurethane foam was developed in the USA (26) and contained hydrated alumina and halogenated flame retardants but was made in a single operation. It was used in institutions, public buildings, hotels etc. but its high density and less than optimum physical properties

Table IV. Ignitibility of PU foams and fabric covered PU foam furniture composites

	Standard PU Foam	High Resilience PU Foam	FR Additive PU Foam	Combustion Modified PU Foam (Al/OH)3	Combustion Modified PU Foam (Melamine)	Combustion Modified PU Foam (Exfoliated Graphite)
BS 4735 tests on PU foams						
Burn Distance mm	150	24-41	93-125	31-38	10-30	-
Burn Date mm/s	1.6-4.2	1.4-2.2	0.9-1.8	0.5-0.6	0.5-2.0	-
BS 5111 tests on PU foams						
Smoke Obscuration %10-20	%10-20	21-42	49-53	94-96	34-43	90-96
BS 5852 ignition tests on fabric covered PU foam						
Acrylic	PC F1	PC F1	PC F1	PC F1/T	PC F1	PC F1
Polypropylene	PC F1	PC F1	PC F1	PC F1	PC F1	- -
Cotton	FC F1	PC F1	PC F1	PC F1/T	PC F1/T	PC F1/T
Polyester	PC F1	PC P1	PC P1	PC P1	PC P1	PC P1
Wool	PC P1	PC P1	PC P1	PC P1	PC P1	PC P1

Table V.   Ignitibility of upholstery with Barrier
Fabrics and Foams

Fabric	Standard PU Foam	High Resilience PU Foam	High Resilience PU Foam & FR Cotton Barrier Fabrics	High Resilience PU Foam & PU Barrier Foam
Nylon	F1	P1/F3	P5	P6
Polyester	F1	P1	P5	P6
Modacrylic	-	P3/4	P5	P7
FR PVC/Cotton	P6	P2/4	P5/	P7
Wool	P2	P5	P5/6	P7

restricted its use for non-high risk applications. This type of
material was available for a limited period in the UK before
being superceeded.
      This was followed both in the UK and USA by a melamine
containing high resilience polyurethane foam. In the UK, the
term combustion modified PU foam is used for PU foams modified by
the addition of melamine or exfoliated graphite and as such
differs from the US use of the term. PU foams containing
hydrated alumina in the UK are prepared by a post impregnation
process and are used as barrier foams and for institution
mattresses) which has improved physical properties when compared
to the US Hypol and CMHR PU foams. In the UK polyurethane foams
containing melamine or exfoliated graphite are both referred to
as "combustion modified polyurethane foam". These two foams are
currently available in the UK and it was the production of these
materials which encouraged the next step towards improving the
fire behaviour of UK domestic furniture since they permitted
significant fire behaviour improvements to be made without the
use of barrier fabrics and foams while using existing methods of
furniture production.
      This is a very important factor because the primary function
of upholstered furniture, mattresses and bed assemblies is to
provide comfortable seating and support which is hardwearing and
durable as well as aesthetically acceptable. The development of
suitable fabrics, foams and other materials is therefore of great
importance and is a factor that can be overlooked in the pursuit
of improved fire performance.

PRINCIPLE OF 1988 UK REGULATIONS FOR DOMESTIC UPHOLSTERED FURNITURE

Cigarette resistance to BS 5852 Part 1 is required by the
primary upholstery composites of the actual furniture and
mattress. This is consistent with all previous UK
regulations and specifications which required that fire
tests should be carried out on the actual upholstery
composite.
      This principle is adhered to for flame ignition tests for
the public area which still requires that the tests are carried
out on the upholstery composite although some specifiers are now
requiring that all PU foam shall additionally comply with the
1988 Consumer Safety Regulations.
      The principle of the 1988 Consumer Safety Regulations is
that all filling materials and covering fabrics shall be ignition
resistant against specified flames. An exception is made for
specified types of fabric which may be used with a flame retarded
interliner.
      The 1988 Consumer Safety Regulations depart from the
principle of testing composites because they essentially test
individual materials in a standard manner although the composite
BS 5852 test is used. Thus flame retarded fabrics and
interliners are tested with a specified standard PU foam and
polyurethane foams and other fillings are tested with a specified
flame retarded polyester fabric. It is understood that this

situation arises because of industrial requests and to simplify the testing of fabrics and foams (note the cigarette test is still carried on the furniture composite).

UK DOMESTIC FURNITURE AND THE CONSUMER SAFETY REGULATIONS 1988(19)

The Consumer Safety Regulations 1988 regulations were recently published for domestic furniture and mattresses. These are a complex series of relatively simple ignition tests designed to restrict the use of easily ignitable fabrics and infill materials. The 1988 regulations replace the earlier regulations of 1980 and 1983.

The Regulations apply to all types of upholstered seating, chairs, sofa beds, sofas, settees, padded stools, furniture with padded areas (head boards), children's furniture, cots playpens, prams etc. garden furniture, caravan seating, mattresses, bed bases, scatter cushions and pillows. Furniture for export, second-hand caravans and furniture made before 1950 are excluded. Various parts of the regulations become effective on different dates between 1st November, 1988 and 1st March, 1993 and the regulations require all furniture to be labelled with the appropriate fixed and swing labels. This paper will only deal with the fire tests and not the time scale of introduction or the labelling system (Tables VI, VII, and VIII).

PERMANENT COVERING FABRICS. These are essentially required to resist small flame ignition and also to protect covering fabrics are therefore tested to ignition source No.1, BS 5852 Part 1 (simulated match) over a standard PU foam. Should the fabric not ignite but merely melt to expose the standard PU foam, the latter will ignite and the fabric will fail the test. Exceptions to this are covering fabrics which comprise at least 75% by weight of cotton, flax, viscose, modal, silk or wool and not coated with a PU system and which must be used with an interliner. FR treated fabrics must be pretreated by a specified water soak test (30 minutes at 40°C) to ensure that the treatments resist typical household spillages etc. Loose covers are tested in the same manner as the permanent covers but stretch covers shall be tested with PU foam which meets the requirements of PU foam blocks and to ignite source 5, Table VI.

INTERLINERS. These are intended to protect the filling against specified flammable cover fabrics. Interliners are tested with ignition source 5 over a standard PU foam to BS5852 Part 2. FR treated interliners shall be subjected to the specified water soak before testing Table VI.

ALL FILLING MATERIALS EXCEPT PU SHEETS OR BLOCKS. These are tested against ignition source 2 which is a butane flame, approximately 120 mm high and applied to the FR polyester fabric covered filling for 40 seconds(Table VII) to BS 5852 part 2. This applies to all loose foam and non-foam filling when tested singly.

Table VI. Summary of Test Requirement for Fabrics

Fabric	Standard Test	Ignition Source to BS5852	Test Modification Criteria
Cover fabric	BS 5852/1	1	A, B, D
Loose covers	BS 5852/1	1	A, B, D
Stretch covers	BS 5852/2	1	C
Non-visible fabrics	BS 5852/2	1	C
Interliners	BS 5852/2	5	B, D, E

A   Except for fabrics containing more than 75% by weight of cotton, flax, viscose, modal, silk or wool when used with an interliner.

B   With non-FR PU foam. BS 3379 Type B hardness Grade 130, 20-22 kg/^{3}m

C   With PU foam, 24-26 kg/m^3 conforming to BS 5852/2 ignition Source 5, 60g mass loss see Table VII.

D   With water soak for FR treated fabrics.

E   Covered with the flame retarded polyester fabric.

F   BS 5852 failure criteria of smouldering and of flame penetration through the specimen depth are waived. (clauses 4.1 e, 4.1 f and 4.9 f of BS 5852 Part 2).

G   Mass loss (mass loss through burning plus mass of drips falling from test rig) shall not be greater than 60 grams.

H   FR Polystyrene beads which consistently fall over and extinguish the gas flame are considered to have passed the test.

The references A to H apply to Tables VI, VII, and VIII.

Table VII. Summary of Test Requirements for Filling
Materials - All Tested with Specified FR
Polyester Fabric

Filling	Standard Test	Ignition Source to BS 5852	Failure Criteria
PU foam	BS 5852/2	IS5	F, G
PU foam crumb	BS 5852/2	IS2	F
PS foam beads	BS 5852/2	IS2	H
Rubber latex	BS 5852/2	IS2	-
Non-foam filling singly	BS 5852/2	IS2	F
Composite fillings for other than mattresses bed bases, cushions, and pillows	BS 5852/2	IS2	
Composite fillings for mattresses and bed bases after removal of outer covering fabric	BS 6807	IS2	

Note 1: Original PU foam blocks to meet test for PU foam.

Note 2: Pillows and cushions with primary covers and "solid" or "loose" fillings and tested as composites. See Table VIII.

See footnotes to Table VI for explanation of criteria.

Table VIII. Summary of Test Requirements for Final Composite
                          Assemblies

Final Composite Assemblies	Standard Test	Ignition Source to BS 5852	Failure Criteria
Upholstered furniture	BS 5852/1	Cigarette	D
Platform composite	BS 9852/1	Cigarette	
Mattresses*	BS 6807*	Cigarette (0) and Cigarette with non-combustible insulation (O/NS)	–
		1	–
Pillows and cushions + with primary cover and non-loose fillings	BS 5852/2	2	D
Pillows and cushions +# with primary cover and loose fillings	BS 5852/2	2	D

*   Not part of Consumer Protection Act but applied by UK
    industry as a voluntary code of practice. This will be
    incorporated into the relevant British Standard and
    used to define "fitness for purpose."

+   Cushions are covered by the specified FR polyester
    fabrics.

For loose fillings, the test rig is lined with the
    specified FR polyester fabric.

See footnotes to Table VI for explanation of criteria.

When the non-foam fillings are used as composite fillings for upholstered seating or for mattresses and bed bases the multiple layer composite is treated as a single material and tested in its final form to BS 5852 part 2 or BS 6807 respectively.

PU foam crumb must also have been made from PU foam block which meet the relevant requirements. Polystyrene foam beads are considered to have passed the test if the beads cascading from the specimen consistently extinguish the gas flame (Table VII).

POLYURETHANE FOAM SHEETS OR BLOCKS. These are required to resist ignition source 5 (17 gram wood crib) of BS5852 Part 2 except that the flames may penetrate the full depth of the specimen and that the mass loss (due to burning and liquid residues falling from the test rig) shall not exceed 60 grams.

These criteria were developed by the UK PU foam industry and were intended to differentiate the melamine or exfoliated graphite containing combustion modified PU foams from the standard, high resilience and flame retarded (chloro and bromo phosphate) containing PU foams (Table IV). This distinction was required because large scale burning tests of real arm chairs and furnished rooms had demonstrated the superiority of the combustion modified polyurethane foams.

CIGARETTE TEST FOR UPHOLSTERED FURNITURE. Is carried out with the fabric (excluding braids and other trimmings but including seams), interliners if used and filling materials to BS5852 Part 1 using the cigarette ignition source. FR treated fabrics are also subjected to the water soak pre-treatment (see interlining) Table VI. This test is also applied to the platform composite in a preposed modification to the regulation.

CIGARETTE TEST FOR MATTRESSES. Is carried out using the full composite to BS 6807 using the cigarette (Source 0) and the cigarette covered with the non-combustible, insulating pad (Source O/NS). NOTE: These tests are currently part of a voluntary code and not part of Consumer Safety Legislation but are part of the British Standard for Mattresses and will be applied as a duty of care within the 1988 regulations. The No.1 gas flame test will be introduced at the same time.

FLAME IGNITION TEST FOR PILLOWS AND CUSHIONS WITH 'SOLID' FILLINGS. This is carried out using ignition to BS 5852 Part 2, Source 2. Cushions are additionally covered with the specified FR polyester fabric (Table VII).

FLAME IGNITION TESTS FOR PILLOWS AND CUSHIONS WITH LOOSE FILLINGS. With primary covers and loose fillings are tested to BS 5858 Part 2 with ignition Source 2 using the specified FR polyester fabric lining to the test rig. Cushions are additionally covered with the specified FR polyester fabric.

TIMESCALE AND LABELLING. It is not proposed to deal with the introduction of legislative requirements in detail but regulations are to be introduced over a period of time. Readers are strongly recommended to consult the Consumer Protection Act for details.

Regulations for the PU foam filling of manufactured items generally come into force on 1st November, 1988 for the sale of such items after 1st March 1989. Regulations for furniture intended for use in the open air and for furniture fixed into caravans is applied on 1st March 1990 and is applied to second hand furniture on 1st March 1993. Furniture made before 1950 or materials for the reupholstery of such furniture are exempt from the regulations.

The regulations require that furniture shall be labelled to indicate its conforming with various parts of the Act. Fire hazard warning labels are required as well as labels which identify the furniture, the person responsible for it in UK law (which may be the manufacturer or importer) as well as information relating to the composition of the furniture. The latter information may be on the label or may be retained by the company but must be available to Trading Standards Officers for a period of 5 years.

PROBABLE EFFECT OF UK CONSUMER REGULATIONS AND FIRE HAZARDS OF FURNITURE

The fire hazards of upholstered furniture have been identified as easy ignition by small sources such as cigarettes and materials and rapid rates of generation of heat, smoke and toxic gases(2)(3)(4). It is therefore logical to assess the UK regulations in terms of these parameters. Upholstered furniture is essentially a finished item but the flammability of bed assemblies is markedly affected by all components. Bedding is excluded from the regulations which apply only to pillows, mattresses and bed bases.

CIGARETTE RESISTANCE. Trims, rouchings, and pipings (which are frequently cellulosic) are specifically excluded from the cigarette test regulations for upholstered furniture. Because of this, it is unlikely that the cigarette test requirement can ever result in upholstered furniture which is completely cigarette resistant.

The platform which can be of a completely different construction to the primary upholstery and may be tested under a proposed modification to the regulations which include testing "invisible" fabrics to Source No.1 over a PU foam resisting Source 5 to the 1988 regulation.

It has been demonstrated that a fabric/foam composite which passes the cigarette test of BS 5852 Part 1 can fail if a strip of viscose rouching is placed along the junction of the vertical and horizontal parts of the test rig.

The paper of McCormack, Damant and Williams (27), describes the results of a total of 5619 tests carried out on 9 test sites on 450 upholstered chairs. The overal number of ignitions was 7.2% although ignition on individual sites varied from zero to 14.4%.

It is arguable that composites of the crevice, top of arm, top of back and smooth surface (34% of ignitions) are tested in the crevice of BS 5852. The important areas of the welt, tufts etc. which includes piping and rouchings which gave 14.2% of ignition are ignored. The platform (6.7% of ignition) is tested separately.

A similar situation exists for mattresses where BS 6807 section 4 specifically excludes the testing of edge piping, tufts, pleats etc. The cigarette is used directly and also covered with a layer of non-combustible insulation representing non-smouldering bedding.

The match test (No.1 gas flame) is currently only applied to upholstered furniture but will be applied to mattress in the future by the same mechanism as the cigarette test.

Defining the performance of fabrics using a standard PU foam may produce a different result (possibly superior) than testing over the actual infill material to be used but will create a simplified system of testing each fabric at source. However, cigarette tests are carried out using the actual composite and match resistance tests at the same time as the cigarette test will only add a few minutes to the test and will produce a technically better result. Although not included as part of the UK regulations, the revised BS 5852 will include ignition test procedures for tests on actual full scale chairs. Full scale ignition tests are already included in BS 6807 for mattresses and bed assemblies.

The use of match ignitable covers over FR interliners is potentially hazardous as flames can spread over real furniture until a flammable item is ignited and this secondary source may then cause ignition of the upholstery. Ignition resistant barrier fabric/HRPU foam composites have been widely used in UK public areas since the 1970's with fabrics comprising wool, FR cotton, FRPVC etc. to give composites which resist Source 5 of BS 5852 part 2. FR barrier systems have hardly ever been applied to domestic furniture in the UK and it is ironic that it is now acceptable because it is the only means of using certain flammable fabrics for domestic furniture.

RATE OF BURNING AND USE OF IGNITION TESTS TO DEFINE FLAMMABILITY

In an ideal situation the parameters used to define furniture should be ignition resistance and the rate of generation of heat, smoke and toxic gases. Tests to do this with actual or mock-up full sized furniture are not yet available as final specifications but the Nordtest (28) and NBS furniture calorimeters (29) represent scientific methods while room/corridor rigs, typically UK DOE PSA FR5 and 6 of 1976 (5)(6) were originally used but are less satisfactory from a scientific point of view. The Californian (30) and Boston tests (31) for public area furniture are essentially simple room tests and are similar in principle to DOE, PSA, FR5 and 6 although the latter do not have pass/fail criteria. Bench scale rate of heat release tests include the NBS cone (29) which, with a code of practice represent a possible alternative but the rate of burning of upholstered furniture is markedly affected by design factors as well as by the fabrics and fillings used. A recent paper characterises the rate of release of PU foams using the OSU calorimeter (32).

In the absence of a readily available Standard rate of fire development test, the UK furniture regulations have used high ignition resistance as an alternative. Although scientifically different from rate of burning (33), high ignition resistance may be a realistic alternative for the next few years. The use of a high ignition resistance filling in conjunction with a flammable fabric in a composite implies that the ignition source replicates or simulates a burning fabric or other covering. If this philosophy were to be followed to its logical conclusion, then each and every infill should be tested in the same way and to the same source. Alternatively, quality assurance type tests may be used in which pass/fail criteria and test conditions have been developed to distinguish between materials known to have acceptable and unacceptable burning behaviour. This approach has been used in the UK regulations and a considerable amount of full scale test data is available concerning PU foam filled furniture. Data for cotton wadding and cotton wadding/PU foam are also available especially from USA. Information concerning polyester fibre and polyester fibre/PU composites is less readily available but limited data indicates that polyester fibres (probably resin bonded) burn in a similar manner to PU foams with a similar overall fire hazard. The use of simple, non-hazard related test data to distinguish between "good" and "bad" materials is potentially hazardous because the test performance has been determined by current materials. Because the tests are simple, cheap and can be carried out in a small fume cupboard, there is a strong possibility that future materials will be developed to meet the small scale tests and will not be evaluated in hazard based tests. This possibility could result in furniture of increased fire hazard.

A further difficulty with small scale tests is that the relative fire performance and even the rank order of materials can change with different fire environments. Small scale tests can rarely reflect real life fire situations and examples already exist where reliance on small scale tests has let to hazardous full scale situations.

The pass/fail criteria of BS 5852 Part 2 have been altered for certain materials to exclude clauses limiting the extent of penetration of flaming or smouldering combustion. A maximum mass loss criteria is introduced for PU foam and is essentially a means of distinguishing between combustion modified and other types of PU foam. In practice, combustion modified PU foam is likely to improve the burning behaviour of upholstered furniture. Although it is arguable that the Consumer Safety Regulations are based on test procedures which are scientifically unsound and that they contradict the essentials of composites testing and hazard based fire tests, large scale hazard based tests involving the newer types of combustion modified PU foams with suitable fabrics have shown

significant improvements in fire behaviour. The time to rapid
burning of a settee or chair may be increased from 3 - 4 minutes
to up to 10-30 minutes by the careful combination of flame
retarded fabrics or fabrics of inherently low flammability with
combustion modified PU foams. The subsequent rates of heat
release may be significantly reduced(21). Clearly this
represents a major improvement although a fire problem can still
exist. Complementary measures, e.g. the installation of smoke
alarms will be advantageous because they will detect the small
initial fire in its early stages.
      Although the ignition resistance of upholstered furniture,
mattresses and bed assemblies are specified by tests on the final
composite specimen, filling materials and foams are additionally
required to conform to the 1988 regulations (Table IX).

RATE OF BURNING TESTS

Technically the recent UK regulations may be viewed as a
short term measure pending the development of proper hazard
based rate of fire development tests. Considerable research
has been published by Babrauskas (29) and others in USA and also
by workers in Scandinavia (25) concerning the measurement of
rates of fire development of full scale furniture (furniture
calorimeter and Nordtest hood/duct) while similar research is
being carried out in the UK by Fire Research Station. The
Californian TB133(30) approach does not measure fundamental
parameters such as rate of heat release but is a relatively
simple approach to limiting the rate of fire growth of
upholstered furniture. The NBS cone calorimeter (29) and OSU
calorimeter (32) are being developed as a means of measuring
rates of heat and smoke generation, as bench scale tests.
However, considerable evidence exists that the design of
upholstered furniture has an important effect and that materials
other than the primary upholster, e.g. viscose rouchings and
trims, platform materials, etc. can not only alter the ignition
resistance but also the mechanisms and rates of burning. At
present the only really satisfactory method of determining the
rates of heat and smoke release from upholstered furniture and
bed assemblies is to burn the actual items. The implications of
specimen size, cost and complexity of test equipment has so far
limited these to type approved tests for critical applications.
A possible solution would be to use a bench scale test to
eliminate the more hazardous composite but, depending on the
outcome of current research, it may still be necessary to use the
full scale hood test approach (Nordtest) to indicate safety.

SUMMARY AND CONCLUSIONS

1)   The Consumer Protection, Furniture and Furnishings (Fire)
     (Safety) Regulations 1988 is a complex application of a
     series of ignition tests based on BS 5852 Parts 1 and 2 and
     BS 6807.

Table IX.   Ignition Test Requirements for Upholstered
Furniture and Mattresses for UK

Upholstered Furniture

BS 5852	Low Hazard	Medium Hazard	High Hazard	Very High Hazard
Ignition Source	Cigarette + No.1 gas flame	Cigarette + No.5 wood cribs	Cigarette + No.7 wood cribs	At discretion of specifier
	Domestic Dwellings Colleges Schools Offices	Hospitals Public Houses & Bars Public Halls Theatres, Cinemas etc Restaurants Public Buildings Offices	Hospitals Hostels Sleeping Accommodation	Prison Cells Locked Psychiatric Units

Mattresses

BS 6807	Section 4	Section 2 Mattress Section 3 Bed Assembly	Section 2 Mattress Section 3 Bed Assembly	DOE FTS 15 Vandalised Mattress 4 x No. 7 Wood Cribs
	Cigarette Cigarette + NC cover No.1 gas flame	Cigarette No.1 gas flame No.5 wood crib	Cigarette No.1 gas flame No.7 wood crib	
	Domestic Dwellings	Public Buildings Halls of Residence at colleges Hostels Hotels	Certain Hospitals Psychiatric Accommodation Hotels Hostels	Prison Cells Locked psychiatric Accommodation

Dimensions of test specimens

Dimensions of Test Specimen	BS 5852 Pt. 1 (mm)	BS 5852 Pt. 2 (mm)	BS 6807 (mm)
Width of back	450	450	−
Height of back	300	450	−
Width of base	150	300	450 square
Depth of filling	75	75	75

2) Actual upholstered composites are required to resist the cigarette source of BS 5852 Part 1 (furniture). Mattresses are also required to resist the cigarette test uncovered and covered with non- combustible insulation of BS 6807.

3) The match (No.1 gas flame) source of BS 5852 Part 1 is only applied to upholstered furniture fabrics over a standard PU foam although exceptions are made for certain fabrics, when used with FR interliners. It may also be applied to mattresses to BS 6807.

4) All filling materials are specified by ignition tests to BS 5852 Part 2 and composite non-PU foam mattress fillings to BS 5852 Part 2 and BS 6807 for seating and mattress applications respectively with a flame retarded polyester fabric. Different ignition sources, constructions and pass/fail criteria are used to differentiate between "acceptable" and "unacceptable" materials.

5) "Acceptable" combustion modified PU foams have been validated in full scale hazard based fire tests.

6) The 1988 UK Regulations will probably be effective for upholstered furniture and mattresses comprising a high proportion of PU foam.

7) Combining ignition resistant fabrics with combustion modified PU foam will significantly improve the fire performance of upholstered furniture and mattresses.

8) The actual composites used for upholstered furniture and mattresses in government and public buildings are required to meet cigarette and crib 5 (for general use) and crib 7 for use in high risk areas to BS 5852 Part 2 and BS6807. Combustions modified PU foam meeting the 1988 regulations is additionally likely to be required although these are outside the outside the 1988 regulations.

8) Because of the lack of suitable standard rate of heat, smoke and toxic gas generation tests, the 1988 UK regulations are based on ignition resistance of individual materials. This contradicts the basic requirements for the fire testing of composites and of hazard related tests and as such it may be possible to develop materials which meet the requirements but which produce hazardous products.

9) The 1988 Regulations should ideally be replaced by performance requirements based on composite specimens tested by hazard related tests e.g. rate of heat and smoke release when they become available.

1. Home Office Statistical Bulletin, HMSO, London 1988.
2. K. N. Palmer, W. Taylor, K. T. Paul, Fire Hazards of Plastics in Furniture and Furnishings BRE Garston. Current papers CP18/74, CP3/75 and CP21/76.
3. Flexible PU Foam, Its Uses and Misuses British Rubber Manufacturers Association, London 1976 revised 1983.
4. F. Prager et al Cellular Polymers 3 (1984).
5. Fire Technical FTS3, 5, 6 and 15. Department of the Environment Crown Suppliers, London.
6. K.T. Paul, J. Cellular Polymers 3 (1984) P.105-132
7. BS 5852, Part 1. Methods of Test for Ignitibility of Upholstered Composites for Seating by smokers materials. BSI London.
8. BS 5852 Part 2, Methods of Test for Ignitability of Upholstered Composites for Seating by Flaming Sources. BSI, London.
9. BS 6807 Methods of Test for the Ignitibility of Mattresses with Primary and Secondary Sources of Ignition. BSI, London.
10. Draft Guide to Fire Precautions in Existing Residential Care Premises. Home Office, Scottish Home and Health Departments 1983.
11. Draft Guide to Fire Precaution in Hospitals. Home Office, Scottish Home and Health Departments. 1983.
12. BS 7176 Method of Test for the Resistance to Ignition of Upholstered Furniture. To be published. BSI, London.
13. BS 7177 Method of Test for the Resistance to Ignition of Mattresses, Divans and Bed Bases. BSI, London 1989.
14. Report of the Technical Sub Committee on the Fire Risks of New Materials. Home Office, London 1978.
15. Consumer Protection, The Upholstered Furniture (Safety) Regulation 1980, HMSO, London.
16. Consumer Protection. The Upholstered Furniture (Safety) (Amendment Regulations) 1983. HMSO, London.
17. Method of Test for the Ignitability of Mattresses in the Form of Bed Assemblies. National Bedding Federation. London 1982.
18. BS 7175 Methods of Test for the Ignitability of Bed Covers and Pillows by Smouldering and Flaming Ignition Sources. BSI, London.
19. Consumer Protection. The Furniture and Furnishings (Fire) (Safety) Regulations 1988. HMSO, London.
20. C.P. Colver, J.C. Colver, J. Fire Sci. Vol.5 Sept/Oct 1987 P.326-337
21. V. Babrauskas, First European on Fire and Furniture, Brussells 1988.
22. J. Fishbein, Oyez Conference, London 1980.
23. G.H. Damant, J. Fire Sci. Vol.4 May/June 1986 p.174-187.
24. K.T. Paul, Urethanes Technology June 1987 p.38-42.
25. BS 4735. Assessment of Horizontal Burning Characteristics of Small Specimens of Cellular Plastics and Cellular Rubber Material When Subjected to a Small Flame, BSI, London.

26. J.F. Szabat, J.A. Gaetane, SPI Conference Polyurethane Division, November 1983, p.326-331 (43C6).
27. J.A. McCormack, G.H. Dament and S.S. Williams, J. of Fire Sciences Vol. 4 March/April 1986.
28. Nordtest No. 410-83, Upholstered Furniture: Burning Behaviour - full Scale Test
29. V. Babrauskas and J. Krasny, National Bureau of Standards Monograph 173, Gaithersburg, USA.
30. California Technical Balletin TB133, California USA 1988.
31. Boston Fire Dept. Procedure BFD 1X-10, USA 1986.
32. J.F. Szabat, W.E. Zirk, D.B. Parrish, SPI Summer Conference USA 1988 p.190-200.
33. K.T. Paul, J. Cellular Polymers 4(1985) p.195-223.
34. K.T. Paul, D.A. King, The Burning Behaviour of Domestic Upholstered Chains Containing Different Types of Polyurethane Foam.
35. B. Sundstrom, I. Kaiser, Technical Report SP-RAPP Bonar Sweden 1986.
36. K.T. Paul, S.D. Christian, J. of Fire Sciences Vol.5, No.3, May/June, 1987.
37. D. Powell, PRI Urethane Conference, Blackpool, UK, April 1989.
38. R. Hurd. PRI Urethane Conference, Blackpool, UK, April 1989.
39. K.T. Paul, D.A. King. To be published.

RECEIVED January 11, 1990

Chapter 31

Heat Release Equipment To Measure Smoke

Marcelo M. Hirschler

BFGoodrich Technical Center, P.O. Box 122, Avon Lake, OH 44012

Smoke has usually been measured in the NBS smoke chamber. Such results cannot be correlated with full scale fire results and do not predict fire hazard. Rate of heat release (RHR) calorimeters (e.g. NBS Cone (Cone) and Ohio State University (OSU)) can be used to determine the best properties associated with fire hazard, as well as smoke. Results from the Cone RHR correlate with full-scale fire results. The best way to determine the fire hazard associated with smoke, for materials which do not burn up completely in a fire, is by using RHR to measure combined smoke and heat release variables, such as smoke parameter or smoke factor.

This work measured smoke and heat released from burning 17 materials, in the Cone and OSU and smoke in the NBS smoke chamber. Results from the RHR calorimeters correlate well with each other while those from the smoke chamber do not. This suggests that the smoke parameter and smoke factor, from either RHR calorimeter, are excellent measures of smoke hazard.

Fire hazard is associated with a variety of properties of a product in a particular scenario [1]. It is determined by a combination of factors, including: product ignitability, flammability, amount of heat release on burning, rate at which this heat is released, flame spread, smoke production and smoke toxicity.

The traditional way of measuring fire properties is to determine each property individually by carrying out small scale tests on materials, in isolation of the fire scenario of interest.   A crude means of fire hazard assessment would then be to establish minimal "passing" standards for each test and require all materials to meet them.

However, the majority of small scale tests actually used to measure fire properties are incapable of determining either more than a single property or combined properties.   Furthermore, there is, often, no attempt to investigate whether the test results are can be related to results to be expected in full scale fires.   This is incompatible, thus, with modern concepts of fire hazard.

It has now been established that the single property which most critically defines a fire is the heat release, in particular its peak value which is indicative of the maximum intensity of the fire[1-3].   This introduction is required in order to understand one of the premises of the present work, viz. the usefulness of measuring rate of heat release and of combining its measurement with that of smoke.

In order for a fire to propagate from the product first ignited to another one, two conditions are necessary.   Firstly, sufficient heat needs to be released to cause secondary ignition.   Secondly, the heat release needs to occur sufficiently fast so that the heat is not quenched in the cooler air surrounding the latter product.

Heat release equipment can be used to measure various parameters on the same instrument, in a manner generally relevant to real fires.   The two most frequently rate of heat release (RHR) calorimeters used are the Ohio State University (OSU calorimeter)[4] and the NBS cone (Cone calorimeter)[5].

The OSU calorimeter [4] has long been used for simultaneously measuring heat and smoke release.   It can also be used to measure release of combustion products. It is the basis of standard tests at both ASTM, the American Society for Testing and Materials (ASTM E906-1983), and FAA, the Federal Aviation Administration [6,7].

The Cone RHR calorimeter [5] is a more modern instrument, designed to meet the same objectives as the OSU calorimeter.   It is now being considered for standardization by ASTM [8] and by the International Organization for Standardization (ISO).   It is a very versatile instrument, which allows simultaneous determinations to be made of release of heat, smoke and other combustion products, and of sample mass loss and soot mass formation.   The Cone RHR calorimeter can, thus, measure the same properties as the OSU RHR calorimeter, plus a number of other ones based on sample and soot mass.

There are a number of differences between both
instruments, both in terms of their basic engineering
design and in terms of the concepts used for measuring
properties. This paper will highlight some of those
differences. The paper will focus on smoke measurement by
means of RHR calorimeters and other instruments.
This paper has the following objectives:

(a)   Describe reasonable small scale smoke measurement
       methods, based on heat release (RHR) equipment
(b)   Define smoke/fire hazard parameters: smoke
       parameter and smoke factor
(c)   Show the good correlations between three
       smoke/fire hazard parameters, measured in two
       different RHR instruments
(d)   Show the lack of correlation between these
       results and those obtained with the NBS smoke
       chamber.

Measurement of Smoke

Smoke obscuration is an essential parameter related to
fire hazard, because it may cause visual impairment both
of the occupants of a fire scenario and of the rescue
team, creating a potential danger.
NBS Smoke Chamber - Static Measurements
The traditional way in which smoke obscuration has
been measured is by determining the maximum smoke density
(or the specific maximum smoke density) by means of a
smoke density chamber developed by the National Bureau of
Standards (NBS smoke chamber, ASTM E662). This instrument
measures the obscuration inside a static 500 L chamber,
after a sample has been exposed, vertically, to a 2.5 W/cm^2
radiant source.
There are many inadequacies in this procedure, both
theoretical and experimental, Table I [9-13]. The most
important deficiency is the lack of correlation of the
results with those from full scale fires. This is
exemplified in Table II, for both obscuration and soot
[14].
Other deficiencies of the NBS smoke chamber will also
be discussed.
When samples are exposed vertically to a flame or
another heat source, some materials melt and drip, and do
not burn up completely. This will cause their smoke
results to be artificially low [9]. Burning samples
horizontally makes material performance comparisons in a
small scale test more logical because the entire sample
will be burnt in every case. This is very relevant when
dealing with fire retarded materials which do not melt or
drip, and will thus, yield similar smoke production
results in the vertical and horizontal modes.

Table I. Deficiencies in the NBS smoke chamber

- Results do not correlate with full-scale fires.
- Vertical orientation leads to melt and drip.
- Time dependency of results cannot be established.
- No means of weighing sample during test.
- Maximum incident radiant flux is 25 kW/m^2.
- Fire self-extinguishes if oxygen level becomes <14 %.
- Therefore, composites often give misleading results.
- Wall losses are significant.
- Soot gets deposited on optics.
- Light source is polychromatic.
- Rational units of m^2/kg are not available.

The next most important aspect is the issue of the oxygen depletion in the chamber. The atmosphere inside the closed compartment often becomes oxygen-deficient before the test ends. Thus, after the oxygen level gets below 14% burning often ceases. This means that, when testing thick samples (typically multi-layered materials or composites) one of the components may not get to burn at all.

Table II. Smoke Generation in a Room Corner Burn Test and in the NBS Smoke Chamber

	Thick	Max Smoke	Soot	Smoke Haz.	
NBS Smoke Material	cm	OD/m	g	MJ/m^3(F)	D_m
No sample	–	1.6	106	7.9	–
Polycarbonate	0.24	>15.1	>2900	>305.9	247
FR ABS	0.23	>15.1	>1460	>160.6	900
Wood panel	0.58	9.6	750	130.6	106
FR acrylic	1.24	7.7	398	> 83.7	435
Generic PVC	0.23	8.3	384	39.9	780
Low smoke PVC	0.12	1.5	93	17.8	94
CPVC	0.12	1.5	75	6.9	53

A 6.3 kg wood crib was used in all room corner burn experiments; total panel area: 6.6 m^2. The PVC materials used in these experiments were rigid.

The fact that it is not possible to weigh the sample during the test means that there is no continuous output of weight loss (and, in fact, normally none of smoke obscuration either, since the output tends to be a single final number). Thus, it is, generally, not possible to relate the sequence of events to the burning

characteristics of the sample. The measurements, even if
modifications were made to make them continuously, would
still reflect a static chamber, where smoke is being
accumulated. Thus, the obscuration values obtained are a
result of the forced combustion of those materials which
are most likely to stay in place, near the heat source,
during the entire burning process.

The incident flux used in the NBS smoke chamber is
only a single value, at 2.5 W/cm^2, which is a relatively
mild flux for a fire, and cannot, thus represent all the
facets of a fire. The light source is polychromatic, which
causes problems of soot deposits and optics cleaning, as
compared to measurements done using a monochromatic (laser)
beam.   Finally, the units of the normal output of this
smoke chamber are fairly arbitrary and the data is of
little use in fire hazard assessment.

<u>Dynamic Measurements - RHR Calorimeters</u>

There is an interesting phenomenon that can be gleaned from
the series of full scale experiments in Table II.   The
amount of smoke produced by a material like PVC, which is
generally regarded as having a high smoke production
tendency (i.e. large amount of smoke produced per unit mass
burnt) is much lower, in the full scale fire, than that
produced by wood, with a lower smoke production tendency.
This apparent contradiction is really a consequence of the
incorrect interpretation of test procedure results.   In the
full scale fire, only a small area of the PVC panel used
was consumed, viz. the area immediately adjacent to the
ignition source.   This occurred because the panel itself
released heat too slowly to cause further flame spread.
This suggests a concept generally applicable to fire hazard
assessment: if a material has a lower rate of heat release,
it will, all else being equal, be associated with lower
fire hazard.   This concept is essential in order to cope
with the results from small scale tests, where the sample
is forced to burn up completely, even for those materials
which would normally self-extinguish.

The Cone calorimeter yields smoke results which have
been shown to correlate with those from full scale fires
[10, 15-18].   The concept of a combined heat and smoke
release measurement variable for small scale tests has been
put into mathematical terms for the cone calorimeter: <u>smoke
parameter</u> (SmkPar) [10].   It is the product of the maximum
rate of heat release and the average specific extinction
area (a measure of smoke obscuration).   The correlation
between this smoke parameter and the smoke obscuration in
full scale tests has been found to be excellent [10].   The
corresponding equation is:

log (SmkPar) = 2.24 * log (full-scale ext coeff) - 1.31,
                                              (Equation 1),

which has a correlation coefficient of 85%.

It is noteworthy to restate that there was no correlation in the series of experiments shown in Table II between the maximum smoke density in the NBS smoke chamber, flaming mode, and the obscuration in the full scale tests.

This parameter, the smoke parameter, is based on continuous mass loss measurements, since the specific extinction area is a function of the mass loss rate. A normal OSU calorimeter cannot, thus, be used to measure smoke parameter. An alternative approach is to determine similar properties, based on the same concept, but using variables which can be measured in isolation from the sample mass. The product of the specific extinction area by the mass loss rate per unit area is the rate of smoke release. A smoke factor (SmkFct) can thus be defined as the product of the total smoke released (time integral of the rate of smoke release) by the maximum rate of heat release [19]. In order to test the validity of this magnitude, it is important to verify its correlation with the smoke parameter measured in the Cone calorimeter.

The same hazard concept could, potentially, be used for full scale tests, multiplying the total heat released, per unit surface exposed, by the maximum smoke obscuration. This is the basis for the magnitude smoke hazard (Smoke Haz.), shown in Table II. It is of interest that smoke hazard results yield the same ranking as mass of soot formed. Cone calorimeter tests are being planned with the same materials used in the full scale tests to investigate the usefulness of this concept.

Experimental

Materials. The materials used in this series of experiments were a variety of plastics, described in Table III in the order in which they appear in all subsequent Tables.

Ohio State University Calorimeter. The OSU calorimeter apparatus measures the heat released from burning samples adiabatically by means of a thermopile, based on either 3 or 5 thermocouples. Glow bars generate radiant input energy and a small methane flame serves as the igniter. Smoke obscuration is measured via the fraction of white (polychromatic) light transmitted horizontally. No load cell is normally available for continuously measuring weight loss of sample.

One of the difficulties inherent in adiabatic calorimetry is that it will, almost inevitably, result in low results, because of heat losses. This is not an intrinsic deficiency of the OSU calorimeter unit, since it can easily be modified to incorporate other methods of heat measurement. However, the traditional detection device used (and recommended in the standards) is the use of a thermopile.

Table III. Materials Used in These Investigations

---

1. Experimental vinyl wire and cable compound
2. Low smoke PVC sheet extrusion compound
3. Weatherable PVC extrusion compound
4. General purpose PVC custom injection moulding
   compound
5. Noryl GFN-3-70 polyphenylene oxide/polystyrene
   with fiberglas
6. FR thermoplastic polyurethane
7. Lexan 141-111 polycarbonate
8. Marlex HXM 50100 polyethylene
9. Cycolac KJT FR acrylonitrile-butadiene-styrene
   (ABS)
10. Dypro 8938 polypropylene
11. Celanex 2000-2 polyester
12. ABS fire retarded with PVC
13. Noryl N-190 polyphenylene oxide/polystyrene
14. Huntsman 351 FR polystyrene
15. Cycolac CTB ABS
16. Rovel 701 EPDM/SAN copolymer
17. Huntsman 333 polystyrene.

Some of these materials are not commercially available.

Samples are normally exposed in a vertical orientation. If samples melt and drip, the heat can be redirected, by means of a system of aluminum foil mirrors, towards a horizontal sample. Many of the materials used for the series of experiments reported here melted excessively, away from the flame. Therefore, vertical burns were impossible for them, without distorting the data. All the materials investigated in the OSU RHR calorimeter, with the exception of the experimental flexible vinyl wire and cable compound, were, thus, exposed horizontally.

The horizontal exposure method is not very adequate for the OSU RHR calorimeter, because the heat reflected from the aluminum foil onto the sample is much lower than the heat generated by the glow bars. Since the OSU calorimeter is based on the adiabaticity of the measurements, any heat losses will represent inaccurate results. The reflection on the aluminum foil is also uneven. Moreover, the use of higher radiant energy causes problems with the mechanical functioning of the instrument (bending and buckling of the back plate).

Data measured, at each of three incident fluxes, include the maximum rate of heat release (Max RHR, in kW/m^2), the total heat released after 15 min (THR@15, in MJ/m^2), the maximum rate of smoke release (Max RSR, in 1/s) and the total amount of smoke released after 15 min

(TSR@15). The data (Tables IV-VI) suggest that this instrument provides a satisfactory method for measuring heat release, even in the horizontal mode. Furthermore, it can differentiate between those materials which are prone to release much heat rapidly and those which perform much better in terms of heat release. The reliability of smoke data is, in principle, lower than that of heat data. In order to establish some criteria, the Tables include SmkFct values at 5 min (in MW/m^2), which will be compared with SmkFct and SmkPar values for the same materials tested in the Cone and with values of specific maximum smoke density measured in the NBS smoke chamber.

National Bureau of Standards Cone Calorimeter. There are three main differences between the Cone and the OSU calorimeters. The Cone calorimeter has:

(a) Heat release based on oxygen consumption
(b) Continuous mass measurements
(c) Improved engineering design

The principle of oxygen consumption is an empirical finding that the rate of heat release is proportional to the decrease in oxygen concentration in the combustion atmosphere [20, 21]. Thus, cone calorimeter heat release measurements do not require adiabaticity of reactions. Therefore, the combustion process can be carried out more openly, and reactions seen with the naked eye. The Cone calorimeter contains a load cell and can, thus, measure any property on a per mass lost basis. This permits determinations of specific variables (e.g. extinction area, gas yields, smoke parameter). The improved engineering design of the Cone calorimeter, with respect to the OSU calorimeter, is a consequence of having been designed more recently with the experience of the advantages and disadvantages of the OSU calorimeter and of other fire tests. Among its features are: the use of an excellent combustion module (a conical heater, for uniform heat distribution on the sample surface), the use of a monochromatic (laser) light for smoke obscuration and the provision of a horizontal burning orientation without compromising heat distribution.

Furthermore, it has been shown that the time period until ignition occurs, in the Cone calorimeter, is proportional to the inverse of the flame spread rate [16]. The Cone calorimeter can also be used to provide the mass loss rate information required for the simplified classification into categories of toxic hazard [1]: quick toxic hazard assessment. Thus, the NBS Cone calorimeter is a very useful tool to overcome some of the disadvantages associated with measuring a single property at a time.

It is of greater interest to this work, however, that the Cone calorimeter can also be used to measure combined

Table IV.   OSU RHR Results at 20 kW/m^2 Flux

	Max RHR	THR@15	Max RSR	TSR@15	SmkFct	
Vinyl wire cpd	17.5	0.7	11.8	46.0	0.2	
Low smoke PVC	39.3	19.8	57.9	271.9	6.4	
PVC Extrusion	40.0	20.9	127.2	652.9	17.3	
Rigid PVC	43.0	16.7	162.2	707.5	25.3	
PPO/PS + FGlas	170.4	55.1	101.9	324.4	42.8	
Thermoplas. PU	158.1	52.8	80.6	292.2	38.6	
Polycarbonate	192.5	51.0	121.7	287.2	36.1	
PE	476.9	80.4	66.8	114.6	49.4	
ABS (FR)	70.7	12.2	384.1	909.0	63.6	
PP	451.2	74.1	85.5	140.8	58.3	
Polyester	316.0	71.1	132.6	246.0	64.6	
ABS (FR with PVC)	152.4	55.1	165.4	538.6	79.1	
PPO/PS	136.4	50.6	308.6	662.3	88.8	
PS (FR)	103.8	16.2	408.4	875.2	90.6	
ABS (non FR)	391.1	77.6	163.3	311.2	121.6	
EPDM/SAN	402.8	80.6	170.1	353.1	141.6	
PS (non FR)	398.9	75.5	252.3	403.5	159.5	

Table V.   OSU RHR Results at 40 kW/m^{2}Flux

	Max RHR	THR@15	Max RSR	TSR@15	SmkFct	
Vinyl wire cpd	42.3	18.2	70.6	588.0	6.5	
Low smoke PVC	37.3	15.0	97.8	325.2	11.9	
PVC extrusion	48.7	15.7	289.4	703.6	33.5	
Rigid PVC	45.6	9.6	245.8	617.5	28.9	
PPO/PS +FGlas	162.3	47.7	145.5	290.3	46.0	
Thermoplas. PU	216.7	57.2	93.8	244.8	49.8	
Polycarbonate	5.0	49.1	139.6	290.2	47.9	
PE	430.2	73.3	131.9	166.2	67.9	
ABS (FR)	140.0	23.5	403.9	569.9	79.7	
PP	400.4	60.8	175.0	177.7	69.2	
Polyester	314.2	60.3	166.4	203.7	63.1	
ABS (FR with PVC)	143.4	47.8	210.0	481.3	69.1	
PPO/PS	104.4	26.2	324.7	574.3	57.8	
PS (FR)	162.7	28.2	453.2	638.8	102.6	
ABS (non FR)	269.0	50.3	186.7	276.2	71.5	
EPDM/SAN	319.7	59.5	232.9	310.1	98.8	
PS (non FR)	320.0	51.7	281.8	371.8	108.3	

Table VI. OSU RHR Results at 70 kW/m^2 Flux

	Max RHR	THR@15	Max RSR	TSR@15	SmkFct	
Vinyl wire cpd	63.5	44.1	187.3	1066.0	19.8
Low smoke PVC	36.0	11.0	143.9	257.8	9.0
PVC extrusion	84.7	17.0	192.7	321.8	26.0
Rigid PVC	86.5	21.8	234.0	320.9	26.4
PPO/PS + FGlas	132.4	45.2	176.8	277.0	35.0
Thermoplas. PU	213.5	59.1	69.7	130.5	24.9
Polycarbonate	116.2	46.7	146.1	234.5	26.6
PE	305.8	53.5	257.7	196.2	55.9
ABS (FR)	118.5	18.7	495.7	375.6	43.8
PP	275.6	48.1	262.6	188.5	50.8
Polyester	225.2	57.1	264.3	155.0	34.2
ABS (FR with PVC)	135.9	47.8	169.7	259.0	35.2
PPO/PS	98.9	18.6	298.2	341.8	33.7
PS (FR)	132.3	30.2	553.1	430.0	56.9
ABS (non FR)	262.5	50.2	249.3	237.9	62.3
EPDM/SAN	257.8	60.8	240.4	282.4	72.4
PS (non FR)	283.4	61.3	283.0	307.1	84.4

properties which can be associated with fire hazard, in
particular smoke parameter and smoke factor.
   It has already been shown that the Cone calorimeter
smoke parameter correlates well with the obscuration in
full-scale fires (Equation 1).   At least four other
correlations have also been found for Cone data: (a) peak
specific   extinction   area   results   parallel   those   of
furniture calorimeter work [12]; (b) specific extinction
area   of   simple   fuels   burnt   in   the   cone   calorimeter
correlates well with the value at a much larger scale, at
similar fuel burning rates [15];   (c)maximum rate of heat
release values predicted from Cone data tie in well with
corresponding full scale room furniture fire results [16]
and (d) a function based on total heat release and time to
ignition accurately predicts the relative rankings of wall
lining materials in terms of times to flashover in a full
room [22].
   For the specific extinction area correlation described
in (a) those materials which did not burn completely in the
full-scale were specifically excluded [12].   The small
scale test always leads to complete consumption of the
sample.   Therefore, more smoke is being produced in the
small scale test than in real fires for those materials
usually associated with lower fire hazards.   This is
exactly the kind of issue that is being remedied by
measurements of smoke parameter or smoke factor.

Results and Discussion

Tables   VII-IX   present   heat   release   data   for   the   same
materials as in Tables IV-VI.   The Tables also present the
same smoke data as measured with the OSU calorimeter, plus
smoke parameter information (SmkPar in MW/kg).
   The principal interest are, of course, real fires.
Since Cone calorimeter results correlate well with those
from full scale fires, it is essential, thus, to check
whether OSU calorimeter results correlate well with Cone
calorimeter results.   There are several aspects of this,
but the present paper will focus mainly on measurement of
smoke.
   Linear correlations were thus attempted for peak rate
of heat release, total heat released after 15 min. and
smoke factor between both calorimeters.   Furthermore,
linear   correlations   were   also   attempted   between   OSU
calorimeter   smoke   factors   and   Cone   calorimeter   smoke
parameters and between Cone calorimeter smoke factors and
Cone calorimeter smoke parameters.   Figures 1-3 show some
of the results.
   In   every   case   there   was   a   very   high   degree   of
correlation, with a statistical significance of over 95%.
Statistical   "rules   of   thumb"   state   that   whenever   a
correlation   shows   a   significance   of   over   90%,   such   a
correlation is statistically significant. The correlation

Table VII. Cone RHR Results at 20 kW/m² Flux

	Max RHR	THR@15	Max RSR	TSR@15	SmkFct	SmkPar
Vinyl wire cpd	15.7	2.5	1.9	0.19	8.9	2.3
Low smoke PVC	89.8	6.2	8.3	0.43	11.5	0.02
PVC extrusion	121.4	3.6	25.7	1.06	38.8	24.0
Rigid PVC	91.8	4.5	14.7	1.10	63.3	47.8
PPO/PS + FGlas	145.5	76.6	7.7	3.85	221.9	204.4
Thermoplas. PU	508.3	115.4	24.6	3.63	548.9	356.6
Polycarbonate	12.9	0.0	0.1	0.01	0.1	0.03
PE	1296.4	270.0	8.4	2.20	1931.6	488.3
ABS (FR)	261.7	47.5	42.8	9.55	1158.6	419.8
PP	1334.4	264.2	11.1	2.70	2628.8	648.6
Polyester	949.0	111.6	12.3	1.74	1603.7	223.0
ABS (FR with PVC)	268.4	38.8	20.8	3.43	401.2	313.5
PPO/PS	242.0	78.4	18.4	6.39	418.6	429.0
PS (FR)	328.2	93.6	39.6	11.49	1924.3	450.4
ABS (non FR)	732.2	190.1	20.2	5.54	2870.4	439.6
EPDM/SAN	882.4	254.7	25.5	7.73	5157.9	933.5
PS (non FR)	861.0	247.0	28.7	8.45	5184.0	1004.9

Table VIII. Cone RHR Results at 40 kW/m² Flux

	Max RHR	THR@A15	Max RSR	TSR@15	SmkFct	SmkPar
Vinyl wire cpd	100.7	17.0	17.2	0.65	47.3	28.3
Low smoke PVC	130.2	67.2	12.0	1.71	' 101.6	0.22
PVC extrusion	217.4	105.6	29.3	6.87	655.4	177.9
Rigid PVC	198.2	90.4	23.2	5.08	476.9	163.4
PPO/PS + FGlas	329.1	142.9	17.6	6.24	1077.9	480.6
Thermoplas. PU	268.1	123.4	17.5	3.75	489.3	169.7
Polycarbonate	514.8	95.9	23.1	3.23	952.3	517.1
PE	1876.6	283.0	12.2	2.19	3021.3	689.7
ABS (FR)	479.8	83.7	63.6	10.66	5094.3	876.1
PP	1826.3	257.8	17.6	2.87	4839.7	997.2
Polyester	1576.2	205.3	35.5	4.43	6162.9	796.1
ABS (FR with PVC)	366.1	127.1	36.1	10.82	1383.7	664.9
PPO/PS	315.8	140.8	28.3	8.74	1455.7	606.7
PS (FR)	413.7	115.9	77.1	18.91	6778.2	1014.9
ABS (non FR)	1126.3	195.2	33.5	5.16	5713.9	963.3
EPDM/SAN	1160.6	242.6	36.3	8.44	7845.6	1322.2
PS (non FR)	1315.1	250.7	43.5	8.44	9091.9	1658.8

Table IX. Cone RHR Results at 70 kW/m^2 Flux

	Max RHR	THR@15	Max RSR	TSR@15	SmkFct	SmkPar
Vinyl wire cpd	123.5	24.9	18.2`	0.54	55.6	25.2
Low smoke PVC	162.5	63.0	18.1	2.44	206.3	78.1
PVC extrusion	234.5	110.4	27.3	9.12	1341.2	419.2
Rigid PVC	244.2	107.2	22.5	7.44	969.4	239.3
PPO/PS + FGlas	501.7	156.1	33.2	8.04	3088.0	866.9
Thermoplas. PU	417.6	129.2	27.3	5.02	1258.9	368.7
Polycarbonate	443.3	119.5	30.6	4.60	1223.6	560.0
PE	3552.9	299.0	24.2	1.85	6466.3	1259.7
ABS (FR)	503.1	73.6	74.3	9.68	4822.0	807.7
PP	3137.7	298.0	29.5	3.00	9146.4	1726.2
Polyester	2189.2	215.5	32.6	2.87	6975.8	984.2
ABS (FR with PVC)	490.7	133.3	38.7	9.85	3655.6	690.2
PPO/PS	391.5	158.4	40.6	10.05	2568.3	824.5
PS (FR)	532.7	98.2	78.5	12.36	6567.6	762.2
ABS (non FR)	1571.4	191.8	47.3	4.52	7039.7	1186.3
EPDM/SAN	1582.4	279.8	55.3	11.13	17525.4	2381.6
PS (non FR)	1861.9	235.2	54.1	6.72	12353.9	1879.7

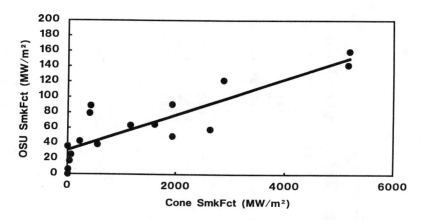

Figure 1. Correlation between the OSU smoke factor and the cone factor at an incident heat flux of 20 kW/m^2.

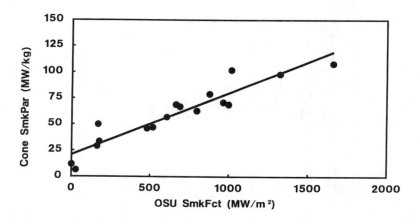

Figure 2. Correlation between the OSU smoke factor and the cone smoke parameter at an incident heat flux of 40 kW/m^2.

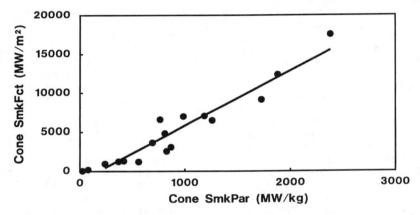

Figure 3. Correlation between the cone smoke factor and the cone smoke parameter at an incident heat flux of 70 kW/m^2.

coefficient and the coefficient of variation (100 times the
ratio of the standard error of the y values and the mean
y values) will, then, reflect the magnitudes of the
differences between the particular linear model chosen and
the experimental results and of the experimental scatter.
In this series of experiments quantities chosen varied over
more than one order of magnitude, indicating that the
correlations are valid over very wide ranges.

A summary of the results of correlation models for
smoke factor and smoke parameter is shown in Table X. For
comparison purposes, correlation models for OSU and Cone
calorimeter peak rates of heat release are also shown in
Table X, together with one of the total heat release
models.

Previous work had shown that the Cone and OSU
calorimeter results were not identical, but was unclear as
to whether the results were correlatable [23]. This work
gives definitive proof of correlation between the OSU
calorimeter and the Cone calorimeter RHR tests.

This work does not give definitive proof, however, that
the results from the OSU calorimeter correlate well with
those from full scale fire tests. It is likely that this
will happen, but the product of two good correlations
cannot be guaranteed to give another good correlation.
Thus, correlation between OSU calorimeter and full scale
fires still remains to be firmly established.

Table XI presents the results of tests on the same
materials in the NBS smoke chamber. It is immediately
clear that these results do not correlate well with those
measured on the RHR apparatuses. Furthermore, an attempt
at a linear correlation between the flaming mode specific
maximum optical density and the Cone calorimeter SmkPar at
20 kW/m^2 yielded a correlation coefficient of ca. 1%, a
coefficient of variation of 217% and statistically invalid
correlations. A comparison between a Cone and OSU
calorimeter correlation and one with the NBS smoke chamber
is shown in Figure 4. This suggests that unrelated
properties are being measured.

The expected discrepancy between NBS smoke chamber
results and those from a good smoke production test were
compounded in this work by the fact that many of the
materials used melt and drip.

Conclusions

Results from the NBS Cone Calorimeter have been shown to
correlate with those from real fires. Moreover, it
measures properties very relevant to fire hazard, in
particular heat release, the most important of them. The
OSU Calorimeter will measure many of the same properties.
Furthermore, the results generated by both instruments have
similar significance because of the good correlation
between them. Smoke measurements are only relevant to fire

Table X. Linear Correlations. Statistical Analysis

	Slope	Intercept	Corr. Coeff. %	Coeff. Var. %
OSU SmkFct vs. Cone SmkPar				
20 kW/m^2	0.129	18.2	76	36
40 kW/m^2	0.059	20.8	88	18
70 kW/m^2	0.028	16.6	78	18
OSU SmkFct vs. Cone SmkFct				
20 kW/m^2	0.023	31.0	74	38
40 kW/m^2	0.009	32.0	77	24
70 kW/m^2	0.004	22.0	79	23
Cone SmkFct vs. Cone SmkPar				
20 kW/m^2	5.1	−371.2	85	49
40 kW/m^2	6.1	−763.1	88	34
70 kW/m^2	7.0	−1225.1	92	27
20 kW/m^2	0.334	49.7	85	31
40 kW/m^2	0.194	55.7	90	22
70 kW/m^2	0.071	91.9	76	27
OSU THR@15 vs. Cone THR@15				
40 kW/m^2	0.210	9.4	65	30
NBS Smoke Chamber F D$_m$g vs. Cone SmkPar (20 kW/m^2)				
	0.114	111.3	1	217

Model used: y = slope * x + intercept

Corr. Coeff.: Correlation Coefficient
Coeff. Var.: Coefficient of Variation

Table XI. Maximum Optical Density in the NBS Smoke Chamber

	F Dm	NF Dm	F Dm/g	NF Dm/g
Vinyl wire cpd	252	242	24.2	24.5
Low smoke PVC	159	117	17.7	13.3
PVC extrusion	607	274	59.3	25.8
Rigid PVC	495	204	61.9	25.8
PPO/PS + FGlas	332	41	36.5	4.6
Thermoplas.PU	491	350	54.8	40.7
Polycarbonate	109	11	11.9	1.1
PE	46	268	7.1	39.4
ABS (FR)	1000	232	1000.0	32.7
PP	56	237	11.6	49.6
Polyester	74	19	7.8	1.9
ABS (FR with PVC)	538	259	70.7	32.6
PS (FR)	1000	409	1000.0	43.3
ABS (non FR)	394	284	54.9	36.5
EPDM/SAN	371	249	36.5	4.6
PS (non FR)	612	251	56.5	29.7

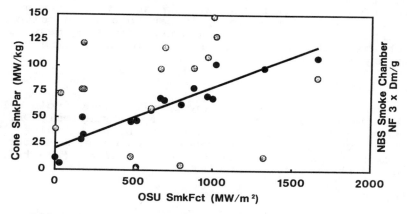

Figure 4. Correlation between the OSU smoke factor and the cone smoke parameter at an incident heat flux of 40 kW/m^2 (dark points) and also NBS smoke chamber results (light points).

hazard if the results obtained correlate with those from real fires. Those criteria mean that both RHR calorimeters are very appropriate smoke measuring devices.

Smoke obscuration, as measured in the NBS smoke chamber, does not correlate well with full-scale fire hazard nor with smoke as measured in heat release calorimeters. In particular, those materials which do not burn up completely in full-scale fires are being treated excessively harshly by this method.

Smoke parameter (in the Cone calorimeter) and smoke factor (in both calorimeters) are combined properties of smoke obscuration and heat release which compensate for the incomplete burning of fire retardant samples and which should predict smoke hazard in real fires.

The NBS Cone calorimeter has been shown to be more versatile than the OSU calorimeter and to allow simultaneous measurements of a large variety of the properties required for a full assessment of fire hazard in real fires. Furthermore, it can be used to calculate combined properties, including those involving mass loss, which are much more useful as indicators of fire hazard than any individual one.

Acknowledgments

The author is grateful to many people. They include Dr. J. W. Summers (who supplied many of the samples), Dr. G. F. Smith (fruitful discussions), Mr. L. A. Chandler, Ms. T. T. Chavez and Ms. C. E. Sauer (who carried out most of the experimental work) and Mr. R. S. Butler (who helped with the statistical analysis).

References

1. Hirschler, M. M., J. Fire Sciences. 5, 289 (1987).
2. Thomas, P.H., Int. Conf. "FIRE: Control the Heat – Reduce the Hazard", Fire Research Station, October 24-25, 1988, London, paper 1.
3. Babrauskas, V., Int. Conf. "FIRE: Control the Heat – Reduce the Hazard", Fire Research Station, October 24-25, 1988, London, paper 4.
4. Smith, E. E., in "Ignition, Heat Release and Non-combustibility of Materials, ASTM STP 502" (Ed. A. F. Robertson), p.119, Philadelphia (1972).
5. Babrauskas, V., Fire Mats. 8, 81 (198).
6. Sarkos, C. P., Filipczak, R. A. and Abramowitz, A., "Preliminary Evaluation of an Improved Flammability Test Method for Aircraft Materials," (DOT/FAA/CT-84/22). Federal Aviation Admin., Atlantic City Airport (1984).
7. Hill, R. G., Eklund, T. I. and Sarkos, C. P., "Aircraft Interior Panel Test Criteria Derived from Full-Scale Fire Tests," (DOT/FAA/CT-85/23). Federal Aviation Admin., Atlantic City Airport (1985).

8. American Society for Testing and Materials, "Proposed Test Method for Heat and Visible Smoke Release Rates for Materials and Products using an Oxygen Consumption Calorimeter," E-5, Proposal P-190, ASTM, Philadelphia, (1986).
9. Breden, L. H. and Meisters, M., J. Fire Flammability. 7, 234 (1976).
10. Babrauskas, V., J. Fire Flammability. 12, 51 (1981).
11. Quintiere, J. G., Fire Mats. 6, 145 (1982).
12. Babrauskas, V., in "PVC - The Issues", SPE RETEC, Sep. 16-17, Atlantic City, Soc. Plast. Engin., p. 41 (1987).
13. Hirschler, M. M., 31st. IUPAC Microsymp. on Macromolecules Poly(Vinyl Chloride)", Prague, 18-21 July (1988) Makromol. Chem., Macromol. Symp. 29, 133-53 (1989).
14. Smith, G. F. and Dickens, E. D., Proc. Eighth Int. Conference on Fire Safety (Ed. C. J. Hilado), Product Safety Corp., San Francisco, p 227 (1983).
15. Mulholland, G. W., Henzel, V. and Babrauskas, V., in 2nd. Int. Fire Safety Science Symp., Tokyo, 13-16 June (1988).
16. Babrauskas, V., "Bench-Scale Methods for Prediction of Full-Scale Fire Behavior of Furnishings and Wall Linings", Soc. Fire Prot. Eng., Boston, Technology Report 84-10 (1984).
17. Babrauskas, V. and Krasny, J. F., in "Fire Safety. Science and Engineering, ASTM STP 882" (Ed. T.Z. Harmathy), p. 268, Philadelphia, 1985.
18. Babrauskas, V., J. Fire Sciences. 2, 5 (1984).
19. Hirschler, M. M. and Smith, G. F., in "Fire Safety Progress in Regulations, Technology and New Products", Proc. Fall 1987 F.R.C.A. Conf., Monterey, CA, October 18-21, p. 133 (1987).
20. Ostman, B.A.-L. and Nussbaum, R. M., in 2nd. Int. Fire Safety Science Symp., Tokyo, 13-16 June (1988).
21. Parker, W.J., "Calculations of the Heat Release Rate by Oxygen Consumption for Various Applications," NBSIR 81-2407, February, 1982.
22. Huggett, C., Fire Mats 4, 61 (1980).
23. Babrauskas, V., J. Fire Sciences. 4, 148 (1986).

RECEIVED November 1, 1989

Chapter 32

Flammability Characteristics of Fiber-Reinforced Composite Materials

D. P. Macaione[1] and A. Tewarson[2]

[1]U.S. Army Materials Technology Laboratory, Watertown, MA 02171–0001
[2]Factory Mutual Research Corporation, Norwood, MA 02162

This paper describes the results of a joint study
undertaken by the U.S. Army Materials Technology
Laboratory (AMTL) and the Factory Mutual Research
Corporation (FMRC) on fiber reinforced composite
materials (FRC) for use in a composite combat vehicle.
The objective of the study was to assess the flammability characteristics of FRC materials using small-
scale experiments.
In the study, five FRC samples, about 3 to 5 mm in
thickness, were examined. The results from the study
show that FRC materials have high resistance to ignition, high heat of gasification and high resistance to
self-sustained fire propagation. These results suggest
that a composite combat vehicle, by virtue of its
construction, does not present an unusual fire hazard.

Fiber reinforced composite materials are used extensively because of
their physicochemical properties and their high strength/weight
ratio. The use of composites in Army vehicles as a means of decreasing weight and enhancing survivability, without reducing personnel
safety, has been considered for some time. The U.S. Army Materials
Technology Laboratory (AMTL) has successfully demonstrated, in an
earlier program, that a ground vehicle turret could be fabricated
from FRC materials; now the technology has been applied to the fabrication of a composite vehicle. The U.S. Navy is also considering the
use of FRC materials for numerous ship and submarine applications,
including use as major structural components. FRC materials are also
finding applications in the aerospace, automobile and other industries. Although FRC materials are very attractive in terms of their
physical properties, one of the major obstacles to their application
is the concern for the hazards expected in fires. In fires, hazardous environments are created as a a result of generation of heat
(thermal) and generation of smoke, toxic and corrosive fire products
(nonthermal).
For the assessment of resistance to heat exposure and generation
of hazardous environments by FRC materials, the following processes
need to be examined: 1) ignition, 2) fire propagation, 3) generation

of vapors, heat, smoke, toxic and corrosive fire products, and 4) fire extinguishment.

The generation of material vapors can be quantified in terms of: 1) relationship between weight loss and sample temperature. AMTL used thermogravimetric techniques for such quantification (1), and 2) the relationship between the generation rate of material vapors and the heat flux. Factory Mutual Research Corporation (FMRC) has developed a technique using FMRC's Small-Scale Flammability Apparatus (2-10), which was used for the quantification.

The ignition process for FRC materials can be described in terms of the relationship between time to ignition and heat flux. A technique developed by FMRC using its Small-Scale Flammability Apparatus (2-6) was used for the quantification.

Techniques have been developed for the quantification of fire propagation using FMRC's Small-Scale Flammability Apparatus (4,6) and the National Institute of Standards and Technology (NIST) Flame Spread Apparatus (11). In this study, the FMRC technique was used. Oxygen Index and its dependency on temperature was used by AMTL to examine the fire propagation behavior of small samples of FRC materials (12).

The heat generated in a fire is due to various chemical reactions, the major contributors being those reactions where CO and CO_2 are generated, and O_2 is consumed, and is defined as the chemical heat release rate (3). Techniques are available to quantify chemical heat release rate using FMRC's Flammability Apparatus (2-6), Ohio State University (OSU) Heat Release Rate Apparatus (13) and the NIST Cone Calorimeter (14). Techniques are also available to quantify the convective heat release rate using the FMRC Flammability Apparatus (2,3) and the OSU Heat Release Rate Apparatus (13). The radiative heat release rate is the difference between the chemical and convective heat release rates (2,3). In the study, FMRC techniques were used.

Techniques are available to quantify the generation of smoke, toxic and corrosive fire products using the NBS Smoke Chamber (15), pyrolysis-gas chromatography/mass spectrometry (PY-GC-MS) (16), FMRC Flammability Apparatus (2,3,5,17,18), OSU Heat Release Rate Apparatus (13) and the NIST Cone Calorimeter (14). Techniques are also available to assess generation of: 1) toxic compounds in terms of animal response (19), and 2) corrosive compounds in terms of metal corrosion (17). In the study, FMRC techniques and AMTL PY-GC-MS techniques were used.

The extinguishment of fire depends on the fire propagation rate, physico-chemical properties of the materials, rate of application and the concentration of extinguishing agents. Water applied through sprinklers is the most widely used liquid extinguishing agent, and Halon 1301 and CO_2 are the most widely used gaseous agents. Techniques are available to quantify fire extinguishment in large-scale fires (20); recently attempts are being made to develop small-scale techniques for fire extinguishment using FMRC's Small-Scale Flammability Apparatus (21). The FMRC technique for flame extinguishment using Halon was used in the study.

AMTL and FMRC have been performing research to quantify the flammability characteristics of FRC materials which are enumerated in this paper.

Concepts

Generation of Material Vapors. The magnitude of heat required to
generate vapors from a material depends on the thermal stability of
the material. As the temperature or heat flux is increased, generation rate of vapors, measured in terms of mass loss of the material,
increases and the following relationship is satisfied (2,3):

$$\dot{m}'' = (\dot{q}_e'' + \dot{q}_f'' - \dot{q}_{rr}'') \, / \, L = \dot{q}_n''/L, \tag{1}$$

where $\dot{m}''$ is the mass loss rate ($g/m^2 s$); $\dot{q}_e''$ is the external heat
flux (kW/m^2); $\dot{q}_f''$ is the flame heat flux (kW/m^2); $\dot{q}_{rr}''$ is the surface
reradiation loss (kW/m^2); $\dot{q}_n''$ is the net heat flux (kW/m^2) and L is
the heat of gasification (kJ/g).
     It has been shown that in small-scale experiments with turbulent
fires, the flame heat flux approaches its asymptotic value for oxygen
concentrations greater than about 30%; the asymptotic value is very
close to the value expected in very large fires (7).
     A condition where $\dot{q}_e'' + \dot{q}_f'' = \dot{q}_{rr}''$, $\dot{m}'' = 0$, and thus $\dot{q}_{rr}''$ represents
the minimum heat flux at or below which the material is not expected
to generate vapors.

Mass Loss Rate as a Function of Temperature. The most commonly used
technique is thermogravimetry (TG). The basic components of modern
TG have existed since the early part of this century (22-24).
Thermal analysis, in the form of TG, has been employed extensively in
the area of polymer flammability to characterize polymer degradation.

Mass Loss Rate as a Function of External Heat Flux. The technique
for the measurement of mass loss rate as a function of heat flux was
developed in 1976 at FMRC using the Small-Scale Flammability Apparatus (8). Several other flammability apparatuses are now available
for such measurements, such as OSU Heat Release Rate Apparatus (13)
and NIST Cone Calorimeter (14).

Ignition. For thermally thick materials, time to ignition is found
to follow the following relationship as external heat flux is varied
(4,6):

$$t_{ig}^{-1/2} \quad \alpha \quad \dot{q}_e'' \, / \, \Delta T_p \, (k \, \rho \, c \,)^{1/2}, \tag{2}$$

where t_{ig} is time to ignition (s); ΔT is the temperature of ignition
above ambient (K); k is the thermal conductivity (kW/m K); ρ is the
density (g/m^3) and c_p is the specific heat (kJ/g K). In Equation
(2), $\Delta T (k\rho c_p)^{1/2}$ is defined as the Thermal Response Parameter (TRP)
of the material and expresses ignition and fire propagation resistance of the material. The minimum value of $\dot{q}_e''$, at or below which
there is no ignition, is defined as the critical heat flux, $\dot{q}_{cr}''$, for
ignition (4,6).
     Experimentally, time to ignition is measured at various heat flux
values and critical heat flux for ignition and TRP are quantified
using techniques such as the one used in the FMRC Small-Scale Flammability Apparatus (4,6).

Heat Release Rate. It has been shown that the heat generated in chemical reactions leading to the generation of CO and CO_2 and depletion of O_2 can be used to calculate the chemical heat release rate using the following relationships (2,3):

$$\dot{Q}''_{Ch} = (\Delta H_T/k_{CO_2})\ \dot{G}''_{CO_2} + [(\Delta H_T - \Delta H_{CO})/k_{CO}]\ \dot{G}''_{CO} \tag{3}$$

and

$$\dot{Q}''_{Ch} = (\Delta H_T/k_{O_2})\ \dot{C}''_{O_2}, \tag{4}$$

where $\dot{Q}''_{Ch}$ is the chemical heat release rate (kW/m^2); ΔH_T is the net heat of complete combustion (kJ/g); ΔH_{CO} is the heat of combustion of CO (kJ/g); k_{CO} and k_{CO_2} are the maximum possible theoretical yields of CO and CO_2, respectively (g/g); k_{O_2} is the maximum possible mass of oxygen consumed per unit mass of material vapors (mass oxygen-to-fuel stoichiometric ratio) (g/g); $\dot{G}''_{CO}$ and $\dot{G}''_{CO_2}$ are the mass generation rates of CO and CO_2, respectively $(g/m^2 s)$; and $\dot{C}''_{O_2}$ is the mass consumption rate of O_2 $(g/m^2 s)$.

It has been shown that the convective heat release rate can be calculated using the following relationship (2,3):

$$\dot{Q}''_{Con} = M\ c_p \Delta T_g/A, \tag{5}$$

where $\dot{Q}''_{Con}$ is the convective heat release rate (kW/m^2); M is the total mass flow rate of fire products and air mixture (g/s), ΔT_g is the gas temperature above ambient (K) and A is the total surface area of the material involved in fire (m^2).

It has been shown that heat release rate satisfies the following relationship (2,3):

$$\dot{Q}''_i = \Delta H_i \dot{m}'' = \chi_i\ (\Delta H_T/L)\ \dot{q}''_n, \tag{6}$$

where i is chemical, convective or radiative and ΔH_T is the heat of complete combustion (kJ/g), χ_i is the combustion efficiency. $\Delta H_T/L$ is a fundamental physico-chemical property of the material.

Experiments can be performed where chemical, convective and radiative heat release rates can be measured at various external heat flux values. Linear relationships should be found for the experimental data, where the slope is equal to $\chi_i\ (\Delta H_T/L)$.

Fire Propagation. It has been shown that the fire propagation behavior can be quantified using a Fire Propagation Index (FPI) (4,6). FPI is expressed as the ratio of the radiative heat release rate to the TRP,

$$FPI\ \alpha\ [(\chi_R \dot{Q}'_{Ch})^{1/3}\ /\ TRP]\ x\ 1000, \tag{7}$$

where $\dot{Q}'_{Ch}$ is the chemical heat release rate per unit width or circumference of the sample (kW/m), and χ_R is the radiative fraction of the chemical heat release rate, assumed to be constant and equal to 0.40. FPI is quantified by measuring $\dot{Q}'_{Ch}$ as a function of time during fire propagation and by measuring TRP in separate ignition experiments.

For small samples, the Oxygen Index (OI) is used to characterize fire propagation behavior (25). OI is defined as the minimum concentration of oxygen, in an oxygen nitrogen atmosphere, necessary to sustain flaming combustion for a specified period of time or specified sample length.

The OI value is affected by the temperature of the environment (26). With increase in temperature, the OI values decrease, and thus in many cases instead of OI a Temperature Index (TI) is used. TI of a material is defined as a temperature of the environment at which its OI becomes equal to the concentration of oxygen in normal air. TI and its profile are found to be more reliable than OI for the assessment of the flammability of materials (12).

Generation of Smoke, Toxic and Corrosive Fire Products. Smoke, toxic and corrosive products are generated in fires as a result of vaporization, decomposition and combustion of materials in the presence or absence of air.

It has been shown that the generation rate of fire products satisfies the following relationship (2,3):

$$\dot{G}''_j = Y_j \, \dot{m}'' = f_i \, (k_j/L) \, \dot{q}''_n \, , \tag{8}$$

where $\dot{G}''_j$ is the generation rate of product j (g/m^{2}s); f_j is the generation efficiency of the product and k_j/L is a fundamental physico-chemical property of the material.

Generation rate of smoke can be quantified by measuring the mass of smoke and/or the optical density of smoke, D, defined as:

$$D = (1/\ell) \, \log_{10} \, (I_o/I) \, , \tag{9}$$

where ℓ is the optical path length (m) and I/I_o is the fraction of light transmitted through smoke. D is expressed as the mass optical density (MOD) in m^2/g, where only fuel mass loss rate, $\dot{m}''$, is taken into consideration (17).

$$MOD = (1/\ell) \, [\log_{10} \, (I_o/I)] \, \dot{V}/\dot{m}'' \, A \, , \tag{10}$$

where $\dot{V}$ is the total volumetric flow rate of the fire product-air mixture (m^3/s). If the generation rate of smoke is considered, then from Equations (8) and (10), specific mass optical density (SMOD) in m^2/g can be defined as:

$$SMOD = (1/\ell) \, [\log_{10} \, (I_o/I) \,] \, \dot{V}/\dot{m}'' Y_s A \, , \tag{11}$$

where Y_s is the yield of smoke (g/g). Thus SMOD = MOD/Y_s.

Fire Extinguishment. The efficiency of fire extinguishment depends on the rate of agent application and the ability of the agent to interrupt the chemical reactions responsible for generating heat in the gas phase or removing heat from the surface of the burning material. No small-scale techniques are currently available for quantifying fire extinguishment; however, recently an attempt has been made to develop them using the FMRC Small-Scale Flammability Apparatus (21).

Experiments

Thermal Analysis. The DuPont Instruments Model 9900, computer controlled thermal analyzer and Model 951 TGA module were used in the experiments, using a gas flow rate of 100 cc/min. Experiments were performed in dynamic and isothermal mode using air and argon.

Oxygen Index and Temperature Index. A Stanton-Redcroft FTA/HFTA Oxygen Index apparatus was used. Experiments were performed using the ASTM D 2 863-77 procedures with temperature variations between ambient and 300°C.

Smoke Density. The smoke density was measured in an NBS Smoke Chamber following the procedures of ASTM E 662.

Pyrolysis-Gas Chromatography-Mass Spectrometry. In the experiments, about 2 mg of sample was pyrolyzed at 900°C in flowing helium using a Chemical Data System (CDS) Platinum Coil Pyrolysis Probe controlled by a CDS Model 122 Pyroprobe in normal mode. Products were separated on a 12 meter fused capillary column with a cross-linked poly (dimethylsilicone) stationary phase. The GC column was temperature programmed from -50 to 300°C. Individual compounds were identified with a Hewlett Packard (HP) Model 5995C low resolution quadruple GC/MS System. Data acquisition and reduction were performed on the HP 100 E-series computer running revision E RTE-6/VM software.

Ignition, Mass Loss Rate, Heat Release Rate, Generation Rates of Fire Products and Fire Extinguishment. Experiments were performed in the FMRC Small-Scale (50 kW) Flammability Apparatus shown in Figure 1. Horizontal, 0.10 x 0.10 m samples with edges covered tightly with heavy duty aluminum foil were exposed to external heat flux up to a maximum of 60 kW/m^2. A 0.01 m long ethylene-air premixed flame located about 0.01 m from the surface, was used as a pilot flame for ignition. Ignition experiments were performed under natural air flow; all other experiments were performed under forced air flow conditions. The sample surface was coated with carbon black to eliminate errors due to differences in the surface absorptivity. In the experiments, the generation rate of material vapors was monitored by measuring the mass loss rate using a load cell. For the determination of heat release rate, generation rates of fire products and optical density of smoke, all the fire products were captured in the sampling duct along with air. In the duct, measurements were made for the total volumetric and mass flow rate of the fire product-air mixture, concentrations of various fire products, optical transmission through smoke and gas temperature.

For the quantification of fire propagation behavior of the FRC materials, 0.10 m wide and 0.61 m long vertical sheets with thickness varying from 3 mm to 5 mm were used. The bottom 0.15 m of the sheet was exposed to 50 kW/m^2 of external heat flux in the presence of a 0.01 m long pilot flame to initiate fire propagation. For the simulation of large-scale flame radiation, experiments were performed in 40% oxygen concentration.

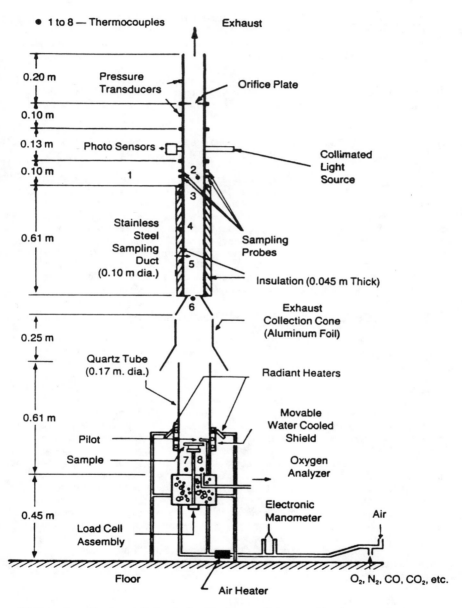

Figure 1.   Factory Mutual Small-Scale (50 kW-Scale) Flammability
Apparatus

Fire extinguishment behavior of the FRC materials using Halon
1301 was quantified with a horizontal 0.10 x 0.10 m sample with edges
covered tightly with heavy duty aluminum foil. The sample surface
was exposed to 60 kW/m^2 of external heat flux. Experiments were performed under forced air flow conditions, where Halon 1301 was added
to the inlet air flow such that fire remained well ventilated.

For the assessment of flame heat flux, expected in large-scale
fires, 0.10 x 0.10 m samples with edges covered tightly with heavy
duty aluminum foil, were burned in 40% oxygen concentration without
the external heat flux. Mass loss rate was measured and Equation (1)
was used to calculate flame heat flux.

The samples used in the study are listed in Table I. Tables II
through VI list the experimental data.

Table I. Fiber Reinforced Composite Materials
Used in the Study

Sample No.	Thickness (mm)	Fiber/Resin Components	Ratio
MTL #1	4.8	S2/Polyester	70/30
MTL #2	4.8	S2/Polyester	70/30
MTL #3	4.8	S2/Polyester	70/30
MTL #4	4.8	Kevlar/Phenolic-PVB*	84/16
MTL #5	3.2	S2/Phenolic	80/20

* Resin is 50/50 Phenolic-PVB.

Table II. Isothermal Thermogravimetric Analysis Data
for FRC Materials

Sample No.	300	400	500	Residue at 500°C (%)
MTL #1	0.73	29.8	32.9	60.1
MTL #2	2.0	24.1	29.1	70.5
MTL #3	3.1	25.1	30.0	67.4
MTL #4	3.1	28.9	54.0	10.4
MTL #5	0.0	3.1	9.1	82.1

Table III. Oxygen Index for FRC Materials Versus Temperature

Sample No.	25	100	200	300
MTL #1	23	23	<10	<10
MTL #2	28	28	13	<10
MTL #3	52	95	77	41
MTL #4	28	30	29	26
MTL #5	53	98	94	80

Table IV. Fire Properties of Fiber-Reinforced

Sample No.	Material	Critical Heat Flux for Ignition (kW/m^2)	Ignition Temperature °C (°F)	Thermal Response (kWs^2/m^2)	Heat of Gasification (kJ/g)	Flame Heat Flux[a] (kW/m^2)
MTL #1	S-2/Polyester (MTL Base line, E-701)	10	355 (671)	382	6.4	51
MTL #2	S-2/Polyester (Owens-Corning Prepreg.)	15	424 (796)	406	5.1	21
MTL #3	S-2/Polyester (American Cyanamide PrePreg.)	10	355 (671)	338[1] 417[2] 476[3]	2.9	37
MTL #4	Kevlar/Phenolic PVB (Russell Corp Spall Liner)	15	424 (796)	403	7.8	21
MTL #5	S-2/Phenolic (Owens Corning Spall Liner)	20	478 (892)	610	7.3	20
	Douglas fir	11	390 (734)	---	1.82	--
	Polypropylene	18	478 (892)	---	2.03	--
	Polystyrene	13	419 (786)	---	1.70	--

[a]For combustion in normal air with a sample area of about 0.01 m^2, exposed to external heat flux in the range of 30 to 60 kW/m^2.
[b]Depletion yield.
[c]CH represents total gaseous hydrocarbons.
[d]Not used in averaging.
[e]Heat flux not used.
[1]Thickness 4.8 mm
[2]Thickness 19 mm
[3]Thickness 45 mm

Composite Materials

External Heat Flux (kW/m^2)	Residue (% of Initial Weight)	Chemical Heat of Combustion (kJ/g)		Yield (g/g)					Mass Optical Density (m^2/g)
		From CO & CO_2	From O_2	CO	CO_2	O_2[b]	Sm	CH[c]	
30	63	17.7	18.5	0.050	1.40	1.47	0.070	0.0086	0.932
45	63	17.7	17.6	0.058	1.45	1.44	0.068	0.0060	0.897
60	62	17.9	18.0	0.053	1.47	1.48	0.073	0.0071	1.070
Average	63	17.9		0.055	1.47	1.44	0.070	0.0072	0.966
30	68	15.0	18.1	0.042	1.23	1.48	0.051	0.0055	0.809
45	67	15.6	15.1	0.038	1.28	1.24	0.051	0.0030	0.804
60	67	14.6	17.3	0.037	1.20	1.42	0.059	0.0031	1.080
Average	67	16.0		0.039	1.24	1.38	0.054	0.0039	0.898
30	69	7.9	9.0	0.122	0.650	0.741	0.066	0.020	0.874
45	70	9.0	9.7	0.085	0.738	0.793	0.066	0.018	0.960
60	69	9.0	11.0	0.099	0.741	0.903	0.072	0.018	1.03
Average	69	9.3		0.102	0.710	0.812	0.068	0.019	0.955
30[d]	24	11.9	--	0.021	1.06	--	0.019	0.0019	0.175
45	26	15.4	14.1	0.031	1.25	1.22	0.034	0.0021	0.550
60	24	14.4	15.3	0.018	1.29	1.32	0.048	0.0012	0.790
Average	25	14.8		0.025	1.27	1.27	0.041	0.0017	0.0670
30[e]	--	--	--	--	--	--	--	--	--
45	91	12.5	--	0.075	1.08	--	0.027	0.0027	0.207
60	90	10.2	13.1	0.057	0.88	1.04	0.019	0.0024	0.332
Average	91	11.9		0.066	0.98	1.04	0.023	0.0026	0.270
--	--	13.0		0.004	1.31	--	0.015	0.001	0.092
--	--	38.6		0.024	2.79	--	0.059	0.007	0.553
--	--	27.0		0.060	2.33	--	0.164	0.014	0.771

Table V. Peak Fire Propagation Index Values for
Fiber Reinforced Composite Materials

Sample No.	Thickness (mm)	Peak FPI
MTL #1	4.8	13.3
MTL #2	4.8	ND
MTL #3	4.8	9.7
	19	7.8
	45	6.6
MTL #4	4.8	7.8
MTL #5	3.2	3.2
Fluorinated Ethylene-Propylene Cable[a]	-	5.0
PE/PVC Cable[b]	-	>20

[a] No self-sustained fire propagation; classified as Group 1 cable (FPI <10)[6];
[b] Very rapid self-sustained fire propagation; classified as Group 3 cable (FPI ≥ 20)[6];
ND: Not determined.

Table VI. Volume % of Halon 1301
Required for Flame Extinction

Sample No.	Volume (%)
MTL #1[a]	4.0
MTL #2[a]	4.0
MTL #3[a]	2.8
MTL #5[a]	4.0
Wood[b]	4.0
Polyethylene[b]	3.9
Polystyrene[b]	3.9
Polyvinylchloride[b]	2.6

a: Fiber reinforced composite materials examined in this study. 0.01 x 0.01 m samples exposed to 60 k/W/m^2 of external heat flux.

b: Data reported by various authors compiled in Ref. 27. Experimental conditions are not defined.

Figure 2 shows an example of the PY-GC-MS data. Many compounds are generated in the pyrolysis of FRC materials (at 900°C in Helium). The total number of compounds identified using the PY-GC-MS technique varied from 19 for MTL #5 to 39 for MTL #2.
     Figure 3 shows plots of FPI values for selected FRC materials.

Discussion

Thermal Analysis. Figure 4 shows a plot of percent weight loss versus temperature for the FRC materials. As can be noted, the FRC materials sustain little thermal damage below 200°C. Major weight loss due to the decomposition of the matrix resin occurs between 300

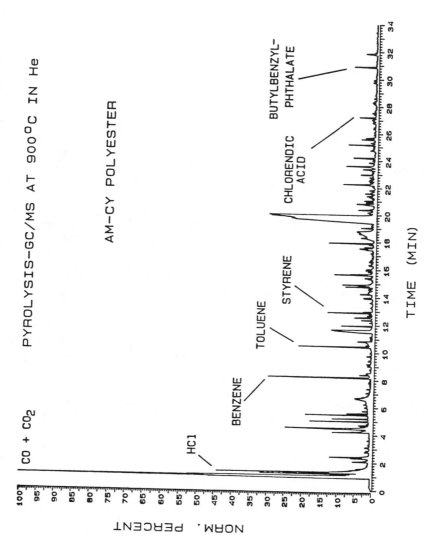

Figure 2.   Pyrolysis Products of MTL #3 Sample

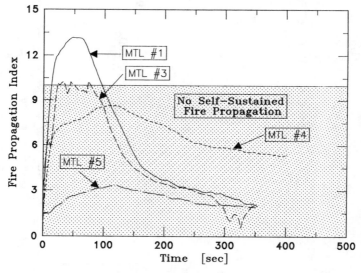

Figure 3. Fire Propagation Index for Fiber Reinforced Composite
Materials. Sheet Length: 0.61 m; Width: 0.10 m; Thickness:
3-5 mm; External Heat Flux: 50 kW/m^2; Oxygen Concentration: 40%.

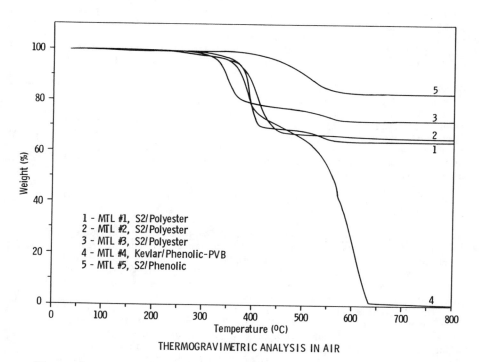

THERMOGRAVIMETRIC ANALYSIS IN AIR

Figure 4. Weight Percent as a Function of Temperature for FRC Materials

to 500°C. For the MTL #4 sample, in which an organic fiber is combined with an organic matrix resin, a second major weight loss occurs
between 500 and 650°C. The maximum weight loss for FRC material
samples MTL #1, #2, #3, #4 and #5 occurs respectively at 385, 423,
350, 390 (second peak 575) and 490°C. These temperatures are very
close to the temperatures estimated from the critical heat flux in
Table IV, determined in the ignition experiments in the FMRC Small-
Scale Flammability Apparatus. It is interesting to note that the TG
data may be related to the ease of ignition of materials and possibly
to the fire propagation behavior of the FRC materials.
     The data (Table II) for the percent residue at 500°C under iso-
thermal thermogravimetric analysis also show reasonable agreement
with the data for the residue from the experiments in the FMRC Small-
Scale Flammability Apparatus where large-scale fire conditions are
simulated. Thus, the TG analysis for flammability assessment of FRC
materials may be more useful than previously considered.

Oxygen Index (OI). The OI data in Table III show that for MTL #1 and
#2 samples OI decreases with temperature and becomes very low above
about 200°C. For MTL #3 and #5, OI increases up to about 200°C and
then decreases rapidly above about 300°C. OI values for MTL #4
appear to decrease very slowly with increase in temperature. The
differences in the OI values at different temperatures may be due to
the differences in the flame radiative heat transfer characteristics
of the FRC materials (flame radiative heat transfer changes with
oxygen concentration) (7). The self-sustained fire propagation behavior of materials in large-scale fires, however, may be very different than indicated by the OI-TI values.

Mass Optical Density of Smoke from NBS Smoke Chamber. Figure 5 shows
the data for the mass optical density measured in the NBS smoke
chamber for the flaming and nonflaming fires under static conditions.
For MTL #1 and #2 samples, the MOD values in flaming fires are lower
than in the nonflaming fires. For the other three samples, for which
the MOD values are smaller, the trend is reversed.
     The MOD values in the flaming fire for the FRC materials have
also been measured in the FMRC Small-Scale Flammability Apparatus
under dynamic conditions (Table IV). A comparison of these values
with the MOD values from the NBS Smoke Chamber is shown in Figure 6.
For flaming fires, it appears that MOD values measured in the NBS
Smoke Chamber are somewhat lower than the values measured in the FMRC
Small-Scale Flammability Apparatus, possibly due to: 1) differences
in the wavelength of light used in the measurements. In the FMRC
Apparatus, MOD values are measured at three wavelengths of light
(0.458, 0.633 and 1.06 micron). The data reported in this paper are
for the 0.633 micron wavelength. In the NBS Smoke Chamber, the wavelength of light is not specified; 2) differences in fire ventilation;
and 3) stratification of fire products within the chamber. The NBS
Smoke Chamber is a closed system and the fires might be under-
ventilated. The MOD values in the FMRC Apparatus are for well-
ventilated flaming fires.
     The MOD values are functions of the yield of smoke as suggested
by Equation (11). Thus a plot of MOD versus Y_s is expected to be a

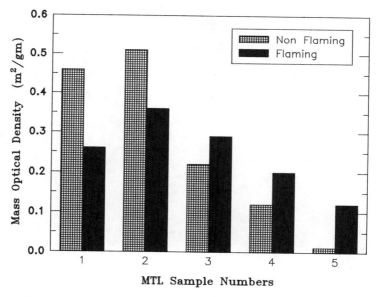

Figure 5. Mass Optical Density of Smoke Measured in the NBS Smoke Chamber for the Fiber Reinforced Composite Materials

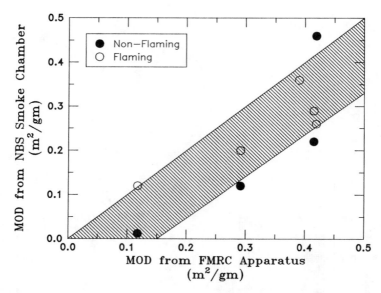

Figure 6.   Correlation Between the Mass Optical Density Data for
Fiber Reinforced Composite Materials Measured in the FMRC Small-
Scale Flammability Apparatus for Flaming Fires and the NBS Smoke
Chamber for Flaming and Non-Flaming Fires

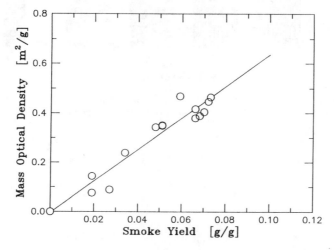

Figure 7.   Relationship Between Mass Optical Density and Yield of
Smoke for Fiber Reinforced Composite Materials

straight line with a slope equal to SMOD. This is shown in Figure 7 for the FRC materials. A first order polynomial regression analysis of the data suggests that for the FRC materials, SMOD = 6.4 m^2/g, a value close to the SMOD values for non FRC materials.

Resistance to Ignition and Fire Propagation. In this study, resistance to ignition and fire propagation was examined in terms of critical heat flux for ignition and TRP. Data for these parameters are listed in Table IV. In general, materials with thicknesses of about 5 mm with TRP values greater than about 300 $kW/s^{1/2}/m^2$, show high resistance to ignition and fire propagation. The FRC materials examined in this study satisfy these conditions and thus are expected to have high resistance to ignition and fire propagation.

Heat Release Rate. Heat release rate has been examined in terms of the ratio of heat of combustion to heat of gasification and the flame heat flux (Equation 6). The ratios are plotted in Figure 8 and the flame heat flux is plotted in Figure 9 for FRC and non-FRC materials. The heat of combustion, heat of gasification and flame heat flux for FRC materials are listed in Table IV. In Figure 8, the ratios of the heat of combustion to heat of gasification within the FRC materials do not vary appreciably and are significantly lower than the ratios for the non-FRC materials, including wood. In Figure 9, flame heat flux values for the FRC materials are significantly lower than the values for the non-FRC materials, with the exception of MTL samples #1 and #3. These data thus suggest that in large-scale fires heat release rates for FRC materials are expected to be significantly lower than the rates for non-FRC materials under similar values of heat flux. Since the flame heat from a burning material for self-sustained fire propagation depends on the heat release rate, the self-sustained fire propagation is expected to be difficult for the FRC materials compared to the non-FRC materials.

Self-Sustained Fire Propagation. The self-sustained fire propagation for the FRC materials has been quantified in terms of the Fire Propagation Index (FPI). The profiles of the FPI values quantified for 0.10 m wide and 0.61 m long vertical sheets of the FRC materials have been shown in Figure 3 and the peak FPI values are listed in Table V. A comparison of FPI values for the FRC materials with values for cables, suggests that the FPI values for FRC materials, with the exception of MTL #1, are comparable to the FPI values for Group 1 cables for which self-sustained fire propagation is not expected and the fire hazard is expected to be limited to the burning of the FRC materials in the ignition zone.

Hazards Due to Thermal and Nonthermal Fire Environments. For FRC materials, the generation of heat and fire products are expected to be limited to the burning in the ignition zone. For relative comparisons of heat release rates and generation rates of fire products and light obscuration, the ratios of heat of combustion to heat of gasification, as shown in Figure 8, yield of individual fire products to heat of gasification and mass optical density to heat of gasification (data listed in Table IV), are useful. Under similar heat flux

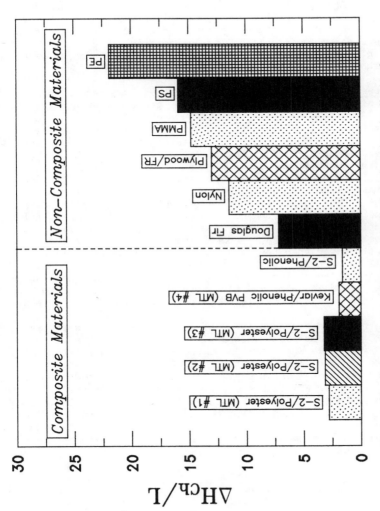

Figure 8. Ratio of the Chemical Heat of Combustion to Heat of Gasification for Fiber Reinforced Composite and Non-Composite Materials

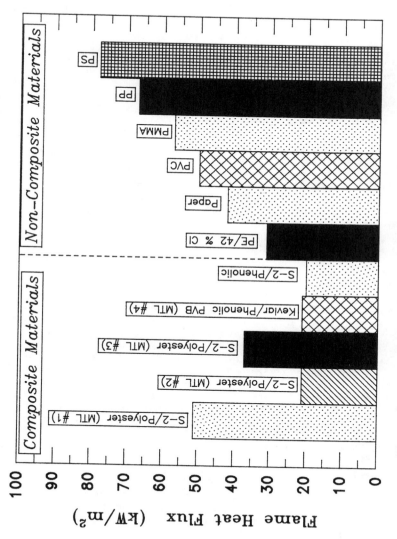

Figure 9. Flame Heat Flux for Fiber Reinforced Composite Materials and Non-Composite Materials Expected in Large-Scale Fires

in the ignition zone, the ratios suggest that for FRC materials, heat release rates, generation rates of fire products, and light obscuration by smoke are expected to be significantly lower than for the non-FRC materials; for non-FRC materials, fire is expected to propagate beyond the ignition zone, and the rates and light obscuration are expected to increase further.

Fire Suppression/Extinguishment by Halon 1301. As discussed previously, for FRC materials the fire propagation is expected to be limited to the ignition zone. Thus, fire suppression/extinguishment by Halon 1301 is required to be effective in this zone. In this study, fire suppression/extinguishment experiments for Halon 1301 were performed with FRC samples exposed to 60 kW/m^2, a very strong ignition source. The heat release rate decreases in the presence of Halon 1301, suggesting that combustion efficiency is decreased. This is supported by the data for the yield of CO (Figure 10). The yield of CO increases with increase in the Halon concentration until flame extinguishment.

The data reported in Table VI for Halon 1301 concentration required for flame extinguishment vary in the range of 3 to 4%, which is comparable to the range found for ordinary combustibles. Thus maintenance of Halon 1301 concentrations in excess of 4% is expected to extinguish fires in the ignition zone for the FRC materials; this concentration limit satisfies the current Halon 1301 requirements for fire suppression systems for tracked vehicles.

Summary

The thermogravimetric (TG) and ignition and mass loss experiments suggest that all of the FRC materials examined in the study are stable below about 200°C with maximum weight loss occurring between 350 and 490°C. The Thermal Response Parameter (TRP), which characterizes the degree of resistance to ignition and fire propagation, is found to be above 300 kWs$^{1/2}$/m^2 suggesting that self-sustained fire propagation is expected to be difficult for these materials.

The estimated vapor generation rates in the absence of external heat flux for MTL #1, #2, #3, #4 and #5 are 5.0, 0.43, 3.0, 1.0 and 0.0 g/m^{2}s, respectively, compared with the values of 10, 24 and 38 g/m^{2}s for wood, PP and PS, respectively. These estimated rates for MTL samples #2, #3, #4 and #5 are close to the critical values for ignition, defined as values at or below which ignition and propagation are not sustained.

For FRC materials, FPI values are less than 10, except for the MTL sample #1, thus fires are not expected to be self-sustained. This is consistent with the critical vapor generation rate for ignition and propagation.

The flame extinguishment data for Halon 1301 measured in this study suggest that for extended ignition zone covering large surface areas, maintenance of a concentration in excess of 4% by volume would be sufficient for the control of fire growth and extinguishment for the MTL samples.

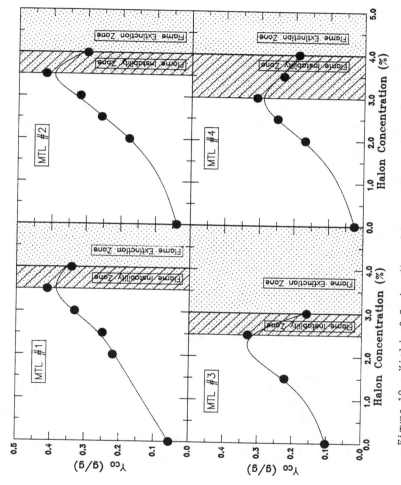

Figure 10. Yield of Carbon Monoxide as a Function of Halon Concentration for the Fiber Reinforced Composite Materials. External Heat Flux of 60 kW/m² in Normal Air was Used in the Experiments.

Acknowledgments

The authors wish to thank the following individuals for performing
the experiments: Mr. David A. Bulpett of AMTL (PY-GC-MS experiments), Ms. Claudette LeCroy of AMTL (Oxygen Index experiments) and
Mr. Stephen D. Ogden of FMRC (experiments in the FMRC's Small-Scale
Flammability Apparatus).

Literature Cited

1. Macaione, D.P., "Flammability of Guardrail IV System Components,"
   1982 Army Materials Technology Laboratory, Watertown, MA,
   Technical Report AMMRC TR 82-37.
2. Tewarson, A., "Experimental Evaluation of Flammability Parameters
   of Polymeric Materials," Flame Retardant Polymeric Materials,
   1982, edited by M. Lewin, S.M. Atlas and E.M. Pearce, Vol. 3,
   Chapter 3, p. 97, Plenum Press, New York, NY.
3. Tewarson, A., "Generation of Heat and Chemical Compounds in
   Fires," Fire Protection Handbook of the Society of the Fire Protection Engineering, 1988, edited by J.P. Dinenno, Chapter 1-13,
   p. 1-179, The National Fire Protection Association Press, Quincy,
   MA.
4. Tewarson, A., and Khan, M.M., "Flame Propagation for Polymers in
   Cylindrical Configuration and Vertical Orientation," Twenty-
   Second International Symposium on Combustion, The Combustion
   Institute, Pittsburgh, PA, (In press).
5. Tewarson, A., "Physicochemical and Combustion Pyrolysis Properties of Polymeric Materials," 1980, National Institute of Stan-
   dards and Technology, Gaithersburg, MD, Technical Report NBS-GCR-
   80-295.
6. Tewarson, A., and Khan, M.M., "Electrical Cables - Evaluation of
   Fire Propagation Behavior and Development of a Small-Scale Test
   Protocol," 1989, Factory Mutual Research Corporation, Norwood,
   MA, Technical Report J.I. OM2E1.RC.
7. Tewarson, A., Lee, J.L., and Pion, R.F., "The Influence of Oxygen
   Concentration on Fuel Parameters for Fire Modeling," Eighteenth
   Symposium (International) on Combustion, 1981, The Combustion
   Institute, Pittsburgh, PA, p. 563.
8. Tewarson, A., and Pion, R.F., Combustion and Flame, 1976, 26, 85.
9. Tewarson, A., and Newman, J.S., "Scale Effects on Fire Properties
   of Materials," Fire Safety Science, Proceedings of the First
   International Symposium, 1986, Hemisphere Publishing Corp., Washington, DC, p. 451.
10. Tewarson, A., and Pion, R.F., "Evaluation of the Flammability of
    a Glass Fiber Reinforced Epoxy Material," 1978, Factory Mutual
    Research Corporation, Norwood, MA, Technical Report J.I.
    OC2N8.RC.
11. Quintiere, J.G., Babrauskas, V., Cooper, L., Harkleroad, M.,
    Steckler, K., and Tewarson, A., "The Roll of Aircraft Panel
    Materials in Cabin Fires and Their Properties," 1985, The U.S.
    Department of Transportation, Federal Aviation Administration,
    Technical Center, Atlantic City, NJ, Technical Report DOT/FAA/CT-
    8/30.

12. Macaione, D.P. and Dowling, R.P., "Flammability Assessment Tests for Organic Materials," 1977, Army Material Technology Laboratory, Watertown, MA, Technical Report AMMRC TR 77-19.
13. "Standard Test Method for Heat and Visible Smoke Release Rates for Materials and Products," 1984, The American Society for Testing and Materials, Philadelphia, PA, ASTM E906-83.
14. Babrauskas, V., "Development of the Cone Calorimeter - A Bench-Scale Heat Release Rate Apparatus Based on Oxygen Consumption," 1984, Fire and Materials, 8, 81.
15. Lee, T.G., "The Smoke Density Chamber Method for Evaluating the Potential Smoke Generation of Building Materials," 1973, National Bureau of Standards (National Institute of Standards and Technology), Gaithersburg, MD, Technical Note, 757, ASTM E-662).
16. Macaione, D.P., "Characterization of FP100 and FP102 Polyester and Polyurethane Materials," 1984, Army Materials Technology Laboratory, Watertown, MA, Technical Report AMMRC TR 84-15.
17. Tewarson, A., "Generation of Smoke from Electrical Cables," International Symposium on Characterization and Toxicity of Smoke, American Society of Testing and Materials, Philadelphia, PA (in press).
18. Tewarson, A., "Relationship between Generation of CO and CO_2 and Toxicity of the Environment Created by Flaming and Non-Flaming Fires and effect of Fire Ventilation," International Symposium on Characterization and Toxicity of Smoke, American Society of Testing and Materials, Philadelphia, PA (in press).
19. Fire and Smoke: Understanding the Hazards, 1986, Committee on Fire Toxicology, Board on Environmental Studies and Toxicology Commission on Life Sciences, National Research Council, National Academy Press, Washington, DC.
20. Yao, C., "The Development of the ESFR Sprinkler System," 1988 Fire Safety Journal, 14, 65.
21. Tewarson, A., "Fire Suppression and Extinguishment in Small-Scale Experiments," Factory Mutual Research Corporation, Norwood, MA, Technical Report (to be published).
22. Nernst, W., and Reisenfield, E.H., Berichte, 1903, 36, 2086.
23. Brill, O., Z. Anorg. Chem., 1905, 45, 275.
24. Honda, K., Science Report Tohoku University, 1915, 4, 97.
25. Fenimore, C.P., and Martin, F.J., Modern Plastics, 1966, 44, 141, (ASTM D-2863-74).
26. Routley, F.F., Central Dockyard Laboratory, H.M. Dockyard, Portsmouth, U.K., Technical Report No. 5/73, 1973
27. Steciak, J., and Zalosh, R.G., "Preliminary Reliability Study of Halon 1301 Extinguishing Systems," 1989, Factory Mutual Research Corporation, Norwood, MA, Technical Report J.I. ON1J7.RU/0Q4R5.RU.

RECEIVED November 20, 1989

Chapter 33

Room Fires and Combustible Linings

Björn Karlsson and Sven Erik Magnusson

Department of Fire Safety Engineering, Institute of Science and Technology, Lund University, Box 118, 221 00 Lund, Sweden

An extensive research program on combustible wall lining materials has been carried out in Sweden. Several lining materials were tested in full scale room tests and 1/3 scale model room tests for two different scenarios, A and B. Scenario A refers to the case where walls and ceiling are covered by the lining material, scenario B where lining mate-- rials are mounted on walls only. A model is presented using material properties derived from standardized bench–scale tests as input data. The model predicts the fire growth in the full or 1/3 scale tests, which includes predicting the rate of heat release, gastemperatures, radiation to walls, wall surface temperatures and downward flame spread on the wall lining material.

This report presents experimental and theoretical results from a study within the project "Fire Hazard — Fire Growth in Compartments in the Early Stage of Development (Preflashover)". The project is carried out jointly by the department of Fire Safety Engineering at Lund University and the Division of Fire Technology at the Swedish National Testing Institute. An outline of the research program is given by Pettersson (1).
    Room fire growth on combustible linings has been a problem of concern to the legislators and authorities since the advent of building fire safety regulations. Work in this area has included development of bench–scale tests to derive basic flammability charac— teristics which could rationally be used as classification criteria. Also, full–scale standard test have been developed to evaluate the fire performance of materials of products under actual in—use situa— tions. The contribution of a specimen to the fire growth within a previously calibrated compartment can then be used to rate materials and to evaluate the validity of existing bench–scale test methods.
    The purpose of the work presented here is to use results from bench–scale flammability tests as input to a mathematical model which could rationally predict full scale fire growth on combustible linings. Only two scenarios are considered here; scenario A, where

0097–6156/90/0425–0566$07.50/0

the lining materials are mounted on compartment walls and ceiling; and, scenario B where the material is mounted on walls only.

Two room sizes were considered; the full scale test room with a single door opening in accordance with methods proposed by ASTM, ISO and NORDTEST; and, a 1/3 scale model of the full scale compartment.

Bench–scale tests

Surface spread of flame test. The 13 materials listed in Table I were tested in the IMO and ISO surface spread of flame tests. The velocity of the flame front, V_f, and the corresponding external heat flux, q''_e, were measured at several positions along the sample. The sample surface temperature was also measured at some of these positions during the test. A detailed description of the test appa–ratus, procedure and results is given in (2).

Determination of the minimum radiant heat flux to sustain piloted ignition. Harkleroad, Quintiere and Walton (3) determined $\dot{q}''_{0,ig}$ from the surface spread of flame test. The method of plotting $1/\sqrt{V_f}$ versus $\dot{q}''_e$ F(t) was introduced, where the intercept on the latter axes gave $q''_{0,ig}$. An expression for the minimum radiant heat flux to sustained piloted ignition was given as

$$\dot{q}''_{0,ig} = h (T_{ig} - T_i) \tag{1}$$

The values of $q''_{0,ig}$ used in this work were reported in (2). Andersson (4) used data from Ref. 2 and the method of plotting $1/\sqrt{V_f}$ versus q''_e F(t) to obtain the flame spread parameter C. Figure 1 shows, as an example, the plot for material no.3 from in Table I. The slope of the line resulting from the plot of $1/\sqrt{V_f}$ versus q''_e F(t) gives this parameter. The values of C reported in (4) and the values obtained in (3) were compared and there seemed to be no large differences in the data except for material no. 12, wood panel, spruce.

Determination of ignition temperature. The ignition temperature, T_{ig}, is an important material parameter when studying opposed flow flame spread over a solid. This temperature can be calculated from equation 1 above but to do this the parameters $q''_{0,ig}$, h and T_i need to be determined.

The values of $q''_{0,ig}$ used here were those derived by Sundström (2). An expression for the heat transfer coefficient, h, is given in (3) as:

Table I. Results from the Surface Spread of Flame Test and the Cone Calorimeter

Material no.	Material name	$\dot{Q}_{max}$ kW/m²	λ s⁻¹	$k\rho c*10^{-3}$ (w/m²K²)	ϕ (m/s)K²
1	Insulating fiberboard	139.8	0.0070	41	63
2	Medium density fiberboard	162.4	0.0027	80	15
3	Particle board	199.8	0.0049	110	14
4	Gypsum plasterboard	27.7	0.0150	100	–
5	PVC cover on gypsum pl. board	107.5	0.0293	75	–
6	Paper cover on gypsum pl. board	105.3	0.0208	100	1
7	Textile cover on gypsum pl. board	222.0	0.0278	80	–
8	Textile cover on mineral wool	246.2	0.0382	4.3	34
9	Melamine-faced particle board	40.9	-0.0032	105	–
10	Expanded polystyren	–	–	–	–
11	Rigid polyurethan foam	130.6	0.0217	4.0	–
12	Wood panel, spruce	149.7	0.0086	85	–
13	Paper cover on particle board	164.1	0.0035	110	19

Flamespread correlations

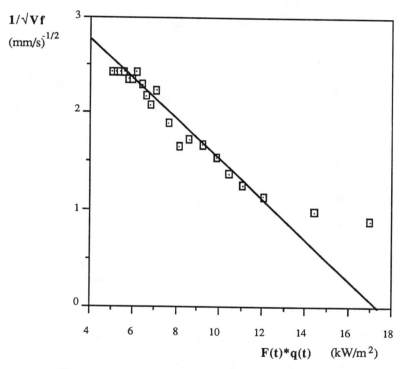

Figure 1. Surface spread of flame test for particle board.

$$h = 0.01 \ (1 + 0.0085 \ (T_s - T_i)) \tag{2}$$

The initial temperature, T_i, was fixed at 20°C for all materials. By assuming a value for T_{ig} and substituting it for T_s when calculating h, T_{ig} could be calculated by an iterative process. The values of T_{ig} obtained in (3) and (4) were compared, the agreement was relatively good.

Determination of the flame spread parameter, φ. The following well known expression has been given ((5), (6), etc) for the velocity of the flame front for a slab initially at the temperature $T = T_s$

$$V_f = \phi/(T_{ig} - T_s)^2 \tag{3}$$

where φ depends on the thermal properties of the solid, the ambient oxygen concentration, flow speed and the heat flux ahead of the advancing flame. V_f is determined from the surface spread of flame test as

$$V_f = dx_f/dt \tag{4}$$

The flame spread parameter φ can be determined from the experimentally-determined parameter C

$$\phi = V_f/(T_{ig} - T_s)^2 = 1/(Ch)^2 \tag{5}$$

The heat transfer coefficient, h, and the flame spread constant, C, were determined as discussed above. Andersson (4) compared the resulting values of φ to those reported in (3) and found the agreement satisfactory. The values of φ used in this work are listed in Table I.

Ignitability test. The test — ISO TC92 TR5657, Fire Tests, Reaction to Fire, Ignitability of Building Products — is described in (28). The main quantitative information from the test is a set of values of t_{ig} for a set of exposure radiation levels q''_e. Data can also be extrapolated to give the minimum level of impressed flux to cause ignition, $q''_{0,ig}$.

With additional thermocouples attached to both sides of the sample the test can be used to derive parameters such as thermal conductivity, k, and thermal capacity, ρc, of the tested specimen. The values of thermal inertia, $k\rho c$, used in this study were derived in this way and are listed in Table I. A full description of the method used to derive these parameters is given in (11).

Rate of Heat Release measurements. The 13 materials were tested in three different RHR apparatuses: the Ohio State University apparatus

(7), an open configuration (8) based on a design originally developed by the National Institute of Standards and Technology (formerly National Bureau of Standards) (9), and the cone calorimeter (10). The measurements referred to here were reported in (8).

The equipment consists of a vertical sampleholder and an electrical radiation panel placed under an open hood. The samples were tested at 5, 3 and 2 W/cm² and some easily ignitable mate—rials also at 1 W/cm². An example of the test output is given in Figure 2 for material no. 3.

An attempt to calculate the mass loss and RHR analytically was not successful, as described in (11). Therefore it was decided to describe and make use of the RHR characteristics of the involved material directly, using a mathematical approximation of the curves shown in Figure 2, primarily the curves valid for external flux equal to 3 W/cm². In the full scale experiments, heat fluxes to the lining material will vary considerably with time and location. A study of available literature indicated that an average value of 3 W/cm² might be more representative than 5 W/cm², but this has not been substantiated.

The experimental curves were idealized as seen in Figure 3, resulting in the expression

$$\dot{Q}''(t) = \dot{Q}''_{max} \, e^{-\lambda(t - t_p)} \tag{6}$$

Equation 6 assumes semi − infinite sample (no returning heatwave) and may have to be changed. The Q_{max} values were taken directly from measurements and are given together with the corre—sponding regression values of λ in Table I. Equation 6 seemed phenomenologically correct except for materials no 9 and 10. Full results are given in (11).

Corner test experiments carried out in Sweden

Full scale tests carried out at the Swedish National Testing Insti—tute, Borås, Sweden. The full scale tests, scenario A, were carried out according to the standard test method NT FIRE 025 (12). The fire test room is 3.6 m long, 2.4 m wide and 2.4 m high with a doorway measuring 0.8 m wide and 2.0 m high. The walls are of lightweight concrete, 150 mm thick. The ignition source was a propane gas burner situated on the floor, in a corner of the room, with an effect of 100 kw. If this effect did not cause flashover in 10 minutes the effect was raised to 300 kw for another 10 minutes. The 13 different lining materials tested are listed in Table I. All tests were terminated after flashover, defined as flames emerging out of the doorway.

This test series will hereafter be referred to as "full scale A". A detailed description of this test series is given by Sundström (13).

For scenario B, the same room, ignition source and procedure were used as described above for scenario A, except that no lining material was mounted on the ceiling. Only 3 materials were tested, namely materials no. 2, 3 and 8 in Table I.

This test series will hereafter be referred to as "full scale B". Ondrus (14) described the test series shortly.

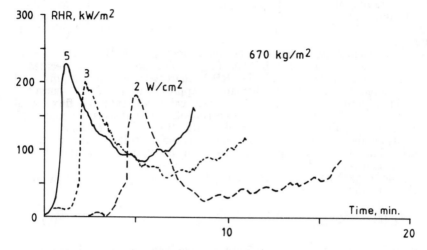

Figure 2. Results from RHR bench-scale test for particle board.

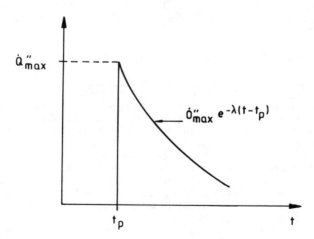

Figure 3. Principle for analytical approximation of experimental RHR curves.

1/3 scale experiments carried out at the Department of Fire Safety
Engineering, Lund University. The experiments were carried out in a
room with a length of 1.2 m, width of 0.8m and height of 0.8 m.
A description of the test procedure and results is given by
Andersson (15). The room is a 1/3 scale model of the full scale
compartment at the National Testing Institute in Borås. The ignition source was a gas burner with an effect of 11 kw for 10
minutes and, if no flashover occurred, 33 kw for another 10 minutes.

For scenario A, where lining material was mounted on both
ceiling and walls, the doorway measured 0.56 m wide and 0.67 m
high. This test series will hereafter be referred to as "1/3 scale
A".

In scenario B, where the lining material was mounted on
walls only, the door opening measured 0.46 m wide and 0.67 m
high. This test series will hereafter be referred to as "1/3 scale
B".

Results from 1/3 scale and full scale tests. The four experimental
series discussed above were carried out over the period of four years
and therefore not very coherent. In some series heat flux was
measured at floor level only, in others at different heights at the
lining material surface as well as at the floor level. Similarly, gas—
temperatures and surface temperatures were measured at different
heights resulting in difficulties when comparing results between series.

Calculation of RHR in room fire experiments

RHR in scenario A. Magnusson (11) developed a model allowing the
RHR in scenario A to be calculated. The method presented in this
chapter is based on that work. Scenario A, as mentioned above,
refers to the case where lining materials are on three walls and
ceiling. Soon after the gas burner is started in the corner test, the
wall material behind it ignites. The time for this to happen must
be evaluated as well as the RHR from the burning wall and ceiling
lining material.

Time to ignition. A quantity $\dot{Q}_{start}$ is defined as the sum of the
heat release from the ignition source and the vertical wall area
behind the burner, assuming complete combustion. The corresponding
time t_{start} denotes the time necessary for the whole of the the
lining material behind the burner to be pyrolysing. t_{start} is taken
directly from the experimental time − RHR curves and thus includes
time delay components such as transportation time in the measure—
ment system.

Comparison of t_{start} values with results from the ISO ignita—
bility test can be done in various ways. It was found that the
simple procedure of correlating ignition time at the 30 kw/m² im—
pressed radiation level with t_{start} seemed to work best. For the
full scale series

$$t_{start} = t_{ign} + 5$$

and for the 1/3 scale series

$$t_{start} = t_{ign} *1.85$$

gives a reasonable approximation of t_{start}. The different dependencies of t_{start} in t_{ISO} reflects the difference in thermal load from the 100 kw and 11 kw gas burner flame respectively.   Observe that the strength of the ignition source means that the total height of the corner is covered by flame from the start of the experiment.

Calculation of RHR. The model is based on the concepts presented in (16), (17) and (18) from which A_p may be considered as a driving force in a process where the rate of increase of A_p is proportional to the quantity A_p itself; i.e. A_p is exponentially in— creasing with time.   In the regression model of ceiling flame spread and combustion presented in (11), which includes the horizontal wall flame propagation along the intersection ceiling − wall,   pyrolysis area A_p was written as

$$A_p(t) = \alpha(e^{at} - 1)^{\beta} \tag{7}$$

where $a = h^2/k\rho c$ and α and β were coefficients to be determined statistically.   The rate of heat release could then be expressed as

$$\dot{Q}(t) = A_p(t) * \dot{Q}''_{av} \tag{8}$$

where $\dot{Q}''_{av}$ denotes a suitable time and space averaged measure of material rate of heat release per unit area.   It was shown in chapter 2.4 that $\dot{Q}''(t)$ for a certain constant impressed heat flux could be written as

$$\dot{Q}''(t) = \dot{Q}''_{max} e^{-\lambda(t - t_p)} \tag{9}$$

Combining Equations 7 to 9 and describing the interaction of flame spread and rate of heat release by a superposition, Duhamel−type integral (19), a final form of the regression equation was given in (11) as

$$\frac{\dot{Q}_{rt} - \dot{Q}_{start}}{\dot{Q}_{cf}} = \alpha\left[e^{at} - e^{-\lambda t}\right]^{\beta}\dot{Q}''_{max}\left[\frac{a}{a + \lambda}\right] \tag{10}$$

where Q_{start} is as defined earlier, Q_{rt} denoting measured RHR (rt meaning room test) and Q_{cf} the non—combustible part of wall corner flame reaching the ceiling ($= Q_{start}$ minus combustion in the verti—cal part of the corner flame). The time t is measured from t = t_{start}. For the relatively short, initial period $0 < t < t_{start}$ the rate of heat release is assumed to grow linearly up to Q_{start}.

The overall average values and coefficients of variations for model eqn (10) were given as:

$$\alpha'_{aver} = -4.58 \ (= \ln \alpha) \quad \sigma_{\alpha'}/\alpha'_{aver} = 0.13$$

$$\beta_{aver} = 1.15 \quad \sigma_{\beta'}/\beta_{aver} = 0.276$$

It remains to be studied how well time—RHR curves from the experiments can be recalculated using average values of α and β. Figure 4 a) shows the results of using the regression equation on material no. 3 in Table 1 for the full scale test. Figure 4 b) shows the same for the 1/3 scale test. The regression equation has been used for 6 materials in both full scale and 1/3 scale tests, showing similar results.

RHR in scenario B. The procedure for calculating rate of heat release in scenario B builds on the same principles as the one described above. In this scenario there is, however, no material on the ceiling.

The total rate of heat release in the room is assumed to come from five sources; the gas burner, the vertical wall area behind the burner, a horizontal strip of material corresponding to the verti—cal height og the ceiling jet at the ceiling—wall intersection, the wall material in the upper layer and, when downward flame spread has started, from the wall linings below the hot gas layer.

The scenario we are considering is the following one: The walls of the test room are lined with the material. The ignition source in the corner ignites the wall corner material and spreads upward on an area, A_w, approximately equal to the width of the burner times the distance from the burner to the ceiling. In this initial period, $0 < t < t_{start}$, the rate of heat release is calculated in the same way as above, i.e. assumed to grow linearly up to Q_{start} at time t_{start}.

The resulting ceiling jet, or flame, spreads along the intersec—tions between the walls and the ceiling in the mode of concurrent flame propagation. After a time t_h the pyrolysing area has propa—gated to the nearest corner in the room and a strip of material at the top of the walls is pyrosying. In the experiments discussed here this strip has a height of around 5% of the room height.

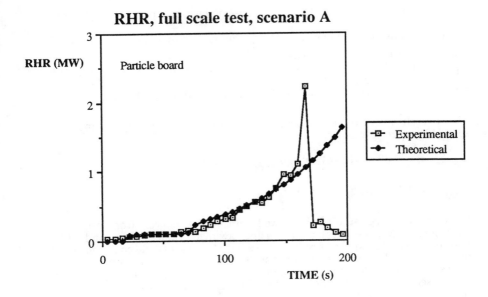

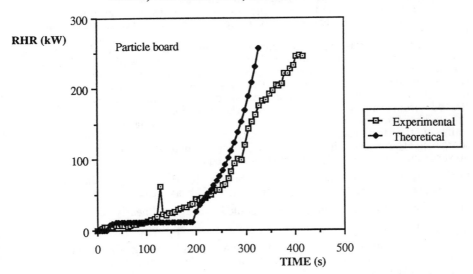

Figure 4. Comparison of experimental and calculated rates of heat release in four different experimental setups. *Continued on next page.*

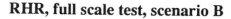

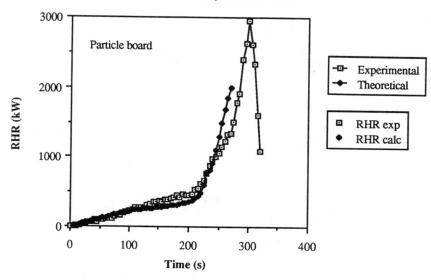

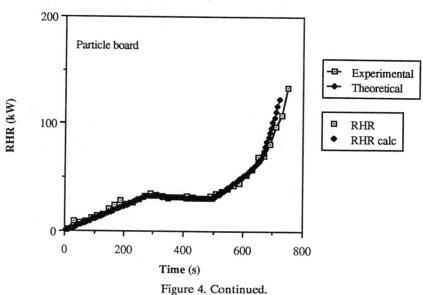

Figure 4. Continued.

We assume we can calculate how long it will take for the pyrolys—ing area to reach the nearest corner. In the period $t_{start} < t <$ t_h there are thus three sources that contribute to the total heat release; the gas burner, the diminishing effect of the pyrolysing area behind the gas burner and the increasing pyrolysis area of the horizontal strip at the wall—ceiling intersection.

It remains to give expressions for t_h, the time at which the horizontal pyrolysing area reaches the an opposite corner. Saito, Quintiere and Williams (20) have given the upward spread velocity of the pyrolysis front as

$$V_p = 4 \ (\dot{q}''_0)^2 \ (x_f - x_p)/[\pi k \rho c (T_p - T_a)^2] \qquad (11)$$

which has been employed in some recent studies by several workers ((21), (22) and others). The flame height x_f has been shown to only depend on the energy release rate per unit wall width ((21), (22), (23) and (24)) and several expressions for x_f have been given.

Efforts are being made to apply the above method to the horizontal concurrent flame spread. Some problems have been encountered in estimating the horizontal flame length x_f since earlier reported flame height correlations may not be valid in the horizontal ceiling—wall intersection configuration. Results from this part of the study will be reported later.

The problem of estimating t_h, the flame spread along inter—section wall — ceiling and the resulting rate of heat release was therefore solved rather crudely as a first effort and is considered to be a temporary solution. Magnusson (25) suggested a simple expression for the calculation of A_{ph}, the horizontal pyrolysing area, assumed to increase linearly with time

$$A_{ph}(t) = k * t * (Q_{start} - Q_{gb})/k\rho c \qquad (12)$$

The factor k was derived from experiments to be $\approx 0.004 \ w/(mK)^2$. For each Δt the pyrolysis area increases by ΔA_{ph}. Equation 12 will be changed, using the methodology described in (24) instead.

Following a similar procedure as for scenario A, the expression for the rate of heat release for the time period $t_{start} < t < t_h$ is then written at time step j (counted from t_{start}) as

$$\dot{Q}(t) = \dot{Q}_{gb} + A_w * \dot{Q}''_{max} * e^{(-\lambda * \Delta t * j)} + \Sigma[\Delta A_{ph} *$$
$$\dot{Q}''_{max} * e^{(-\lambda * \Delta t * i)}] \qquad (13)$$

The first term is the effect from the gas burner, the second the contribution from the wall behind the burner and the third from the part of the horizontal strip which is pyrolysing.

For the period $t > t_h$ the downward flame spread contributes to the increasing rate of heat release. To equation 13 is then added the contribution of the downward flame spread in the upper layer and, once the flames reach the intersection of the hot gas layer and the lower ambient layer, the downward flame spread below the hot layer.

Figure 4 c) shows the results of using this procedure on material no. 3 in Table 1 for the full scale test. Figure 4 d) shows the same for the 1/3 scale test. The procedure has been used on 3 other materials, the results are similar but not quite as good.

The following sections describe how the gastemperatures, surface temperatures in the hot layer and below the hot layer and the downward flame spread are calculated, thus adding to the rate of heat release.

Calculation of gastemperatures

The basic principle used to calculate the temperature in a compart—ment fire is the conservation of mass and energy. Since the energy release rate and the compartment temperature change with time, the application of the conservation laws will lead to a series of differential equations.

By making certain assumptions on the energy and mass trans—fer in and out of the compartment boundaries, the laws of mass and energy conservation can result in a relatively complete set of equations. Due to the complexity and the large number of equa—tions involved, a complete solution of the set of equations would usually only be obtained from computer programs.

However, now there exist regression formulae which, with a number of limiting assumptions, allow the gastemperature in a naturally or mechanically ventilated compartment to be calculated by hand.

McCaffrey, Quintiere and Harkleroad ($\underline{26}$) used a simple conservation of energy expression and a correlation of a relatively wide range of data to develop a hand— calculation formula for the hot layer temperature in a naturally ventilated compartment.

The upper layer temperature was written as a function of two dimensionless groups

$$\frac{\Delta T}{T_0} = C \cdot X_1^N \cdot X_2^M \tag{14}$$

The constants C, N and M were determined from a wide range experimental data, the final form of the regression equation in ($\underline{26}$) was given as

$$\frac{\Delta T}{T_0} = 1.63 \left[\frac{Q}{\sqrt{g}\ c_p\ \rho_0\ T_0\ A_0\ \sqrt{H_0}} \right]^{2/3} \left[\frac{h_k\ A_w}{\sqrt{g}\ c_p\ \rho_0\ A_0\ \sqrt{H_0}} \right]^{-1/3}$$

$$\tag{15}$$

The heat transfer coefficient, h_k, depends on the duration of the fire and the thermal characteristics of the compartment boundaries. The thickness of the lining materials treated here and the short duration of the corner test is such that the outer boundaries of the test compartment do not have an effect on the heat transfer coefficient. It can therefore be written as:

$$h_k = \sqrt{k\rho c/t} \qquad (16)$$

Calculation of gastemperatures in the current paper. The energy released within the compartment in flame and upper layer combus—tion is restricted by the availability of oxygen. The heat release measured in the experiments includes the energy released in the flames coming out through the opening. This part of the heat release does not influence the gastemperature within the compartment. The availability of oxygen in the upper layer can be approximated by using a simple flame formula (27), calculating the entrainment of air into the corner flame

$$m'_{air} = 2/3 \; \alpha \; \rho_a \; \sqrt{2g \; (1 - T_a/T)} \; X_f^{3/2} \qquad (17)$$

where X_f is the effective entrainment flame height. Thus, the maximum rate of heat release inside the compartment was found to be approximately 0.6 Mw for the full scale room and 40 kw for the 1/3 scale room.

The method developed in (26) was followed to calculate the upper layer gastemperatures in the corner test experiments. An attempt was made to determine the constants C, N and M by regression analysis but results were not satisfactory. The two constants, N and M, as they appear in equation 4 did however seem to describe the slope and shape of the experimental curves well. The constant C was then determined for each of the experi—mental series with the following results:

Full scale, scenario A	C_{aver}	=	2.048
1/3 scale, scenario A	C_{aver}	=	2.237
Full scale, scenario B	C_{aver}	=	2.700
1/3 scale, scenario B	C_{aver}	=	2.240

The experimental gastemperature was measured by thermocouples 5 cm from the ceiling in both 1/3 scale tests series and 30 cm from the ceiling in full scale test, scenario A. But in the full scale B test series the gastemperature was measured only 10 cm from the ceiling, resulting in relatively much higher gastemperature values than in the other test series. This accounts for the much higher C_{aver} in the last mentioned test series.

The procedure of limiting the RHR and finding a pre—exponential factor for each test series proved to be very robust and showed good agreement with experimental results. Figures 5 a), b), c) and d) show results of using the above procedure on material 3 in Table I for both scenarios and both compartment sizes. This has also been done for most of the materials in Table 1, showing similar results.

Heat transfer to walls

When the fire in the corner starts and the lining material in the corner ignites, combustion products and plume entrained air are transferred to the ceiling. The hot gas layer forms, descends and increases in temperature with time. Relatively early in the test the layer reaches the top of the opening, stabilizes and hot gases start flowing out through the opening.

Classical heat transfer provides expressions for quantities such as view factors, radiation and temperature fields in semi—infinite bodies. The lining materials studied here were treated as semi—infinite bodies since the test duration is relatively short.

One long side of the compartment wall was split into a large number of thin, horizontal strips and the heat flux from the gas layer to the center of each strip calculated using the well known expression

$$q'' = \epsilon \ F \ \sigma \ (T_g^4 - T_0^4) \tag{18}$$

The emission coefficient was taken to be a constant value close to unity. The configuration factor, F, was calculated in a conventional way, treating the center of each strip as a point. Once the down—ward flame spread started the radiation from the wall flames and the pyrolysing lining material behind the flames was added to the smoke layer radiation. The heat flux to the walls was then cal—culated from the expression

$$q'' = \epsilon_g \ F_g \ \sigma \ (T_g^4 - T_0^4) + \epsilon_f \ F_f \ \sigma \ (T_f^4 - T_0^4) +$$
$$\epsilon_p \ F_p \ \sigma \ (T_p^4 - T_0^4) \tag{19}$$

where the subscript g refers to the gaslayer, f to the flame and p to the pyrolysing wall material. The view factors from the flame and the pyrolysing wall material were assumed to be identical and equal to the total burning area of lining material. The flame temperature was taken to be $\simeq 1100^0$ K and the pyrolysing material surface temperature was assumed to be $\approx 750^0$ K. Further, the flame emission coefficient was taken to be $= 0.5$ as was the surface of the pyrolysing material. A sensitivity analysis is necessary.

Wall surface temperatures

As explained above, one long side of the compartment wall was split into a large number of thin strips and the heat flux to the center of each strip calculated. For a constant heat flux, assuming the wall material to be semi—infinite, the wall surface temperature

Hot gas temperatures, full scale test, scenario A

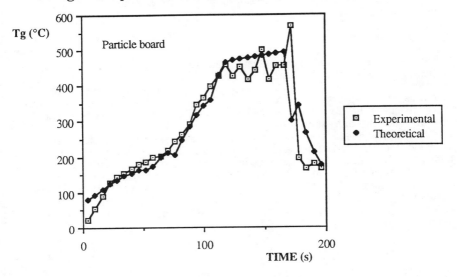

HOT GAS TEMPERATURES
1/3 scale room test, scenario A

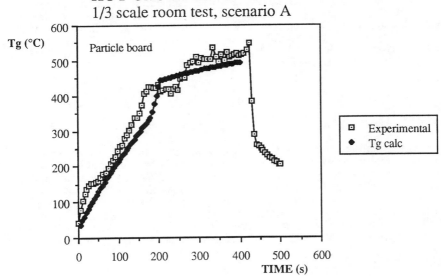

Figure 5. Comparison of experimental and calculated hot gas temperatures in four different experimental setups. *Continued on next page.*

Hot gas temperatures, full scale test, scenario B

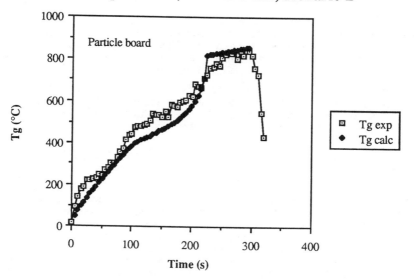

Hot gas temperatures, 1/3 scale test, scenario B

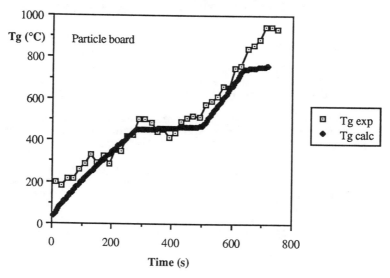

Figure 5. Continued.

can be calculated from

$$T_s - T_0 = (\dot{q}''/h) * (1 - e^{h^2\,t/k\rho c} * \text{erfc}(\sqrt{h^2t/k\rho c}))$$
(20)

Knowing the heat flux as a function of time and using the super—position principle, the surface temperature of the lining material was calculated at the center of each strip. The surface temperature was therefore be calculated as a function of time and height from the floor. Here, the Newtonian cooling coefficient was assumed to be a constant value, h = 30 w/m^2, throughout the test. An explicit form of the superposition integrals can be found in (29).

The surface temperature of the wall material emerged in the hot layer was approximated by the expression

$$T_s - T_0 = (T_g - T_0) * (1 - e^{h^2\,t/k\rho c} * \text{erfc}(\sqrt{h^2t/k\rho c}))$$
(21)

using the superposition principle and assuming a constant heat transfer coefficient. Putting the Newtonian cooling coefficient and the heat transfer coefficient equal to a constant value is of course an oversimplification and is seen only as a temporary measure.

No wall surface temperatures were measured in the full scale test series, scenario A and in the 1/3 scale test series, scenario B. Figure 6 a) shows the experimental and calculated wall surface temperatures, at a height of 0.45 m from the floor, for material no. 3, 1/3 scale test, scenario A. Figure 6 b) shows the same, but at a height of 1.2 m from the floor, for the full scale test, scenario B.

Downward flame spread

The RHR in scenario A is calculated from a regression formula, equation 10, where the only dependent variable is time. Calculation stops before downward flamespread has become significant or dominant.

In scenario B, however, both the horizontal concurrent flame spread and the downward flame spread, in and below the hot gas layer, are directly linked to the rate of heat release.

Downward flame spread for scenario A. Relatively early in the test the hot gas layer reaches the top of the opening, stabilizes and hot gases flow out through the opening. We have assumed that at the beginning of the test the hot layer is already stabilized at the height of the opening, this has a relatively small influence on the radiation from the hot layer to the walls since the the layer is relatively cold to begin with.

No attempt is made to predict what happens within the hot smoke layer for scenario A. The smoke is quite thick in this scenario and it is difficult to visually see what happens there. When the smoke layer has been heating the wall surfaces for some time, occasional flames start appearing at the interface of the smoke layer and the walls. Shortly after, a thin, horizontal line of flames has been established on the lining material at this interface. The downward flame spread is quite slow to begin with but accelerates with time and can be calculated from a similar expression to equation 3

$$V_f = \phi/(T_{ig} - T_{fl.f})^2 \qquad (22)$$

where T_{ig} and ϕ are obtained from the bench–scale tests discussed earlier in this paper. $T_{fl.f}$ is the material surface temperature just ahead of the flame front. If the position of the flame front is known, this surface temperature at a certain wall height and certain time, can be extrapolated from the surface temperatures calculated at the center of each wall strip at each time step. The downward flame spread can thus be calculated.

Since what happens in the gaslayer is not treated in scenario A, the wall flames are assumed to start at the intersection of the walls and the hot layer, i.e. 40 cm from the ceiling in the full scale tests and 13 cm from the ceiling in the 1/3 scale tests.

In the full scale test series, scenario A, the test was terminated at flashover so no data is available for the downward flame spread in this series. Figure 6 c) shows the experimental and calculated downward flame spread for material no. 3, 1/3 scale test, scenario A.

Downward flame spread for scenario B. Once the horizontal, con–current flame spread along the wall ceiling intersection has reached an opposite corner in the compartment the downward flame spread in the upper layer starts. In reality, this could possibly start happening during the concurrent flame spread time interval. In the current version of the model, no account is taken of the relatively low oxygen concentration in the upper layer. The flame spread is quite slow at first since the wall material has a relatively low sur–face temperature. It then accelerates until it reaches the interface of the smoke layer and walls.

At this point the flame spread slows down since the walls beneath the smoke layer have a lower surface temperature than the walls immersed in the hot layer. The downward spread then accelerates again. The flame spread is calculated from equation 22 as above.

Figure 6 d) shows the experimental and calculated downward flame spread for material no. 3, full scale test, scenario B.

Remarks on the results

No sensitivity testing has so far been carried out with respect to the different assumptions and procedures just enumerated. Changes will certainly be introduced, especially regarding the horizontal con–current flame spread which will follow the methods outlined in (21), (22) and (24). The Newtonian cooling coefficient and the heat transfer coefficient in equations (20) and (21) will be calculated as

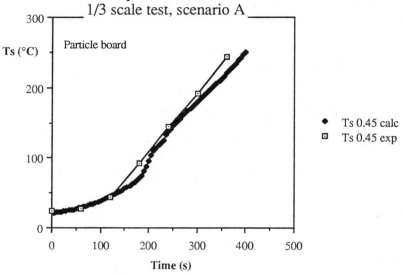

Wall surface temperature 0.45 m from the floor
1/3 scale test, scenario A

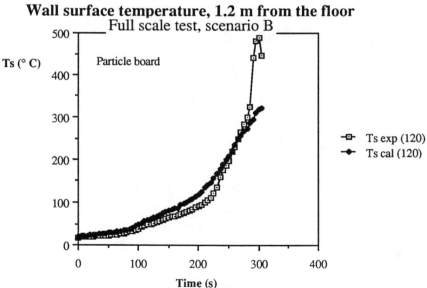

Wall surface temperature, 1.2 m from the floor
Full scale test, scenario B

Figure 6. Comparison of experimental and calculated data (wall surface temperatures and downward flame spread) in two different experimental setups. *Continued on next page.*

Downward flamespread, 1/3 scale test, scenario A

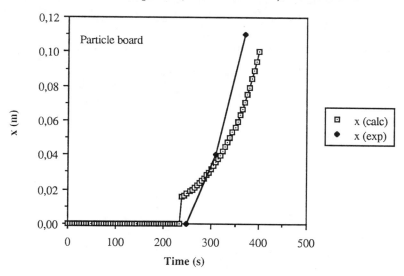

Downward flamespread, full scale test, scenario B

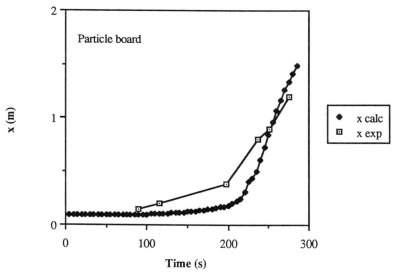

Figure 6. Continued.

functions of surface temperatures.   Certain other areas in the proce—
dure need to be looked at more closely, a sensitivity analyses with
regards to emission coefficients and heat transfer coefficients, as well
as other input parameters, will need to be carried out.
        The ISO surface spread of flame test seems to correlate well
with room test behavior when directly comparing times to ignition
and rates of opposed flow flame spread.   For the RHR—test the
correlation is more implicit. A simple model incorporating data from
this test and the ignitability test is capable of predicting the first
phases of room fire growth in scenario A.   The basic structure of
the model for predicting fire growth in scenario B seems acceptable
although it needs improving.   This is valid for both full scale and
the 1/3 scale test room.

Legend of Symbols

A	=	Area
C	=	Flame spread constant in eqn. 5 and constant in eqn. 14
c_p	=	Heat capacity (of air unless otherwise stated)
F	=	Configuration factor
g	=	Gravitational acceleration
h	=	Convective and radiative heat transfer coefficient
h_k	=	Convective heat transfer coefficient
H_o	=	Height of opening
k	=	Constant in eqn. 12
$k\rho c$	=	Thermal inertia
$\dot{q}''$	=	Radiative heat transfer per area
$\dot{Q}''$	=	Energy release rate per fuel area
$\dot{Q}_{start}$	=	Total heat release from gas burner and lining material behind burner
t_h	=	Time for pyrolysis front to move horizontally along wall—ceiling intersection to opposite corner
t_{start}	=	Time to reach Q_{start}
T	=	Temperature
x_f	=	Position of the flame front, flame height in eqn. 17
α	=	Entrainment coefficient in eqn. 17, correlation coefficient in eqn. 10
β	=	Correlation coefficient
λ	=	Decay coefficient
ρ_0	=	Density of ambient air
σ	=	Stefan—Boltzmann constant
ϕ	=	Flame spread parameter
ϵ	=	Emissivity

Subscripts:

av	=	average
e	=	external
f	=	flame
fl.f	=	flame front
g	=	gas
gb	=	gas burner
i	=	initial
max	=	maximum
o	=	opening
p	=	pyrolysis
s	=	surface
w	=	wall

Literature Cited

[1]   Pettersson, O., "Fire Hazards and the Compartment Fire Growth Process – Outline of a Swedish Joint Research Program", FoU–brand No 1, 1980.

[2]   Sundström, B. "Results and Analysis of Building Materials Tested According to ISO and IMO Spread of Flame Tests", SP–RAPP 1984:36, Swedish National Testing Institute, Borås, 1984.

[3]   Harkelroad, M., Quintiere, J., Walton, W., "Radiative Ignition and Opposed Flow Flame Spread Measurements on Materials", Report No. DOT/FAA–CT–83/28, FAA Technical Center, Atlantic City Airport, N.J., 1983

[4]   Andersson, B., Internal Report, Department of Fire Safety Engineering, Lund University, Lund 1988.

[5]   DeRis, J.N., "Spread of Laminar Diffusion Flame", 12th Symposium (International) on Combustion, the Combustion Institute, Pittsburgh, PA, 1969.

[6]   Quintiere, J. G., "An Approach to Modeling Wall Fire Spread in a Room", Fire Safety Journal Vol 3, 1981.

[7]   Blomqvist, J., "RHR of Building Materials – Experiments with an OSU Apparatus Using Oxygen Consumption", Report LUTVDG/(TVBB–3017), Division of Building Fire Safety and Technology, Lund University, Lund, Sweden, 1983.

[8]   Svensson, G., Östman, B., "Rate of Heat Release By Oxygen Consumption Testing of Building Materials", Meddelande Serie A Nr 812, Swedish Institute for Wood Research, Stockholm, 1983.

[9]   Sensenig, D. L., "An Oxygen Consumption Method for Determining the Contribution of Interior Wall Finishes to Room Fires" NBS Technical Note 1128, National Bureau of Standards, Washington DC, 1980.

[10]   Babrauskas, V., "Development of the Cone Calorimeter – A Bench–Scale Heat Release Apparatus Based on Oxygen Con–sumption" NBSIR 82–2611, National Bureau of Standards, Washington DC, 1982.

[11]   Magnusson, S.E., Sundström, B., "Combustable Linings and Room Fire Growth – A First Analysis", Report LUTVDG/(TVBB 3030), Department of Fire Safety Engineering, Lund University, Lund 1985.

[12]   "Room Fire Test in Full Scale for Surface Products", NORDTEST Fire Test Method. NT FIRE 025.

[13]   Sundström, B., "Full Scale Fire Testing of Surface Materials", Technical Report SP–RAPP 1986:45, Swedish National Testing Institute, Borås 1986.

[14]   Ondrus, J., Internal Report, Department of Fire Safety Engineering, Lund University, Lund 1988.

[15]   Andersson, B., "Model Scale Compartment Fire Tests With Wall Lining Materials", Report LUTVDG/(TVBB–3041), Department of Fire Safety Engineering, Lund University, Lund 1988.

[16]   Orloff, L., DeRis, J., Markstein, G. M.,Upward Turbulent Fire Spread and Burning of Fuel Surfaces", 15th Symposium (International) of Combustion, 1975.

[17]   Fernadez–Pello, A. C., Mao, C. P., Comb. Sci. Tech., Vol. 26, Gordon and Breach Science Publishers, London, England, 1981, pp 147 – 155.

[18]   Parker, W. J., "An Assessment of Correlations Between Laboratory and Full–Scale Experiments for the FAA Aircraft Fire Safety Program, Part 3: ASTM E 84", NBSIR 82–2564, Center for Fire Research, Washington DC, August 1982.

[19]   Wickström, U., Internal Working Memorandum, Swedish National Testing Institute, Borås, Sweden, 1983.

[20]   Saito, K., Quintiere, J.G., Williams, F.A., "Upward Turbulent Flame Spread", Int. Assoc. for Fire Safety Science. Fire Safety Science Proceedings. 1st Int. Symposium. October 7–11 1985. Grant, C. E., and Pagni, P. J., Editors, Gaithersburg, MD, Hemisphere Publishing Corp., New York, 1985.

[21]   Quintiere, J. G., Harkleroad, M., Hasemi, Y., Comb. Sci. Tech., 48, p. 191, 1986.

[22]   Hasemi, Y., "Thermal Modeling of Upward Wall Flamespread", Int. Assoc. for Fire Safety Science. Fire Safety Science Proceedings. 1st Int. Symposium. October 7–11 1985. Grant, C. E., and Pagni, P. J., Editors, Gaithersburg, MD, Hemisphere Publishing Corp., New York, 1985.

[23]   Delichatsios, M. A., "Flame Heights in Turbulent Wall Fires with Significant Flame Radiation", Comb. Sci. Tech., 39, p. 195, 1984.

[24]   Kulkarni, A. K., Fisher, S., "A Model for Upward Flame Spread on Vertical Wall", National Bureau of Standards, Center for Fire Research, Gaithersburg, 1988.

[25]   Magnusson, S. E., Personal communication, Dept. of Fire Safety Eng. Lund University, Lund, 1989.

[26]   McCaffrey, B. J., Quintiere, J. Q., Harkleroad, M.F., "Estimating Room Temperatures and the Likelyhood of Flashover Using Fire Test Data Correlations". Fire Technology, Vol 17, pp 98–119; 1981.

[27]   Quintiere, J. G., "An Assessment of Correlations Between Laboratory and Full–Scale Experiments for the FAA Aircraft Fire Safety Program, Part 2: Rate of Energy Release in Fire", NBSIR 82–2536, National Bureau of Standards, Washington DC, 1982.

[28]   Östman, B., "Ignitability as Proposed by the International Standards Organisation Compared with Some European Fire Tests for Building Panels", Fire and Materials, Vol. 5, no. 4, 1981.

[29]   Mitler, H.E., "The Physical Bases for the Harvard Computer Fire Code", Division of Applied Sciences, Harvard University, Home Fire Project Technical Report No. 34, Oct. 1978.

RECEIVED February 22, 1990

Chapter 34

Fire Hazard in a Room Due to a Fire Starting in a Plenum

Effect of Poly(vinyl chloride) Wire Coating

F. Merrill Galloway and Marcelo M. Hirschler

BFGoodrich Technical Center, P.O. Box 122, Avon Lake, OH 44012

An issue of interest is the contribution to fire hazard in a room from products in a plenum space above it. This contribution can result from two scenarios: fire in the room or fire in the plenum. The products being addressed here are PVC electrical products contained in a plenum.

The first scenario involves a fire starting in the room.

Three room dimensions and two ceiling materials were analysed; the products were PVC conduit (rigid and ENMT, semi-rigid; 100 m of either) and PVC wire coating (400 m). It was found that the amount of energy needed for the room fire to cause thermal decomposition of the PVC products in the plenum was larger than that needed to take the room to flashover. Furthermore, if the PVC products did eventually decompose or burn, somehow, they would cause a lethal smoke concentration only significantly later than a lethal (by toxicity) atmosphere had already been created by the fire itself. Thus, the PVC products did not add any significant fire hazard to that caused by the room fire.

The next scenario is more complex: it involves a fire starting in a plenum and has been analysed only for wire coating.

Calculations were made, for many fire scenarios, in which the fire hazard model F.A.S.T. was used to simulate hazard to occupants of a standard room following a fire starting in a plenum above it. A total of 400 m of PVC wire coating was assumed to be present in the plenum. Its decomposition was made a function of the plenum temperature achieved. The fire ranged between RHR of 50 and 500 kW; the heat of combustion of the fire varied between a standard value (20 kJ/g) and that of methane (57 kJ/g). Various vent connections between compartments and surroundings were used. The plenum temperature was never enough to decompose all the PVC wire coating.

If the plenum was vented to the surroundings, almost no smoke entered the room. In an unvented plenum smoke entered the room but the fire burnt for a short period only: the level of oxygen was not enough for full combustion. In extreme cases the fire generated enough heat for an untenable atmosphere in the room. In almost all single plenum cases studied the smoke flowing into the room was insufficient to generate a concentration lethal to man.

Therefore, using such low heat release PVC wire coating products did not cause a significant increase in the fire hazard to occupants.

In office buildings it is very common to have plenums, i.e. spaces above rooms where the air handling system is located, together with electrical wires and cables, as well as abundant wood and other construction materials. These concealed spaces are usually ca. 1 m (3 ft) high and are invisible from the room below.

In the 1980's years there has been some controversy about the effect of fires involving combustible products contained in such concealed spaces. This addresses the room-plenum scenario both when a fire starts in the room and when a fire starts in the plenum and investigates its spread into the room below.

In recent years there has been much controversy surrounding the impact of smoke toxicity following a fire. This has included discussions regarding means to measure toxic potency, by one of a variety of small-scale methods, and how to use these results to evaluate fire hazard. There has been, in particular, much speculation regarding the hazards due to certain plastics, typically poly(vinyl chloride) (PVC).

The present paper will deal with this issue in several stages.

(1) Address the issue of PVC fire properties, including smoke toxicity and hydrogen chloride decay.

(2) Describe measurements of mass loss rates of various electrical PVC products, by thermoanalytical experiments.

(3) Address the concealed space scenarios used.

(4) Investigate the fire hazard in a room due to the burning of PVC electrical products in a plenum space

above it, when a fire starts in the compartment underneath.
(5) Investigate the fire hazard in a room due to the burning of PVC wire coating in a plenum space above it, when the fire starts in the plenum.

Fire Properties of Poly(Vinyl Chloride)

Other papers in this volume address the importance of a variety of fire properties on fire hazard, in particular the relative importance (or lack of it) of toxic potency of smoke (e.g. Ref. [1]).

PVC is unique among commodity materials in that it contains chlorine and, thus, produces hydrogen chloride (HCl) when it decomposes or burns [2, 3].

The fire properties of PVC have been put into perspective recently [4, 5]. They show that PVC is a polymer with a high ignition temperature and low flammability. Furthermore, PVC products are associated with a low rate of heat release as well as little total heat released [4-9]. This will depend, clearly, on the type of product, since plasticised PVC products are obviously more flammable than rigid ones.

Undoubtedly, fire hazard is partially associated with the toxicity of the smoke itself. The smoke of a variety of common materials, as measured e.g. by the NBS cup furnace toxicity test [10], has recently been compared with the intrinsic toxic potency of other poisons and of toxic gases, as well as with toxicity categories [11]. It has been shown that toxicity is a relatively minor factor because there is very little difference between the intrinsic toxic potency of the smoke of the majority of common materials, with very few exceptions.

Detailed studies have also been made on the toxicity of HCl, an irritant gas often present in fires. It does not cause baboon or rat incapacitation up to very high exposure doses which are sufficient (or very close) to cause eventual death [12]. Furthermore, a recent study has shown that the effects of irritants are heavily dependent on the animal model used [13].

Interestingly, the mouse is much more sensitive to HCl than the rat [13-16]. In turn, however, the rat works as a good model for a primate, as far as lethality due to HCl is concerned [17, 18]. This is important because all rodents (mice and rats) are obligate nose breathers, while primates can also breathe through their mouths, and it has been speculated that this would make rodents less sensitive to HCl than primates [19].

The relevance of all this to the present paper is that the toxic potency of PVC smoke or of HCl are fairly similar to those of other smoke or of carbon monoxide (CO) respectively.

A large number of studies have also been done to investigate the lifetime of HCl in a fire atmosphere [20-24]. These studies have shown that HCl reacts very rapidly with most common construction surfaces (cement block, ceiling tile, gypsum board, etc.) so that the peak atmospheric concentration found in a fire is much less than would have been predicted from the chlorine content of the burning material. Furthermore, this peak concentration soon decreases and HCl disappears completely from the atmosphere.

These considerations are included here because in the discussion that follows HCl decay will be ignored, to facilitate the calculations. HCl decay is also important when measuring PVC toxic potency, because the walls of exposure chambers are made of non-sorptive materials, where such decay is minimised. In this connection it is worth pointing out that the highest concentration of HCl found when fire fighters entered buildings actually on fire was ca. 280 ppm [25, 26].

Mass Loss Rates of PVC Products

Table I presents the results of "isothermal" simultaneous thermoanalytical (STA) runs, at 573 K and 773 K, for all three products. Similar data, at a fixed heating rate is shown in Table II. One of the crucial parameters is the temperature of maximum weight loss rate, corresponding to the time when dehydrochlorination of PVC starts becoming important. This temperature is close to 573 K in all cases. In fact, at a relatively fast heating rate, almost no decomposition occurs at temperatures under 563 K. If the materials are heated at 573 K for a prolonged period, complete dehydrochlorination takes place, but no further stages of PVC decomposition occur. None of the three materials investigated decomposes completely until a temperature of ca. 773 K is attained. Even then only a certain fraction of the entire mass of the samples is volatilised, due to the presence of inorganic fillers in their composition.

The average rate of mass loss is calculated from the amount of mass lost and the corresponding time period. The calculations in Table I at 573 K represent the average mass loss of isothermal dehydrochlorination. Thus, the values in Table I (3.4 %/min for blue conduit, 2.9 %/min for grey conduit and 2.3 %/min for wire coating) represent a reasonable estimate of the mass loss rate of the PVC products in a fire, at a temperature not exceeding 563 K.

Concealed Space Scenarios

A few recent events make it particularly interesting to valuate the fire hazard resulting from the burning of PVC materials, when they are present in a plenum. These include the recent regulations promulgated in New York State regarding the creation of a data base for smoke

Table I. Results of "isothermal" STA runs

	ENMT Conduit		Rigid Conduit		Wire Coating	
	573	873	573	873	573	873
Top Temperature K	573	873	573	873	573	873
Total wt loss (%)	51.0	85.9	50.1	85.5	49.9	62.0
First wt loss (%)	49.0	56.0	50.1	47.7	49.9	46.6
$T_{1\%}$(K) 255	290	279	290	269	282	
Max DTG (mg/min)	2.6	7.5	3.1	6.1	1.9	7.8
Max DTG (%/min)	10.8	34.2	13.1	29.3	7.9	35.6
Average DTG (%/min) (first wt loss)	3.4	14.3	2.9	12.5	2.3	10.5
Second wt loss (%)	2.0	20.5	–	18.7	–	8.8
Max DTG (mg/min)	–	1.8	–	0.2	–	0.9
Third wt loss (%)	–	14.6	–	22.9	–	6.5
Max DTG (mg/min)	–	0.3	–	0.06	–	0.04

toxicity of building products [27]. The products that were tested in the first year (electrical) are often found behind fire rated walls or ceilings. An example of a calculation of this type has been made for fluorinated wire coatings [28] and for PVC electrical materials [29] and, another one, following a different procedure and a different scenario, for PVC-based electric non-metallic (ENMT, semi-rigid) conduit [30]. The philosophy used in references [28] and [29] is that fire hazard is higher if the time to reach a lethal atmosphere is lower.

The scenarios investigated here involve various room-plenum configurations. The room has a standard opening corresponding to the size of a normal door (2.03 x 0.74 m). The rooms in the first part of the study (fire in the room) were of dimensions which might represent, approximately, a small warehouse (10.0 x 10.0 x 3.0 m), a bedroom (3.7 x 3.7 x 2.7 m) and an office (2.5 x 2.5 x 2.5 m) and the plenum height is 1.0 m throughout. In each case, two different ceiling materials were considered (viz. gypsum wallboard (GB) and ceiling tile (CT)). Furthermore, the intensity of heat of combustion of the fuel involved in causing the fire was set at both 20 kJ/g and 40 kJ/g, thus covering the ranges typical for most materials.

It was assumed that the smoke was instantaneously distributed among either the room and plenum or the room and plenum plus another three rooms identical in size to the burn room. All room walls were assumed to be made of gypsum wallboard and all floors of concrete. This part will present, for each case, an assessment of the time required to achieve an untenable atmosphere, as a consequence of the exclusive presence, in the corresponding plenum, of PVC-coated electrical wire (400 m), of PVC-based rigid conduit (100 m) and of PVC-based electrical non-metallic ENMT, semi-rigid, conduit (100 m). These times will be compared with the times at which such an untenable atmosphere is generated due to the toxicity of the materials burning in the room, assuming them to be of normal toxic potency, similar to that of an ordinary wooden product (e.g. Douglas fir).

The rooms in the second part of the study are all of the same size, viz. 3.7 x 3.7 x 2.7 m. The plenums being considered are either one with the same floor size and 1.0 m height or one with 3 times the floor size (3 plenum configuration) or 10 times the floor size (10 plenum configuration). The only ceiling material considered is gypsum board and the only product being investigated is a PVC wire coating fire retarded to give very low heat release and flame spread. In this case, the fire starts in a plenum and the work investigates whether it spreads into the room below in terms of its effects on temperature, smoke layer levels and concentrations of toxic gases, mainly carbon monoxide, in both room and plenum.

Table II.　Results of STA runs at 5 and 20 K/min heating rate

Heating rate K/min	ENMT Conduit		Rigid Conduit		Wire Coating	
	5	20	5	20	5	20
Total wt loss (%)	92.0	91.9	86.9	87.6	77.6	75.4
First wt loss (%)	57.6	57.8	52.9	53.2	50.7	50.2
$T_{1\%}$(K) 522	546	500	528	575	595	
Max DTG (%/min)	13.5	33.3	8.0	31.1	8.6	25.3
Temp max DTG (K)	546	578	555	574	568	584
Average DTG (%/min) (first wt loss)	1.7	7.4	1.3	6.1	1.1	15.6
Second wt loss (%)	22.9	16.8	11.1	17.0	15.1	12.1

Max DTG (%/min)	1.5	7.2	2.7	8.0	—	3.7
Temp max DTG (K)	696	713	712	725	—	727
Average DTG (%/min) (second wt loss)	0.9	3.0	0.9	2.8	0.4	1.3
Average DTG (%/min) (first and second wt losses)	1.3	5.6	1.2	4.8	0.8	4.4
Third wt loss (%)	13.9	24.2	22.9	17.4	11.8	13.2
Max DTG (%/min)	1.0	2.4	1.9	6.0	1.6	3.9
Temp max DTG (K)	727	818	787	857	962	981
Average DTG (%/min) (third wt loss)	0.5	1.1	1.0	2.7	1.1	1.5
Average DTG (%/min) (first to third wt losses)	1.1	2.3	1.1	4.1	0.9	2.7

Fire starting in the Room

For the analysis, a steady-state fire was assumed. A series of equations was thus used to calculate various temperatures and/or heat release rates per unit surface, based on assigned input values. This series of equations involves four convective heat transfer and two conductive heat transfer processes. These are:
- (a) convective transfer from fire to upper room layer
- (b) conductive transfer through suspended ceiling to plenum floor
- (c) convective transfer from suspended ceiling to plenum air
- (d) convective transfer to plenum ceiling
- (e) conductive transfer through concrete plenum ceiling slab
- (f) convective transfer to air above plenum (ambient temperature)

The heat release rate necessary for flashover was calculated, from the equation given by Quintiere et al. [31]. The series of equations is then solved, with the assumption that the temperature increase for flashover is 500 K (leading to an upper level temperature of T_{UL}: 795 K) and the plenum temperature for decomposition of the PVC products is 573 K. The results in Table III show that a much more intense fire is required, in all cases, to cause the PVC products to undergo dehydrochlorination than to take the room to flashover. Thus, the heat released by this fire at flashover is insufficient to dehydrochlorinate the PVC products in the plenum, for any of the scenarios. Therefore, the occupants of the room will succumb before there is an effect due to the plenum PVC products.

It is of interest to calculate, too the time required for both the fire itself and the thermal decomposition of the plenum PVC products to produce a lethal atmosphere. Table III presents such results for the fire, for heats of combustion of 20 kJ/g and 40 kJ/g, a range typical of most fires. In order to carry out this calculation it is assumed that the smoke is distributed instantaneously throughout the volume being considered, one or four room-plenums. The barriers represented by walls or transport processes are ignored. The toxic potency used for the fire is a minimal value, an LC_{50} of 40 mg/l for a 30 min exposure in the NBS smoke toxicity test, in the non-flaming mode. This could be representative of a variety of materials (e.g. wood) and is within the normal range of toxic potencies.

In order to calculate the "time to lethal concentration" the concentration of smoke (per unit time) is first calculated. Then the total amount of smoke (in concentration per unit time) is calculated from the mass of material (and, in the case of the PVC products, the percentage of the weight of the product that can be volatilised, as seen from the STA results). To the ratio of the toxic potency to the amount of smoke is added the time for thermal penetration of hot gases through the ceiling and the 30 min exposure time that the toxic potency refers to. The calculations are all repeated for transport of the smoke over four room-plenums of the same size.

Table III presents the results of calculating the "time to lethal concentration" for each one of the PVC products investigated. The toxic potency values used for all the materials are based on 30 min exposures in the NBS cup furnace toxicity test, in the Non-Flaming mode, the one most relevant to this scenario.

It is clear that the "time to lethal concentration" for the smoke from any of the PVC products in the plenum, in all the six scenarios considered, is much longer than the time required for the fuel in the room itself to cause a lethal concentration in the same scenario.

This indicates clearly that <u>these PVC plenum products will not cause a serious fire hazard concern until well after the fire itself has reached flashover conditions and has long since caused lethal concentrations, both in the room of origin and in other rooms</u>.

It is worth stressing that the calculations done in this work have ignored HCl decay. This is very important since the rate of HCl decay in sorptive surfaces (such as concrete or ceiling tile) is extremely high (half lives of HCl of less than 1 min have been calculated for a plenum with such surfaces [32]).

The same calculation procedure has also been applied to other products in the same scenario. In particular, it has been used for PTFE wire coating in one of the scenarios being considered here [28, 29]. The results showed that, even if the toxic potency of the product in the plenum is extremely high, it is extremely unlikely to contribute significantly to fire hazard in the habitable areas if it has very good fire performance.

Fire starting in the Plenum

In this case a completely different approach was taken. It was decided to use a fire model, of zonal type, to predict smoke flows, temperatures and gas concentrations. The model chosen for these calculations was the NBS Fire and Smoke Transport model (F.A.S.T.), version 18.3 [33]. This model requires that the transport between rooms be in a horizontal manner. In order to achieve this, a virtual room is needed and a vent is needed in both the room and the plenum. In order, therefore, to analyse a broad variety of different fires and scenarios, the only product used was a low heat release wire coating.

The product used for these calculations was a fire retarded plasticized PVC wire coating material, which does not spread flame or continue burning unless an external source of heat or flame is directed at it. This material was chosen because PVC represents the most common cable

Table III. Fire in Room

Scenario	A	B	C	D	E	F
Room length (m)	10.0	10.0	3.7	3.7	2.5	2.5
Room width (m)	10.0	10.0	3.7	3.7	2.5	2.5
Room height (m)	3.0	3.0	2.7	2.7	2.5	2.5
Susp. ceil. material	GB	CT	GB	CT	GB	CT
T ambient (K)	295	295	295	295	295	295
T_{UL} at flashover (K)	795	795	795	795	795	795
RHR flashover (MW)	2.09	2.01	0.89	0.86	0.64	0.62
T plenum (PVC dec) (K)	573	573	573	573	573	573
T_{UL} (PVC dec) (K)	902	978	902	978	902	978
RHR reqd (PVC dec) (MW)	2.86	3.27	1.21	1.40	0.88	1.02

Effect of Room Fire Fuel Alone

If fuel has Delta H_{comb} of 20 kJ/g and LC_{50} (30 min) of 40 mg/l

	A	B	C	D	E	F
Time to lethal conc. (s)	112	98	42	37	20	17

If the volume considered includes four rooms

	A	B	C	D	E	F
Time to lethal conc. (s)	448	391	170	147	80	69

If fuel has Delta H_{comb} of 40 kJ/g and LC_{50} (30 min) of 40 mg/l

	A	B	C	D	E	F
Time to lethal conc. (s)	224	196	85	73	40	34

If the volume considered includes four rooms

	A	B	C	D	E	F
Time to lethal conc. (s)	895	782	339	293	160	138

PVC wire coating (400 m; LC$_{50}$: 31.6 mg/L)

Time to lethal conc. (s)	1928	1948	639	659	475	495

If the volume considered includes four rooms

Time to lethal conc. (s)	8339	8356	3180	3200	2527	2547

PVC rigid conduit (100 m; LC$_{50}$: 37.0 mg/L)

Time to lethal conc. (s)	1465	1485	564	584	450	470

If the volume considered includes four rooms

Time to lethal conc. (s)	4687	4707	1082	1102	626	646

PVC ENMT conduit (100 m; LC$_{50}$: 32.4 mg/L)

Time to lethal conc. (s)	1962	1982	644	664	477	497

If the volume considered includes four rooms

Time to lethal conc. (s)	6675	6695	1402	1422	735	755

coating material used overall, although the fire characteristics of the particular example chosen are better than those of the average plasticized PVC. Plenum cables are, of course, very often also made of fluorinated materials. A total of 400 m of the PVC wire were assumed to be present in a plenum. Calculations were made for many fire scenarios. These varied in fire size and in the burning characteristics of the material burning in the fire. In view of the good fire performance of the wire coating itself, it was possible to ignore the minute probability of it being the item first ignited.

A fractional factorial design was used to examine the effects of 9 variables which were thought to be significant. The variables were:

- Fire heat release rate
- Fire heat of combustion

A set of five variables relating to size and orientation of vents connecting plenum and room (tops of plenum vent and plenum and of room vent and room coincide)

- Width of a duct connecting room and plenum
- Width of vent in plenum
- Location of bottom of vent in plenum
- Width of vent in room
- Location of bottom of vent in room

A set of two variables relating to size and orientation of single vent connecting room and surroundings (bottoms of vent and room coincide)

- Vent width
- Vent height

Three fire sizes were chosen: 50, 275 and 500 kW, and the heats of combustion picked, viz. 20, 40 and 57 kJ/g, represent a spread between the normal heat of combustion of most common materials (20 kJ/g) and that of methane (57 kJ/g). This covers a very wide range of fires and of combustible materials starting the fire.

In order to cover these nine variables adequately, a statistical experimental design was calculated. The statistical experimental design requires the use of 15 simulations for each plenum size. Simulations were repeated using 3 and 10 plenums.

The National Bureau of Standards (NBS, now National Institute for Standards and Technology, NIST) fire and smoke transport model, F.A.S.T., version 18.3, was used to generate the information concerning the temperatures and gas concentrations. This is a zone model which predicts the formation of two layers in each compartment.

Once the conditions generated by each fire were known, decisions were taken as to which fires would cause significant decomposition of these cables. Some examples

of such cases were then run, where it was assumed, for
simplicity, that the PVC generated, initially, only
hydrogen chloride (HCl). The rate of HCl generation
incorporated into each example was calculated based on the
temperatures achieved in the plenum, as predicted by
F.A.S.T. HCl decay was ignored, as a first approximation,
just as it had been in the other set of cases.

Results And Discussion of Final Case

Tables IV-VI present the main results of the simulations
carried out with 1, 3 and 10 plenums. Most cases can be
divided in two categories:

(i) the fire is very intense but runs out of oxygen
and self-extinguishes fairly quickly;

(ii) the fire continues burning for a long time, but
plenum temperatures are fairly low.

A number of additional cases were also tried, in
which there was a direct opening between the plenum and
the surroundings. None of them produced any significant
amount of smoke flowing into the room: the net flow
through the opening was always outward, so that no air
entered the system to replenish the oxygen. These cases
are not being reported in detail here, in the interest of
space economy.

Only those fires in category (i) cause sufficiently
high plenum temperatures to allow decomposition of the
PVC. PVC will start decomposing at ca. 473 K, and will
decompose rapidly at temperatures above 523 K only.

In all the cases studied with ten plenums, which
represent a heating, ventilating and air conditioning
system, the fire was of category (ii). Even in those
cases were the upper level plenum temperature exceeded 523
K, this never occurred for a period of more than 2 min.

Virtually all the fires resulted in a CO
concentration in the room upper level which was
sufficiently high to cause serious concern. However, in
all single plenum cases, the size of the lower level (cold
layer) in the room and its CO concentrations were such
that escape was virtually always possible.

A total of ca. 60 simulations were run and in the
vast majority of them PVC decomposition plays a
negligible, if any, role. In only two of the single
plenum simulations was there a high enough plenum
temperature for PVC decomposition to take place over a
period of more than 1 min. Those worst cases, viz. # 2,
and # 13, were analysed further, by considering various
rates of PVC decomposition (HCl generation), depending on
upper level temperatures.

Table IV.   Simulations with a single plenum

Sim. #	Time1 >473K min	Time2 >523K min	Depth in room m	maxRm CO ppm	maxRm T max K
1 (1)	1	0	1.7 (11)	11,280 (11)	338
2 (1)	2	1 (614)	1.8 (11)	4,987 (11)	334
3 (1)	1	1 (621)	1.1 (11)	6,150 (11)	346
4 (2)	0	0	0.9 (6)	12,150 (8)	301
5 (1)	1	1 (618)	1.1 (10)	6,121 (11)	341
6 (1)	1	1 (622)	1.2 (9)	10,960 (11)	344
7 (1)	1	1 (618)	1.2 (11)	2,297 (11)	342
8 (1)	1	1 (573)	1.3 (10)	4,808 (11)	349
9 (2)	0	0	0.9 (7)	1,849 (9)	303
10 (2)	0	0	1.8 (11)	5,724 (8)	300
11 (2)	0	0	1.2 (8)	1,837 (9)	304
12 (1)	1	1 (630)	1.6 (11)	6,983 (11)	346
13 (1)	2	1 (563)	1.9 (11)	4,101 (11)	335
14 (2)	0	0	1.0 (7)	1,122 (9)	303
15 (2)	0	0	1.4 (10)	3,846 (8)	303

Legends: Time1: period upper level plenum temperature exceeds 473 K; Time2: idem for 523 K (maximum, in K); Depth max: maximum smoke layer depth (time reached, in min); Rm CO max: maximum room upper level [CO] (time reached, in min); Rm T max: maximum room upper level temperature (time reached, in min).

The dehydrochlorination rates of PVC considered were [3, 29]:

(a) for the range 473 - 523 K: 0.3 %/min
(b) for the range 523 - 563 K: 1.0 %/min
(c) for the range above 563 K: 2.3 %/min

Furthermore, for lower level temperatures well below mininum PVC decomposition temperature, it was assumed that no more than 20 % of the cable length, viz. 80 m, was

Table V. Simulations with three plenums

Sim. # —	Time1 >473K min	Time2 >523K min	Depth in room m	maxRm CO ppm	maxRm T max K
1 (1)	2	2 (541)	1.6 (2)	15,470 (11)	364
2 (1)	2	1 (614)	1.8 (11)	4,987 (11)	334
3 (1)	1	1 (621)	1.1 (11)	6,150 (11)	346
4 (2)	0	0	0.9 (6)	12,150 (8)	301
5 (1)	1	1 (618)	1.1 (10)	6,121 (11)	341
6 (1)	1	1 (622)	1.2 (9)	10,960 (11)	344
7 (1)	1	1 (618)	1.2 (11)	2,297 (11)	342
8 (1)	1	1 (573)	1.3 (10)	4,808 (11)	349
9 (2)	0	0	0.9 (7)	1,849 (9)	303
10 (2)	0	0	1.8 (11)	5,724 (8)	300
11 (2)	0	0	1.2 (8)	1,837 (9)	304
12 (1)	1	1 (630)	1.6 (11)	6,983 (11)	346
13 (1)	2	1 (563)	1.9 (11)	4,101 (11)	335
14 (2)	0	0	1.0 (7)	1,122 (9)	303
15 (2)	0	0	1.4 (10)	3,846 (8)	303

Legends as in Table IV.

decomposed simultaneously. The linear density of cable coating used is 70 g/m. In both of these simulations, worst case scenarios, (Table VII) it is clear that the HCl concentration does not introduce much additional hazard to that due to the fire itself, since the lethal potencies of HCl and of CO are very similar [1, 11, 13, 34, 35]. An investigation of those cases, among the 3 plenum simulations, with the highest potential for effects by PVC yields the same implications.

The main reason for this is that the products concerned have good fire performance. They have very low heat release characteristics, so that they do not add significantly to the energy of the fire and, furthermore, will not spread flame in the absence of an external energy source, so that they hardly increase the fuel supply for the fire.

Table VI. Simulations with ten plenums

Sim. #	Time1 >473K min	Time2 >523K min	Depth in room m	maxRm CO ppm	maxRm T max K
1 (4)	0	0	2.1 (4)	15,050 (8)	3 3 6
2 (1)	0	0	1.8 (11)	4,987 (11)	3 3 4
3 (1)	0	0	1.1 (11)	6,150 (11)	3 4 6
4 (2)	0	0	0.9 (6)	12,150 (8)	3 0 1
5 (1)	0	0	1.1 (10)	6,121 (11)	3 4 1
6 (1)	0	0	1.2 (9)	10,960 (11)	3 4 4
7 (1)	0	0	1.2 (11)	2,297 (11)	3 4 2
8 (1)	0	0	1.3 (10)	4,808 (11)	3 4 9
9 (2)	0	0	0.9 (7)	1,849 (9)	3 0 3
10 (2)	0	0	1.8 (11)	5,724 (8)	3 0 0
11 (2)	0	0	1.2 (8)	1,837 (9)	3 0 4
12 (1)	0	0	1.6 (11)	6,983 (11)	3 4 6
13 (1)	0	0	1.9 (11)	4,101 (11)	3 3 5
14 (2)	0	0	1.0 (7)	1,122 (9)	3 0 3
15 (2)	0	0	1.4 (10)	3,846 (8)	3 0 3

Legends as in Table IV.

Table VII. Results of Some Simulations with PVC

Simulation #	1,2	1,13	3,2	3,13
	Room Upper Layer Results			
CO @ 2 min (ppm)	4,325	1,221	3,314	2,957
HCl @ 2 min (ppm)	20	0	0	93
CO @ 5 min (ppm)	4,325	1,593	5,204	3,156
HCl @ 5 min (ppm)	20	108	112	126
CO @ 10 min (ppm)	4,564	3,828	5,204	5,282
HCl @ 10 min (ppm)	202	668	112	696
Max Temp (K)	335	335	327	353
Max Smoke layer (m) 2.2		1.8	1.9	2.0
Max Plen Temp (K)	615	562	497	583

Conclusions

Fires starting in a room may eventually get transferred to a plenum above it. However, by the time the effects of such a fire cause PVC products (rigid conduit, ENMT conduit and wire coating) in the plenum to burn, the room has already reached flashover conditions. Furthermore, the smoke generated by the room fire fuel causes much faster toxic concern than that from the PVC products in the plenum.

Fires starting in a plenum communicated to the outside are unlikely to cause concern in habitable areas. If the plenum is isolated from the outside, a fire starting in it is more likely to cause a hazardous situation in the room below if the plenum is communicated with other plenums.

The use of fire safe PVC wire coating products in a plenum, did not contribute, in virtually any of the simulations reported here, to a significant increase in the fire hazard due to the fire itself. This conclusion is valid both for the cases where the fire starts in the room and for the cases where the fire starts in the plenum.

References

1. Hirschler, M.M. General Principles of Fire Hazard and the Role of Smoke Toxicity, This volume.
2. Cullis, C.F. and Hirschler, M.M. The Combustion of Organic Polymers, Oxford University Press, Oxford, 1981.
3. Hirschler, M.M., Europ. Polymer J., 22, 153 (1986).
4. Hirschler, M.M., Fire Prev. 204, November, p. 19 (1987).
5. Smith, E.E.,in Ignition, Heat Release and Non-combustibility of Materials, A.S.T.M. STP 502 (Ed. A.F. Robertson), p.119, Amer. Soc. Testing Mater., Philadelphia, PA (1972).
6. Hilado, C.J., Flammability Handbook for Plastics, 3rd.Edn, Technomic, Lancaster, 1982.
7. Hirschler, M.M. and Smith, G.F. Determination of Fire Properties of Products by Rate of Heat Release Calorimetry: Use of the National Bureau of Standards Cone and Ohio State University Instruments, in Proc. Fire Retardant Chemicals Assoc. Fall 1987 Conf., Fire Safety Progress in Regulations, Technology and New Products, Monterey, CA, Oct. 18-21, p.133 (1987).
8. Hirschler, M.M. and Smith, G.F., Eastern States Comb. Inst. Fall Tech. Mtg, Nov. 2-6, 1987, pap. 63, Gaithersburg, MD, Combustion Institute, Pittsburgh, PA (1987).

9. Hirschler, M.M., _31st. IUPAC Microsymp. on Macromolecules Poly(Vinyl Chloride)_, Prague, 18-21 July (1988), _Makromol. Chem., Macromol. Symp._ 29, 133-53 (1989).

10. Levin, B.C., Fowell, A.J., Birky, M.M., Paabo, M., Stolte, A. and Malek, D., _Further Development of a Test Method for the Assessment of the Acute Inhalation Toxicity of Combustion Products_, Nat. Bur. Stands., Gaithersburg, MD, NBSIR 82-2532 (1982).

11. Hirschler, M.M., _J. Fire Sci._ 5, 289 (1987).

12. Kaplan, H.L., Grand, A.F., Switzer, W.G., Mitchell, D.S., Rogers, W.R. and Hartzell, G.E., _J. Fire Sci._ 3, 228 (1985).

13. Hinderer, R.K. and Hirschler, M.M. _J. Vinyl Technology_, 11(2), 50 (1989).

14. Higgins, E.A., Diorca, V., Thomas, A,A. and Davis, H.V., _Fire Technol._ 8, 120 (1972).

15. Darmer, K.I., Kinkead, E.R. and DiPasquale, L.C., _Am. Ind. Hyg. Assoc. J._ 35, 623 (1974).

16. Kaplan H.L., Hirschler, M.M., Switzer, W.G. and Coaker, A.W., _Proc. 13th. Int. Conf. Fire Safety_ (Ed. C.J. Hilado), p. 279, Product Safety, San Francisco, CA (1988).

17. Hartzell, G.E., Packham, S.C., Grand, A.F. and Switzer, W.G., _J. Fire Sci._ 3, 195 (1985).

18. Hinderer, R.K. and Kaplan, H.L., _Dangerous Properties of Industrial Materials Report_, p. 2, Mar-Apr (1987).

19. Alarie, Y.R., _Ann. Rev. Pharmacol. Toxicol._ 25, 325 (1985).

20. Beitel, J.J., Bertelo, C.A., Carroll, W.A., Gardner, R.A., Grand, A.F., Hirschler, M.M. and Smith, G.F., _J. Fire Sci._ 4, 15 (1986).

21. Bertelo, C.A., Carroll, W.F., Hirschler, M.M. and Smith, G.F., _Proc. 11th. Int. Conf. Fire Safety_ (Ed. C.J. Hilado), p. 192, Product Safety, San Francisco, CA (1986).

22. Bertelo, C.A., Carroll, W.F., Hirschler, M.M. and Smith, G.F., _Fire Safety Science, Proc. 1st. Int. Symp._ (Ed. C.E. Grant and P.J. Pagni), p. 1079, Hemisphere, Washington (1986).

23. Beitel, J.J., Bertelo, C.A., Carroll, W.A., Grand, A.F., Hirschler, M.M. and Smith, G.F., _J. Fire Sci._ 5, 105 (1987).

24. Galloway, F.M. and Hirschler, M.M., _Mathematical Modeling of Fires, A.S.T.M. STP 983_ (Ed. J.R. Mehaffey), p. 35, Amer. Soc. Testing. Mater., Philadelphia, PA (1987).

25. Burgess, W.A., Treitman, R.D. and Gold, A., _Air Contaminants in Structural Firefighting_, N.F.P.C.A. Project 7X008, Harvard School Public Health, Cambridge, MA (1979).

26. Grand, A.F., Kaplan, H.L. and Lee, G.H., _Investigation of Combustion Atmospheres in Real Fires_, U.S.F.A. Project 80027, Southwest Research Institute, San Antonio, TX (1981).

27. New York State Uniform Fire Prevention and Building Code - Art. 15, Part 1120 Combustion Toxicity Testing and Regulations for Implementing Building Materials and Finishes. Fire Gas Toxicity Data File, Albany , NY (1987).
28. Bukowski, R.W., Fire Technol. 21, 252 (1985).
29. Hirschler, M.M., J. Fire Sci. 6, 100 (1988).
30. Benjamin, I.A., J. Fire Sci. 5, 25 (1987).
31. McCaffrey, B.J., Quintiere, J.G. and Harkleroad, M.F., Fire Technol. 17, 98 (1981).
32. Galloway, F.M. and Hirschler, M.M. Fire Safety J. 14, 251 (1989).
33. Jones, W.W., A Model for the Transport of Fire, Smoke and Toxic Gases (FAST), NBSIR 84-2934, Natl Bur. Stands., Gaithersburg, MD (1984).
34. Babrauskas, V., Levin, B.C. and Gann, R.G., Fire Journal. 81(2), 22 (1987).
35. Hartzell, G.E., Packham, S.C., Grand, A.F. and Switzer,
    W.G., J. Fire Sci. 3, 195 (1985).

RECEIVED November 1, 1989

Author Index

Affiliation Index

Subject Index

Production: Donna Lucas
Indexing: Deborah H. Steiner
Acquisition: Robin Giroux

Elements typeset by Hot Type Ltd., Washington, DC
Printed and bound by Maple Press, York, PA